The Polar Cusp

NATO ASI Series

Advanced Science Institutes Series

A series presenting the results of activities sponsored by the NATO Science Committee, which aims at the dissemination of advanced scientific and technological knowledge, with a view to strengthening links between scientific communities.

The series is published by an international board of publishers in conjunction with the NATO Scientific Affairs Division

A	Life Sciences	Plenum Publishing Corporation
B	Physics	London and New York
C	Mathematical and Physical Sciences	D. Reidel Publishing Company Dordrecht, Boston and Lancaster
D	Behavioural and Social Sciences	Martinus Nijhoff Publishers
E	Engineering and Materials Sciences	The Hague, Boston and Lancaster
F	Computer and Systems Sciences	Springer-Verlag
G	Ecological Sciences	Berlin, Heidelberg, New York and Tokyo

Series C: Mathematical and Physical Sciences Vol. 145

The Polar Cusp

edited by

Jan A. Holtet

and

Alv Egeland

Institute of Physics, University of Oslo, Blindern, Oslo, Norway

D. Reidel Publishing Company

Dordrecht / Boston / Lancaster

Published in cooperation with NATO Scientific Affairs Division

Proceedings of the NATO Advanced Research Workshop on
The Morphology and Dynamics of the Polar Cusp
Lillehammer, Norway
May 7-12, 1984

Library of Congress Cataloging in Publication Data

NATO Advanced Research Workshop on the Morphology and Dynamics of the Polar
    Cusp (1984: Lillehammer, Norway)
    The polar cusp.

    (NATO ASI series.   Series C, Mathematical and physical sciences; v. 145)
    "Proceedings of the NATO Advanced Research Workshop on the Morphology
and Dynamics of the Polar Cusp, Lillehammer, Norway, May 7–12, 1984"-T.p. verso.
    "Published in cooperation with NATO Scientific Affairs Division."
    Includes index.
    1. Polar cusp–Congresses. I. Holtet, J. A. (Jan A.), 1939–
II. Egeland, Alv, 1932–   . III. North Atlantic Treaty Organization. Scientific
Affairs Division. IV. Series.
QC994.75.N28   1984   551.5'14'0911   84–27528
ISBN-13: 978-94-010-8838-1   e-ISBN-13: 978-94-009-5295-9
DOI: 10.1007/978-94-009-5295-9

Published by D. Reidel Publishing Company
P.O. Box 17, 3300 AA Dordrecht, Holland

Sold and distributed in the U.S.A. and Canada
by Kluwer Academic Publishers,
190 Old Derby Street, Hingham, MA 02043, U.S.A.

In all other countries, sold and distributed
by Kluwer Academic Publishers Group,
P.O. Box 322, 3300 AH Dordrecht, Holland

D. Reidel Publishing Company is a member of the Kluwer Academic Publishers Group

CONTENTS

PREFACE

These proceedings are based upon introductory talks, research reports
and discussions from the NATO Advanced Workshop on the "Morphology and
Dynamics of the Polar Cusp", held at Lillehammer, Norway, 7-12 May,
1984.
   The upper atmosphere at high latitudes is called the "Earth's window to outer space". Through various electrodynamic coupling processes as well as through direct transfer of particles many geophysical
effects displayed there are direct manifestations of phenomena occurring
in the deep space. The high latitude ionosphere will also exert a
feedback on the regions of the magnetosphere and atmosphere to which
it is coupled, acting as a momentum and energy source and sink, and a
source of particles. Of particular interest are the sections of the
near space known as the Polar Cusp. A vast portion of the earth's
magnetic field envelope is electrically connected to these regions.
This geometry results in a spatial mapping of the magnetospheric processes and a focusing on to the ionosphere. In the Polar Cusps the
solar wind plasma has also direct access to the upper atmosphere. The
polar regions are thus of extreme importance when it comes to understanding the physical processes in the near space and their effect on
our environment.
   The Introductory Talks given at this workshop provided a common
background for discussing and understanding the physics of the Polar
Cusp. By this book we will make the information which thus was provided to the participants of the workshop accessible to a wider audience.
   We will take this opportunity to thank Mrs A.-S. Andresen who was
in charge of the Workshop Secretariate. The assistance received from
The Royal Norwegian Council for Scientific and Industrial Research's
Space Activity Division (NTNFR) in organizing the meeting is also ack-
nowledged.
   In addition to the NATO grant, financial support was also received
from Office of United States Naval Research, London Branch Office.
This support has been of particular value in preparation of these
Proceedings.

OSLO, October 1984

J.A. Holtet   A. Egeland

PARTICIPANTS

Baron, M.J.	EISCAT Scientific Association Box 705 S-981 27 Kiruna Sweden
Baumjohan, W.	Max-Planck-Institut für Extraterr. Physik D-8046 Garching Germany
Berthelier	LGE - CNRS 4 Avenue de Neptune F-94 100 St. Maur France
Blix, T.	Norw. Defence Researc:. Establishment P.O.Box 25 N-2007 Kjeller Norway
Boström, R.	Uppsala Ionospheric Observatory S-75 590 Uppsala Sweden
Brekke, A.	University of Tromsø The Auroral Observatory P.O.Box 953 N-9001 Tromsø Norway
Broom, S.M.	British Antarctic Survey High Cross Madingley Road Cambridge CB3 OET England
Carlson, H.C.	Department of the Air Force Air Force Geophysics Laboratory (AFSC) Hanscom Air Force Base, Massachusetts 01731 USA
Carovillano, R.L.	Office of Naval Research - London Branch 223 Old Marylebone Road, London NW1 5TH England
Craven, J.D.	Department of Physics and Astronomy The University of Iowa Iowa City, Iowa 52242 USA
Doyle, D.	St. Patric College, Maynooth, Kildare, Ireland

Eather, R.H.   Boston College,
                                Department of Physics
                                Chestnut Hill, Massachusetts 02167
                                USA

Egeland, A.   University of Oslo
                                Institute of Physics
                                P.O.Box 1038 Blindern
                                N.0315 Oslo 3
                                Norway

Evans, D.S.   Space Environment Laboratory
                                U.S. Department of Commerce
                                NOAA
                                Boulder, Colorado 80302
                                USA

Feldman, P.   Physics Department
                                Johns Hopkins University
                                Baltimore, Maryland 21210
                                USA

Foster, J.C.   Massachusetts Institute of Technology
                                Haystack Observatory
                                Westford, Massachusetts 01886
                                USA

Fälthammar, C.-G.   Royal Institute of Technology
                                Institution for Plasmaphysics
                                S-100 44 Stockholm
                                Sweden

Greenwald, R.A.   Applied Physics Laboratory
                                Johns Hopkins University
                                Johns Hopkins Road
                                Laurel, Maryland 20707
                                USA

Gustafsson, G.   Uppsala Ionospheric Observatory
                                S-755 90 Uppsala
                                Sweden

Haldoupis, C.   University of Crete,
                                Physics Department
                                Iraklion
                                Crete
                                Greece

Heelis, R.A.   The University of Texas at Dallas
                                Box 688
                                Richardson, Texas 75080
                                USA

Heikkila, W.J.   The University of Texas at Dallas
                                Box 688
                                Richardson, Texas 75080
                                USA

Henriksen, K.   University of Tromsø
                                The Auroral Observatory
                                P.O.Box 953
                                N-9001 Tromsø
                                Norway

Holmgren, G.	Uppsala Ionospheric Observatory S-75 590 Uppsala Sweden
Holtet, J.A.	University of Oslo Institute of Physics P.O.Box 1038 Blindern N-0315 Oslo 3 Norway
Johnstone, A.	University College London Department of Physics Mullard Space Science Laboratory Holmbury St Mary. Dorking, Surrey RH5 6NT England
Jørgensen, T.S.	Meteorological Institute Lyngbyvej 100 2100 København Ø Denmark
Kelly, J.D.	SRI International 333 Ravenswood Avenue Menlo Park, California 94025 USA
Killeen, T.L.	The University of Michigan Space Physics Research Laboratory College of Engineering Dept. of Atmospheric and Oceanic Science Ann Arbor, Michigan 48109 USA
Lemaire, J.	I.A.S. 3 Avenue Circulaire B-1180 Bruxelles Belgium
Link, R.	York University Centre for Research in Exp. Space Science 4700 Keele Street, Downsview, Ontario M3J 1P3 Canada
Lundin, R.	Kiruna Geophysical Institute P.O.Box 704 S-981 27 Kiruna Sweden
Lybekk, B.	University of Oslo Institute of Physics P.O.Box 1038 Blindern N-0315 Oslo 3 Norway
Machard, C.	LGE - CNRS 4 Avenue de Neptune F-94100 St Maur France

Marklund, G.   Royal Institute of Technology
                                Institution for Plasma physics
                                S-100 44 Stockholm
                                Sweden

Maynard, N.C.   NASA/Goddard Space Flight Center
                                Code 696
                                Greenbelt, Maryland 20771
                                USA

McCormac, F.G.   Ulster Polytechnic
                                Shore Road
                                Newtownabbey
                                CO Antrim BT37 0QB
                                United Kingdom

McEwen, D.J.   Inst. of Space and Atmospheric Studies
                                University of Saskatchewan
                                Saskatoon, Saskatchewan, S7N 0W0
                                Canada

Meng, C.I.   Applied Physics Laboratory
                                Johns Hopkins University
                                Johns Hopkins Road
                                Laurel, Maryland 20707
                                USA

Muldrew, D.B.   Communications Research Centre
                                Department of Communications
                                Ottawa, Ontario K2H 8S2
                                Canada

Myrabø, H.K.   Norw. Defence Research Establishment
                                P.O.Box 25
                                N-2007 Kjeller
                                Norway

Mæhlum, B.N.   Norw. Defence Research Establishment
                                P.O.Box 25
                                N-2007 Kjeller
                                Norway

Neubert, T.   Danish Space Research Institute
                                Lundtoftevej 7
                                DK-2800 Lyngby
                                Denmark

Nielsen, E.   Max-Planck-Institut für Aeronomie
                                Postfach 20
                                D-3411 Katlenburg-Lindau
                                West Germany

Peterson, W.K.   Lockheed Missiles and Space Comp, Inc.
                                Research and Development
                                3251 Hanover Street
                                Palo Alto, California 94304
                                USA

Potemra, T.A.   The Johns Hopkins University
                                Applied Physics Laboratory
                                Johns Hopkins Road
                                Laurel, Maryland 20707
                                USA

Primdahl, F.	Danish Space Research Institute Lundtoftevej 7 DK-2800 Lyngby Denmark
Ranta, H.	Geophysical Obs. of the Finnish Academy of Science and Letters SF-99600 Sodankylä Finland
Rosenberg, T.J.	University of Maryland Inst. for Physical Science and Technology College Park, Maryland 20742 USA
Sandholt, P.E.	University of Oslo Institute of Physics P.O.Box 1038 Blindern N-0315 Oslo 3 Norway
Sckopke, N.	Max-Planck-Institut für Extraterr. Physik D-8046 Garching West-Germany
Shepherd, G.G.	York University Faculty of Science 4700 Keele Street Downsview, Ontario M3J 1P3 Canada
Sivjee, G.G.	University of Alaska, Fairbanks Department of Physics 224 Duckering Building Fairbanks, Alaska 99701 USA
Smith, R.W.	Ulster Polytechnic Shore Road Newtownabbey CO Antrim BT37 0QB United Kingdom
Stamnes, K.	University of Tromsø The Auroral Observatory P.O.Box 953 N-9001 Tromsø Norway
Stauning, P.	Meteorological Institute Lyngbyvej 100 DK-2100 København Denmark
Steen, Å.	Kiruna Geophysical Institute P.O.Box 704 S-981 27 Kiruna Sweden

Svenes, K.   University of Oslo
                               Institute of Physics
                               P.O.Box 1038 Blindern
                               N-0315 Oslo 3
                               Norway
Sørnes, F.   University of Bergen
                               Department of Physics
                               Allégaten 55,
                               N-5000 Bergen
                               Norway
Søraas, F.   University of Bergen
                               Department of Physics
                               Allégaten 55,
                               N-5000 Bergen
                               Norway
Tsunoda, R.   SRI International
                               333 Ravenswood Avenue
                               Menlo Park, California 94025
                               USA
Tulunay, Y.   58 Sorak No 3
                               EMEK MAHALLESI
                               Ankara
                               Turkey
Vasyliunas, V.M.   Max-Planck-Institut für Aeronomie
                               Postfach 20
                               D-3411 Katlenburg-Lindau
                               West-Germany
Weber, E.J.   Department of the Airforce
                               AFSC
                               Hanscom Air Force Base
                               Massachusetts 01731
                               USA
Winningham, J.D.   Southwest Research Institute
                               P.O.Drawer 28510
                               San Antonio, Texas 78294
                               USA
Witt, G.   University of Stockholm
                               Department of Meteorology
                               Arrhenius Laboratory
                               S-106 91 Stockholm
                               Sweden.

PLASMA AND FIELD OBSERVATIONS IN THE EXTERIOR CUSP, ENTRY LAYER, AND PLASMA MANTLE

Norbert Sckopke
Max-Planck-Institut für Physik und Astrophysik
Institut für extraterrestrische Physik
8046 Garching
Federal Republic of Germany

ABSTRACT.   This report briefly summarises the basic properties of the magnetopause boundary layer, in particular those of its high-latitude portions. Emphasis is placed on the results of the Heos 2 studies while those derived from the more recent Prognoz 7 data are not treated in great detail since they are the subject of a separate paper in this volume.

INTRODUCTION

The Earth's magnetopause is not impermeable to the solar wind. As a consequence, a plasma boundary layer forms inside of and adjacent to the magnetopause, a major contribution to which comes from the magneto-sheath. First observations in the polar region were carried out over 10 years ago by means of the Heos 2 spacecraft, and it is these high-latitude observations with which this report is mainly concerned. The entire subject has recently been reviewed in more detail by Haerendel and Paschmann (1982).

MORPHOLOGY

The magnetopause boundary layer extends over the entire known part of the magnetopause, but its properties vary substantially from one region to the other. These spatial regimes are conveniently distinguished by a number of different names. Figure 1 shows the boundary layer as it appears in the noon-midnight cross section. Three regions have been identified: the low-latitude boundary layer (LLBL), the entry layer (EL), and the plasma mantle (PM) (e.g. Hones et al., 1972; Rosenbauer et al., 1975; Paschmann et al., 1976; Haerendel et al., 1978; Eastman and Hones, 1979). The exterior cusp (EC), though located outside the magnetopause, is likely to influence the plasma and field behaviour in its neighbourhood, that is, in the entry layer. Its properties have been investigated by Mencke Hansen et al. (1976), and by Vasyliunas et al. in an unpublished study part of whose results have been reported by Sckopke (1979).

1

J. A. Holtet and A. Egeland (eds.), The Polar Cusp, 1–7.
© 1985 by D. Reidel Publishing Company.

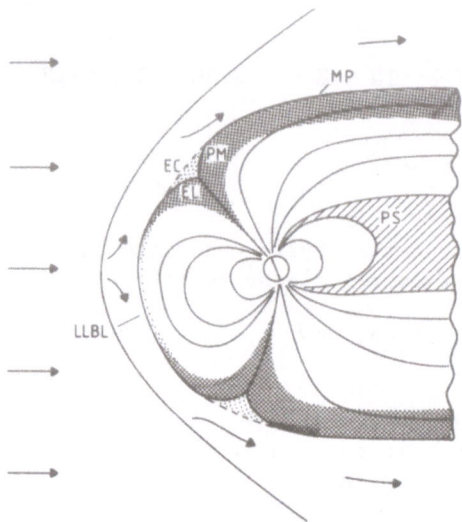

Figure 1. Noon-midnight meridional cross section of the magnetosphere
illustrating schematically the plasma boundary layer inside of and
adjacent to the magnetopause (MP). The layer is contiguous but different portions exhibit different plasma properties, and are referred
to by different names: the low-latitude boundary layer (LLBL), entry
layer (EL), and plasma mantle (PM). EC is the exterior cusp, and PS
denotes the plasma sheet.

Complementary to the noon-midnight cross section, Figure 2 shows a
frontside view of the northern magnetopause in which the regions
occupied by the low-latitude boundary layer, entry layer, and plasma
mantle are indicated. The classification is based on magnetopause
crossings collected during the 2 1/2 years of Heos-2 operation. The
crossings are represented by different symbols which indicate different
properties of the boundary layer plasma (for an explanation, see the
figure caption).
     This morphology is valid only in an average sense as all the data
have been lumped together irrespective of external factors such as the
direction of the interplanetary magnetic field (IMF) which is known to
influence quite strongly the magnetospheric electric and magnetic
fields. In particular, it is conceivable that the region labelled EL
is in fact an envelope of the area occupied by the entry layer at
various instances. At any given time, the latitudinal and azimuthal
extent of the entry layer might be smaller than that suggested by the
figure. Also, the entry layer need not be symmetrical to the noon
meridian if there is a strong IMF y component present.

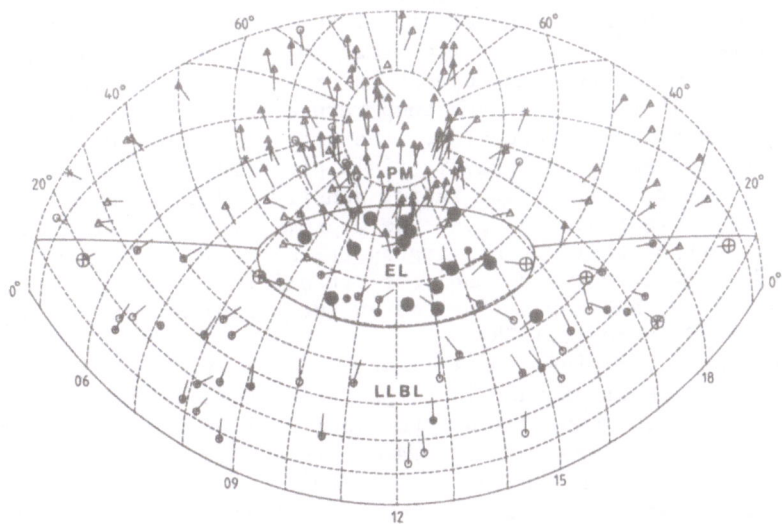

Figure 2. Front view of a model magnetopause indicating the average
extent of the entry layer (EL), low-latitude boundary layer (LLBL), and
plasma mantle (PM). Locations of Heos 2 magnetopause crossings are
marked by various symbols which indicate the type of plasma encountered:
triangles represent the plasma mantle, open circles indicate the absence
of boundary layer plasma, and the other symbols refer to the EL and
LLBL. Here, filled-in circles and encircled crosses distinguish between
large and small density jumps at the magnetopause, and the symbol size
indicates the thickness of the layer. Short lines emanating from the
symbols indicate the interior field component tangential to the model
surface (from Haerendel et al., 1978).

PLASMA AND FIELD PROPERTIES

The *exterior cusp* is bounded on the inside by the cusp-like indentation
of the magnetopause (see Figure 1), and on the outside by the innermost
free-flow stream line of the magnetosheath flow which detaches from the
magnetopause somewhere upstream, and re-attaches again downstream of the
cusp, as was originally suggested by Spreiter and co-workers (e.g.
Spreiter et al., 1968). The properties of the plasma in the exterior
cusp are distinctly different from those encountered in the free-flow
magnetosheath: the temperature is higher while the density and Mach
number are lower, and the flow is also more irregular, both in magnitude
and direction. A qualitative explanation was given by Haerendel et al.
(1978) who noted the similarity of the situation near the cusp to hydro-
dynamic flow around a corner, in which vortex formation and separation
are known to occur and to initiate some level of turbulence.
    The *entry layer* is located on magnetospheric field lines just
equatorward of the cusp. It has been so termed because it appears to

be the region of dominant plasma entry into the magnetosphere (Haerendel
and Paschmann, 1975; Paschmann et al., 1976). The strongest support for
this view comes from the observation that the plasma density often stays
nearly constant when the magnetopause is crossed from the outside into
the entry layer. Also, the plasma temperature is very similar to that of
the exterior cusp. As a consequence, the plasma β, i.e. the ratio of
plasma thermal pressure to magnetic field pressure, is of the order $\beta \sim 1$
in the entry layer. Fairly flat radial density and temperature profiles
across the layer indicate the presence of some very effective inward
plasma transport mechanism. This transport is likely to be achieved
through eddy convection which manifests itself in the irregular, low-
speed plasma flow, and may be incited by the turbulence in the adjacent
exterior cusp (Haerendel, 1978).

In contrast, the *plasma mantle* is located on field lines just tailward of the cusp. Most of the time, the field lines appear to be open,
and the mantle plasma expands into the open tail. Hence, the plasma
mantle is a region of predominant plasma efflux from the magnetosphere,
with an estimated average number flux of 10^{26} s^{-1} (Pilipp and Morfill,
1978). On an outward directed profile across the plasma mantle, the
density rises by several orders of magnitude towards the magnetosheath
density level, and the plasma temperature (or mean energy) and bulk
speed increase as well, although less strongly, towards their respective
magnetosheath levels. The flow direction is tailward and very closely
aligned with the magnetic field. The pressure ratio is $\beta << 1$ throughout
most of the mantle with the exception of the outermost region where this
value, too, approaches its magnetosheath level, i.e. $\beta \sim 1$. These
features have been successfully explained by Haerendel (1974) to result
from the anti-sunward convection of entry layer plasma across the polar
cusp. The combination of the energy-independent cross field convection
speed, and the energy- (and pitch-angle-) dependent particle motion
parallel to the field, together with particle mirroring in the low
altitude cusp, establishes a velocity filter effect from which all the
observed plasma properties follow in a consistent way.

On rare occasions, however, the plasma encountered by Heos 2 in
the plasma mantle regime had radically different properties: the bulk
speed was low, and the temperature was more than one order of magnitude
higher than usual. This *stagnant plasma mantle* (Rosenbauer et al., 1975;
Sckopke et al., 1976) is associated with periods of northward IMF. In
fact, a re-examination of the data has shown that, for the examples of
the Heos 2 data set, the IMF had had a northward component for a prolonged period of time prior to the observations. Unfortunately, the
number of such cases was low, and a detailed analysis has yet to be
carried out.

The *low latitude boundary layer*, finally, forms a broad belt from
the subsolar region past the flanks toward the magnetotail (which was
the region from which the first observations of boundary layer plasma
had been reported (Hones et al., 1972). Magnetosheath plasma enters the
LLBL on the dayside, as witnessed by the ion energy spectra (Haerendel
et al., 1978), and, more directly, by the results of the more recent
plasma composition measurements (see, for example, Lundin, 1985 (this
volume)). However, the plasma entry mechanism seems to be less effective

than that operating in the entry layer, as the density experiences a
strong gradient at the magnetopause and/or across the layer. Similar to
the density, the temperature, too, assumes values in the layer which are
about half way between their respective inner (ring current) and outer
(magnetosheath) levels. Moreover, the form of the ion and electron
spectra indicates quite clearly that the LLBL plasma is a mixture of
plasma originating on both sides of the layer. The flow speed is sometimes irregular, most notably near the stagnation point. At moderate-
to-large distances from the subsolar region, the flow is fairly steady
and similar to the external flow both in magnitude and direction. However, this latter statement holds only in an average sense: high-resolution plasma and field data from ISEE 1 and 2 have shown that the LLBL is
often not uniform, but consists of a series of plasma blobs sliding
tailward along the magnetopause with about the magnetosheath flow velocity (Sckopke et al., 1981). Two implications arise from the observed
LLBL flow direction: (i), the flow has a strong cross-field component,
and (ii), most of the magnetosheath plasma that enters the magnetosphere
through the dayside LLBL bypasses the high latitude region of the entry
layer and polar cusp, and flows directly into the tail, again at an
estimated rate of 10^{26} ions/s (Eastman et al., 1976).

COUPLING TO THE IONOSPHERE

At ionospheric heights, the entire magnetopause boundary layer maps into
a small region adjacent to the area usually identified with the cusp
(see Figure 1 of Vasyliunas (1979)). Transfer of magnetosheath plasma
into the dayside boundary layer is necessarily accompanied by transfer
of momentum. This momentum is coupled to the polar ionosphere via the
magnetic field, and manifests itself in the usual convection pattern at
low altitudes. The precise nature of the momentum transfer process as
well as the question about the location of the main transfer site are
still a matter of debate. An even greater enigma is the mode of mass
transfer across the magnetopause, i.e. the initial mass loading of magnetospheric field lines with external plasma. Both topics are discussed
in greater detail by Haerendel and Paschmann (1982). Of the magnetosheath plasma that has crossed the magnetopause, only that populating
the entry layer has access to the low altitude cusp, as is evident from
the foregoing description.
      Finally, transfer of mass and momentum into the ionosphere implies
also the possibility of the reverse process, namely the contribution of
ionospheric plasma to the boundary layer. This contribution is most
readily apparent in the plasma mantle. The earlier (indirect) reports
on the presence of ionospheric oxygen in the plasma mantle (e.g. Frank
et al., 1977) have been confirmed more recently through plasma composition measurements (Lundin et al., 1982; see also Lundin, 1985 (this
volume)).

REFERENCES

Eastman, T.E., E.W. Hones, Jr., S.J. Bame, and J.R. Asbridge, 'The mag-
    netospheric boundary layer: site of plasma, momentum and energy
    transfer from the magnetosheath into the magnetosphere', Geophys.
    Res. Lett., 3, 685-688, 1976.
Eastman, T.E., and E.W. Hones, Jr., 'Characteristics of the magneto-
    spheric boundary layer and magnetopause layer as observed by Imp 6',
    J. Geophys. Res., 84, 2019-2028, 1979.
Frank, L.A., K.L. Ackerson, and D.M. Yeager, 'Observations of atomic
    oxygen (O$^+$) in the Earth's magnetotail', J. Geophys. Res., 82, 129-
    134, 1977.
Haerendel, G., 'Die Spur der Magnetopause in der Magnetosphäre', Mitt.
    Astron. Ges., 35, 165-181, 1974.
Haerendel, G., and G. Paschmann, 'Entry of solar wind plasma into the
    magnetosphere', in: *Physics of the Hot Plasma in the Magnetosphere*
    (B. Hultqvist and L. Stenflo, ed.s), pp. 23-43, New York: Plenum,
    1975.
Haerendel, G., 'Microscopic plasma processes related to reconnection',
    J. Atmos. Terr. Phys., 40, 343-353, 1978.
Haerendel, G., G. Paschmann, N. Sckopke, H. Rosenbauer, and P.C. Hedge-
    cock, 'The frontside boundary layer of the magnetosphere and the
    problem of reconnection', J. Geophys. Res., 83, 3195-3216, 1978.
Haerendel, G., and G. Paschmann, 'Interaction of the solar wind with
    the dayside magnetosphere', in: *Magnetospheric Plasma Physics*,
    (A. Nishida, ed.), pp. 49-142, Dordrecht: D. Reidel, 1982.
Hones, E.W., Jr., J.R. Asbridge, S.J. Bame, M.D. Montgomery, S. Singer,
    and S.-I. Akasofu, 'Measurements of magnetotail plasma flow made
    with Vela 4B', J. Geophys. Res., 77, 5503-5522, 1972.
Lundin, R., B. Hultqvist, N. Pissarenko, and A. Zackarov, 'The plasma
    mantle: Composition and other characteristics observed by means
    of the Prognoz-7 satellite', Space Sci. Rev., 31, 247-345, 1982.
Lundin, R., 'Plasma composition and flow characteristics in the mag-
    netospheric boundary layers connected to the polar cusp', 1985,
    (this volume).
Mencke Hansen, A., A. Bahnsen, and N. D'Angelo, 'The cusp-magnetosheath
    interface', J. Geophys. Res., 81, 556-561, 1976.
Paschmann, G., G. Haerendel, N. Sckopke, and H. Rosenbauer, 'Plasma and
    field characteristics of the distant polar cusp near local noon:
    The entry layer', J. Geophys. Res., 81, 2883-2899, 1976.
Pilipp, W.G., and G. Morfill, 'The formation of the plasma sheet result-
    ing from plasma mantle dynamics', J. Geophys. Res., 83, 5670-5678,
    1978.
Rosenbauer, H., H. Grünwaldt, M.D. Montgomery, G. Paschmann, and
    N. Sckopke, 'Heos 2 plasma observations in the distant magneto-
    sphere: The plasma mantle', J. Geophys. Res., 80, 2723-2737, 1975.
Sckopke, N., G. Paschmann, H. Rosenbauer, and D.H. Fairfield, 'Influence
    of the interplanetary magnetic field on the occurrence and thick-
    ness of the plasma mantle', J. Geophys. Res., 81, 2687-2691, 1976.
Sckopke, N., 'External plasma flow', in: *Magnetospheric Boundary Layers*,
    (B. Battrick, ed.), pp. 37-42, Paris: ESA, SP-148, 1979.

Sckopke, N., G. Paschmann, G. Haerendel, B.U.Ö. Sonnerup, S.J. Bame, T.G. Forbes, E.W. Hones, Jr., and C.T. Russell, 'Structure of the low latitude boundary layer', J. Geophys. Res., 86, 2099-2110, 1981.

Spreiter, J.R., A.A. Alksne, and A.L. Summers, 'External aerodynamics of the magnetosphere', in: *Physics of the Magnetosphere*, (R.L. Carovillano et al., eds.), pp. 301-375, Dordrecht: Reidel, 1968.

Vasyliunas, V.M., 'Interaction between the magnetospheric boundary layers and the ionosphere', in: *Magnetospheric Boundary Layers*, (B. Battrick, ed.), pp. 387-393, Paris: ESA, SP-148, 1979.

PLASMA COMPOSITION AND FLOW CHARACTERISTICS IN THE MAGNETO-
SPHERIC BOUNDARY LAYERS CONNECTED TO THE POLAR CUSP

Rickard Lundin
Kiruna Geophysical Institute
Box 704
981 27 Kiruna
Sweden

ABSTRACT. A summary of recent magnetospheric boundary layer
results from the PROMICS-1 ion composition experiment on
Prognoz-7 will be presented. Particular emphasis will be put on
observations which are relevant for the entry of magnetosheath
plasma into the magnetosphere and the energy and momentum
transfer associated with this process.
     Observations in the low-latitude boundary layer and the
entry layer demonstrate that magnetosheath plasma may penetrate
into the boundary layer and appear as high density plasma blobs
having excess momentum as compared to the "halo" of the boundary
layer. Similar penetration structures are frequently found in
the exterior cusp and the plasma mantle.
     The first observations of individual flow vectors of
different ion species in the different boundary layers are also
described and discussed.
     Ions of solar wind and ionosphere origin frequently show
different flow vectors. This observation is discussed in
relation to the energy and momentum transfer in the magnetosheath penetration structures.
     Finally, the Prognoz-7 plasma mantle observations are
summarized and the results are related to the open magnetic
field model.

1. INTRODUCTION

In this report I will present some recent results of the hot
plasma composition and flow properties in the magnetospheric
boundary layers connected to the dayside cusp region. The results have been obtained by means of the PROMICS-1 experiment on
board the Prognoz-7 satellite. A detailed description of this
experiment has been given by Lundin et al. (1982).

J. A. Holtet and A. Egeland (eds.), The Polar Cusp, 9–32.
© 1985 by D. Reidel Publishing Company.

     The magnetospheric boundary layers are the interface
regions between the hot solar wind/magnetosheath plasma and the
hot and cold magnetosphere/ionosphere plasmas. These interface
regions thus control the energy and momentum transfer from the
solar wind into the magnetosphere. Ultimately the energy
transferred into the magnetosphere manifests itself as e.g.
auroral processes in the Earth's upper atmosphere and
ionosphere.
     The plasma which populates the magnetospheric boundary
layers - the low-latitude boundary layer (LLBL), the entry layer
(EL), the exterior cusp and the high-latitude boundary layer/
plasma mantle (PM) - is dominated by particles of solar wind
origin (eg. H^+ and He^{2+}). However, a small admixture of particles of terrestrial origin may be found in the boundary layers
as well. Occasionally this admixture is also so strong that the
terrestrial plasma component may considerably affect the energy
and momentum transfer processes. In most cases, however, the
terrestrial plasma component serves as an important tracer for
local as well as remote processes associated with the solar wind
energy and momentum transfer process.
     It seems now a common belief that a full dynamical descrip-
tion of the boundary layer plasma processes requires mainly
integral quantities of the plasma and fields. To some extent
this has been successfully applied on reconnection observations
in the dayside LLBL - i.e. one has found that reconnection is
evidenced by the observations of local stress balance (magnetic
vs plasma flow, see e.g. Paschmann et al. 1979).
     I will here present both observations and theoretical con-
siderations which show that partial quantities of the plasma
distribution function may indeed complicate the dynamical picture of the local energy and momentum transfer in the boundary
layer. Substantial differences observed between the drift velocities for ions of magnetosheath and magnetosphere origin is for
example here interpreted as the signature of a local MHD-genera-
tor process. Such a generator process - driven by e.g. either
the inertia of the drifting boundary layer plasma or by local
pressure gradients - will also transfer energy to the Earth's
ionosphere and upper atmosphere via field aligned currents and
induced electric fields. Thus, we are here also offering an
alternative view of the transfer of energy from the shocked
solar wind plasma into the magnetosphere, as compared to the
merging and reconnection hypothesis where the free energy comes
from the ambient electric and magnetic field. A similar view of
the boundary layer as a generator region driven by the plasma
flow has also been favoured by e.g. Eastman et al. (1976),
Lemaire and Roth (1978, 1979), Heikkila (1979) and Akasofu
(1983).
     Previous measurements of the plasma momentum properties in
the boundary layers near the exterior cusp have entirely been
based on experiments without mass-resolving capabilities. (see
e.g. Sckopke and Paschmann, 1978 and Paschmann, 1979). Except

for a few "anomalies" (e.g. Sckopke et al., 1976) the plasma
momentum characteristics have, according to these measurements,
agreed with the generally accepted "open" magnetic field model
as introduced by Dungey (1961) and later discussed by Cowley
(1980). The HEOS-2 results reviewed by Sckopke in these proceedings, have for instance emphasized the difference between the
low-latitude boundary layer (LLBL) on one hand and the entry
layer (EL) and the plasma mantle (PM) on the other - the latter
two being the only boundaries which are topologically connected
to the cusp.

Although the results from Prognoz-7 to a large extent
agree with the classification inferred from HEOS-2, the composition measurements have added a completely new dimension into
the results that calls for a substantial revision of previous
interpretations. We have already mentioned the flow aspect and
the generator concept. Other important aspects are for instance
the plasma sources, the plasma circulation and the degree of
"openness" of the magnetosphere. The test of "openness" of the
magnetosphere is an example of how ion composition measurements
can be used for identifying plasma boundaries. In the open
magnetic field model the nightside (plasma mantle) magnetopause
is at times believed to be open in the sense that the earth
magnetic field lines are connected to the interplanetary magnetic field lines. Plasma should thus have free access to either
side of the magnetopause (solar wind plasma streaming in and
magnetosphere plasma streaming out). Although previous measurements from e.g. HEOS-2 were not able to actually confirm "steady
state" reconnection (e.g. Haerendel et al., 1978), the HEOS-2
results were not in conflict with the open model. On the
contrary, very good arguments for open field lines in the plasma
mantle were found. For instance, a much thicker and denser
mantle during a southward IMF (e.g. Sckopke et al. 1976), frequently with a smooth transition characteristic (increasing
temperature, density and velocity towards the magnetopause). The
Prognoz-7 observations (e.g. Lundin et al., 1982), however, indicated a more closed magnetopause than was anticipated from the
HEOS-2 results. The arguments for this and the related conse-
quencies for the open magnetic field model will be reviewed
here.

This report will be ordered in such a way that we start by
summarizing the Prognoz-7 observations in the low-latitude
boundary layer/entry layer near the noon meridian. We continue
with some observations made near the exterior cusp and then go
on by summarizing the rather extensive material from the plasma
mantle. This is also the order in which we expect the penetrating solar wind plasma to encounter the boundary regions discussed.

2. THE ENTRY LAYER/LOW LATITUDE BOUNDARY LAYER

The reason for blending these two topologically different regions (see e.g. Paschmann et al. 1976 for a definition of the entry layer) is because the plasma characteristics pertaining to both boundary regions may not be much different at higher latitudes. We will demonstrate that narrow "entry regions" can be observed in what appears to be the LLBL but which may as well be an equatorial extension of the EL. These entry regions in the boundary layer, hereafter referred to as "injection structures", are all characterized by a high "magnetosheath-like" plasma density, a large fraction of solar wind ions (H^+ and He^{2+}) and usually an enhanced cross-field flow.

The PROMICS-1 experiment on board Prognoz-7 has not only provided the first ion composition measurements in the high-latitude boundary layers, but also the first determination of individual flow vectors for different ion species (Lundin and Dubinin, 1984 a, b; Lundin, 1984). On a few occasions data were taken in the magnetospheric boundary regions and transmitted to ground at a rate of about 4 kbits/s for the PROMICS-1 experiment. For those few cases quite detailed information about the distribution of densities, temperatures, pressure and flow vectors of the four major ion constituents has been obtained. An example of such a dayside magnetopause crossing and boundary layer encounter is shown in Figure 1. The magnetopause crossing is indicated by the vertical bar in Fig. 1 and the magnetosheath is characterized by the region of continuous high density prior to the magnetopause crossing. Notice that the magnetopause marks a distinctive discontinuity with respect to both the magnetic field vector (dashed line in the two lowermost panels) and the ion composition (O^+ density close to the measurement threshold on the sheath side, see uppermost panel).

The most obvious features inside the magnetopause are the three high density, low temperature and enhanced flow velocity regions. These structures are very magnetosheath-like (with respect to density, temperature and composition) but have also features which are different from those found in the magnetosheath (e.g. flow vector). Notice for instance the more tailward and dawnward flow of H^+ inside the magnetopause as compared to outside of it. Notice also a remarkably strong difference between the flow vectors for H^+ and O^+ (marked by + and o respectively) and the strong cross field flow component for all ions inside the boundary layer. Figure 2 depicts data from the same boundary layer encounter but now in terms of the cross-field flow vectors (-vxB-vectors). We can here see that not only does a significant cross-field flow component exist in the boundary layer, but even more pronounced is the fact that we have a marked difference between the H^+ and O^+ cross-field flow vector (hatched area).

How can we then understand the observation of different cross-field flow vectors for H^+ and O^+ ions? The ion drift can

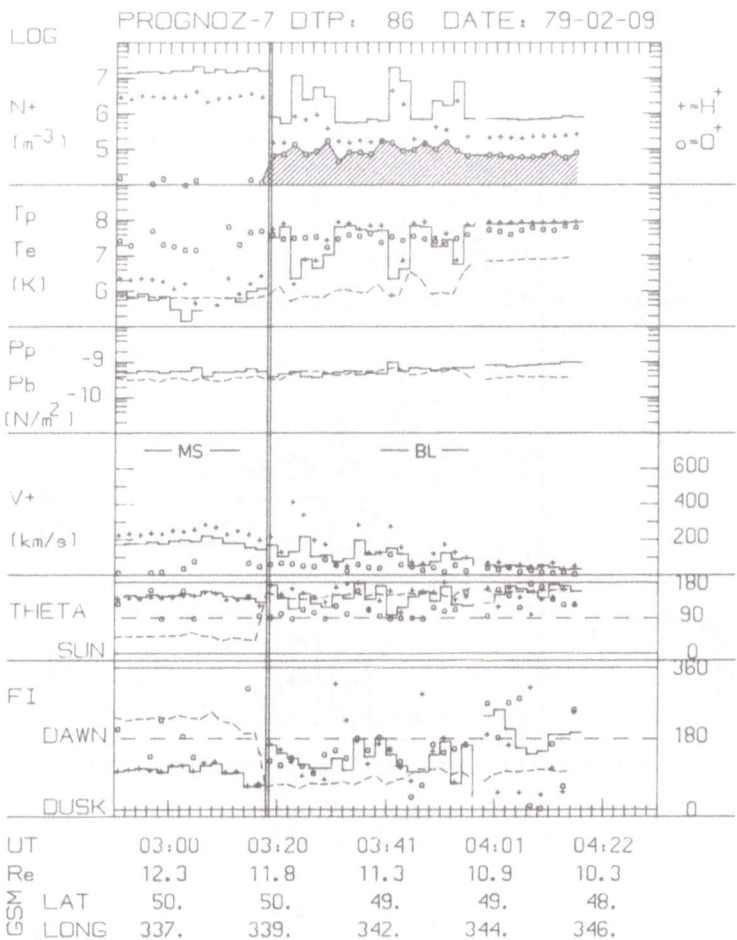

Figure 1. Plasma momentum parameters and magnetic field data for a PROGNOZ-7 inbound dayside magnetopause crossing near noon. The magnetopause crossing is marked by the vertical bar, magneto-sheath is marked "MS" and the boundary layer "BL" in the bulk velocity panel (V_+). The O^+ number density in the density panel (N_+) is indicated by the hatched area. The electron temperature in the second panel from the top is represented by the broken curve. The magnetic pressure (P_b) and the direction of the magnetic field vector in polar coordinates is given by dashed curves in the third, fifth and sixth panels respectively. Solid histogram curves represents plasma parameters deduced from the E/q-ion spectrometers under the assumption that all ions are protons. (+) and (o) represents plasma parameters deduced for H^+ and O^+ ions respectively using the Ion Composition Spectro-meters.

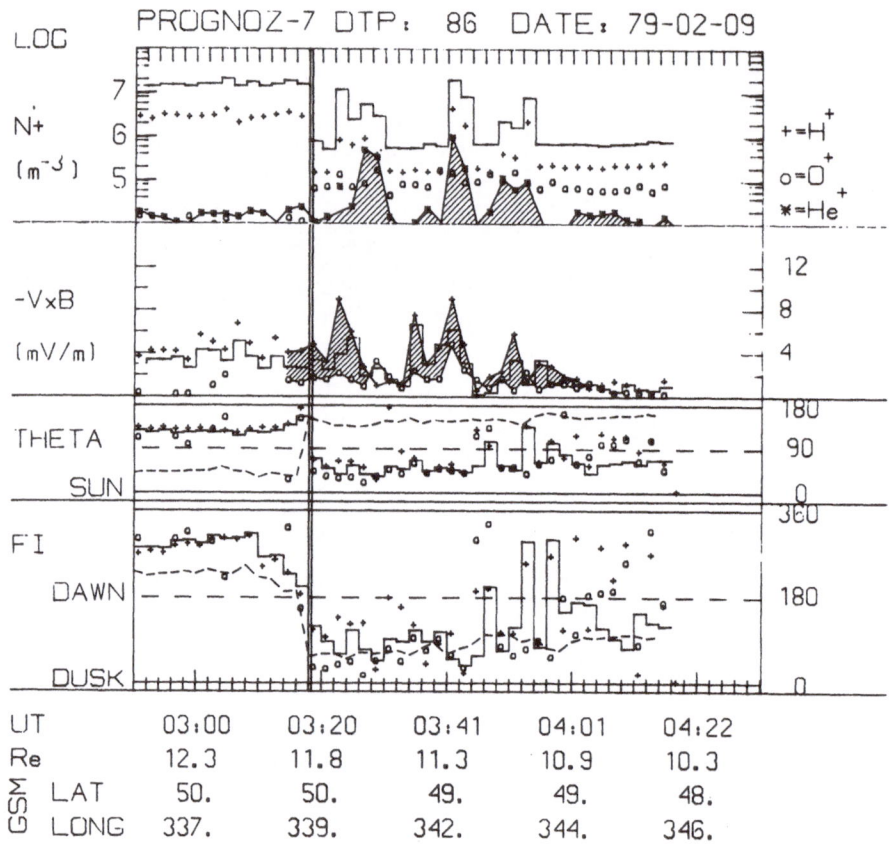

Figure 2. The magnitude and direction of the cross-field flow for the dayside magnetopause crossing contained in Figure 1. The -vxB vector has been deduced for H⁺ (+), O⁺ (o) and all ions (E/q-spectrometers assuming the mass of protons). The He⁺ density in the upper panel is marked by the hatched area. Hatched area in the second panel indicates that the H⁺ ions have a cross-field component that exceeds the O⁺ cross-field component. Broken curve in the two bottom panels gives the direction of the magnetic field vector in polar coordinates.

be associated with a convection electric field which gives rise to ExB drift of ions and electrons. This electric field cannot be different for different ion species, and hence one needs other drift-terms to account for this difference.

Before going into details about the drift differences we may note from Fig. 2 that the convection field inferred from the ion drift is pointing predominantly radially outward in this dawnside boundary layer encounter. This is of course an expected consequence of a plasma flow being tangential to the magneto-pause surface in the boundary layer. Likewise, the convection field should point radially inward in the duskside boundary layer - which the second example by Lundin and Dubinin 1984 a also demonstrates. The complete reorientation of the convection field vector at the magnetopause in Fig 2. suggests, however, that the convection field is not an externally applied but a locally induced one by the plasma flow. The plasma is basically flowing in the same direction on both sides of the magnetopause independent of the magnetic field orientation. Notice also from Figure 2 the very high He$^+$ density inside the region of en-hounced cross-field flow. Lundin and Dubinin (1984) demonstrated that such high He$^+$ percentages are expected if "cold" plasma from the high altitude ionosphere/the plasmasphere have become locally accelerated in the boundary layer. Such local accelera-tion regions with high He$^+$ densities have predominantly been found in the LLBL and EL.

A general expression for the difference between the flow vectors of two different ion species 1 and 2, when all time variations are much slower than the ion gyrofrequencies is the following:

$$\underline{V}_1 - \underline{V}_2 = \frac{B}{B^2} \times \left[\left(\frac{m_1}{q_1} - \frac{m_2}{q_2} \right) \frac{d\underline{V}_o}{dt} + \frac{\nabla p_{\perp 1}}{q_1 n_1} - \frac{\nabla p_{\perp 2}}{q_2 n_2} \right]$$

$$+ \left[\frac{(W_\parallel - W_\perp)_1}{q_1} - \frac{(W_\parallel - W_\perp)_2}{q_2} \right] \left(\frac{B \times \nabla B}{B^3} + \frac{\mu_o}{B^2} \underline{j}_\perp \right) \tag{1}$$

where $\underline{V}_o = \frac{E \times B}{B^2}$, $\frac{d}{dt} = \frac{\partial}{\partial t} + (\underline{V}_o \cdot \nabla)$,

$$p_\perp = nW_\perp = \frac{nmv_\perp^2}{2} , \quad p_\parallel = nW_\parallel = n\frac{mv_\parallel^2}{2} ,$$

(e.g. Volkov 1966 and Stasiewicz et al. 1984).

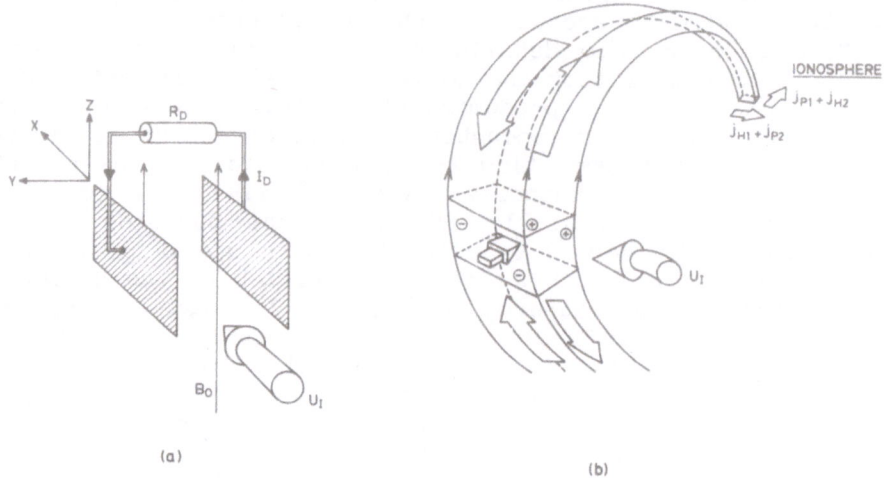

(a)   (b)

Figure 3. Circuit model of an MHD generator where plasma kinetic
energy is converted into electric energy.
(a) Diagrammatic representation of a linear MHD generator with
continuous electrodes and a load R_D.
(b) Qualitative picture of a dawn current system resulting from
a local penetration of magnetosheath plasma moving perpendicular
to B. Open arrows mark currents.

From this equation we see that the flow velocities of different
ion species with different mass, charge, magnetic moment or
magnetic field aligned velocity are always different. A time
and/or space variation of the ExB velocity may also contribute
to the difference between the species in terms of flow velocity. The question is, how big may the velocity difference between two ion species be in the boundary layer? Very rough esti-
mates show that the dV_g/dt, $\nabla p_\perp$ and $j_\perp$ terms easily can reach
values of some hundreds of kilometres per second (Stasiewicz et
al. 1984).
    Lundin (1984) suggested that the regions with different
flow direction and magnitudes of solar wind protons and "local"
magnetospheric plasma (e.g. O^+) are generator regions which
drive Birkeland currents through a dissipating ionosphere. The
situation is illustrated in Figure 3. In the same generator
region the "local" plasma, which have been found to flow with
the same velocity vector, is likely to have been accelerated by
induced electric fields to the direction and magnitude of the
ExB drift vector (Lundin and Dubinin 1984 b). They also

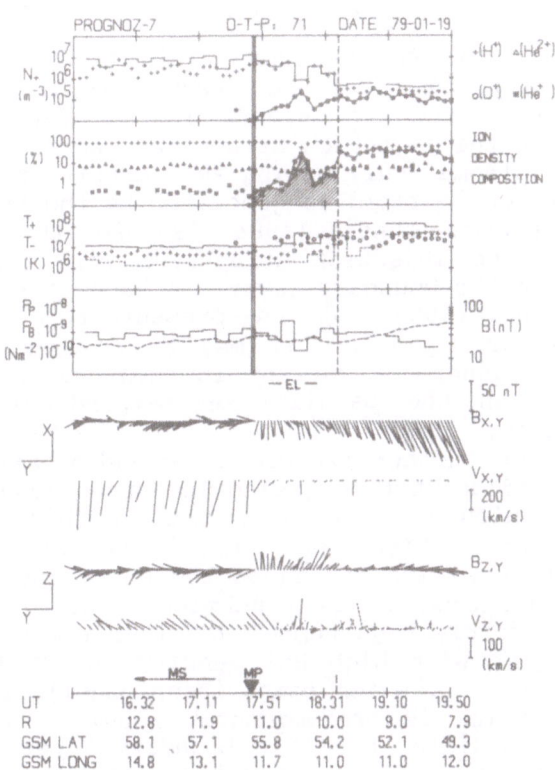

Figure 4. Low time resolution plasma and magnetic field parameters for an inbound crossing of the magnetopause and entry
layer near noon. The magnetopause is marked by "MP" and the vertical solid bar. The extension of the entry layer is indicated
by the vertical broken line and the hatched area in the ion density composition panel of the He$^+$ percentage. The magnitude and
direction of the plasma flow and magnetic field is given by
vector-projections onto the xy and yz solar ecliptic planes.
Velocity vectors are given for H$^+$ (solid lines) and O$^+$ (broken
lines).

demonstrated that whenever the "local" plasma density is sufficiently low, the "local ions" (e.g. O$^+$) may be used as test particles which feel the electric field produced by the penetrating
solar wind plasma. From this and on basis of the generalized
Ohm's law, Lundin (1984) derived the following simplified
expression for the dynamo current:

$$j_y \approx \sigma_o \ (V(O^+) - V(H^+)) \ B_z \tag{2}$$

where $V(O^+)$ and $V(H^+)$ are the cross-field flow vectors for O^+ and H^+ respectively and σ_O is the conductivity in the dissipator (the ionosphere, see e.g. Figure 3). The fact that j_y was always negative in the injection structures was held as evidence for the injected plasma truly acting as a generator plasma. This means that the electric field is generated by the braking action of the injected plasma. Lundin (1984) favoured the inertia term (dV_O/dt) because the injection structures were flowing faster than the "halo" in the boundary layer. As have been demonstrated by Stasiewicz et al. (1984) both the pressure term $(\nabla p_\parallel)$ and the anisotropy terms in equation (1) may be as important as the inertia term. The generator current is then driven by the free energy available in the parallel or perpendicular (cycloid) motion of particles.

Figure 4 gives another example of an entry layer encounter near the noon meridian, but now using low-speed mode data with a time resolution which is four times smaller than that in e.g. Figure 1. The two vertical bars mark the encounter of the boundary layer that we identified as the entry layer, according to the definition introduced by Paschmann et al. (1976). Notice for instance the fairly high number density, the reduced plasma flow and the region with high He^+ density inside the magnetopause (hatched area in the ion composition panel). The latter is again a feature which is predominantly found in the LLBL and EL. It is quite possible that this boundary layer passage contained as much structure as the previous example. However, if it was so, this is probably squashed out by the limited time resolution.

3. THE EXTERIOR CUSP

Hansen et al. (1976) identified from three HEOS-2 passes a region of reduced flow magnetosheath which separates the distant cusp plasma and the "free flow" magnetosheath plasma. This reduced flow region was later named "exterior cusp" after Vasyliunas et al. (1977). Sckopke (1979) also discussed the HEOS-2 measurements in the outer cusp region and concluded that the exterior cusp topology and plasma characteristics agreed with the predictions of Spreiter et al. (1968) - the exterior cusp being a magnetopause indentation that separates a pocket of hot and stagnant plasma from the magnetosheath flow. Hansen et al. (1976) and Sckopke (1979) also discussed the gasdynamic model by Walters (1966) and concluded that the HEOS-2 results did not corroborate the predictions of a second standing shock attached to the magnetopause near the cusp-magnetosheath interface as suggested by Walters. Sckopke (1979), however, noted

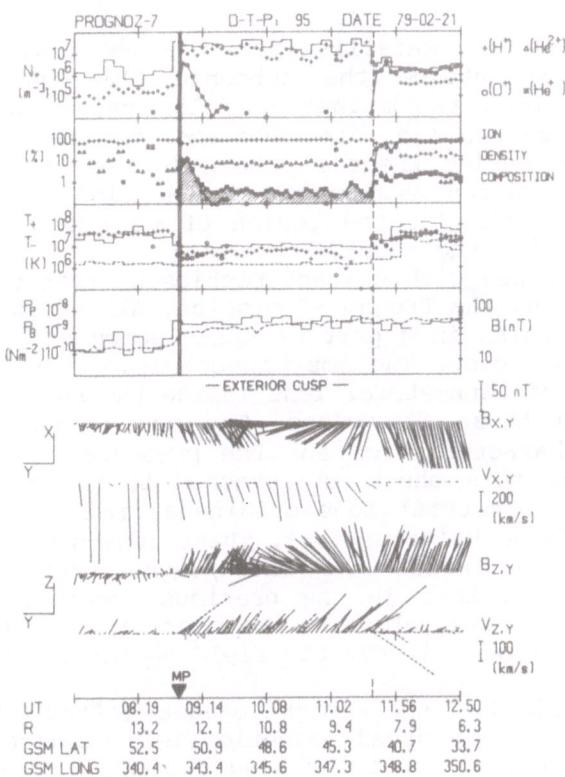

Figure 5. Low time resolution plasma and magnetic field parameters for an exterior cusp crossing near the noon meridian. The
exterior cusp extension is given the vertical bars (solid and
broken), where the solid bar marks the location of the shock
front/magnetopause and the broken bar represents the outer
boundary of the ring current. The format is the same as in
Figure 4.

that cases of supersonic magnetosheath flow could indeed be
found as well.
     We will here present exterior cusp passes which corrobora-
tes both the Spreiter model with a deep pocket of stagnant
plasma and the model of Walters with a supersonic magnetosheath
and a second shock attached to the magnetopause near the outer
cusp.
     Figure 5 shows an inbound crossing of the magnetopause and
boundary layer near the noon meridian during the Feb. 21, 1979
magnetic storm discussed by e.g. Johnson et al. (1983) and

Hultqvist (1983). This pass is unusual in many ways, but I will here only focus on a few of the flow and magnetic field properties.

Around 0850 UT the satellite passed a shock with a supersonic plasma flow outside the subsonic flow inside of it. Notice, however, that the magnetic field inside the shock is very magnetosphere-like (in direction and magnitude) and has a plasma flow which during the first hour is predominantly field aligned and much reduced as compared to the flow outside of the shock. Notice also the limited region of very high O^+ and He^+ plasma density inside the shock.

Only a few passes with characteristics similar to this have been observed during the Prognoz-7 mission. All of them occurred near the noon meridian in a part of space which one would refer to as the exterior cusp. The most spectacular characteristic – the high amounts of terrestrial ions inside the shock – has only been observed in those few cases. The much reduced flow the magnetic field characteristics and the presence of terrestrial ions suggest that this shock is attached to the magnetopause. Most likely the terrestrial ions originate from the high altitude ionosphere (high He^+ fraction). These ions have then become accelerated locally inside the shock/magnetopause in a way similar to that described in the previous section. Notice for instance that the O^+ flow close to the shock is directed perpendicular to the H^+ flow and magnetic field vector in the YZ-velocity projection.

Further inside the shock/magnetopause (after ≈1000 UT) a region with strongly perturbed magnetic field occurred – a perturbation that remained until the ring current encounter around 1130 UT.

This pass is therefore an example of that a second shock may indeed develop in the outer cusp part of the magnetosheath as suggested by Walters (1966) – a shock which is also attached to the magnetopause.

Figure 6 gives another example of an inbound crossing of the magnetopause, now in the dawn-flank part of the cusp. The magnetic field orientation inside the magnetopause (≈0410 UT) shows that the crossing occurs on field lines that are located slightly poleward of the cusp. Notice that the flow is much reduced and essentially field aligned inside the magnetopause, yet the plasma density in the outer part is similar to the plasma density in the magnetosheath. The innermost part is characterized by strong fluxes of upgoing O^+ ions which appear to be injected directly from the low altitude ionosphere (strong variations in O^+ temperature as well as flow velocity, the O^+ density exceeding the He^+ density). At about 0920 UT the satellite encounters the ring current, characterized by the lack of plasma flow, a strong increase in plasma temperature and a reorientation of the magnetic field vector.

This exterior cusp encounter gives an indication of the plasma characteristics which are expected to be present in the

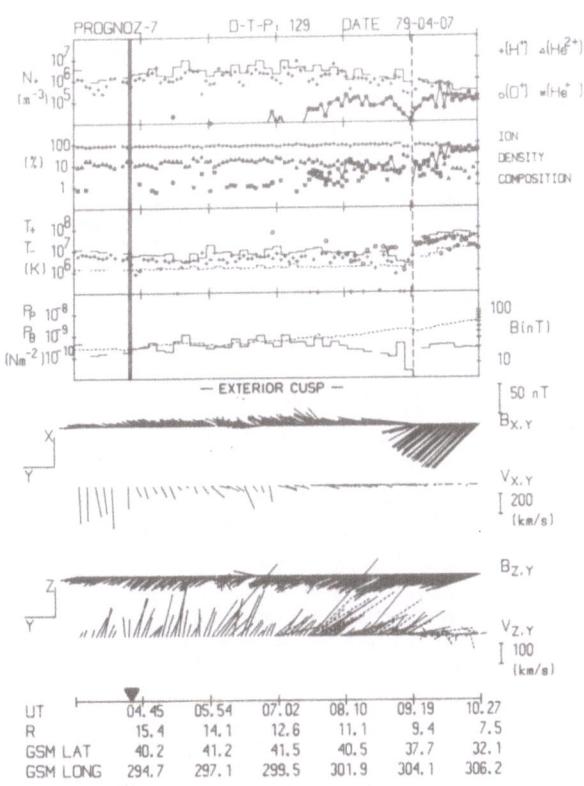

Figure 6. Low time resolution plasma and magnetic field parameters for an exterior cusp crossing near dawn slightly poleward
of the cusp (as indicated by the magnetic field direction). The
format is the same as in Figure 5.

plasma mantle further tailward of the cusp. The main difference
between the exterior cusp here and the mantle is the lack of a
well defined magnetopause magnetic field signature, the much
higher density and to some extent lower flow velocity in the
exterior cusp. Finally the mantle usually has an "empty" lobe
inside, instead of a region of hot ring current plasma.
      The exterior cusp characteristics in this example agree
with those found from HEOS-2 (Skopke, 1979) i.e. a magnetopause
indentation with stagnant magnetosheath plasma in the indentation region as suggested by Spreiter et al. (1968). Notice, however, that we infer at least the innermost part of this "pocket"
to be an extension of the plasma mantle. It is not topologically
connected to the magnetosheath as will become obvious in the
next section. Even if no clear magnetopause signature is evident
from this pass, we may yet identify a magnetospheric boundary

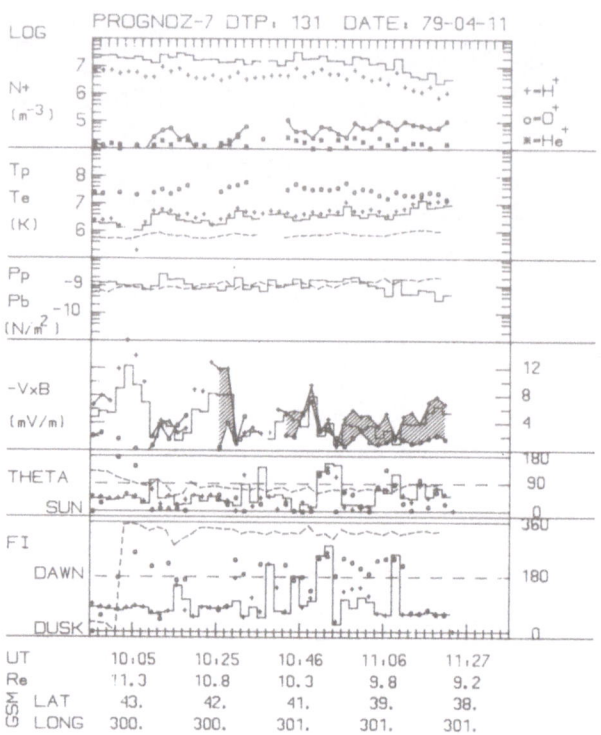

Figure 7. High time resolution plasma and magnetic field parameters for an exterior cusp encounter in almost the same local
time and latitude sector as the encounter in Figure 6. The
format is the same as in Figure 2, but with the temperature and
pressure panels included.

(at ≈0430 UT). Inside this boundary the plasma seems to be at
least partly attached to terrestrial field lines.

Figure 7 gives another example of an exterior cusp crossing, but now in considerable more detail as deduced, from the
high speed data. The plasma characteristics are here very similar to those found in the previous exterior cusp crossing of
Fig. 6. The density of the magnetosheath plasma in the boundary
layer is, however, almost an order of magnitude higher, but the
flow velocity is yet not very high (≲200 km/s). Also in this
time scale, the transition between magnetosheath and magnetosphere plasma is very smooth, opposite to what was found in the
LLBL/EL (e.g. Fig. 1).

Notice that a significant cross field flow component may be
observed in the exterior cusp as well, especially from the
magnetosheath plasma flow. It is possible that Prognoz-7 encountered the magnetosheath around 1005 and 1025 UT. The main argu-

ment for magnetosheath encounter is the lack of O^+ and the very strong H^+ cross field flow in these regions. A weak magnetic field signature may also be observed, especially before 1008 UT when the B_x vector is positive. A marked dynamo signature, as evidenced by the difference in cross-field flow vectors for O^+ and H^+ respectively (see equation (2)), can be found in the innermost part of the boundary layer. Except for a few polarity reversal (at e.g. 1050 - 1056 UT) the electric field is dominated by a sunward component (and upward for H^+). This is very similar to the LLBL/EL observation in Fig 1. However, considering the magnetic field orientation, this means that the injected magnetosheath plasma is now escaping outward instead of inward as in the LLBL/EL-observation.

4. THE PLASMA MANTLE

As we have mentioned already, the plasma mantle may be defined as the high latitude/tailward extension of the exterior cusp. This does not mean that a simple geometrical connection to the individual source regions for the mantle plasma exists. For instance, we expect most of the ionospheric plasma component (eg. O^+) to be injected into the mantle from the dayside high latitude ionosphere, whilst magnetosheath plasma may be injected directly into the mantle via the exterior cusp. An extensive review of the composition and other characteristics of the mantle has been given by Lundin et al. (1982). Some of their results, plus a few new highspeed data observations, are summarized here.

The plasma mantle density and composition depend strongly on the magnetospheric disturbance level. This is illustrated by Figure 8. In quiet conditions the mantle is frequently "weak", i.e. thin and/or has low plasma density. The relation with the disturbance level is, however, not very clear. This can be seen in Figure 8a, in which the product of the thickness of the mantle and the peak proton flux recorded during the passage is plotted for 32 mantle passages of Prognoz-7. The peak proton flux was then the integral flux (0.2-17 keV) taken with spectrometers pointing perpendicularly to the Sun-Earth line (spin axis of the satellite). As magnetic activity parameter the sum of K_p over a 12 hour period centered on the magnetopause crossing has been plotted. The circles represent the nightside observations and the triangles flank ones. A positive correlation can be seen. It is, however, not a very strong one. The scatter is large and fairly good mantles are found also at low activity level.

A similar scatter diagram for O^+ ions is contained i Figure 8b. The correlation between K_p and the product of mantle thickness and peak O^+ flux is fairly good. The amount of O^+ ions in the mantle thus depends strongly on the magnetospheric disturbance level. In very quiet situations no O^+ ions are found

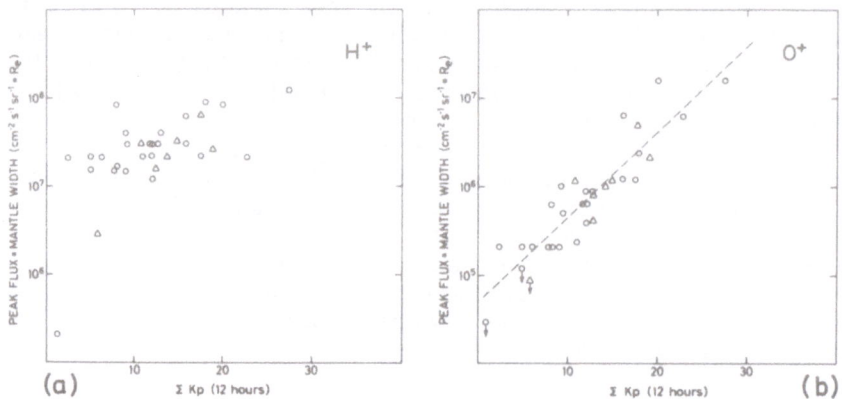

Figure 8. Diagram showing the peak flux of ions (H$^+$ and O$^+$) times the estimated width of the plasma mantle versus magnetic activity for 32 mantle passages (after Lundin et al., 1982a). The broken line in Figure 8b represents an exponential least squares fit with a correlation coefficient of 0.82.

in the mantle. The difference between the point distributions in Figure 8 is obviously consistent with the existence of a H$^+$ source fairly independent of magnetic activity in addition to the activity dependent one(s).

The mantle plasma generally flows along the magnetopause and along the magnetic field lines with velocities up to that of the solar wind in the adjacent magnetosheath. An example of such a mantle is shown in Figure 9.

Near the inner edge of the mantle in Figure 9 there is a region with β>1, no or only very little O$^+$ content, and different magnetic field vector than in the vicinity of the mantle or in the magnetosheath. The magnetopause appears to have been crossed near UT 1450. This high β-region is an example of a large number of observations of what is believed to be a solar wind plasma cloud that has penetrated into the mantle (Lundin and Aparicio, 1982, Lundin et al., 1982). In some of these cases one may dispute whether the satellite was in the mantle or had made an excursion into the magnetosheath. Because of a number of arguments, including the presence of O$^+$ in some regions, special flow and magnetic field characteristics, occurrence of such regions at the inner edge of the mantle and the occurrence of a variety of degrees of "magnetosheath likeness" in both particles and fields among the observed β>1 events we believe that these regions were not magnetosheath but part of the mantle, obviously special parts frequently associated with strong currents.

At the edges of these dense magnetosheath-like regions in the mantle the density of ionospheric ions frequently peak (see Figure 9), indicating that particularly strong Birkeland currents flow there.

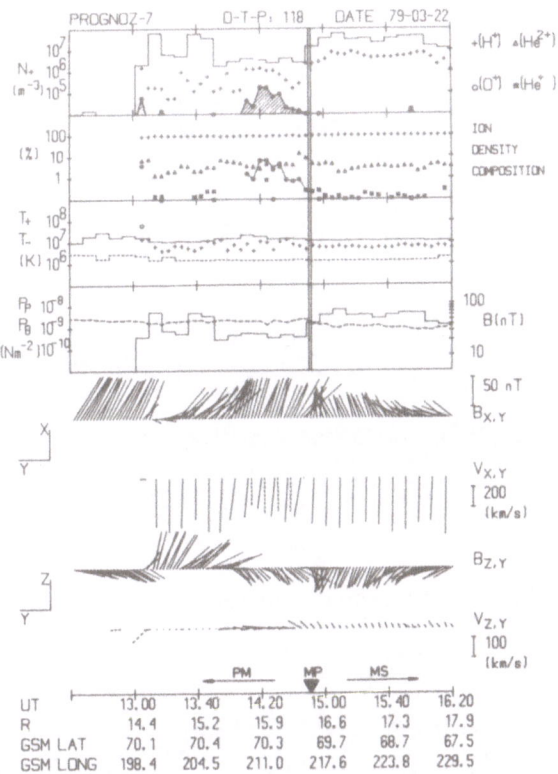

Figure 9. Low time resolution plasma and magnetic field parameters for an outbound crossing of the plasma mantle and magnetopause (marked by the vertical bar and "MP"). Hatched area in
the number density panel marks O^+ density. The data format is
the same as in Figure 5.

We may from this conclude that newly injected magnetosheath
plasma may be present in the plasma mantle as well. This is of
course not surprising, considering the frequent observations of
newly injected plasma structures in the dayside LLBL and EL. The
observation suggests, however, that injection may also take
place directly into the mantle.

An example of a penetration structure which has been injected much further upstream in e.g. the entry layer, and which
thus is expected to have lost most of its excess momentum, is
shown in Figure 10. This penetration structure was located in
the innermost part of the mantle around 2220 UT. The magnetopause was crossed some ten earth radii further out (at ≈1040 UT
on Jan 4) suggesting that this was an extremely thick mantle.
The plasma density and β-value in the penetration structure is
quite high which results in a strong diamagnetic effect. So in

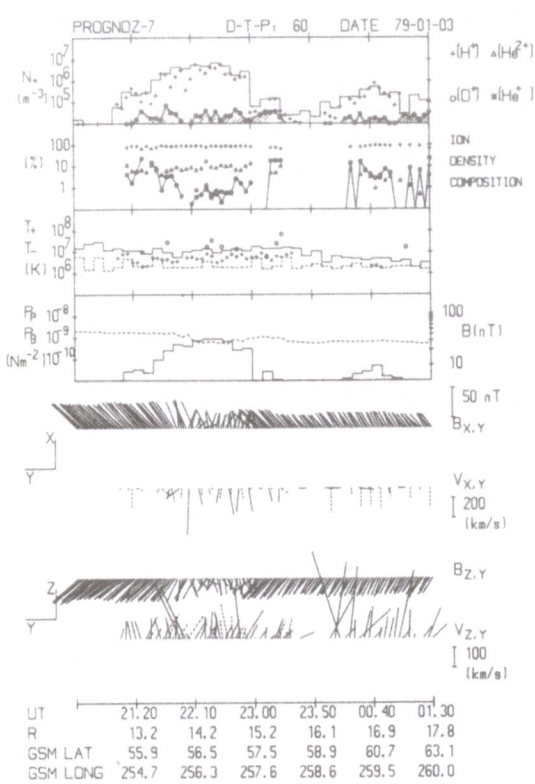

Figure 10. Low time resolution plasma and magnetic field parameters for an outbound encounter of the plasma mantle with an
injection structure (≈2210 UT). The magnetopause was crossed
some 12 hours later (at ≈24.5 R_e).

this respect the injection structure is quite similar to that in
Fig. 9. Except for the difference in flow, the penetration
region in Fig. 10 also differs by the presence of a substantial
amount of O^+ ions in the center of the structure. This is a
further indication of a far upstream penetration region of this
structure, i.e. the injected plasma has already released most of
its available free energy.

Figure 11 shows another mantle penetration structure, now
obtained from the high speed data. This structure differs from
the previous two in that it represents an "intermediate" stage.
Except for the lack of O^+ and an enhanced flow velocity in the
center, the plasma momentum parameters are yet quite similar to
those found in Fig. 10.

Notice from the plasma flow panels that a pronounced difference in flow vectors between O^+ and H^+ is again present. A

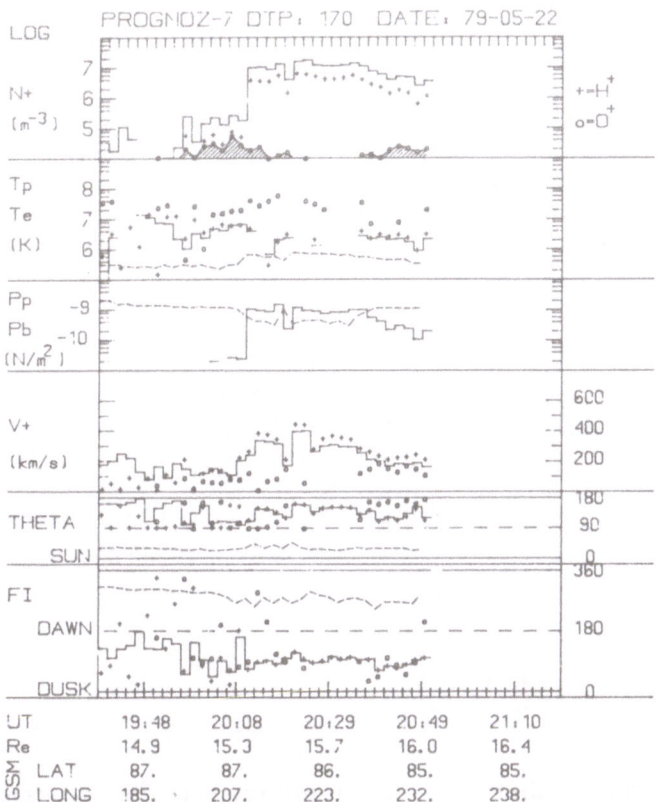

Figure 11. High time resolution plasma and magnetic field parameters for an outbound encounter of a plasma mantle injection
structure. The format is the same as in Figure 1.

peculiar wave-like motion of both O^+ and H^+ ions may also be
observed. This inward/outward motion of particles is quite similar to the "plasma blob" observation by Sckopke et al. (1981),
only that we can here see an out-of-phase characteristics between O^+ and H^+.
    Figure 12 finally shows the cross-field flow for the
example contained in Fig. 11. Notice that we may again detect
dynamo characteristics ($j_y<0$) in these penetration structures
(marked by hatched areas). The cross-field flow vectors are very
variable in direction, but a more careful analysis reveals a
rotational characteristic versus time which is due to the quasi-
periodic inward/outward motion of the plasma. Notice also that
the weakest cross-field flow is present in the region dominated
by magnetosheath plasma ($\approx$2029 UT). This suggests that the
energy transfer is concentrated to the edges of the penetration
structure.

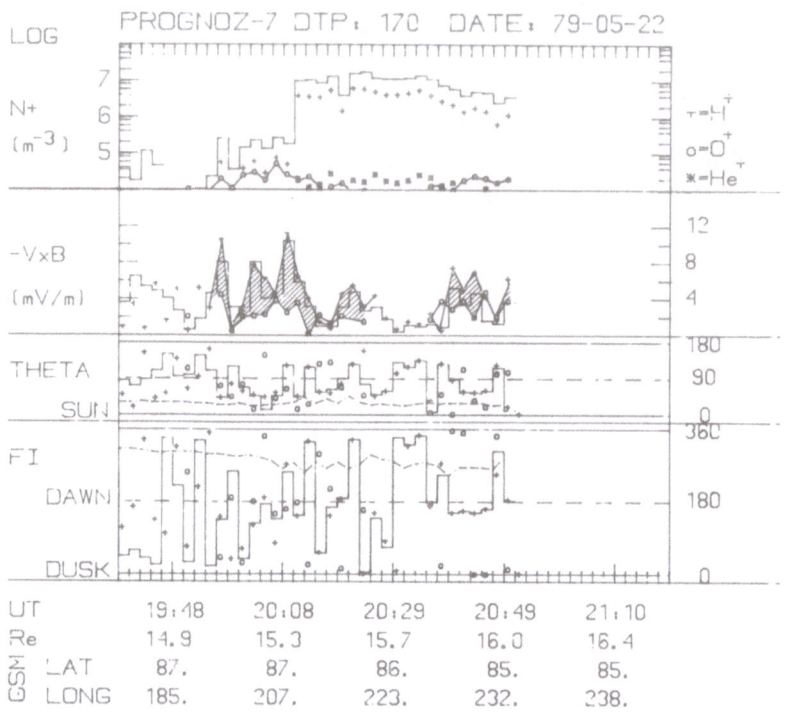

Figure 12. The magnitude and direction of the cross-field flow for the plasma mantle encounter contained in Figure 11. The format is the same as in Figure 2.

Notice also from Fig. 11 that the flow of H^+ ions has a tendency of being directed outward, except in the central part of the injection structure where it is more field aligned than in other parts of the mantle. This tendency of outward cross-field motion is similar to that found in the exterior cusp.

The reason for concentrating on plasma mantle examples with magnetosheath injection structures is because they usually occur together with enhanced fluxes of O^+ ions and hence are attributed to regions of local energy transfer and upward acceleration of ionospheric ions. There are, however, also mantle crossings of Prognoz-7 in which the mantle represents a smooth transition region between the lobe and magnetosheath as well as mantles in which hardly any ordered flow has been observed. An example of a non-flowing "stagnant" mantle can be seen in Figure 13. The stagnant mantles have also frequently, as in the case shown in Fig. 13, much higher plasma temperatures than normal and fairly large O^+ content (Lundin et al., 1982). The processes responsible for their production are not understood. One hint

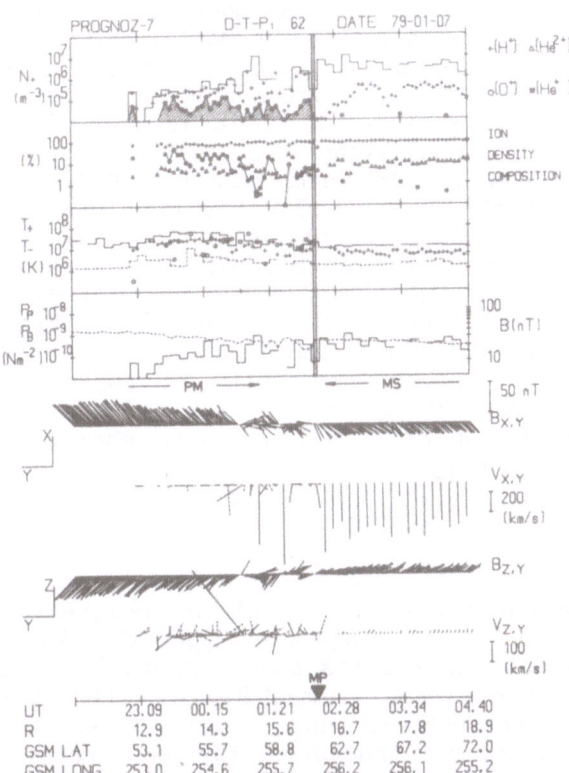

Figure 13. Low time resolution plasma and magnetic field parameters for an outbound crossing of a "stagnant" plasma mantle.
The format is the same as in Figure 5.

may be the fact that most of the stagnant mantles are observed
whenever the magnetic field vectors on either sides of the
magnetopause are more or less antiparallel. However, this condition does not seem to be a sufficient one. The antiparallel condition has an interesting bearing to the "Theta" aurora (Frank
et al., 1982), and it may also be that the stagnant mantle and
the "Theta" aurora are actually related phenomena. One interesting interpretation is that closed field lines during e.g. northward IMFs may reach extremely far north (above 80° in magnetic
latitude), and solar wind plasma injected into these field lines
may become locally trapped.

Another significant outcome of the plasma mantle study was
that the nightside magnetopause appears to be a more solid
boundary for the heavy ions of ionospheric origin, than is expected from the open magnetosphere model. This is illustrated by
Figures 11 and 13. In fact, whenever strong O^+ fluxes were found

in the plasma mantle – a common observation during strong magnetic disturbances – the composition boundary at the magnetopause
was more pronounced. The mantle passages of Prognoz-7 with significant amounts of O^+ ions in the mantle also constituted a
majority of the total set of 32. Even if not all field lines are
supposed to be open according to the model, the openness of the
magnetosphere seems questionable if it is so highly improbable
to see open field lines along the mantle magnetopause that
Prognoz-7 should have missed them all. The O^+ ions observed in
the mantle are expected to be slightly decelerated in passing
through the magnetopause current layer into the magnetosheath
(see e.g. Cowley 1980). The observable O^+ ion density on the
magnetosheath side should thus be higher, rather than lower,
than in the mantle and the ion energy should certainly be in the
energy range of the PROMICS-1 instrument. We thus expect to see
a continuous change of the O^+ density at the magnetopause in
going from inside to outside if the magnetic field lines pass
through it. Instead the observational data show a quite strong
composition discontinuity.

5.   CONCLUSIONS

The Prognoz-7 measurements of the plasma composition and flow
characteristics   in   the   cusp-magnetosheath   interface   have
revealed interesting new results about processes responsible for
the transfer of energy and momentum into the magnetosphere.
Observations in the LLBL/EL have demonstrated that high β,
magnetosheath injection structures may penetrate the dayside
magnetopause and generate electric fields which locally accele-
rates the pre-existing boundary layer plasma (e.g. "cold" high
altitude ionospheric plasma). The observed difference in vxB
motion for ions of different origin is believed to be the signature of an MHD-dynamo process driven by injected clouds of
magnetosheath plasma.
      The LLBL/EL is usually quite structured in the high time
resolution data as opposed to the exterior cusp, where the
mixing of magnetosheath and magnetosphere plasma seems much
smoother. The exterior cusp is also characterized by a slow
transition from reduced flow magnetosheath into a region with
strong fluxes of upgoing ionospheric ions. Significant cross-
field flow components and differential (mass-dependent) ion
drift motions indicate that the magnetosheath plasma in the
exterior cusp acts as a generator plasma. Observations of a
second standing shock attached to the magnetopause in the
exterior cusp indicates that the gasdynamic model of Walters
(1966) at times may apply.
      High β magnetosheath injection structures have been
observed in the plasma mantle as well. This suggests that plasma
injection may take place on the poleward as well as on the
equatorward side of the cusp. Also the plasma mantle injection

structures are associated with dynamo signatures, but now the terrestrial ions originate mostly from the low-altitude iono-sphere (O^+ more aboundant than He^+). Terrestrial ions in the plasma mantle comprises, usually less than 10% of the total plasma population - the percentage being highly magnetic acti-vity dependent.

Injection structures are also observed very "deep" in the plasma mantle - sometimes adjacent to the lobe. Whether these structures are attached to the magnetopause or not ("groves" or isolated plasma clouds) is difficult to say from these single-point measurements. However, there are energy dispersion charac-teristics in these structures which favours a "temporal" injec-tion of isolated plasma structures that are well separated from the magnetopause.

"Stagnant" plasma mantles, i.e. mantles which are charac-terized by high, plasma sheet-like, temperatures and the lack of plasma flow are also observed. The external magnetic field orientation associated with "stagnant" mantles suggests that it is connected to the "Theta aurora" (Frank et al. 1982), but the relation is not clear.

Finally we conclude that a large majority of the nightside magnetopause crossings represents a distinct plasma composition boundary. This observation contradicts the general belief that the plasma mantle field lines are threaded to the solar wind magnetic field lines.

REFERENCES

Akasofu, S-I., 1983, Planet. Space Sci., **31**, 25.
Cowley, S.W.H., 1980, Space Sci. Rev., **26**, 217.
Dungey, J.W., 1961, Phys. Rev. Letter, **6**, 47.
Eastman, T.E., Hones, E.W., Jr., Bame, S.J. and Asbridge, J.R., 1976, Geophys. Res. Lett., **3**, 685.
Frank, L.A., Craven, J.D., Burch, J.L., and Winnignham, J.D., 1982, Geophys. Res. Lett., **9**, 1001.
Haerendel, G., Paschmann, G., Sckopke, N., and Rosenbauer, H., 1978, J. Geophys. Res., **83**, 3195.
Hansen, A.M. Bahnsen, A., and D'Angelo, N., 1976, J. Geophys. Res., **81**, 556.
Heikkila, W.J., 1979, Proceedings of Magnetospheric boundary layers conference, Alpach, ESA SP-148.
Hultqvist, B., 1983, Planet. Space Sci., **31**, 173.
Johnson, R.G. (Ed.), 1983, Energetic Ion Composition in the Earth's Magnetosphere, Terra Scientific Publish. Co., Tokyo,D. Reidel Publish. Co., Dordrecht.
Lemaire, J., and RotH, M., 1978, J. Atmos. Terr. Phys., **40**, 331.
Lemaire, J., 1979, Proceedings of Magnetospheric boundary layers conference, Alpach, ESA SP-148.

Lundin, R., Hultqvist, B., Pissarenko, N., and Zakharov, A., 1982, Space Sci. Rev., 31, 247.

Lundin, R., Hultqvist, B., Pissarenko, N., Zakharov, A., 1982a, Space Sci. Rev., 31, 247.

Lundin, R., and Aparicio, B., 1982, Planet. Space Sci., 30, 93.

Lundin, R., 1984, Planet. Space Sci., in print.

Lundin, R., and Dubinin, E., 1984a, Planet. Space Sci., in print.

Lundin, R., and Dubinin, E.M., 1984b, submitted to Planet. Space Sci.

Paschmann, G., Sonnerup, B.N.Ö., Papamastorakis, I., Sckopke, N., Haerendel, G., Bame, S.J., Asbridge, J.R., Gosling, J.T., Russel, G.T., and Elphic, P.C., 1979, Nature, 282, 243.

Sckopke, N., Paschmann, G., Rosenbauer, H., and Fairfield, D.H., 1976, J. Geophys. Res., 81, 2687.

Sckopke, N., and Paschmann, G., 1978, J. Atmosph. Terr. Phys., 40, 261.

Sckopke, N., 1979, Proceedings of Magnetospheric boundary layers conference, Alpach, ESA SP-148.

Sckopke, N., Paschmann, G., Haerendel, G., Sonnerup, B.U.Ö., Bame, S.I., Forbes, T.G., Jones, E.W. Jr., and Russel, C.T., 1981, J. Geophys. Res., 86, 2099.

Spreiter, J.R., A.Y. Alksne, and A.L. Summers, 1968, in Physics of the Magnetosphere (ed R.L. Carovillano) Dordrecht, D. Reidel.

Stasiewicz, K., Lundin, R., and Hultqvist, B., 1984, to be submitted to Planet. Space Sci.

Walters, G.K., J. Geophys. Res., 1966, 71, 1341.

Volkov, T.F., 1966, Reviews of Plasma Physics, Vol.4, 1, Consultants Bureau, New York.

SIMULATION OF SOLAR WIND–MAGNETOSPHERE INTERACTION

J. Lemaire
Institut d'Aéronomie Spatiale de Belgique
3, Avenue Circulaire
B-1180 Brussels
Belgium

ABSTRACT. According to Bostik's (1956) generic definition, any plasma density irregularity observed in the solar wind and in the magneto-phere could be called a "plasmoid".
    The motion of plasmoids across different non-uniform magnetic field configurations has been discussed from a theoretical point of view. When the dielectric constant of the streaming plasma is large enough for collective polarization effects to become important, an electric field develops which permits cross-B motions of all charged particles as a whole plasma entity. The intensity and direction of this E-field has been given for a plasmoid traversing a magnetic "tangential discontinuity".
    It is reemphasized that the value of the integrated Pedersen conductivity is a determining factor in cross-B plasma motion, although it has been neglected in most reconnection theories. The proper scaling of the value of the integrated Pedersen conductivity is therefore also essential in Terrella simulation experiments.
    It is argued that the magnetopause is located at the surface where the geomagnetic strength has a critical value and where the value of the integrated Pedersen changes abruptly from a typical solar wind value to magnetospheric value which is determined by the transverse conductivity in the Earth ionosphere.
    The thickness of the Plasma Boundary Layer, where 'viscous-like' interaction is taking place as a result of Impulsive Penetration of solar wind plasmoids, is indeed determined by the Pedersen conductivity in the ionosphere and by the degree of inhomogeneity of the impinging solar wind stream. Consequently, the Impulsive Penetration model shares a feature in common with conventional "closed" magnetospheric models where 'viscous-like interaction' is essential to drive magnetospheric convection.
    On the other hand, interconnection of interplanetary magnetic field lines and geomagnetic field lines, as in conventional "open" magnetospheric models results, in the Impulsive Penetration model, from collective diamagnetic effects produced by magnetized plasmoids injected into the magnetosphere. But in this more recent model, interconnection is an intermittent (i.e. a time dependent) and a

33

J. A. Holtet and A. Egeland (eds.), The Polar Cusp, 33–46.
© *1985 by D. Reidel Publishing Company.*

patchy (i.e. localized in space) process.

Simple laboratory simulations have been suggested to verify the theoretical expectations discussed in this article.

INTRODUCTION .

A variety of names have been used to describe the plasma "inclusions", vortices, eddies, filaments, magnetic holes, irregularities or "flux transfert events" that are currently observed in the vicinity of the magnetopause. All these events are 'plasma-magnetic entities' like those studied by Bostik (1956) in his laboratory experiments. He has used the word 'plasmoids' to describe these plasma-magnetic entities. To avoid further proliferation of additional new synonyms for the same type of physical phenomenon or 'event', one should once and for all agree on a generic name : 'plasmoid' originally introduced in 1956 by Bostik, seems to be a most appropriate choice.

To illustrate the interaction of solar wind plasmoids with the geomagnetic field, and to model their impulsive penetration into the magnetosphere, let us first recall the important results deduced from Baker and Hamel's (1965) laboratory experiments.

BAKER AND HAMEL'S EXPERIMENTS .

Fig. 1 illustrates schematically these experiments and their results : a non-diamagnetic collisionless plasma stream can move with a constant velocity v_o across a uniform magnetic field if the magnetic field lines pass through insulating walls : i.e. if Σ_p, the Pedersen conductivity integrated along the magnetic field lines, is equal to zero or at least very small. Polarization charges of opposite sign accumulate respectively at the upper and lower surface of the injected plasma stream as a result of drifts acting in opposite directions on the ions and electrons. This action results in the production of a net positive charge on the upper boundary of the stream and an equal amount

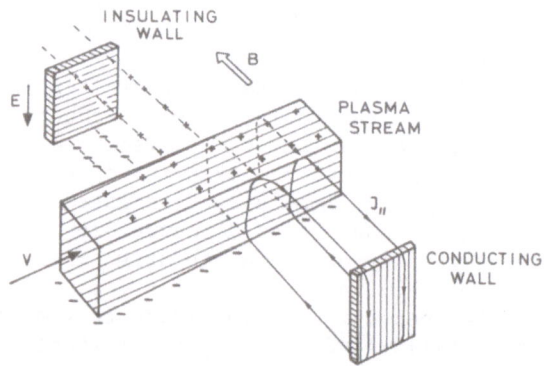

Figure 1.   Simplified sketch of a non-diamagnetic plasma beam crossing a uniform magnetic field;   a) the charging up effect of an insulating wall and   b) the depolarizing effect of a conducting wall (Baker and Hamel 1965) are illustrated.

of negative charge on the lower boundary as shown in fig. 1. Such a
charge separation will continue until the resulting electric field is
strong enough to cancel the deflecting effect of the magnetic field,
i.e. until

$$\underline{E} = - \underline{v}_o \times \underline{B} \qquad\qquad (1)$$

The plasma stream then continues to move across the uniform magnetic
field with the velocity

$$\underline{v} = \underline{E} \times \underline{B}/B^2 = \underline{v}_o \qquad\qquad (2)$$

Since the plasma is collisionless, its electric conductivity
parallel to $\underline{B}$ is very large, and the polarization charge will be re-
distributed such that the magnetic field lines become lines of equal
electric potential. Furthermore, in a collisionless plasma the transverse Pedersen conductivity σ_p is almost equal to zero. But when the
magnetic field lines pass through a good conductor, the polarization
charge will be neutralized by the short-circuiting effect in the plate
as fast as it is produced by the action of the unbalanced cross-B drift
currents. Depolarizing currents will continue to flow in the conducting
walls as long as the plasma stream has forward momentum (Baker and
Hamel 1965; Baum et al. 1984).

The above model of a polarized plasma stream will be extended to
cases of non-uniform magnetic field distributions as well as to cases of
non-uniform wall conductivities. When these extensions are applied to a
diamagnetic plasma stream of finite length (i.e. a plasmoid), they
become especially useful in explaining some of the phenomena observed
near the magnetopause : e.g.(i) 'viscous-like interaction' in the
Plasma Boundary Layers; (ii) intermittent and patchy interconnections
of interplanetary and geomagnetic field lines.

PLASMA FLOW WITH A PARALLEL COMPONENT.

A first extension of Baker and Hamel's plasma experiment would be
to inject a plasmoid with initial velocity components both perpen-
dicular and parallel to the uniform B-field : i.e. $v_{o_\perp} \neq 0$ and $v_{o_\parallel} \neq$
0. This situation is illustrated in fig. 2a. It simulates the motion of
solar wind plasma elements across the interplanetary magnetic field
which is not perpendicular to the radial solar wind velocity.

When the integrated Pedersen conductivity is zero (i.e. for non-
conducting walls or, equivalently, for a collisionless plasma unbounded
along the magnetic field lines) the electric field which builds up
within the plasmoid is perpendicular to $\underline{B}$ and is given by Eq. (1). The
perpendicular component of the streaming velocity is given by Eq. (2)
and is equal to the initial perpendicular velocity. In the absence of
external forces the parallel component of $\underline{v}$ is conserved. As a
consequence, the total velocity $\underline{v}$ of the plasma element remains also
equal to the injection velocity $\underline{v}_0$, as in Baker and Hamel's experiment
where $v_{o_\parallel}$ was equal to zero.

The field-aligned electric conductivity is large in collisionless
plasmas, so that parallel electric fields are almost everywhere equal

to zero. However at the edge of a plasmoid, where magnetic field lines
traverse its surface, a double potential layer is formed to prevent
electrons from escaping out of the plasmoid (Lemaire and Scherer 1978).
    Furthermore, when the total plasma pressure inside the plasmoid
does not match the exterior field and kinetic pressure, the plasmoid
expands or contracts. Field-aligned expansion of the plasmoid or cross-
B expansion ceases when the net momentum flux across all parts of the
surface of the plasmoid is equal to zero.
    The double potential layers form all around at the surface, and
produce electric fields in a sheath whose thickness ranges between a

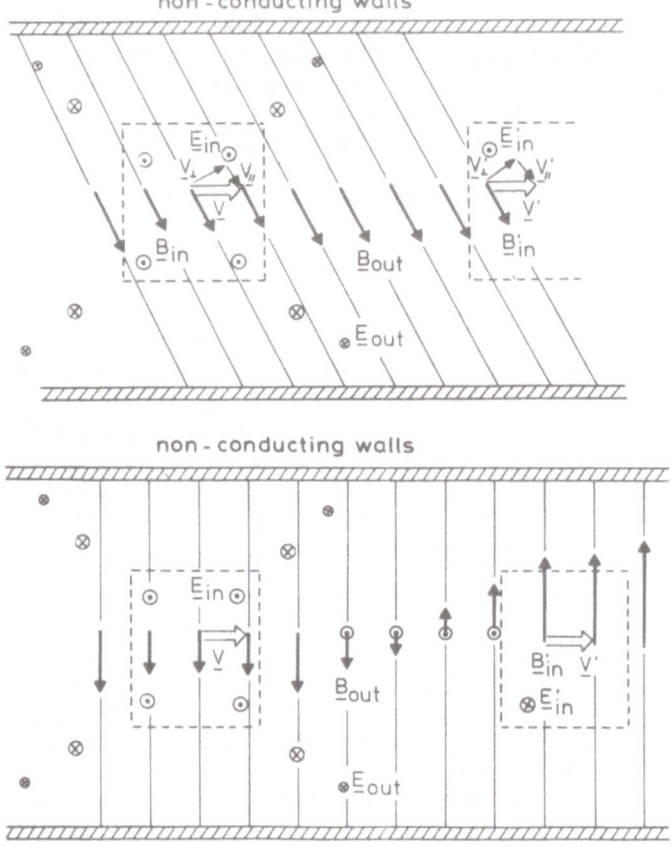

Figure 2. Schematic representation of plasmoids moving across magnetic field lines which pass through insulating walls. In 2a) the magnetic field is uniform, the initial bulk velocity of the injected
plasmoid has components perpendicular and parallel to $\underline{B}$; its bulk
velocity is conserved as well as the magnetic flux through a co-
moving surface. In 2b) the magnetic field direction and strenght
change along the ox-axis.

few Debye lengths and a few ion Larmor radii. These microscopic edge E-
fields must be added to the large scale-polarization field (1) induced
by the plasma motion.

Bostik (1956) observed that particles from the lateral periphery
of the plasmoid are left behind. This result is obvious considering
that particles in the surface sheaths find themselves in a weaker
electric field than those inside. These particles experience smaller
electric drifts than does the bulk of the plasma and are consequently
left behind (Schmidt, 1960).

NON UNIFORM MAGNETIC FIELDS.

Another possible extension of Baker and Hamel's experiment would
be to inject a non diamagnetic plasmoid into an external magnetic field
whose direction and/or intensity change along its trajectory. A
tangential discontinuity like that illustrated in fig. 3 corresponds to
such a magnetic field distribution. Let us then assume that the initial
injection velocity v_o is parallel to the x-axis; furthermore let us
consider that v_o has no component along the magnetic field lines which
are all perpendicular to ox.

Following a demonstration similar to that given by Schmidt (1960)
for the two-stage plasma accelerator, Lemaire (1984) has shown how a
charge separation electric field is produced in this case, inside the
moving dielectric plasmoid, by the gradient-B and inertial drifts.
Indeed ions and electrons drift in opposite directions toward the
lateral surfaces of the plasma element. The resulting electric field
direction remains always perpendicular to the $\underline{B}(x)$. The components of
the electric field inside the plasmoid are then given by

$$E_x(x) = v_z B_y - v_y B_z \; ; \quad E_y(x) = v_x(x) \, B_z(x) \; ; \quad E_z(x) = - \, v_x(x) \, B_y \, (x)$$

$$(3)$$

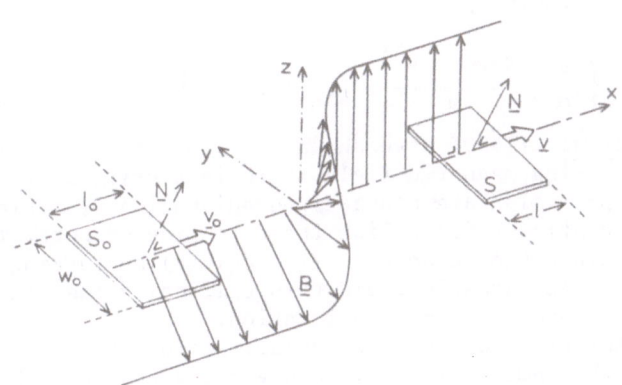

Figure 3. Magnetic field distribution across a tangential discontinuity.
A plasmoid injected into such an external B- field can move through with
a constant bulk velocity when the strength B is independent of x.

It can be verified that the bulk velocity of such a plasmoid is constant and equal to $\underline{v}_o$, when the magnetic field strength is independent of x :

$$\frac{d\underline{v}}{dt} = \frac{d(\underline{E} \times \underline{B})/B^2}{dt} \simeq 0 \quad \text{when} \quad \frac{dB}{dx} = 0 \qquad (4)$$

This result is slightly changed when $dB/dx \neq 0$. Indeed when the magnetic field strength as well as its direction are function of x, Lemaire (1984) has shown that

$$\frac{dv}{dt} = \frac{d|\underline{E} \times \underline{B}|/B^2}{dt} = -\frac{m^+\overline{(v_\perp^+)^2} + m^-\overline{(v_\perp^-)^2}}{2m^+B} \cdot \frac{dB}{dx} \qquad (5)$$

Furthermore, taking into account conservation of the magnetic moment of charged particles

$$\mu = m(V_\perp)^2/2B \qquad (6)$$

one obtains by integrating eq. (5)

$$v^2 + (V_\perp^+)^2 + (V_\perp^-)^2 = Cst = v_o^2 + (V_\perp^+)_o^2 + (V_\perp^-)_o^2 \qquad (7)$$

This means that the sum of the translational, $1/2\ mv^2$, and thermal (gyration) energy, $1/2\ m(V_\perp^+)^2 + 1/2\ m(V_\perp^-)^2$, is a constant of motion. A similar result was first demonstrated by Schmidt (1960) but for non-rotating magnetic field distributions.

A plasmoid penetrating into a region of lower magnetic field intensity will therefore be accelerated due to adiabatic loss of thermal energy. Conversely, when a solar wind plasma element moves from the magnetosheath into the magnetosphere where the B-field intensity is often higher, it can be seen by combining eqs. (6) and (7) that the velocity of the intruding plasmoid decreases as

$$v(x) = \left\{ v_o^2 + \frac{2(\overline{\mu^+} + \overline{\mu^-})}{m} \left[B_o - B(x) \right] \right\}^{1/2} \qquad (8)$$

Demidenko et al. (1969) have shown experimentally that in a monetonically increasing magnetic field there is a critical field strenght $B(x_1)$ where the plasmoids are stopped and adiabatically reflected. This holds also for rotating B-field like those considered here. The value of the critical field can be deduced from eq. (8) by setting $v(x_1) = 0$. In the case of the geomagnetic field this critical value $B(x_1)$ determines the average magnetopause location.

As the vector $\underline{B}(x)$ rotates, the polarization electric field $\overline{\underline{E}}(x)$ rotates through the same angle but, both vectors remain orthogonal to each other : see eq. (3).

An especially simple case (but often considered in reconnection theories) is when $B_y = 0$, and when the remaining component $B_z(x)$ changes sign at some distance x_2. It can be seen from eq. (3) that the only non-zero component of the electric field, $E_y(x)$, does then also change sign at x_2.

In stationary reconnection models where $B_z(x)$ changes sign at a neutral line, one has traditionally forced the electric field to have non-zero component in the y direction at $x = x_2$. This assumption has been introduced by Petschek (1964) to make the flow velocity $\underline{v}$ singular at the $x = x_2$; indeed when $B_z(x)$ changes sign at x_2 but not $E_y(x)$, it results then that $v(x) = E_y(x)/B_z(x)$ must tend to infinity as $x \to x_2$. To resolve this question in the framework of a steady-state ideal MHD flow theory, the 'reconnectionists' suggested that jetting should take place at x_2 in directions parallel to the oyz plane. On both sides the plasma is <u>supposed</u> to converge toward this plane of discontinuity which is then imagined to coincide with the magnetopause; the converging plasma flows are assumed to be deflected through a right angle, and to diverge away from a narrow "diffusion region" surrounding the neutral point or the X-line (Petschek 1964; Semenov <u>et al</u>. 1983).

However, it must certainly not be assumed, a priori, that this very particular type of flow pattern is necessarily imposed on nature by any pressure from any external physical factors - except for pressure from the MHD community. The flow stream described by eqs. (3-8) has the new advantage of being non-singular at x_2. It is based on simple plasma kinetic and electromagnetic theory without the need to invoke anomalous or undefined physical processes ('some kind of instability') anywhere in space, not even in a small "diffusion region".

DISCONTINUOUS VALUES FOR THE INTEGRATED PEDERSEN CONDUCTIVITY.

Let us once again return to Baker and Hamel's simple experimental set up. However, consider now that only the first section of the tank is made of insulating walls ($\Sigma_p = 0$), while the second part is coated with highly conducting material such that $\Sigma_p = \infty$ (see fig. 4a). The magnetic field $\underline{B}$ is uniform, and a plasmoid is injected with an initial

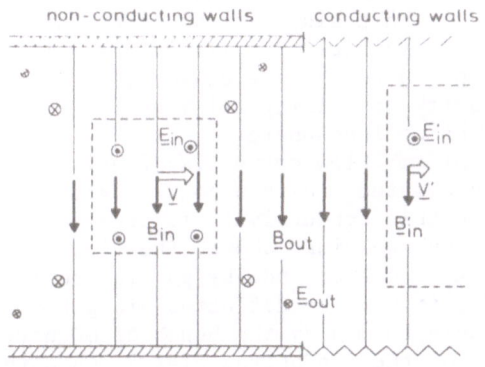

Figure 4a.   Schematic representation of plasmoids moving across magnetic field lines of which some pass through insulating walls and others through conducting plates. In 4a) the applied magnetic field is uniform; the injected plasmoid is braked in the conducting section of the tank like a bullet in a viscous fluid.

velocity $\underline{v}_o$ parallel to the ox axis. It has been shown above and
demonstrated experimentally that, in the first portion of the tank, the
velocity $\underline{v}$ of the plasmoid remains constant and equal to $\underline{v}_o$ as a
consequence of the vanishingly small value of Σ_p and of the high value
of the dielectric constant   (Schmidt 1960).

But as soon as the front edge of the plasmoid penetrates into the
region where the field lines are connected to the conducting walls, the
forward motion of the plasma is stopped as a consequence of the
depolarization currents which flow through the walls and neutralize all
space charges and all interior potential differences in the plasmoid
(Baker and Hamel 1965; Baum et al. 1984;see also the Introduction of
this article).

Because of the strong electromagnetic coupling been the collisionless plasma and the conducting walls, its forward momentum is
transferred to the walls. By reaction, the incident plasmoid can be
deflected or reflected backwards when the walls are superconductors, and
when they have much larger inertia than the plasmoid itself. In this
ideal MHD case, no energy dissipation is involved during the bouncing
back.

When the conductivity of the walls is not too large, and consequently when Σ_p has intermediate values the plasmoid does not bounce
back but it is braked over a finite distance like a bullet being
decelerated in a viscous fluid. The penetration depth, d, is almost
inversely proportional to the integrated Pedersen conductivity
(Lemaire 1977). Therefore, when $\Sigma_p = 0$ the penetration depth is
infinitely large (i.e., there is no braking like that in the first
section of the tank where magnetic field lines pass through insulating
walls); on the contrary, when $\Sigma_p = \infty$, one has d = 0 as expected in the
ideal MHD approximation, and as already discussed in the previous
paragraph.

Consider now that the second portion of the plasma tank is also
constructed with poorly conducting wall material, but that two conducting plates are placed on both sides of the plasma stream, perpendicular to the uniform magnetic field $\underline{B}$ as illustrated in fig. 4b. A
parabolic shape has been given to the plates so as to simulate the
Earth's magnetospheric cavity containing all magnetic field lines
passing through the conducting ionosphere.

The non-diamagnetic plasma flow convecting freely across magnetic
field lines with small integrated Pedersen conductivity eventually
impinges on a region where this conductivity is significantly enhanced.
For physical reasons already explained above, the flow will lose
momentum in the forward ox direction and, by partial reflection at the
interface of both regions, it will be deflected along the surface of
the electromagnetic obstacle formed by the bunch of magnetic field
lines passing through one of the conducting plates. When the value of
Σ_p in the conducting plates is adjusted to vary from a very large to
a very small value, the penetration depth changes from almost zero
(i.e. the MHD limit) to infinity (i.e. free cross-B flow).

Note that in the simulation experiment proposed in fig. 4b, the
magnetic field intensity is "northward" in both sections of the tank.
But the conclusions reached above do not directly depend on the

direction of the magnetic field in the incident plasma flow relative to
the magnetic field orientation in the conducting section. Indeed it has
been indicated already that the cross-B bulk velocity of a non-
diamagnetic plasmoid does not depend on the rotation angle of the
external B-field distribution.

Fig. 4c illustrates the streamlines and magnetic field line
topology for the case in which the magnetic field in the source region

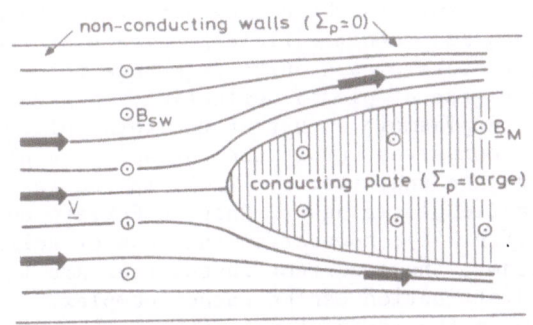

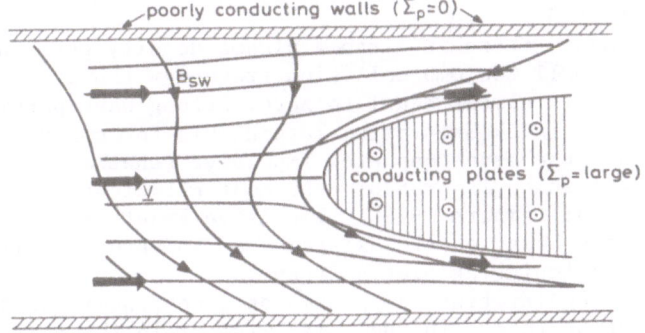

Figures 4b and c. Schematic representation of plasmoids moving across
magnetic field lines of which some pass through insulating walls and
others through conducting plates. In 4b) the applied magnetic field
is uniform, but the distribution of wall conductivities is disconti-
nuous; the non-diamagnetic plasma beam moving across the magnetic
field is deflected sideward around the conducting electromagnetic
obstacle. In 4c) the magnetic field direction in the source region
of the diamagnetic plasma beam is no longer parallel to the applied
field in the conducting region; diamagnetic effects produced by Chapman-Ferraro currents determine the final magnetic field line distribution; impulsive penetration of small scale plasma irregularities
result in the formation of an Plasma Boundary Layer whose thickness
is a function of the value of the integrated Pedersen conductivity in
the conducting region of the simulation experiment.

where the plasmoid is formed, is neither parallel nor antiparallel to
the B-field inside the conducting section of the tank.

The incident plasma flow in fig. 4c is supposed to simulate a high-
β solar wind drifting across geomagnetic field lines linked to the
Earth's ionosphere where Σ_p is of the order of 1 Siemens and is quite
different from that in the interplanetary medium. When the value of β
in the streaming plasma is of the order of unity, we will see below
that diamagnetic effects are important.

Chapman-Ferraro currents circulate parallel to the average
magnetopause and plasmoids surface where they are driven by kinetic
plasma pressure gradients and magnetic field gradients. These magne-
tization, grad-B and curvature plasma currents change the magnetic
field intensity in the region of plasma deflection; they enhances the
strength of B outside the plasmoid such that the total plasma pressure
(kinetic + magnetic) is conserved across the surface of a plasmoid or
of the magnetopause (Burlaga and Lemaire, 1978). In the case of Baker
and Hamel's non-diamagnetic plasma beams, Chapman-Ferraro current
intensities are small (fig. 4b); however in the case of solar wind,
diamagnetic plasma streams, these current intensities are large (Roth
1978, 1979) and their distribution can be rather complex.

It is clear that when the incident velocity of the plasma is
supersonic a shock wave must form in front of the obstacle.

VISCOUS LIKE INTERACTION .

At the magnetopause there is often a plasma density gradient.
Because of the very small Coulomb collision frequency the cross-B
diffusion coefficient is almost equal to zero. Strong wave-particle
interactions have been invoked as being responsible for solar wind
plasma diffusion into the magnetosphere. Impulsive Penetration is a
third entry mechanism which is probably the most efficient of all
(Lemaire and Roth 1978; Lemaire 1979b). Impulsive Penetration is a
time-dependent process based on the existence of momentum density
inhomogenities in the impinging plasma stream.

These plasma irregularities are plasmoids according to Bostik's
definition. They can be formed in the solar wind or in the solar corona
by unstable flow regimes. Kelvin-Helmholz instability at the magnetopause or in the interplanetary medium can perhaps also produce such
plasma inhomogeneities in the impinging solar wind flow. But whatever
their origin may have been, once such plasmoids are present in the
solar wind they penetrate deeper into the geomagnetic field when they
have larger momentum density. Indeed the critical magnetic field
strength $B(x_1)$ for which such more impulsive plasmoids are stopped by
adiabatic heating is larger than for the average background solar wind
plasma; their bulk velocity given by eq. (8) goes to zero deeper into
the geomagnetic field than average solar wind plasma elements which are
stopped at the average magnetopause surface.Their penetration depth,
d, has also been shown to be larger when the integrated Pedersen
conductivity of polar cusp geomagnetic field lines is smaller (Lemaire
1977, 1979a).

Considering a distribution of plasmoids with a broad range of
momentum densities, geometrical shapes and dimensions, one can account

the existence of a rather wide Plasma Boundary Layer at the outer edge
of magnetosphere. Magnetospheric plasma and solar wind plasma are
intermixed in this transition layer (Roth 1978, 1979; Lee and Kan 1982;
Lemaire 1979b; Eastman and Hones 1979).

It has been suggested by Heikkila (1982) that the hypothetical
'viscous-like' interaction mechanism introduced to drive magnetospheric
convection is a consequence of the impulsive penetration of plasmoids
into the magnetospheric Plasma Boundary Layer; the solar wind momentum
of the penetrating plasmoids is deflected eastwardly (Lemaire, 1984),
and can indeed be transfered, via electromagnetic coupling, to both
the magnetospheric and the ionospheric plasma in the dayside cusp
region (Lemaire 1979a).

It can be concluded, therefore that Impulsive Penetration theory
fits well in the context of "closed" magnetospheric models where
viscous-like interactions play an important role (Piddington 1960a,b;
1979; Johnson 1960; Axford and Hines 1961).

DIAMAGNETIC EFFECTS.
In Baker and Hamel's and Demidenko et al's experiments as well as
in the solar wind, the plasma density, n, is large enough for
collective dielectric effects to be important (i.e. $K \gg 1$) :

$$K = 1 - \varepsilon = \frac{nm^+}{\varepsilon_o B} = \frac{nm^+ c^2}{B^2/\mu_o} = \left(\frac{c}{v_A}\right)^2 = \left(\frac{\omega_p^+}{\Omega^+}\right)^2$$

where ε is the plasma dielectric constant; v_A its Alfvèn speed; ω_p^+ the
ion plasma frequency and Ω^+ the ion Larmor frequency. But in these
laboratory experiments, n is not large enough for diamagnetic effects
to be as important as in the solar wind. Indeed, Baker and Hamel's
plasma streams have a low β-value; β is much smaller than in interplanetary space plasmas where $\beta = .1-10$.
When the condition

$$\beta \ll 1 \tag{9}$$

is not satisfied, the magnetic fields generated by plasmoids are
appreciable as compared with the external B-field; the plasma then
exhibits "diamagnetic" properties (Schmidt 1960). As a consequence of
plasma pressure gradient drifts, magnetic gradient drifts, and curvature drifts, collective currents $j_p(x,y,z,t)$ are driven within the
plasmoid and at its surface. They determine an additional magnetic
(induction) field $\underline{B}_p$ such that

$$\nabla \times \underline{B}_p = \underline{j}_p \tag{10}$$

The fields $\underline{B}_p(x,y,z,t)$ that are generated by these local plasma
currents can be rather complex, but have to be added (vectorially) to
the external magnetic fields $\underline{B}_e(x,y,z,t)$ produced by distant (exterior)
currents or magnetized bodies; the external field satisfies the
equation

$$\nabla \times \underline{B}_e = 0 \tag{11}$$

and the total field $\underline{B} = \underline{B}_p + \underline{B}_e$ verifies also eq. (10) with $\underline{B}_p$ replaced by $\underline{B}$.

When $\overline{\beta} \sim 1$, $\underline{B}_p$ is of the order of $\underline{B}_e$ and the field line distribution of the external magnetic field ($\overline{\underline{H}} = \underline{B}_e/\mu_o$) is drastically perturbed. This is illustrated in fig. 5a and 5b showing the magnetic field line distribution obtained when a cylindrical current density system (axial and toroidal) is superimposed on an external dipole which simulates the Earth's magnetic field; the filamentary current system simulates that created by an ideal cylindrical solar wind plasmoid penetrating into the magnetosphere. Magnetopause and tail currents have been ignored for the sake of simplicity. These simplifications, however, are not essential for the aims of this demonstration.

In fig. 5a the filament is located at 10 Earth radii. One can see

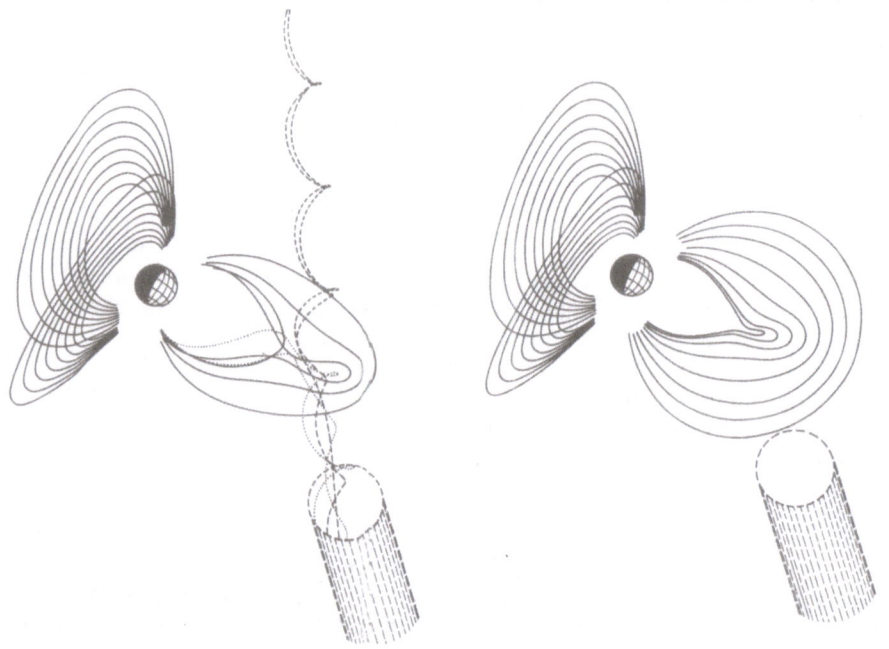

Figure 5. The magnetic field line distribution resulting from superposition of a dipole field and the perturbation fields generated by a cylindrical system of given axial and toroidal current density distributions. The currents simulate those of filamentary solar wind plasmoids penetrating into the geomagnetic field; they produce helicoidal diamagnetic fields of which bundles of field lines interconnect with those of the dipole; a) on the right panel the axis of the current system is located at 10 Earth's radii, b) on the left panel it is closer to Earth at 8 R_E where the local dipole field intensity is significantly larger than the diamagnetic field perturbation.

that, in addition to "closed" geomagnetic field lines and interplanetary magnetic field lines, there are bundles of field lines,
originating in the polar cusp regions, which are interconnected to
those of the solar wind. These interconnected field lines ressemble
those often drawn to represent approximatively the magnetic field
topology in 'flux transfer events'. It should be pointed out that the
calculated field topologies represented in fig. 5a and b, satisfy
precisely Maxwell's equation

$$\underline{\nabla} \cdot \underline{B} = 0 \ , \tag{12}$$

and that there exists at least a current system producing them.

The figs. 5a and 5b have been reproduced from a video-montage
illustrating the time-dependent interconnection of geomagnetic field
lines and interplanetary magnetic field lines induced by diamagnetic
solar wind plasmoids (Lemaire 1982b). Such motion pictures show more
clearly and convincingly than a two-dimensional diagram how these
interconnections proceed from one latitude to another, and from one
polar cusp longitude to another as a plasmoid penetrates deeper into
the geomagnetic field. They indicate that the Impulsive Penetration
theory, when properly understood, does predict the existence of
interconnected (some would probably prefer to say 'reconnected' or
'merged') magnetic field lines inherent in earlier steady state 'open'
magnetospheric models.

CONCLUSIONS.

The theory of Impulsive Penetration of Solar Wind plasmoids into
the geomagnetic field contains aspects of both the "open" and "closed"
magnetospheric models. Indeed, it incorporates the existence of open
(interconnected) magnetic field lines, as well as the notion of
'viscous-like' interaction as discussed above.

Baker and Hamel's experiments have been used here as a basis for
discussion. Several possible modifications and qualifications of these
experiments have been suggested in order to simulate more accurately
the interaction of an inhomogenous 'gusty' solar wind plasma with the
Earth's magnetosphere.

It has also been shown that it is not only the topology of
magnetic field lines which determines the position of a magnetopause
and thickness of the Plasma Boundary layer, but also the value of their
integrated Pedersen conductivity. Consequently, even in 'Terrella'
simulation experiments like those of Baum and Bratenahl (1982) or
Podgorny et al. (1978) it appears to be essential to scale properly the
values of the integrated Pedersen ionospheric conductivity by making
appropriate adjustments to the thickness of the insulating wall
coatings that are supposed to carry the depolarization Pedersen
currents. To imitate correctly the solar wind magnetosphere interaction,
it is important to scale carefully the different boundary conditions,
not only in the interplanetary medium but also in the ionosphere of a
planet.

ACKNOWLEDGMENTS.
    I wish to thank Prof. F.S. Johnson, W.J. Heikkila and Dr. M. Roth
for stimulating discussions on the problem of Impulsive Penetration of
solar wind irregularities. M. Roth is also acknowledged for his
assistance in the editing of the video-film. I have appreciate the
collaboration of M. G. Minnis and of the staff of the Institute for
Space Aeronomy for editing this manuscript. This work was supported by
the Fonds National de la Recherche Scientifique in Belgium. NATO
support to attend the Lillerhamer conference has been appreciated.

REFERENCES.

Axford, W.I. and Hines, C.O 1961, Canad. J. Phys., 39, 1433-1464.
Baker, D.A. and Hammel, J.E. 1965, Physics of Fluids, 8, 713-722.
Baum, P.J. and Bratenahl, A. 1982, Geophysical Research Letters, 9,
435-438.
Baum, P.J., Bratenahl, A., Lemaire, J., and Roth, M. 1984, (submitted).
Bostik, W.H. 1956, Phys. Rev., V. 104, 292-299.
Burlaga, L.F. and Lemaire, J.F. 1978, J. Geophys. Res., 83, 5157-5160.
Demidenko, I.I., Lomino, N.S., Padalka, V.C., Rutkevich, B.N. and
Sinel'Nikov, K.D. 1969, Soviet Physics-Technical Physics, 14, 16-22.
Eastman, T.E., and Hones E.W. Jr. 1979, p. 401-411 in Quantitative
Modeling of Magnetospheric Processes, (ed. W.P. Olson) AGU Geophysical
Monograph 21, Washington D.C..
Heikkila, W.J. 1982, Geophys. Res. Letters, 9, 159-162.
Johnson, F.S. 1960, J. Geophys. Res., 65, 3049-3051.
Lee, L.C. and Kan, J.R. 1982, J. Geophys. Res., 87, 139-143.
Lemaire, J. 1977, Planet. Sp. Sci., 25, 887-890.
Lemaire, J., and Roth, M. 1978, J. Atm. Terr. Phys., 40, 331-335.
Lemaire, J. 1979a, p. 365-373 in Proceedings of Magnetospheric
Boundary Layers Conference, Alpbach, 11-15 June 1979, ESA SP. 148.
Lemaire, J. 1979b, p. 412-422 in Quantitative Modeling of the Magneto-
spheric Processes, Geophys. Monograph. 21 (ed. W.P. Olson), AGU,
Washington D.C.
Lemaire, J. 1982b, Video-cassette, IASB, Brussels.
Lemaire, J. 1984, (submitted).
Petschek, H.E. 1964, p. 425 in Proceedings of AAS-NASA Symposium on the
physics of solar Flares, (ed. W.H. Hess), NASA SP-50.
Piddington, J.H. 1960a, Geophys. J. Roy. Astron. Soc., 3, 314-332.
Piddington, J.H. 1960b, J. Geophys. Res., 65, 93-106.
Piddington, J.H. 1979, J. Geophys. Res. 84, 93-100.
Podgorny, I.M., Dubinin, E.M. and Potanin, Yu. N. 1978, Geophys.
Res. Lett., 5, 207-210.
Roth, M. 1978, J. Atmosph. and Terrestrial Physics, 40, 323-329.
Roth, M. 1979, p. 295-309 in Proceedings of Magnetospheric Boundary
Layers Conference, Alpbach, 11-15 June, ESA SP-148.
Schmidt, G. 1960, Phys. Fluids, 3, 961-965.
Semenov, V.S., Kubryshkin, I.V., Heyn, M.F., and Biernat, H.K. 1983,
J. Plasma Physics, 30-2, 321-344.

ELECTRON INJECTION IN THE POLAR CUSP

A. D. Johnstone
Mullard Space Science Laboratory
Department of Physics & Astronomy
University College London
Holmbury St. Mary, Dorking,
Surrey, U.K.

ABSTRACT.   Observations of the entry of electrons through the
magnetopause and their travel down field lines to the low-altitude
polar cusp are reviewed.   The evidence is strongly in favour of there
being a small region, located poleward of 80° magnetic latitude near
noon in quiet times within which electrons have direct access from the
magnetosheath.   It is more difficult to determine the source of the
electrons precipitating in a ring just equatorward of the direct access
electrons.   While they have some characteristics in common with
electrons observed in the boundary layer of the low latitude
magnetopause, they also show the signature of acceleration, probably
occurred at low altitudes.   It is not clear whether these electrons,
observed on the dayside, should be distinguished from those coming from
the nightside boundary-layer of the plasma sheet.   More detailed
observations are required to determine the origin of this type of
precipitation.

1.   INTRODUCTION

When Chapman and Ferraro (1931) described the 'hollow' which the
geomagnetic field carves out of solar plasma streams they recognised
that particles from the stream would be able to enter the hollow at the
two neutral points on the surface and then travel down the magnetic
field lines to the atmosphere.   The two entry paths, which they called
horns from their appearance in figure 1, are what we now call the polar
cusp.   They could not locate the latitude at which the field line
through the neutral point reaches the atmosphere accurately, but it
would obviously be on the dayside so they realised these solar
particles could not be responsible for the nightside aurora.   They
would be a source of plasma in the magnetosphere but were unlikely to
make a significant contribution to the formation of the ring current
during magnetic storms.
        Much has been learned about the neutral points, and the
magnetopause since Chapman and Ferraro wrote their important papers.

47

J. A. Holtet and A. Egeland (eds.), The Polar Cusp, 47–65.
© 1985 by D. Reidel Publishing Company.

but we are still trying to answer the same questions.   Where, and how, do solar wind particles cross the dayside magnetopause, and how important are they as a source of magnetospheric plasma?

In reviewing electron observations in the polar cusp I shall have to address a more elementary question.   Have the soft electrons observed at high latitudes in the dayside ionosphere come directly from the magnetosheath and, if so, can we use the low-altitude observations of these particles to give us a better overall view of plasma entry into the magnetosphere?   If we can they may help us to answer the more fundamental questions posed above.

Although cusp is descriptive of the field configuration near the neutral points, I use the term here, as I believe it is being used generally, to refer empirically to the low-altitude region where soft electron fluxes are detected.   This does not necessarily even mean that the magnetic field lines from the so-called cusp pass through the neutral point.

By limiting the review to electron observations I am accepting a constraint because it is now clear that other observations (e.g. positive ions, and electric fields) are also necessary to place the electron observations in the correct context and to identify the position of the magnetic field through the neutral points.

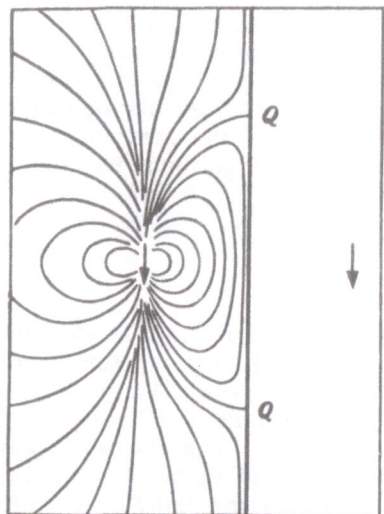

Figure 1.   Sketch of the combined field of the geomagnetic dipole and the electric currents induced in the plane surface layers of the advancing neutral ionised stream.   Particles can enter the geomagnetic field at the neutral points Q and follow the 'horns' to the ionosphere (Chapman and Ferraro 1931).

Nevertheless I believe that a review of electron observations is useful for two reasons. First the higher velocity of the electrons compared with the protons ensures that they follow magnetic field lines from the magnetopause to the ionosphere without being significantly dispersed in energy by the electric field. Therefore they trace the magnetic field lines more accurately. Secondly, most of the energy flux into the ionosphere is carried by electrons so they are responsible for the ionospheric effects (Shepherd et al 1976). Thus we shall concentrate here on what can be learned from the electron observations.

2. TRACING PARTICLE MOTION IN THE MAGNETOSPHERE

Since it is impossible literally to follow particles along field lines we link the measurements made in different places by Liouville's theorem, which effectively states that the particle density remains constant along a trajectory.

In detail this means comparing the high-altitude electron phase space density over the entire range of energies and at the pitch angles which have access to the ionosphere with the densities observed at low altitude. The sharpness of this tool is blunted by the fact that the measurements at the two separate points are rarely, if ever, made simultaneously and the particle velocity distributions vary with time in both shape and intensity. Therefore it is really only possible to compare typical, or characteristic, spectral shapes and statistical distributions of the intensity levels, usually without regard to the pitch angle.

If the characteristic spectral shapes at high altitude, and in the ionosphere are similar then there is evidence for magnetic conjugacy between the regions. However the spectral shapes may be different even though there is magnetic conjugacy if the electrons are accelerated on the journey along the field line. Electron acceleration parallel to the magnetic field produces a recognisable signature in the spectral shape. It is still possible, in theory, to deduce the source spectrum of the electrons in the presence of acceleration but the process is more difficult and requires a much more detailed approach. Most of the work so far is simply based on a comparison of spectra.

The comparison is made more difficult by two other aspects of the physics. First, the mechanisms of particle energy to the magnetosphere are much more complicated than was initially thought, leading to more than one characteristic spectrum, and second, the path that the particles take through the magnetosphere varies continuously as its configuration changes with the solar wind and the interplanetary magnetic field.

3. EARLY OBSERVATIONS OF THE CUSP

The first observations of the cusp at low altitude were reported by Burch (1968), Heikkila and Winningham (1971) and Frank and Ackerson

(1971). They found intense fluxes of soft (E < 500 eV) electrons
poleward of the latitude at which the flux of higher energy electrons,
presumably trapped on closed field lines, decreased sharply. Since
their spectrum was very similar to the spectrum of magnetosheath
electrons (fig 2) the results were interpreted as showing that solar
wind particles had direct access to the ionosphere through the neutral
points on the magnetopause. Similar fluxes were intercepted by IMP-5,
approximately half way between the magnetopause and the ionosphere,
confirming the interpretation of the low altitude data (Frank 1971).
At low altitudes the zone of soft electron precipitation was found to
extend over a fairly wide range of local times (0800-1600 MLT) giving
rise to the idea that there was a neutral line on the magnetopause
rather than a neutral point, or a cleft rather than a cusp (Heikkila
1972). One feature of the data evident in many of these observations
is the variation of the electron fluxes by more than two orders of
magnitude in seconds of time, or hundreds of metres spatially depending
on whether the structure is temporal or spatial (Maynard and Johnstone
1974).

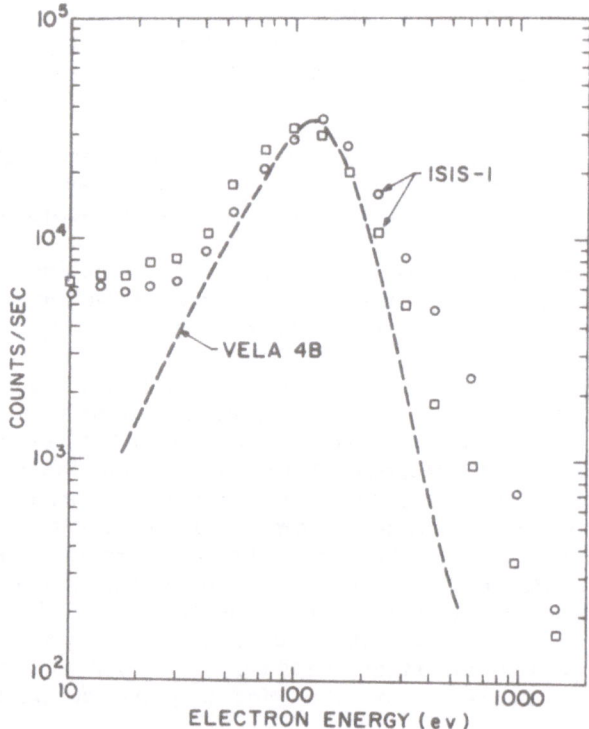

Figure 2. Relative comparison of Vela 4B electron spectrum
obtained in the magnetosheath and ISIS-1 electron spectra
obtained in the dayside high latitude region (Heikkila and
Winningham 1971).

Thus far the concept of direct access seemed successful.   A more detailed examination of the data showed that the situation was actually more complicated.   The existence of loss-cone type pitch-angle distributions, where the intensity peaks at 90° pitch angle, is evidence of a trapped particle population and hence closed magnetic field lines.   Such field lines could not pass through a neutral point or bend back over the polar cusp into the tail.   McDiarmid et al (1976) showed that part of the magnetosheath-like electron distributions thought to be coming from the neutral points were to be found on closed field lines (figure 3) as judged by simultaneous observations of the pitch-angle distribution of high energy electrons (Kintner et al 1978).

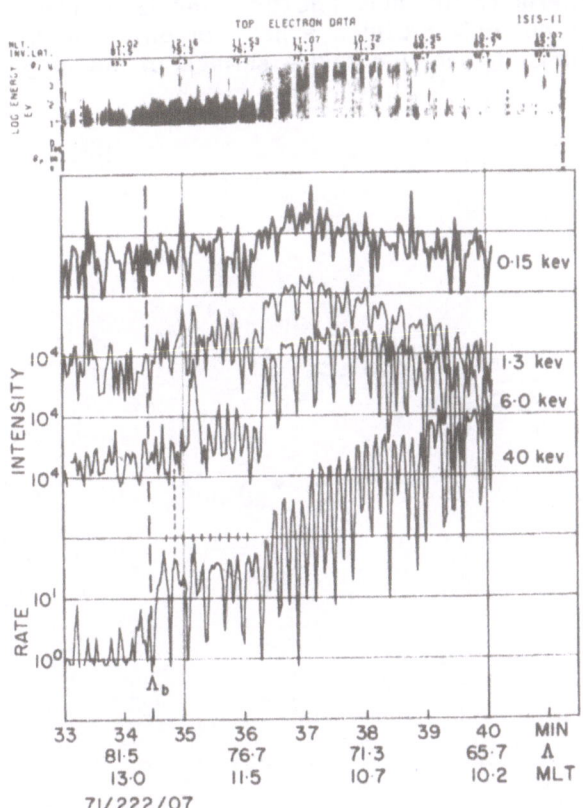

Figure 3.   Example of a prenoon pass of the ISIS-2 satellite showing the SPS spectrogram and intensity plots of four electron energies from the EPD experiment.   The 40 keV electron background boundary Λ_b is given by the dashed vertical line.   The other dashed vertical line corresponds to a pitch angle of 90° and indicates the highest latitude at which anisotropic distributions are observed (McDiarmid et al.   1976).

Furthermore, on afternoon passes, the cusp electrons had a much more energetic spectrum than magnetosheath electrons.

So it was clear that, while magnetosheath electrons might have direct access to some parts of what had been defined as the polar cusp, not all the observations of soft electron precipitation at low altitude could be explained by that mechanism.

4.   OBSERVATIONS AT HIGH ALTITUDES

The topology of the field lines on the surface of the magnetosphere is such that they all pass through both neutral points on the surface.   Therefore the field line from the low-altitude cusp to the neutral point not only provides a path for direct entry, it is also connected magnetically with the entire surface of the magnetopause.

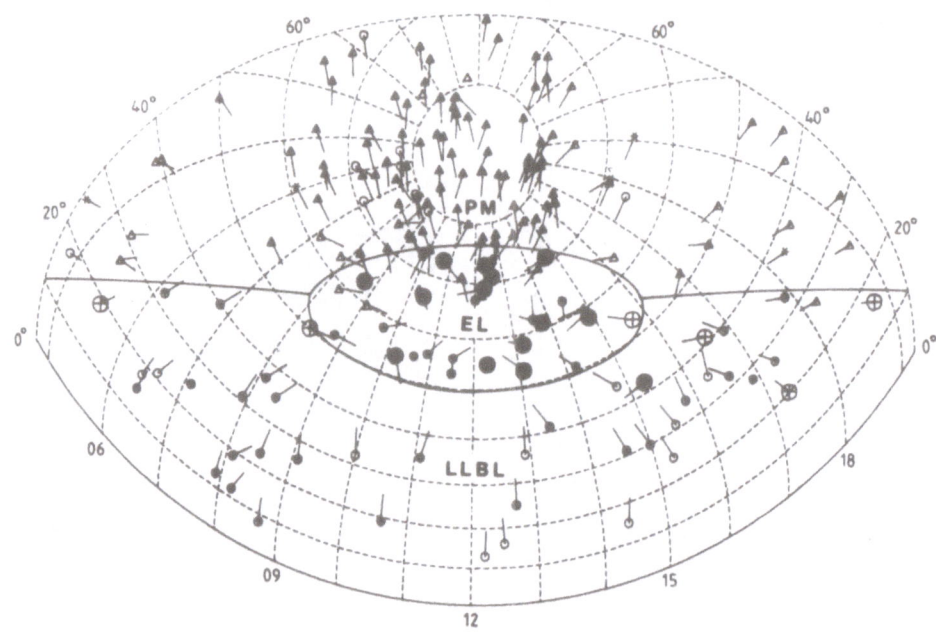

Figure 4.   View of section of model magnetopause.   Dashed lines indicate constant latitudes and local times in solar magnetic coordinates.   Each symbol marks the location of a HEOS-2 magnetopause crossing with the short line indicating the direction of the tangential magnetic field inside the magnetosphere.   Circle and triangle symbols indicate plasma conditions across the boundary.   The heavy solid lines divide the surface into the three distinct regions found: EL the entry layer, PM the plasma mantle, LLBL the low latitude boundary layer (Haerendel et al.   1978).

That means that whenever electrons cross the magnetopause and by whatever means, some effect of the process might be observed near the polar cusp. Rather than consider specific mechanisms such as reconnection or cross-field diffusion I shall approach the problem empirically by asking what types of electron distribution observed near the magnetopause are thought to be associated with entry into the magnetosphere and by attempting to trace these distributions to the low-altitude polar cusp.

Four spatial regions with different plasma populations have been distinguished in the vicinity of the magnetopause. They are the magnetosheath which is the source of all the plasma entering the magnetosphere; the entry layer in the neighbourhood where we would expect to find the indentation in the magnetopause caused by the neutral points; the plasma mantle found tailward and the low-latitude boundary layer found equatorward of the entry layer (figure 4).

The distinction between these regions is made mostly on the basis of the proton population (Haerendel et al 1978). Unfortunately there are relatively few samples of the electron spectra which concern us reported in the literature. I shall give three examples. The first, figure 5 (Formisano 1980), was obtained from HEOS-2 and although not detailed has the advantage for these observations that the minimum energy band starts at 10 eV. It shows that the energy spectra in the magnetosheath, entry layer, plasma mantle, and cusp are all similar in shape although the two latter ones tend to be lower in intensity. The

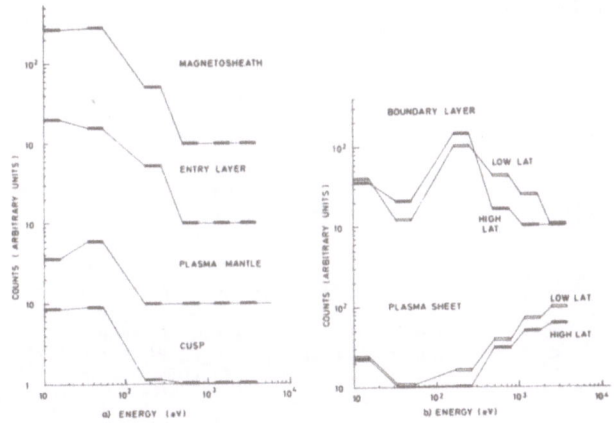

Figure 5. Average low energy electron spectra in the magnetosphere observed by HEOS-2. (a) magnetosheath-like electron spectra (from the top to the bottom) in the magnetosheath, in the entry layer, in the plasma mantle, and in the cusp. (b) boundary layer-like electron spectra (on the top) and plasma-sheet-like electron spectra (on the bottom); two spectra are shown in each case, one from low latitude (SM) observations, the other from high latitude observations (Formisano 1980).

boundary layer is quite different and clearly more energetic, though
not as energetic as the plasma sheet. The second example (fig 6), also
from HEOS-2, (Haerendel et al 1978) shows a succession of spectra
obtained as the spacecraft crossed from the entry layer into the
magnetosheath. The minimum energy in these electron spectra is 100 eV
and therefore does not cover the lowest energy range where the spectra
are most intense. There is no noticeable difference in the electron
spectra as the boundary is crossed confirming that the entry layer
electrons are very much like magnetosheath electrons. The boundary can
only be identified in the proton data.

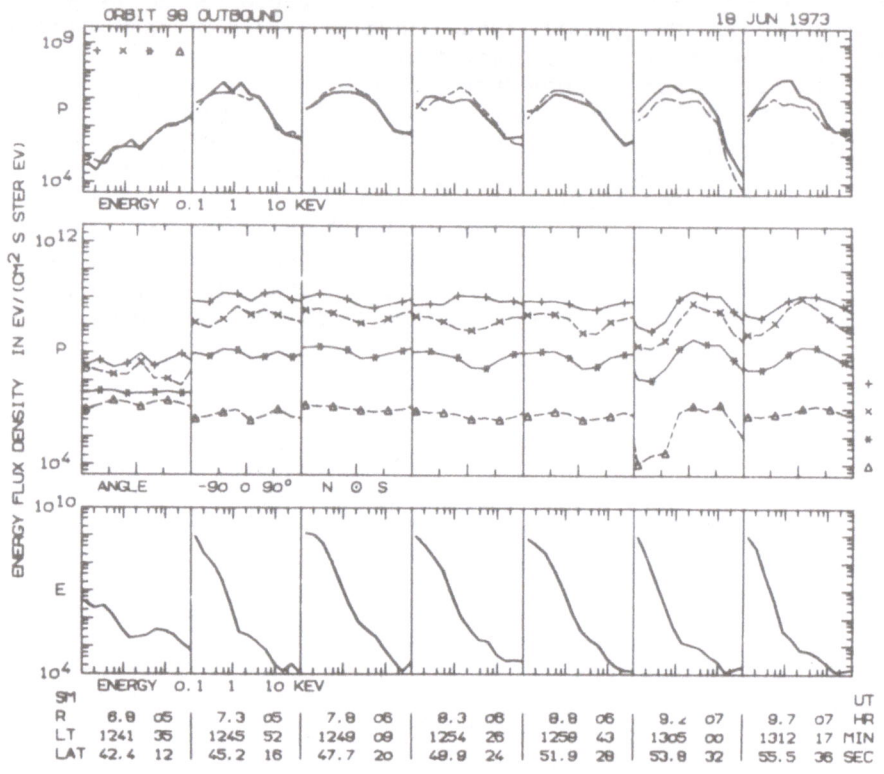

Figure 6. Sequence of seven proton and electron
distributions measured during an outbound traversal of the
entry layer (first five frames) and the magnetosheath (last
two frames). The electron spectra, in the bottom panel, are
averages over all directions. Little change is visible in
the electron spectra in going from the entry layer to the
magnetosheath. The boundary can only be detected as a change
in the proton measurements in the top and middle panels
(Haerendel et al. 1978).

The third example (fig 7) is taken from GEOS-2 which only crosses the magnetopause at times of the highest solar wind pressures. Although the conditions are not then typical there does not appear to be anything especially different about the results. This detailed spectrum (Rodgers et al 1984) which covers the lowest energies shows the boundary layer, close to the equator, detected between plasma sheet and magnetosheath. The boundary layer was found to be completely inside the magnetosphere on geomagnetic field lines.

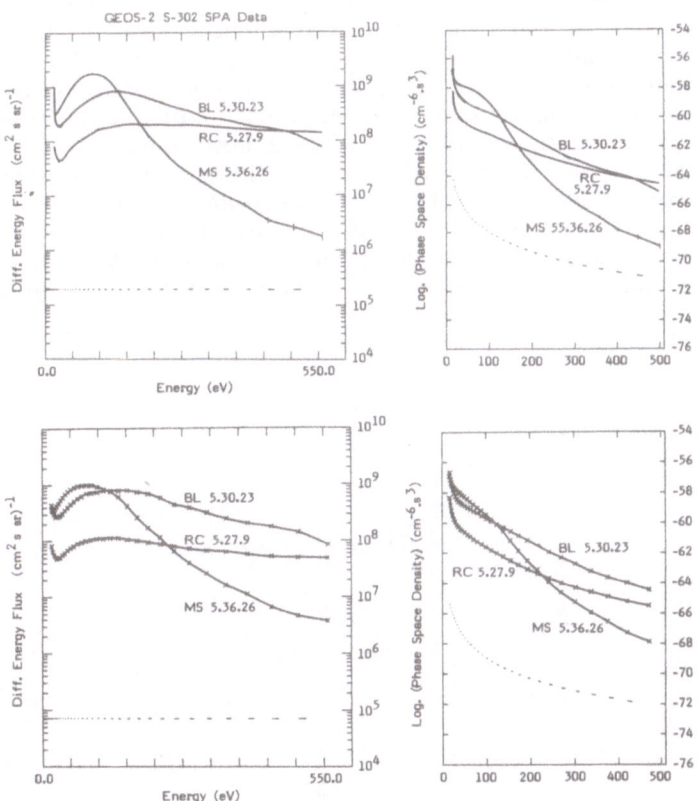

Figure 7. Comparison of 6s average electron spectra observed in the ring current region (RC), boundary layer (BL) and magnetosheath (MS) on 28 August 1978. The same data are presented as differential energy flux and phase space density. Counts for the A detector (top panels) perpendicular to the ecliptic plane, and the B detector (bottom panels) at 10⁰ to the ecliptic plane are shown (Rodgers et al. 1984)

This is confirmed by observations of higher energy electrons and ions made during the same crossing by Korth et al. (1979) figure 8. The figure shows the pitch angle distribution of electrons E > 22 keV in the same structures as the data of figure 7, though not exactly coincident in time. The first three frames 5.28.00 UT to 5.29.11 UT are in the ring current; the next two 5.30.01 UT to 5.30.34 UT are in the boundary layer and the final one is in the magnetosheath. In the boundary layer the intensity is reduced and the pitch angle distribution is no longer symmetric but these data show that these are high energy electrons from the trapped electron population on boundary layer field lines near the magnetopause.

In summary, there are two main types of electron spectra. The magnetosheath-type spectrum with a peak at energies below 100 eV, usually found to be close to Maxwellian with a characteristic energy of the order of 50 eV. The second type is the boundary-layer spectrum which is more energetic than the magnetosheath spectrum and is intermediate in shape between the magnetosheath and the plasma sheet. The protons in the boundary layer have a flow velocity which takes them tailward parallel to the solar wind flow. The electrons in the boundary layer have a near isotropic spectrum as the data in figure 7 show. Although they will drift in the electric field associated with the ion flow, those with sufficiently small pitch angles to reach the ionosphere will precipitate before travelling far from their entry point.

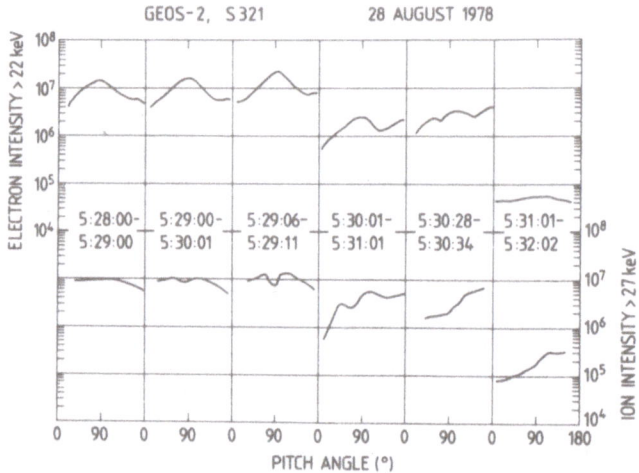

Figure 8. These pitch angle distributions of electrons E > 22 keV and ions E > 27 keV were obtained during the same magnetopause crossing as the data in figure 7. Counting from the left, the first three are in the ring current and show distributions symmetrically peaked at 90° typical of a trapped population. The next two were obtained in the boundary layer and the last one in the magnetosheath (Korth et al. 1979).

5.   OBSERVATIONS AT INTERMEDIATE ALTITUDES

There have only been a few observations in the mid-altitude region between the magnetopause and the cusp.   The first, already mentioned, were reported by Frank (1971) from IMP-5.   The most comprehensive study is by Formisano and Bavassano-Cattaneo (1978) who have traced the path of the cusp electrons from the magnetopause down towards the ionosphere using the HEOS-2 data.   Figure 9 shows the portions of orbit on which cusp-like spectra were detected.   The route from the neutral point towards the ionosphere is clearly shown although there is great variability in the magnetospheric conditions as demonstrated by the variation in the detected location of the magnetopause.

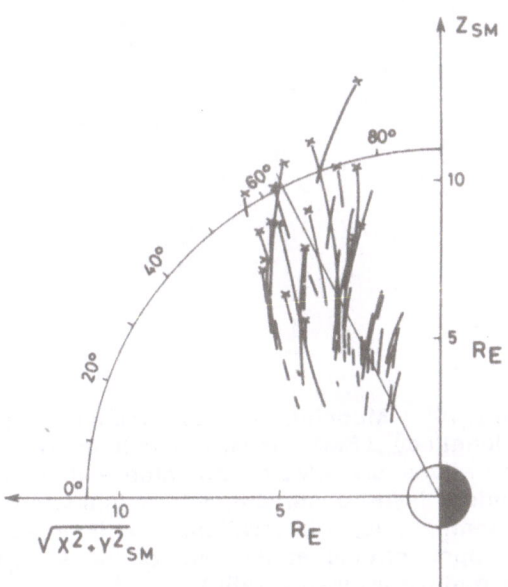

Figure 9.   Portions of the HEOS-2 orbits during which cusp electrons were observed.   The orbits have been rotated into the noon meridian plane around the Z_{SM} axis.   The crosses at the end of trajectories at larger radial distance indicate the magnetopause positions (Formisano and Bavassano-Cattaneo 1978).

These data on the location of cusp electrons, have been compared with similar data on the location of boundary layer electrons and have been plotted in the magnetic local time-invariant latitude coordinate system in figure 10 (Formisano 1980).   Effectively this projects the high-altitude measurements down magnetic field lines to the polar-cap ionosphere.   The plot shows plasma-sheet and boundary-layer electron spectra in concentric rings around the pole while the magnetosheath-

like electrons are only found at high latitudes on the dayside.   This
suggests that boundary layer electrons are on different field lines
from the magnetosheath-like spectra.

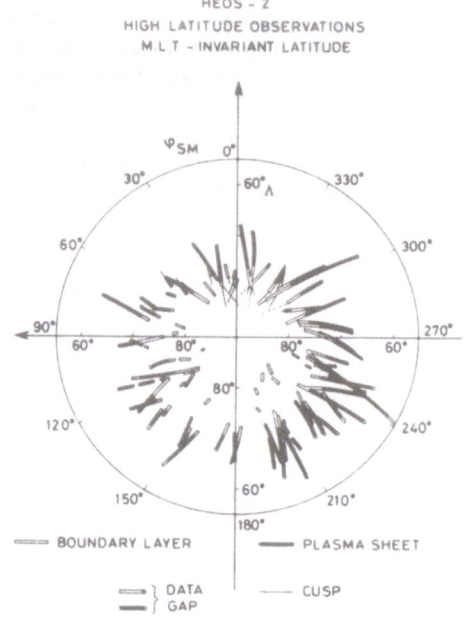

Figure 10.   Mapping of magnetospheric low energy electrons in
the longitude (SM), invariant latitude plane.   Plasma sheet
observations are always indicated with thick dark segments.
Boundary layer observations are indicated with a thick open
segment.   Cusp observations are indicated by thin segments.
Little open circles at the end of the segment indicates a
data gap (Formisano 1980).

The next three diagrams figures 11, 12 and 13 show possible magnetic
field configurations relating the boundary layer and the entry layer to
the cusp.   In the first, fig 11, (Haerendel et al 1978) the boundary
layer, the entry layer and the cusp are on the same field lines.   Thus
at low altitudes the boundary layer electrons would be detected on the
same field lines as the cusp electrons.   The second, fig 12, (Formisano
1981, Frank 1971) places the boundary layer on closed field lines at
lower latitudes than the open entry layer and cusp field lines but does
not differentiate between the boundary layer on the nightside, where it
lies between the plasma sheet and the tail lobes, and the dayside
boundary layer, linked with the magnetopause.   This matches the data of
figure 10 where the day and nightside boundary layers have not been
distinguished.   The third diagram, fig 13, (Vasyliunas, 1979)
introduces the distinction but does not support it with data.   Whether
such a distinction should be introduced is still an open question

(Vasyliunas, private communication)'. The data from low altitudes have some bearing on this question.

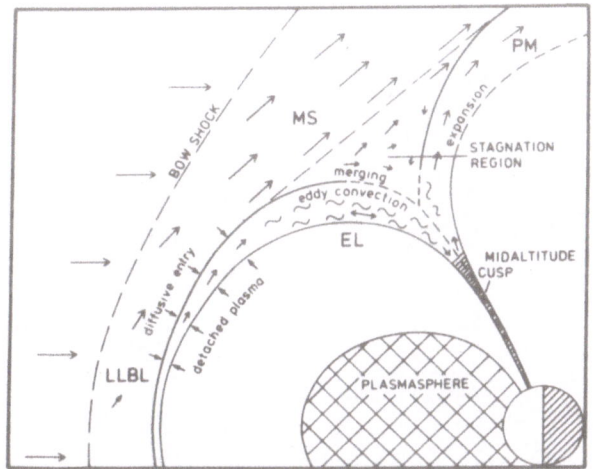

Figure 11. This figure and the two following compare three models of the magnetic configuration near the magnetopause and neutral points. This figure (Haerendel et al. 1978) shows the boundary layer on field lines which pass through the entry layer before reaching the ionosphere.

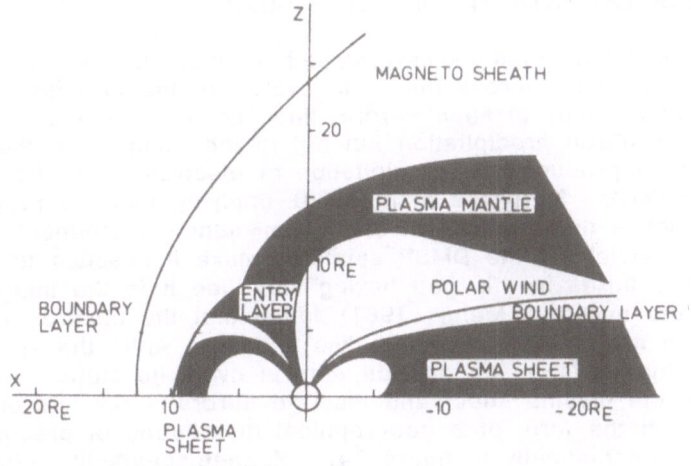

Figure 12. In this configuration the boundary layer is on closed field lines separate from the entry layer field lines but no distinction is made between the nightside and dayside boundary layer (Formisano 1981).

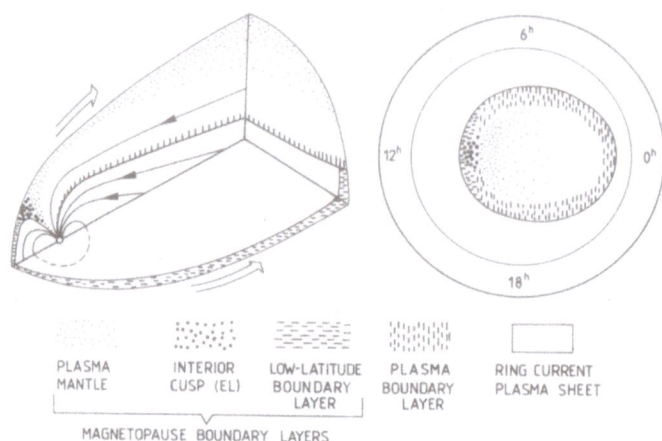

PLASMA   INTERIOR   LOW-LATITUDE   PLASMA   RING CURRENT
MANTLE   CUSP (EL)   BOUNDARY   BOUNDARY   PLASMA SHEET
                            LAYER   LAYER

MAGNETOPAUSE BOUNDARY LAYERS

Figure 13.   Here the field lines from the magnetopause
boundary layer are distinguished from the plasma sheet
boundary layer although the boundary between these two
regions shown in the polar plot on the left is highly
uncertain.   Otherwise the configuration is essentially the
same as in figure 12.   The great extent of the low-latitude
boundary layer is speculative (Vasyliunas 1979).

6.   MEASUREMENTS IN THE POLAR CUSP

It has been noted by several authors that discrete auroral forms
are not continuous across the noon sector on the dayside.   There is a
gap in the intensity of such auroral lines as 557.7 nm associated with
energetic electron precipitation but not in the intensity of the 630.0
nm emission produced by precipitation of electrons E < 100 eV (Dandekar
and Pike 1978;   Murphree et al 1980) implying that the precipitation in
the gap has a different character.   The imaging instrument and particle
detectors carried by the DMSP satellites make it possible to measure
the particle fluxes in this gap having identified it in the images of
the discrete aurora.   Meng (1981) found that the electron precipitation
in the gap had a magnetosheath-like spectrum while the spectra in the
neighbouring parts of the dayside auroral oval, he states, were more
typical of the plasma sheet and discrete aurora.   His interpretation of
the data, in the form of a geographical distribution of precipitation,
is shown schematically in figure 14.   Magnetosheath-like electrons are
limited to the small region of the gap in a very similar way to the
cusp electrons in the HEAO-2 data of Formisano (1980) shown in figure
10.
Potemra et al (1977) also find two different spectral types with a
similar spatial distribution in data from AE-C and AE-D (figure 15).

In a pass across a region, all of which might be called the cusp from
its low energy characteristics, there is a clear boundary between a
poleward region which has the spectra corresponding most closely to
magnetosheath electron spectra and a second region which is more
energetic and coincides with a region of field-aligned currents
detected magnetically.   The authors do not present detailed spectra
from the second region but note that inverted V forms are sometimes
found.   Similar events are also apparent in the data of McDiarmid et al
(1976) and the rocket data of McEwen (1977).   The inverted V is thought
to be the signature of low-altitude acceleration.

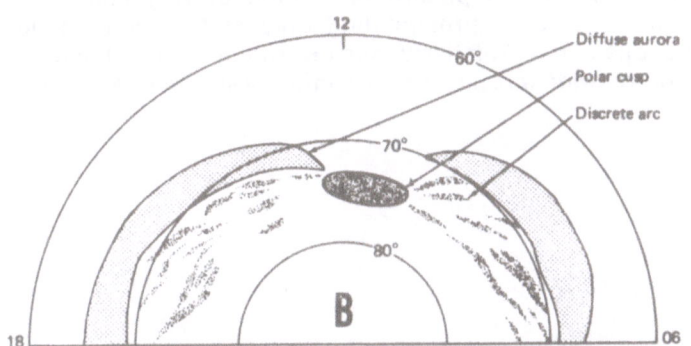

Figure 14.   Electrons with a magnetosheath-like spectrum are found in
the region marked "Polar cusp" in the gap in the discrete auroral
structures found in the dayside auroral oval (Meng 1981).

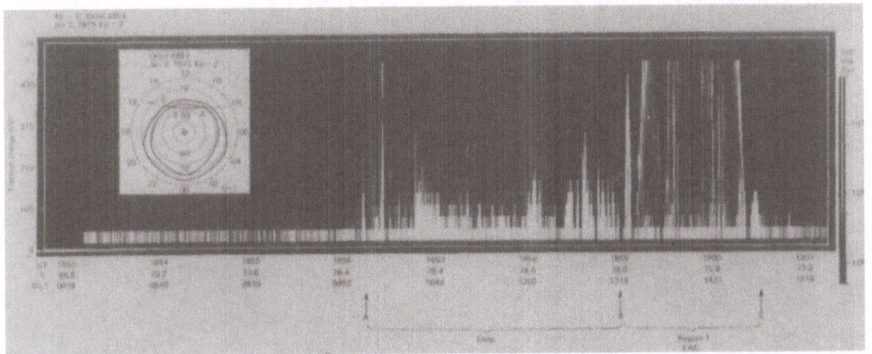

Figure 15.   In this pass across the auroral oval (see insert) there is
a marked difference in the electron spectrum between the two sections
AB, BC marked.   In the section AB which corresponds to the highest
dayside latitudes the spectrum is magnetosheath-like, and variable in
intensity.   The section BC, which maps to region 1 field-aligned
currents, has a more energetic spectrum like the boundary layer, but
also with indications of inverted V structures (Potemra et al.   1977).

Data from the DMSP satellites have also been used by Candidi et al
(1983) to compile contour plots of the intensity of precipitation
electrons at a number of fixed energies (fig 16) averaged over a large
number of passes.   It is impossible to reconstruct typical spectra from
the plot but even so the features already mentioned are also obvious in
these data.   Notice in particular the peak flux zone in the three
lowest energy channels 50 eV, 77 eV and 110 eV extends into a region
above 80° latitude between 0900 and 1400 MLT.   There is almost a
complete absence of electrons in the higher energy channels, 434 eV and
1.06 keV in this region though they overlap the softer electrons
equatorward of 80° latitude and spread out in a ring towards the
nightside.   This confirms the picture of a zone of very soft
magnetosheath-like electrons precipitating only at the high latitude
boundary of the cusp in a relatively narrow range of local times and
bounded by a more widespread zone of more energetic electron spectra.

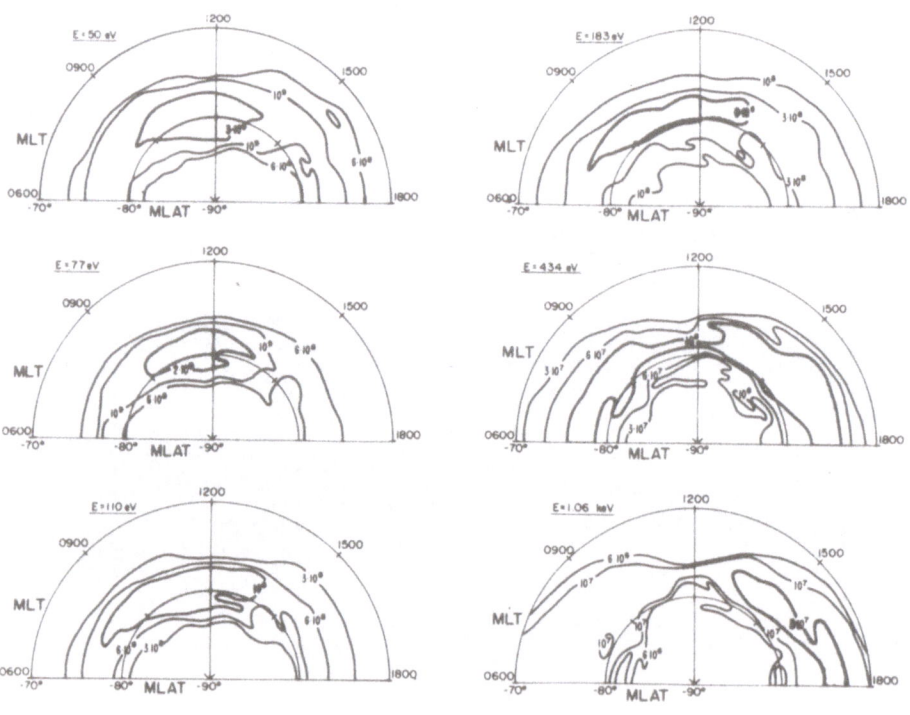

Figure 16.   The maps show contours of the average intensity
of electrons at the six fixed energies marked on the plots,
compiled from a large number of polar passes of the DMSP
satellites (Candidi et al.   1983).

The information from a comparison of spectra is consistent on the existence of a small, high latitude, dayside region of magnetosheath-like spectra which can be attributed to direct access. It is also consistent in showing another ring of precipitation, at lower latitudes with more energetic spectra. The electron spectra in the outer ring, associated by Meng (1981) with discrete aurora could be derived from the precipitation of boundary layer electrons as the observations of Formisano (1980), fig 10 , suggest but it is by no means certain. The boundary layer electrons are more energetic than magnetosheath electrons but the existence of inverted V's in the data mean that an acceleration mechanism is operating at low altitudes on these field lines as in the nightside auroral oval thereby distorting the spectra of the source electrons. The source could then be either entry layer or boundary layer particles because the present data are not good enough to distinguish.

So far the argument has relied on a comparison of spectra. Further information can be obtained from pitch angle distributions and the intensity levels. Very little has been published on these topics. Both Craven and Frank (1976) and Maynard and Johnstone (1974) found that the pitch angle distributions were peaked strongly along the downward field-aligned direction but little is known about the reasons for this.

Recently Candidi and Meng (1984) have found that the intensity of the precipitating electrons in what can be called the direct access cusp is correlated with the solar wind density. There is a lot of scatter in the data showing that other parameters also exert control over the relationship but the dependence on density is unmistakeable and non-linear.

7.   USE OF LOW ALTITUDE DATA FOR MAGNETOPAUSE STUDIES

The use of low-altitude particle data to track the position of the cusp as a function of solar wind parameters (Burch 1972) is well-known and will not be reviewed here. Figure 17 (Candidi et al. 1983) shows the influence of the By component of the interplanetary magnetic field. on the zone of 50 eV electron precipitation which includes both the magnetosheath spectra and the more energetic spectra just equatorward. This diagram is interesting for two reasons. It shows that the path from the magnetopause varies considerably as the solar wind conditions change making the tracing of such trajectories difficult even on a statistical basis. More importantly perhaps it supports the view that these measurements might well be able to provide a great deal more information about the nature of electron entry to the magnetosphere.

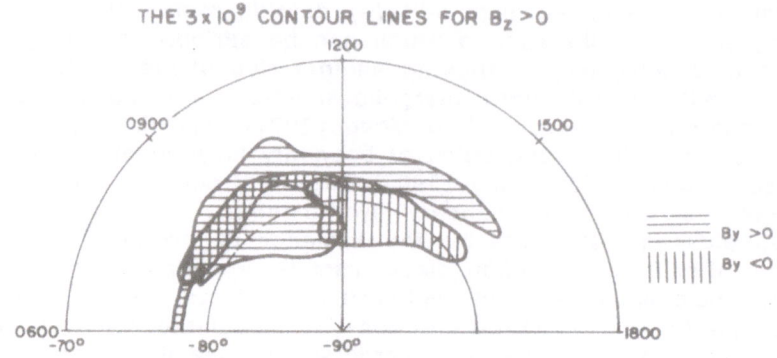

THE 3×10^9 CONTOUR LINES FOR $B_z > 0$

Figure 17. The contours of the average 50 eV flux compiled using the data included in figure 16, but separated into two groups according to the sign of the By component of the interplanetary magnetic field (Candidi et al. 1983).

8.   CONCLUSIONS

The evidence is quite strong that direct entry of magnetosheath electrons does take place through the entry layer on the magnetopause and then down magnetic field lines to a relatively small region at latitudes above 80° (in quiet times) near noon. Within this region electron fluxes are very variable in intensity suggesting that the entry layer – cusp region is not a wide throat down which the magnetosheath electrons are poured, but a region which allows a limited, and fluctuating access of electrons to the magnetosphere. A more energetic region of precipitation, traditionally also included as part of the cusp is on neighbouring and probably closed field lines at lower latitudes. The source of this precipitation is not yet clear but it may map to the low–latitude boundary just inside the magnetopause and it is also possible that these electrons are accelerated en route. Further studies, involving a comparison of electron spectra and pitch angle distributions up to energies of tens of keV, may be able to resolve these questions. When this is done then the structure of the low–altitude cusp and its variation in position should be able to provide a great deal more information about magnetospheric entry processes.

9. REFERENCES

Burch, J.L. 1968, J. Geophys. Res., 73, 3585-3591.
Burch, J.L. 1972, J. Geophys. Res., 77, 6696-6707.
Candidi, M., Kroehl, H.W., and Meng, C-I. 1983, Planet. Space Sci., 31, 489-498.
Candidi, M., and Meng, C-I. 1984, The relation of the cusp precipitating electron flux to the solar wind the interplanetary magnetic field, (preprint).
Chapman, S., and Ferraro, V.C.A., 1931, Terr. Magnetism & Atmos. Elec., 36, 77-97, 171-186.
Craven, J.D., and Frank, L.A. 1976, J. Geophys. Res., 81, 1695-1699.
Dandekar, B.S., and Pike, C.P. 1978, J. Geophys. Res., 83, 4227.
Formisano, V. 1980, Planet. Space Sci., 28, 245-257.
Formisano, V. 1981, Adv. Space Res., 1, 207-224.
Formisano, V., and Bavassano-Cattaneo, M.B. 1978, Planet. Spac Sci., 26., 993-1006.
Frank, L.A., 1971, J. Geophys. Res., 76, 5202-5219.
Frank, L.A., and Ackerson, K.L., 1971, J. Geophys. Res., 76, 3612-3643.
Haerendel, G., Paschmann, G., Sckopke, N., Rosenbauer, H., and Hedgecock, P.C. 1978, J. Geophys. Res., 83, 3195-3216.
Heikkila, W.J., 1972, P67 in Critical Problems of Magnetospheric Physics (Ed. E.R. Dyer), Inter-Union Commission on Solar Terrestrial Physics Secretariat, Washington DC.
Heikkila, W.J., and Winningham, J.D., 1971, J. Geophys. Res., 76, 883-891.
Kintner, P.M., Ackerson, K.L., Gurnett, D.A., and Frank, L.A., 1978, J. Geophys. Res., 83, 163-168.
Korth, A., Kremser, G., Wilken, B., Amata, E., and Candidi, M., 1979, in Proceedings of Magnetospheric Boundary Layers Conference, ESA, SP-148, 157-159.
McDiarmid, I.B., Burrows, J.R., and Budzinski, E.E., 1976, J. Geophys. Res., 81, 221-226.
McEwen, E.J., 1977, Planet. Space Sci., 25, 1161-1166.
Maynard, N.C., and Johnstone, A.D., 1974, J. Geophys. Res., 79, 3111-3123.
Meng, C-I., 1981, J. Geophys. Res., 86, 2149-2174.
Murphree, J.S., Cogger, L.L., Anger, C.D., Ismail, S., and Shepherd, G.D., 1980, Geophys. Res. Lett., 7, 239.
Potemra, T.A., Peterson, W.K., Doering, J.P., Bostrom, C.O., McEntire, R.W., and Hoffman, R.A., 1977, J. Geophys. Res., 82, 4765-4776.
Rodgers, D.J., Wrenn, G.L., and Cowley, S.W.H., 1984, Planet. Space Sci., in press.
Shepherd, G.G., Whitteker, J.H., Winningham, J.D., Hoffman, J.H., Maier, E.J., Brace, L.J., Burrows, J.R., and Cogger, L.L., 1976, J. Geophys. Res., 81, 6092-6102.
Vasyliunas, V.M., 1979, in Proceedings of Magnetospheric Boundary Layers Conference, ESA, SP-148, 387-393.

ION INJECTION AND ACCELERATION IN THE POLAR CUSP

W. K. Peterson

Lockheed Palo Alto Research Laboratories
Palo Alto, California 94304

ABSTRACT

This paper discusses cusp ion observations and their relationship to the morphology and dynamics of the polar cusp. Ion
mass spectrometer data are used to illustrate the mixing of solar
wind, ionospheric, and magnetospheric plasmas that occurs in the
cusp.

INTRODUCTION

The existence of the magnetospheric cusp was suggested by
magnetometer data from early satellites. The first high resolution ion and electron measurements made in the dayside auroral
region confirmed the existence of the cusp (see Winningham and
Heelis, 1983 for a historical review). Cusp ion observations, in
particular the observation of predicted energy-latitude ion
dispersion signatures, provided key insights into the structure of
the magnetosphere as well as the morphology and dynamics of the
cusp. In the past few years a new generation of high sensitivity
and high temporal resolution instruments have returned data from
the cusp region. In particular ion mass spectrometer data over
the thermal energy range of magnetosheath plasma from the S3-3,
Dynamics Explorer, and Prognoz satellites and both ion mass
spectrometer and three dimensional total ion measurements from
Dynamics Explorer -1 satellite are now available. The S3-3 and
Dynamics explorer data (Shelley, 1979, Shelley et al. 1982,
Gurgiolo and Burch 1982) have shown that the escape of ionospheric
plasma from the polar cap and cusp region is very much larger than
that predicted by early polar wind models. Data from the Prognoz
satellites (Lundin et al. 1982, Lundin, 1984 and 1985, and Lundin
and Dubinin, 1984) show the complicated interaction of this
ionospheric plasma in the entry layer and how it relates to the
generation of field aligned currents.

J. A. Holtet and A. Egeland (eds.), The Polar Cusp, 67–84.

In this paper we will discuss the entry and subsequent drift of solar wind ions, acceleration of ionospheric ions in the cusp, and the observations of magnetospheric ions in the cusp. Since the subject of plasma and field observations in the external cusp, entry layer, and mantle regions are discussed earlier in this volume (Lundin, 1985, Sckopke, 1984) we will emphasize the low to mid altitude region (i. e. below about 5 earth radii).

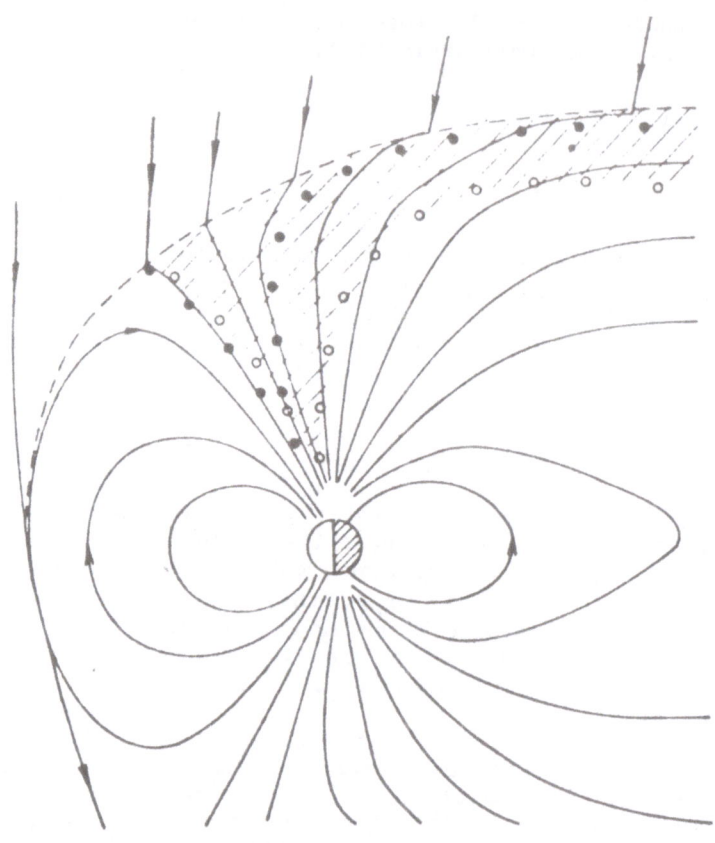

Figure 1. Model of entry of magnetosheath ions into the magneto-sphere and subsequent antisunward drift. Lines with arrows represent magnetic field lines. Open circles represent particles with relatively slow velocities parallel to the magnetic field and solid circles represent faster particles. From Rosenbauer et al. (1975).

ENTRY AND SUBSEQUENT DRIFT OF SOLAR WIND IONS

    The general behavior of ions in the low and mid altitude cusp
(i. e. below about ~5 earth radii, geocentric) is most often
discussed using a model that was introduced by Rosenbauer et al.
(1975) to explain the observed spatial distribution of the plasma
mantle or entry layer. Figure 1 is reproduced from their paper.
It shows the combined effects of of localized entry of magnetosheath plasma and subsequent antisunward drift.   Magnetosheath
plasma, including ions with a wide energy distribution, simultaneously gain access to the cusp along newly interconnected field
lines which are then convected antisunward.   In this figure the
solid circles show the trajectory of ions with higher field
aligned velocities and the open circles represent the trajectories
of ions with lower field aligned velocities. The differences in
field aligned velocities follow from the velocity spread in
magnetosheath plasma and the different path lengths traversed by
ions with various injection pitch angles.   The key features of
this model are the entry being restricted spatially, temporally or
both, and the convection electric field. This model leads to the
prediction of a latitude-energy dispersion for ions in the cusp
(Shelley et al. 1976, Reiff et al. 1977), and an energy-pitch
angle dispersion for ions above the mirror altitude and below the
injection point on field lines connected to the cusp (Burch et
al., 1982).   The illustration in Figure 1, is based on the
assumption that the cusp injection occurs at high latitudes as
suggested by Harendel et al. (1978) and others.   It must be
pointed out that Figure 1 needs only slight modifications to be
consistent with impulsive injection (Lemaire, 1977, Heikkila,
1980), flux transfer events (Russell and Elphic, 1979), or
reconnection at low latitudes (Paschmann et al. 1979).

    Shelley et al. (1976) and Reiff et al. (1977) reported
observing the latitude-energy dispersion predicted by this 'time
of flight' model from satellites at altitudes of less than 1000 km
in the polar cusp. Figure 2, which is reproduced from Shelley et
al. (1976), shows the low-latitude limits of observation for
particles of different masses and energies as a function of the
inverse velocity in units of $(amu / keV)^{1/2}$. The horizontal bars
represent the six second time resolution of the measurements. The
solid line is a fit to the observed points but is identical to one
calculated assuming impulsive injection at a distance of 10 (or
18) Re (earth radii) and convection in a dawn-dusk electric field
of 60 (or 33) mV/m if the convection was constant in time and
approximately parallel to the satellite track.

    In addition to observing the energy-latitude ion dispersion,
Reiff et al. (1977) had available simultaneous ion drift measurements from the AE - C spacecraft and were able to infer an
injection distances in two cases (18 +30/-7 Re and 26 +/- 2 Re).

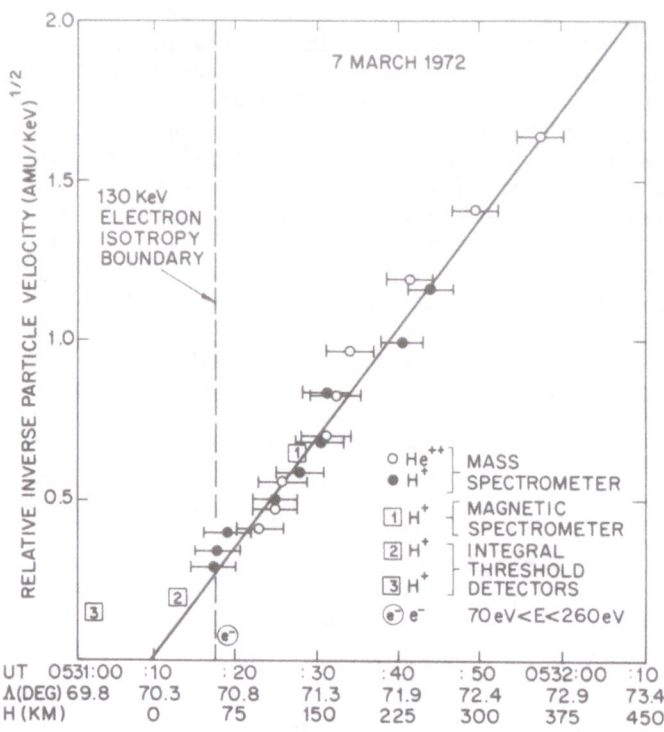

Figure 2. Lowest latitude of detection of particles of different mass and energy vs. inverse velocity for a poleward cusp crossing by the satellite 1971-089A. The horizontal bars represent the time resolution of the measurements. The satellite position at various times (UT) is given in both invariant latitude (λ) and relative displacement (H). The diagonal solid line represents the dependence of inverse velocity on latitude for antisunward convection in the cusp. From Shelley et al. (1976).

Carlson and Torbert (1980) were able to make more precise estimates of the injection distance using data from a rocket between 300 and 600 km. Figure 3 is reproduced from their paper. The format is similar to that used by Shelley et al. (Figure 2) with three important differences: 1) Since the rocket was not moving rapidly with respect to magnetic field lines, the dispersion in ion time of arrival arises from motions of the solar wind entry point or temporal variations in the entry process; 2) There was no mass analysis of the ions and there is a separate velocity dispersion curve for each ion species observed in the electrostatic analyzer used; and 3) The observed ion drifts are sunward in this case. Carlson and Torbert inferred that the injections occurred at a distance of 12 +/- 1 Re from the data in Figure 3

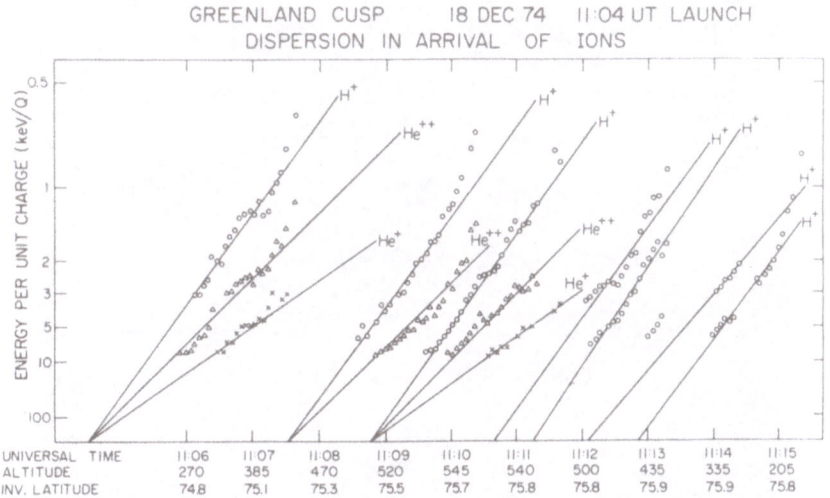

Figure 3. Inverse velocity vs. arrival times of ions observed by a rocket borne ion spectrometer. From Carlson and Torbert (1980).

and 7 and 19 Re from data obtained in another rocket flight. They also used the time width of the individual injections (~ 20 seconds) and the 100 second difference in arrival time of the 10 keV and 1 keV ions to infer that injection occurred nearly simultaneously over a flux tube with a diameter greater than 1000 km at the source. They explained the sunward drift by noting that the measurements were probably made equatorward of the flow reversal boundary. We will discuss below the relationship of the polar cap convection patterns to the observed energetic ion measurements.

Burch et al. (1982) have shown that in addition to the energy-latitude dispersion of particles in the polar cusp there is also a energy-pitch angle dispersion at higher altitudes. This effect is illustrated in Figure 4 using ion mass spectrometer data from the Dynamics Explorer -1 Satellite (Shelley et al. 1981). Seven minutes of data acquired on October 15, 1981 are presented in the form of modified energy-time spectrograms (from top to bottom) for H^+, He^{++}, and a channel sensitive to all ions. The Energetic Ion Composition Spectrometer (EICS) has several operating modes with varying cycle times, energy ranges, and masses sampled. During this period the instrument cycle time was 24 seconds or approximately four spin periods. In each instrumental cycle the EICS samples 15 logarithmically spaced energies covering the energy per charge range 10 eV/e to 17 keV/e at 24

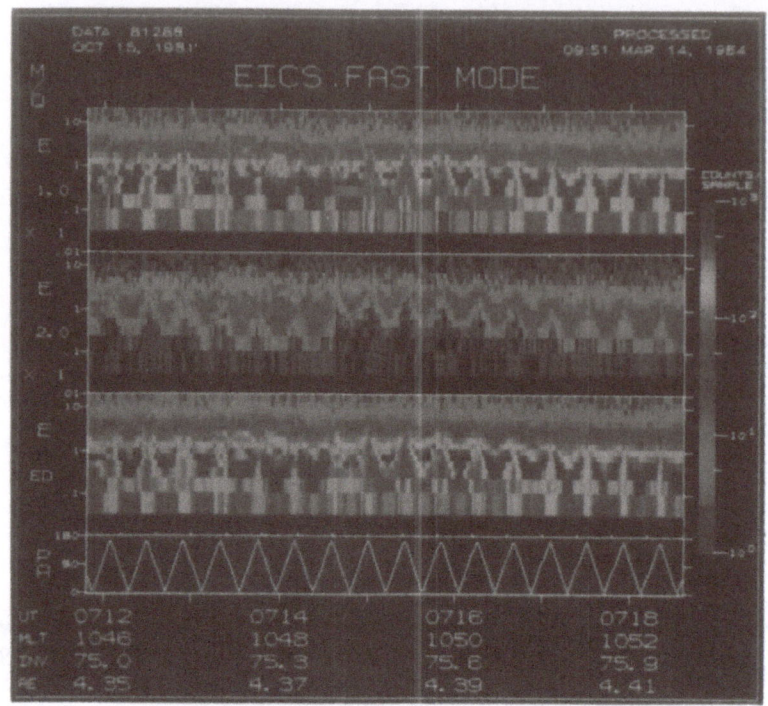

Figure 4.   Energy-time spectrogram of ions observed in the cusp from Dynamics Explorer -1.   The top three panels are H^+, He^{++} and non mass analyzed data respectively.   See text for a description of the format.

angles for both H^+ and He^{++} in one channel and a separate detector sensitive to all ions in another channel.   Since the instrument samples only one mass species at a time, the abscissa in Figure 4 has been modified to present the 24 energy spectra for each species in angle order (rather than time order) for each instrument cycle.   Pitch angle is indicated in the bottom panel.   Note that this display convention results in an apparent satellite spin period equal to the instrumental cycle time.   The instrumental counting rate, which is nearly proportional to number flux is encoded using the color bar on the right.   The universal time (UT) and orbital parameters geocentric distance in earth radii (Re), invariant latitude (INV) and magnetic local time (MLT) are also indicated.

Both the energy-time and energy-pitch angle dispersion signatures are clearly discernable in all three energy-time spectrograms.   In fact two injection events are shown overlapping during the period from ~ 0712 to ~ 0715 in Figure 4.   Detailed examination of the relative He^{++} and H^+ energy-pitch angle data show that the high energy He^{++} and H^+ data from the same instrument cycle have similar distributions in particle velocity, not energy, consistent with the predictions of the time-of-flight model. The low energy data for the two species are, however, dissimilar throughout the interval presented in Figure 4.

The EICS H^+ and He^{++} data from one instrument cycle centered on 0715 UT are presented in Figure 5 as phase space density contour plots. The contours of constant phase space density are displayed in a coordinate system fixed with respect to the satellite velocity direction. The origin and the 500 km/sec grids in this system, as well as the direction of the magnetic field (B) are indicated. The labels are the $\log_{10}$ of the phase space density in units of sec^3/km^6. The energy-pitch angle distributions of the high energy H^+ and He^{++} ions coming down magnetic field lines are similar as noted above. It is also apparent from Figure 5 that the difference between the low energy H^+ and He^{++} data is the low energy H^+ ions flowing up the magnetic field line from the ionosphere below. We defer discussion of upward accelerated ionospheric ions in the cusp to the next section.

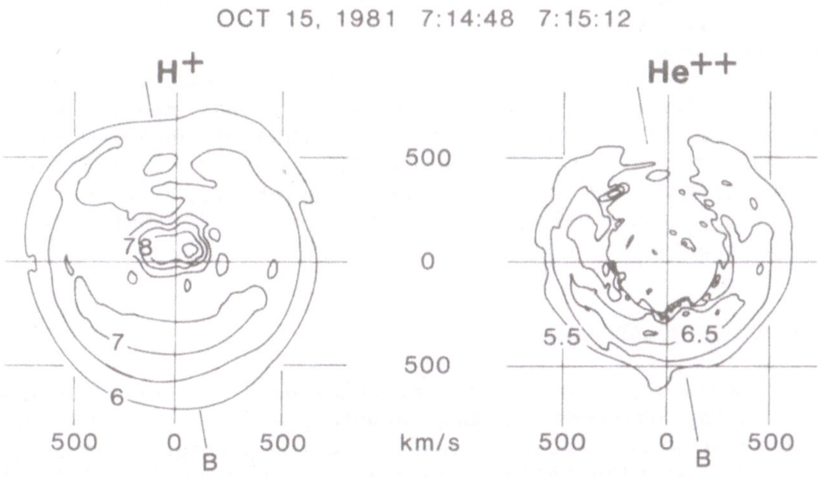

OCT 15, 1981   7:14:48   7:15:12

Figure 5. Contours of constant phase space density for a 24 second interval from the period shown in Figure 4. The coordinate system and contour units are indicated in the text. The direction of the local magnetic field is indicated by the lines labeled 'B'.

As with all geophysical models, the steady drift of plasma after a localized injection model discussed above is not consistent with all of the experimental observations. Reiff et al. (1977 and 1980) and Burch et al. (1980) have identified and discussed in detail events with ' V ' shaped ion energy-latitude dependences. Injection followed by drift in a constant direction leads to a steady increase or decrease of ion energy with latitude. The ' V ' type signatures have both an energy decrease and increase with latitude and a minimum in the observed ion energy in the middle. These events are not rare and occur during periods of weak or northward interplanetary magnetic field (Reiff et al., 1980).

Reiff et al. (1977 and 1980) proposed a very high rate of 'cross field diffusion' to explain these events. Burch et al. (1980) introduced a model of the polar ionospheric electric potential distribution and showed that the observed low altitude Atmosphere Explorer ion energy-latitude dependance of both the normal and ' V ' type could be qualitatively explained by injection followed by drift under the influence of the electric fields that were included in their model.

The relationship between convection direction, altitude, and assumed magnetic topology in the dayside magnetosphere is complex as pointed out by Paschmann et al. (1976), Crooker (1979), Heelis et al. (1980), Winningham and Heelis (1983), Heelis (1985) and others. Measurements of cusp ion composition, energy and angular distributions under various geophysical conditions continue to provide both insights and limitations on global models of the magnetosphere. In particular the dependence of observed plasma flows, and Birkeland currents on the 'y' component of the interplanetary magnetic field as reported by Burch et al. (1984) have lead Reiff and Burch (1984) to propose a modified gobal model of the magnetosphere.

ENTRY OF MINOR SOLAR WIND IONS

There are only a few reports of solar wind ions other than H^+ and He^{++} being observed in the magnetosphere. The third most abundant solar wind species, oxygen, is a few percent of the solar wind helium. As with helium, solar wind oxygen is identified by its charge state. Solar wind oxygen is typically 6 or 7 times ionized. For a discussion of minor ions in the solar wind see Bame et al. 1983, Schmidt et al., 1980, Kunz et al. 1983 and references therin.

Detection of minor solar wind ion species in the earth's magnetosphere at the level of a few parts in 10,000 is experimentally challenging. Lynch et al. (1976) observed that solar wind oxygen was less than 20% of the observed He^{++} or 0.15% of the H^+ in the evening auroral zone. Hovestadt et al. (1978) have

reported observing solar wind helium, oxygen, carbon, and heavier
ions at extremely high energies in the earth's radiation belts.
Peterson et al. (1983) have reported some preliminary correlated
observations of thermal oxygen ions in the solar wind and cusp.
The planned releases of lithium at the subsolar point by AMPTE
(Active Magnetospheric Particle Tracer Explorers) in the fall of
1984 (Krimigis et al, 1982) will provide a unique opportunity to
study the transfer of mass from the solar wind to the magnetosphere and its further transport and energization within the
magnetosphere.

ACCELERATION OF IONOSPHERIC IONS IN THE CUSP

The first direct observation of upward accelerated ions in
the cusp was reported by Sharp et al. (1977). Torbert and Carlson
(1980) reported electron and ion spectra at low altitudes in the
cusp that are best explained by an electrostatic acceleration
parallel to the magnetic field at an altitude of several thousand
kilometers. Recently several statistical studies of the magnetic
local time dependence of upward accelerated ions have been made
(Ghielmetti et al. 1978, Klumpar, 1979, Shelley, 1979, Gorney et
al. 1981, and Yau et al. 1984).   Only one of these studies
(Shelley, 1979) was limited to the cusp region. All of these
studies report high occurrence probabilities of upward accelerated
ions in the cusp region.

Shelley (1979) in his study of 100 orbits of S3-3 ion mass
spectrometer data acquired above 5000 km between 0900 and 1500
magnetic local time used simultaneous electron data to select 57
'clearly identified cusp crossings'. Upstreaming O^+ ions with
energies greater than 500 eV were unambiguously observed on 33 of
these crossings (58%). Shelley pointed out that this fraction
is a lower limit because of instrumental sensitivity, the limited
energy range, and limited temporal resolution. He suggested that
ionospheric ions are being accelerated out of the mid-altitude
cusp into to boundary layer regions of the magnetosphere on a
nearly continuous basis. The later studies are consistent with
this suggestion.

The cusp, then, contains a mixture of recently injected solar
wind plasma and accelerated ionospheric plasma. We will show in
the next section that it also contains some energetic magnetospheric plasma from the dayside magnetosphere. The mass spectrometer data from Dynamics Explorer -1 provide a clear picture of
the relative importance of these three plasma sources.

Energetic ion composition data from a cusp crossing on
October 8, 1981 are presented in Figure 6 and 7. Figure 6 is a
set of energy-time spectrograms similar to Figure 4, except that

the data have been averaged over all pitch angles and the time
scale compressed. The instrumental cycle time during the period
these measurements were made was 96 seconds which is the temporal
resolution in Figure 6 and 7. The three color panels are, from
top to bottom, H^+, He^{++}, and O^+. The color bar on the right
indicates the average flux in units of $(cm^2 -sec - sr - keV/e)^{-1}$.
The energy-latitude dispersion of the solar wind ions (H^+ and
He^{++}) is the most prominent feature in this presentation. The
origin of this dispersion was discussed above. Figure 7 is a set
of three spin phase angle-time spectrograms for the same interval
presented in Figure 6. This type of presentation is particularly
useful for defining magnetospheric boundaries. To produce Figure
7 the fluxes were sorted into 24 spin phase angle bins (zero
degrees corresponds to the spacecraft velocity direction) and then
integrated from 10 eV/e to 1 keV/e. The flux units encoded by the
color bar on the right are $(cm^2-sec-sr)^{-1}$. Over the polar cap the
magnetic field makes an angle of about 90 degrees with the

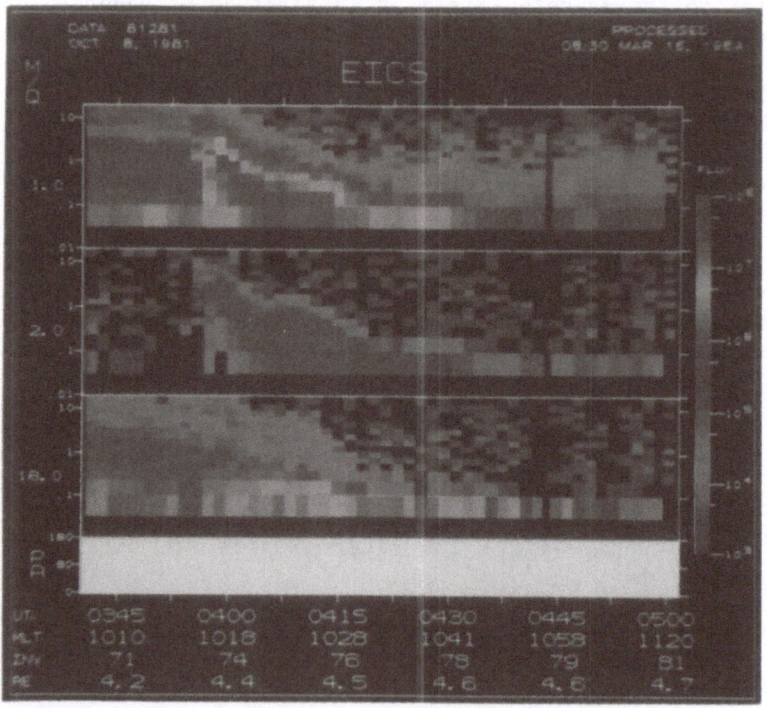

Figure 6. Energy-time spectrogram for a Dynamics Explorer cusp
crossing on October 8, 1981. The three top panels are data for
H^+, He^{++} and O^+ respectively. The data have been averaged over all
pitch angles. The flux units encoded by the color bar on the
right are $(cm^2-sec-sr-keV/e)^{-1}$.

satellite velocity direction, so ions flowing up magnetic field lines from the ionosphere are near 90 degrees in Figure 7. The equatorward boundary of solar wind ions (H^+ and He^{++}) is crossed in one 96 second instrumental cycle period. This instrument cycle is also the onset of a prominent feature in the O^+ panel.

The O^+ feature in Figure 7 starting at about 0355 is centered on the magnetic field direction. The low energy (less than 1 keV) ions that form it have been accelerated away from the ionosphere below. At the onset they are observed at pitch angles close to 90 degrees. This is the same type of velocity distribution as that shown for the low energy H^+ ions in Figures 4 and 5 above. It is commonly called a ' conical ' distribution. Ions must be accelerated normal to the magnetic field to produce such distributions (Sharp et al. 1977). Ion distributions that are narrowly collimated about the magnetic field direction are called ion ' beams '.

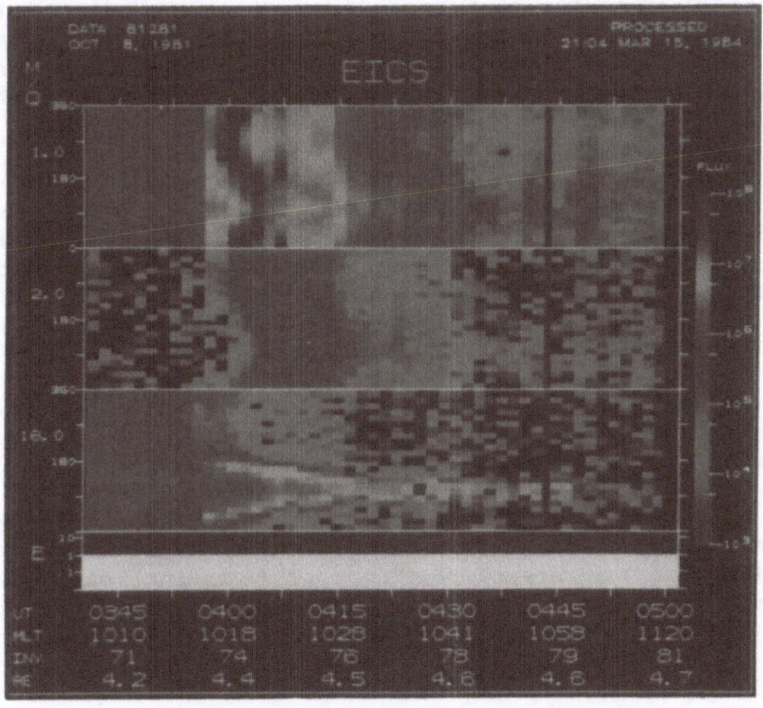

Figure 7. Spin phase angle-time spectrogram for the same period displayed in Figure 6. The data have been collected in 24 angular bins and integrated from 10 eV/e to 1 keV/e. The flux units encoded by the color bar on the right are $(cm^2-sec-sr)^{-1}$. Low energy O^+ ions flowing up magnetic filed lines from the ionosphere below are visible from ~0415 to ~0450.

Ion beams can be formed in two ways: acceleration of ions parallel to the magnetic field direction; or as the result of the conservation of magnetic moment which will cause the 'cone' angle of conic distributions to become smaller than the instrumental resolution as they move upward into regions with weaker magnetic fields.

In the segment of data shown if Figures 6 and 7, the O^+ distribution changes from conical at the equatorward edge of the cusp region to beam-like over the polar cap. It must be noted that this is the only example found to date in the Dynamics Explorer data set of such a smooth transition between types of distribution. The more typical case has multiple variations between conical and beam-like distributions, sometimes on a time scale short compared to the instrumental cycle time. Beam and conical events are sometimes not experimentally distinguishable. Gorney et al. (1981), Yau et al. (1984), and Shelley (1979) have published relative occurrence frequencies of the two types of accelerated ion distributions in the cusp local time region. All of these reports show that beam like and conical distributions have roughly equal probabilities of occurrence in the cusp. Klumpar et al. (1984) have also shown that acceleration parallel and perpendicular to the magnetic field direction can occur on the same magnetic field lines.

As noted above, the reason for distinguishing between different types of ion distributions is that different acceleration mechanisms are thought responsible for them. Unfortunately identifying the acceleration as being perpendicular or parallel to the magnetic field direction is not enough information to uniquely identify the processes involved -- there are many more than two acceleration mechanisms to choose from. Lennartsson (1983) has provided a short introduction into types of acceleration mechanisms that have been considered relevant to the magnetosphere. It is not the intention here to survey the vast number of reports discussing the problem of acceleration of ionospheric ions. For cusp studies, however, the work of Ungstrup et al. (1979) and Gorney (1983) are particularly relevant. Ungstrup et al. (1979) focused attention on electrostatic ion cyclotron waves as a mechanism for transversely heating ions (creating conic distributions). Gorney (1983) pointed out that downflowing ion beams (i.e. the cusp type injection events discussed above and by Burch et al., 1982) form 'rings' of energetic ions similar to those shown in Figure 5, and that these rings have characteristics that make them unstable and are therefore a possible energy source for the waves that in turn cause particle acceleration.

Another possible source of very low energy upflowing iono-spheric ions in the cusp region is the 'classic polar wind' described by Axford (1968), Banks and Holzer (1969) and others. Recent reports of the polar wind have been made by Chappell et al.

(1982), Gurgio and Burch (1982), and Sojka et al. (1983). The classic polar wind is made up of cold (1 eV or less) H^+ ions with a H^+ to O^+ flux ratio of about 100, although some recent work by Barakat and Schunk (1983) shows how the original models of the polar wind may be modified to get larger O^+ fluxes. Gurgiolo and Burch (1982) and Shelley et al. (1982) have noted that there is an additional component in the upward flowing low energy ion obser-vations that was not predicted by models of polar wind. Shelley et al. (1982) observed O^+ to H^+ ratios of 10 and energies of 10's of eV. Gurgiolo and Burch (1982) noted a 'heated' component of the polar wind. The observations of Gurgiolo and Burch (1982), and those shown in Figures 6 and 7 above, show that at least some of the warm O^+ beams observed streaming up magnetic field lines in the polar cap can originate in the cusp. This is possible because the antisunward convection velocities are comparable with the upward O^+ flow velocities. For example, a 10 eV O^+ ion travels ~13 km/sec or ~8.5 minutes per earth radius. There are, however, other physical processes that can accelerate ionospheric ions over the polar cap and outside the cusp region, such as those respon-sible for polar cap aurora. Independent of source, cusp, polar wind, polar aurora, or a combination of these the flux of iono-spheric ions out of the polar cap is considerable, and can make a non-negligible contribution to the plasma sheet (Shelley et al. 1982, Yau et al. 1984).

MAGNETOSPHERIC IONS IN THE CUSP

It is remarkable that energetic O^+ fluxes only gradually decrease in intensity poleward of the point where the satellite enters onto cusp field lines in the data presented in Figures 6 and 7. Equatorward of the cusp (i. e. prior to about 0355 UT), the dayside magnetosphere energy and mass composition is consis-tent with that reported at lower magnetic latitudes on dayside magnetic field lines (Young, 1980, Lennartsson et al. 1981, and Lennartsson and Sharp 1982). Poleward of the first observation of solar wind ions (i. e. after about 0355 UT), the flux of energetic (greater than 1000 eV) O^+ ions at ALL pitch angles only slowly decreases. This point is illustrated in Figure 8 which is a spin phase angle-time spectrogram similar to Figure 7 except that the integral fluxes displayed are from 1 keV/e to 17 keV/e. Signif-icant fluxes of downcoming energetic O^+ are observed until almost 0415 UT. Evidence of energetic magnetospheric ions on cusp field lines has also been noted by Shelley et al. (1976).

There have been several recent reports in the literature that have called into question the meaning and detectability of 'closed' and 'open' magnetic field lines. McDiarmid et al. (1976) and others have noted 'trapped' electron distributions in the cusp region at low altitudes; Eastman and Frank (1982) have noted a similar situation on field lines where magnetic reconnection has been reported (Paschmann et al. 1979); Frank et al. (1982) have reported long lived, cross polar cap auroral structures (Theta Aurora); Peterson and Shelley (1984) have detected isotropic energetic O^+ on field lines over the polar cap; Foster and Burrows (1977) have reported trapped electron fluxes over the polar cap; and Huang et al. (1983) have reported isotropic keV ion distributions in the magnetotail lobes. In Figures 6, 7 and 8 above we see both freshly injected solar wind ions (H^+ and He^{++}) and energetic magnetospheric ions (greater than 1 keV isotropic O^+) simultaneously on the same magnetic field lines.

Several different physical explanations of these surprising observations have been offered. Daly and Fritz (1982) have noted that electrons can be reflected (i.e. trapped) at the minimum in magnetic field strength on 'open' magnetic field lines at the magnetopause and that ions are, at best, only partially reflected at such a minimum. Peterson and Shelley (1984) have noted that

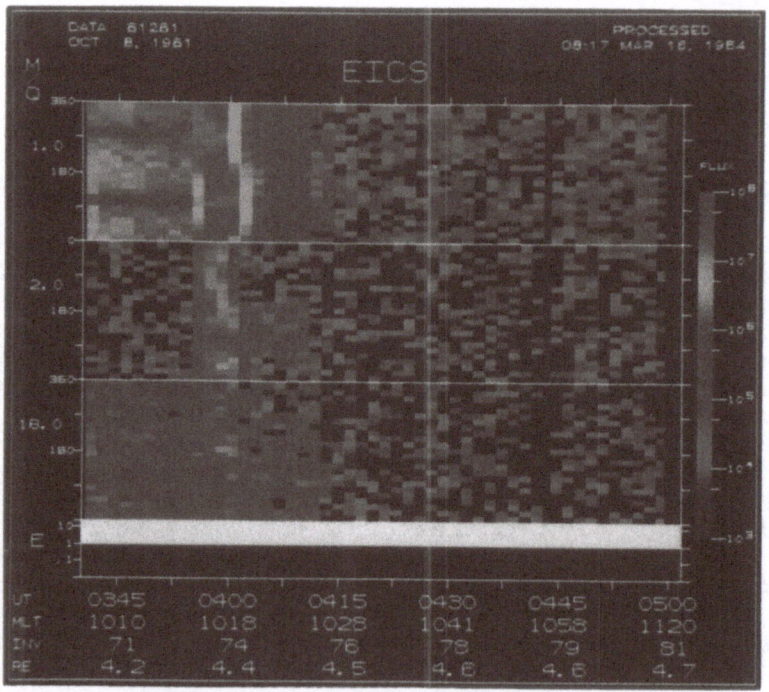

Figure 8. Spin phase angle-time spectrogram for the same interval presented in Figures 6 and 7. The data have been integrated from 1 keV/e to 17 keV/e in Figure 8. The flux units encoded by the color bar on the right are $(cm^2\text{-sec-sr})^{-1}$.

substantial pitch angle scattering of energetic O^+ from magneto-spheric field lines that have been recently 'opened' (i. e. merged with solar wind magnetic field lines) and convected over the polar cap could account for their observations. Diffusion of ions out of the magnetosphere into the magnetosheath or boundary layers has also been discussed by several authors (e.g. Freeman et al. 1977, and Haerendel et al. 1978). Huang et al. (1983) explained their surprising observations in the magnetotail lobes as filaments of plasma sheet plasma convecting 'away' from the plasma sheet during times when the interplanetary field was northward.   Foster and Burrows (1977) inferred the infrequent existence of a fluctuating potential barrier far down the magnetotail lobes to explain their observations. Frank et al. (1982a and 1982b) have explained the Theta Aurora as arising from bifurcation of the magnetotail lobes, a process that most certainly would involve cusp plasmas.

Not all of the phenomena listed above involve the cusp, but the proposed mechanisms can all occur on cusp field lines. Because the energetic O^+ ions do not immediately decrease in intensity after ~0355 in Figures 6,7 and 8, they may well arise from magnetospheric plasma that is on recently 'opened' field lines that are drifting through the cusp. They could also arise from a very strong diffusion process. In any case, the above discussion shows that particle signatures alone are not always an unambiguous indicator of 'open' or 'closed' field lines.

CONCLUDING REMARKS

The polar cusp was originally studied because it provided information and insights on solar wind plasma entering the magnetosphere. With the data from modern high sensitivity and high temporal resolution instruments it is now possible to separate ionospheric and magnetospheric components from the solar wind plasma on cusp field lines. We have briefly discussed the results so far obtained in the study of this mixture: ionospheric plasma is accelerated upwards into the cusp; ionospheric plasma accelerated in the cusp is observed over the polar caps; and this plasma could make a non-negligible contribution to the maintenance of the plasma sheet. We have also noted the numerous reports that show the difficulty in using particle pitch angle signatures alone as an indicator of 'open' or 'closed' magnetic field lines.

ACKNOWLEDGMENTS

I would like to thank Dr. W. Lennartsson for his helpful comments and Drs E. G. Shelley and R. D. Sharp for the opportunity to pursue this research. This research was supported by Lockheed Independent Research funds and NASA contract NAS5-25694.

REFERENCES

Axford, W.I.,   J. Geophys. Res. 73, 1968, 1968.
Barakat, A.R., and R.W. Schunk,   J. Geophys. Res. 88, 7887, 1983.
Bame, S.J., W.C. Feldman, J.T. Gosling, D.T. Young, and R.D. Zwickl,
    p. 73, in Energetic Ion Composition in the Earth's Magnetosphere,
    (Ed. R.G. Johnson). Terra Scientific Publishing Company, Tokyo,
    1983.
Banks, P.M., and Holzer,   J. Geophys. Res. 74, 6317, 1969.
Burch, J.L., P.H. Reiff, R.W. Spiro, R.A. Heelis, and S.A. Fields,
    Geophys. Res. Lett. 7, 393, 1980.
Burch, J.L., P.H. Reiff, R.A. Heelis, J.D. Winningham, W.B. Hanson,
    C. Gurgiolo, J.D. Menietti, R.A. Hoffman, and J.N. Barfield,
    Geophys. Res. Lett. 9, 921, 1982.
Burch, J.L., P.H. Reiff, J.D. Menietti, R.A. Heelis, W.B. Hanson, D.S.
    Shawhan, E.G. Shelly, M. Sugiura, and J.D. Winningham, Submitted
    to J. Geophys. Res., 1984.
Carlson, C.W., and R.B. Torbert,   J. Geophys. Res. 85, 2903, 1980.
Chappell, C.R., J.L. Green, J.F.E. Johnson, and J.H. Waite, Jr., Geo-
    phys. Res. Lett. 9, 933, 1982.
Crooker, N.U., J. Geophys. Res. 84, 951, 1979.
Daly, P.W., and T.A. Fritz, J. Geophys. Res. 87, 6081, 1982.
Eastman, T.E., and L.A. Frank, J. Geophys. Res. 87, 2187, 1982.
Foster, J.C., and J.R. Burrows, J. Geophys. Res. 82, 1977.
Frank, L.A., J.D. Craven, J.L. Burch, and J.D. Winningham, Geophys.
    Res. Lett. 9, 1001, 1982a.
Frank, L.A., J.D. Craven, and A.J. Smith, EOS Trans. Amer. Geophys.
    U. 63, 1056, 1982b.
Freeman, J.W., H.K. Hills, T.W. Hill, P.A. Reiff, and D.A. Hardy,
    Geophys. Res. Lett. 4, 185, 1977.
Ghielmetti, A.G., R.G. Johnson, R.D. Sharp, and E.G. Shelley, Geophys.
    Res. Lett. 5, 59, 1978.
Gorney, D.J., A. Clarke, D.Croley, J. Fennell, J. Luhmann, and P.
    Mizera, J. Geophys. Res. 86, 83, 1981.
Gorney, D.J., Geophys. Res. Lett. 10, 417, 1983.
Gurgiolo, C. and J.L. Burch, Geophys. Res. Lett. 9. 945, 1982.
Haerendel, G., G. Paschmann, N. Sckopke, H. Rosenbauer, and P.C. Hed-
    gecock, J. Geophys. Res. 83, 3195, 1978.
Heelis, R.A., J.D. Winningham, W.B. Hanson, and J.L. Burch, J. Geophys.
    Res. 85, 3315, 1980.
Heelis, R.A., This volume, 1985.
Heikkila, W.J., Geophys. Res. Lett. 9, 159, 1982.
Hovestadt, D., G. Gloeckler, C.Y. Fan, L.A. Fisk, F.M. Ipavich, B.
    Klecker, J.J. O'Gallagher, and M. Scholer, Geophys. Res. Lett. 5,
    1055, 1978.
Huang, C.Y., L.A. Frank, W.K. Peterson, G.K. Parks, W. Lennartsson, and
    R.J. Decoster, EOS, Trans. Amer. Geophys. U. 64, 812, 1983.
Klumpar, D.M., J. Geophys. Res. 84, 4229, 1979.
Klumpar, D.M., W.K. Peterson, and E.G. Shelley, Submitted to J. Geo-
    phys. Res., 1984.

Krimigis, S.M., G. Haerendel, R.W. McEntire, G. Paschmann, and D.A. Bryant, EOS Trans. Amer. Geophys. U. 63, 843, 1982.

Kunz, S., P. Bochsler, J. Geiss, K.W. Ogilvie, and M.A. Coplan. Solar Physics 88, 359, 1983.

Lemaire, J., Planet. Space. Sci. 25, 887, 1977.

Lennartsson, W., R.D. Sharp, E.G. Shelley, R.G. Johnson, and H. Balsiger, J. Geophys. Res. 86, 4678, 1981.

Lennartsson, W., and R.D. Sharp, J. Geophys. Res. 87, 6109, 1982.

Lennartsson, W., p. 23 in Energetic Ion Composition in the Earth's Magnetosphere (Ed. R.G. Johnson). Terra Scientific Publishing Company, Tokyo, 1983.

Lundin, R., B. Hultqvist, N. Pissarenko, and A. Zackarov, Space Sci. Rev. 31, 247, 1982.

Lundin, R., to appear in Planet. Space Sci. 32, 757, 1984.

Lundin, R., this volume, 1985.

Lundin, R., and E. Dubinin, to appear in Planet. Space Sci., 1984.

Lynch, J., D. Pulliam, R. Leach, and F. Scherb, J. Geophys. Res. 81, 1264, 1976.

McDiarmid, I.B., J.R. Burrows, and E.E. Budzinski, J. Geophys. Res. 81, 221, 1976.

Paschmann, G., G. Haerendel, N. Sckopke, H. Rosenbauer, and P.C. Hedgecock, J. Geophys. Res. 81, 2883, 1976.

Paschmann, G., B.U.O. Sonnerup, I. Papamastorakis, N. Skopke. G. Haerendel, S.J. Bame, J.R. Asbridge, J.T. Gosling, C.T. Russell, and R.S. Elphic, Nature 282, 243, 1979.

Peterson, W.K., E.G. Shelley, W.K.H. Schmidt, P. Bochsler, and H. Balsiger, p. 413 in Bulletin no. 48, International Association of Geomagnetism and Aeronomy, Hamburg, 1983.

Peterson, W.K., and E.G. Shelley, to appear in J. Geophys. Res., 1984.

Reiff, P.H., T.W. Hill, and J.L. Burch, J. Geophys. Res. 82, 479, 1977.

Reiff, P.H., J.L. Burch, and R.W. Spiro, J. Geophys. Res. 85, 5997, 1980.

Reiff, P.H., and J.L. Burch, Submitted to J. Geophys. Res., 1984.

Rosenbauer, H., H. Gruenwaldt, M.D. Montgomery, G. Paschmann, and N. Skopke, J. Geophys. Res. 80, 2723, 1975.

Russell, C.T., and R.C. Elphic, Geophys. Res. Lett. 6, 33, 1979.

Schmidt, W.K.H., H. Rosenbauer, E.G. Shelley, R.D. Sharp, R.G. Johnson, and J. Geiss, 1980, Geophys. Res. Lett. 7, 697, 1980.

Sckopke, N., this volume, 1984.

Sharp, R.D., R.G. Johnson, and E.G. Shelley, J. Geophys. Res. 82, 3324, 1977.

Shelley, E.G., R.D. Sharp, and R.G. Johnson, J. Geophys. Res. 81, 2363, 1976.

Shelley, E.G., in Proceedings of Magnetospheric Boundary Layers Conference, Alpach 11-15 June 1979, ESA SP-148, 1979.

Shelley, E.G., D.A. Simpson. T.C. Sanders, E. Hertzberg, H, Balsiger, and A. Ghielmetti, Space Sci. Instrumentation, 5, 443. 1981.

Shelley, E.G., W.K. Peterson, A.G. Ghielmetti, and J. Geiss, Geophys. Res. Lett. 9, 941, 1982.

Sojka, J.J., R.W. Schunk, J.F.E. Johson, J.H. Waite, and C.R.   ·
    Chappell, J. Geophys. Res. 88, 7895, 1983.
Torbert, R.B., and C.W. Carlson, J. Geophys. Res. 85, 2909, 1980.
Ungstrup, E., D.M. Klumpar, and W.J. Heikkila, J. Geophys. Res. 34,
    4289, 1979.
Winningham, J.D., and R.A. Heelis, in High-Latitude Space Plasma
    Physics (Eds. B. Hultqvist and T. Hagfors). Plenum Press, New
    York, 1983.
Yau, A.W., B.A. Whalen, W.K. Peterson, and E.G. Shelley, to appear
    in J. Geophys. Res., 1984.
Young, D.T., Habilitationsschrift, University of Bern, 1980.

AVERAGE ELECTRON PRECIPITATION IN THE POLAR CUSPS, CLEFT AND CAP

M.S. Gussenhoven*, D.A. Hardy*, and R L Carovillano**

*Space Physics Division
Air Force Geophysics Laboratory
Hanscom Field, MA 01731 U.S.A.

**Department of Physics
Boston College
Chestnut Hill, MA 02167 U.S.A.

ABSTRACT. Results are given of the statistical properties of electron precipitation measured at low altitudes by polar orbiting satellites. The morphology of high latitude electron precipitation is determined in the polar cap, the polar cusp, and the polar cleft as a function of magnetic latitude, local time, and activity, Kp. Results show that a polar cusp region highly confined in latitude and local time can be identified by a minimum in the precipitating electron average energy. This region lies very close to local noon, but is not spatially coincident with the maximum in precipitating electron flux which occurs several hours earlier. Surrounding the cusp is a well-defined region of low energy precipitation whose contours of constant integral flux have crescent shapes centered about the flux maxima. This is best seen in cases of low magnetic activity when boundary plasma sheet electron precipitation is not strong. We refer to this region as the cleft. The average energy of the cleft electrons increases steadily as one moves away from the cusp in local time and in latitude. The polar cap appears to have two states: an active state, when the IMF is northward; and a quiet state, characterized by polar rain precipitation and occurring when the oval is active. Two-dimensional maps of integral flux and average energy for polar rain occurrence show that the basic variation in this precipitation is from dayside to nightside. The axis of symmetry for the variation is pre-noon to pre-midnight with the integral flux (average energy) increasing (decreasing) from day to night. These variations in number flux and average energy exhibit a symmetry that is an extension of that seen in the cleft.

1. INTRODUCTION

How particles access the high latitude magnetosphere from the solar wind and magnetosheath is still a fundamental and unanswered question. Possible entry processes are addressed mainly using in situ

85

J. A. Holtet and A. Egeland (eds.), The Polar Cusp, 85–97.
© *1985 by D. Reidel Publishing Company.*

measurements of the transition region. In our approach, statistical
studies were made of low altitude precipitating particles whose sources
are the dayside entry regions. Results from these studies provide a
spatial overview which must be compatible with operative entry
processes. Such a study provides information unobtainable from event
studies. Both statistical and event studies have advantages and
disadvantages. Event studies can give great detail, but typically are
restricted to a few cases that are often difficult to interpret. In
the averaging method, small spatial and temporal variations are, by
necessity, smoothed over, but a global picture can be obtained. The
number of large-scale statistical studies done in the past has been
restricted by requirements of large data sets (millions of samples) and
the availability of computer time.

2. INSTRUMENTATION

The data used for this analysis were from the SSJ/3 detectors on
the F2 and F4 satellites of the Defense Meteorological Satellite
Program and the CRL-251 experiment on the P78-1 satellite of the Space
Test Program. All detectors consisted of sets of curved-plate
electrostatic analyzers capable of measuring the flux of precipitating
electrons in 16 energy channels between 50 eV and 20,000 eV and of
determining a spectrum in one second or less.
All three satellites had circular, sun-synchronous orbits. The F2
orbit plane was initially in the dawn-dusk meridian and precessed
towards the 0830-2030 MLT meridian during the satellite's lifetime.
The F4 and P78-1 satellites' orbit planes were near noon-midnight. The
periods for which data are available are: F2:9/77 to 2/80; F4:6/79 to
8/80; P78-1:3/79 to present. The F2 and F4 satellites were at an
altitude of 840 km; the P78-1 was at 600 km. The DMSP sensors were
operated continuously and approximately 80% of the data were available
for this study. For the P78-1 satellite, approximately 1000 polar
passes were available from 1979. For the determination of the average
characteristics for the cusp and cleft studies, the F2, F4 and P78-1
data were used. For the polar rain studies only F2 data were used.

3. METHOD OF ANALYSIS

In our studies two different spatial griddings were used to
determine the average two-dimensional patterns of high-latitude
electron precipitation. In the first, which used all of the electron
data, the high latitude region was divided into zones in magnetic local
time (MLT) and corrected geomagnetic latitude. The MLT divisions were
48 one-half-hour sections. In latitude there were 30 divisions: two
degree increments between 50 and 60 degrees, one degree increments
between 60 and 80 degrees, and two degree increments between 80 and 90
degrees. Seven such maps were created, one for the Kp 0, 0+ values,
one for Kp 1-, 1, 1+ values, and so on up to Kp 5-, 5, 5+. The last
map included all cases greater than Kp 6-. Fifteen months of data were

used from the F2 and F4 satellites. The fifteen months were chosen to
give an even distribution of the data over the seasons of the year and
to provide sufficient coverage at high activity. Altogether the data
provided 13.6 million spectra. All orbits of the P78-1 satellite in
the interval from February, 1979 to January, 1980 were used. This
comprised approximately 0.52 million additional spectra. The 14.1
million spectra distributed among the 7 levels of Kp as follows: Kp 0,
7.85%; Kp 1, 23.82%; Kp 2, 26.9%; Kp 3, 21.9%; Kp 4, 10.5%; Kp 5, 5.3%;
Kp 6, 3.6%.

In the second, a subset of the data in which only polar rain was
observed was used. Here a two-dimensional cartesian grid was
superposed on MLT-CGM coordinates with each bin having a one degree by
one degree dimension along the dawn-dusk and noon-midnight axes.
Separation by Kp was made into a low activity case (Kp 0 to 2+) and a
moderate activity case (Kp 3- to 5+). All polar rain intervals
observed by the F2 satellite between September, 1977 and August, 1978
were used. This provided about 400,000 spectra.

In both gridding methods in each zone, the average and standard
deviation of the differential number flux for each of the 16 energy
channels of the detector were calculated using all spectra that fell
within that zone. The final product is, therefore, the average
spectrum in each zone at each level of activity. From the average
spectra we calculated integral quantities over the entire energy range
of the spectrum. The three quantities calculated were the integral
number flux in units of $(cm^2-sec-ster)^{-1}$; the integral energy flux in
$keV/(cm^2 -sec-ster)$; and average energy in keV. These three quantities
are displayed in isocontour and color coded maps in the next sections.

4. CUSP AND CLEFT RESULTS

Figures 1a-d and 2a-d are color-coded plots of the integral
number flux and average energy, respectively, for all precipitating
electrons and for Kp = 0, 1, 2, 3. The figures show that the high
latitude region separates naturally into two regions based upon the
average energy of the electrons, i.e., the dark blue and the
yellow-orange regions in Figure 2. The hotter plasma is confined to a

Figure 1a-d. Polar color spectrograms of the average integral number
flux of electrons with energies between 50 eV and 20 keV as a function
of magnetic local time and latitude at four levels of geomagnetic
activity as measured by Kp. The number flux is in units of
$(cm^2-sec-ster)$ and the color is assigned on a logarithmic scale.

Figure 2a-d. Polar color spectrograms of the average evergy of
electrons with energies between 50 eV and 20 keV as a function of
magnetic local time and latitude at four levels of geomagnetic activity
as measured by Kp. The average energy is in keV. The sixteen colors
refer to energies in keV at .05, .10, .20, .30, .40, .60, .80, 1.0,
1.5, 2.0, 3.0, 4.0, 5.0, 7.0, 10.0, and 12.0.

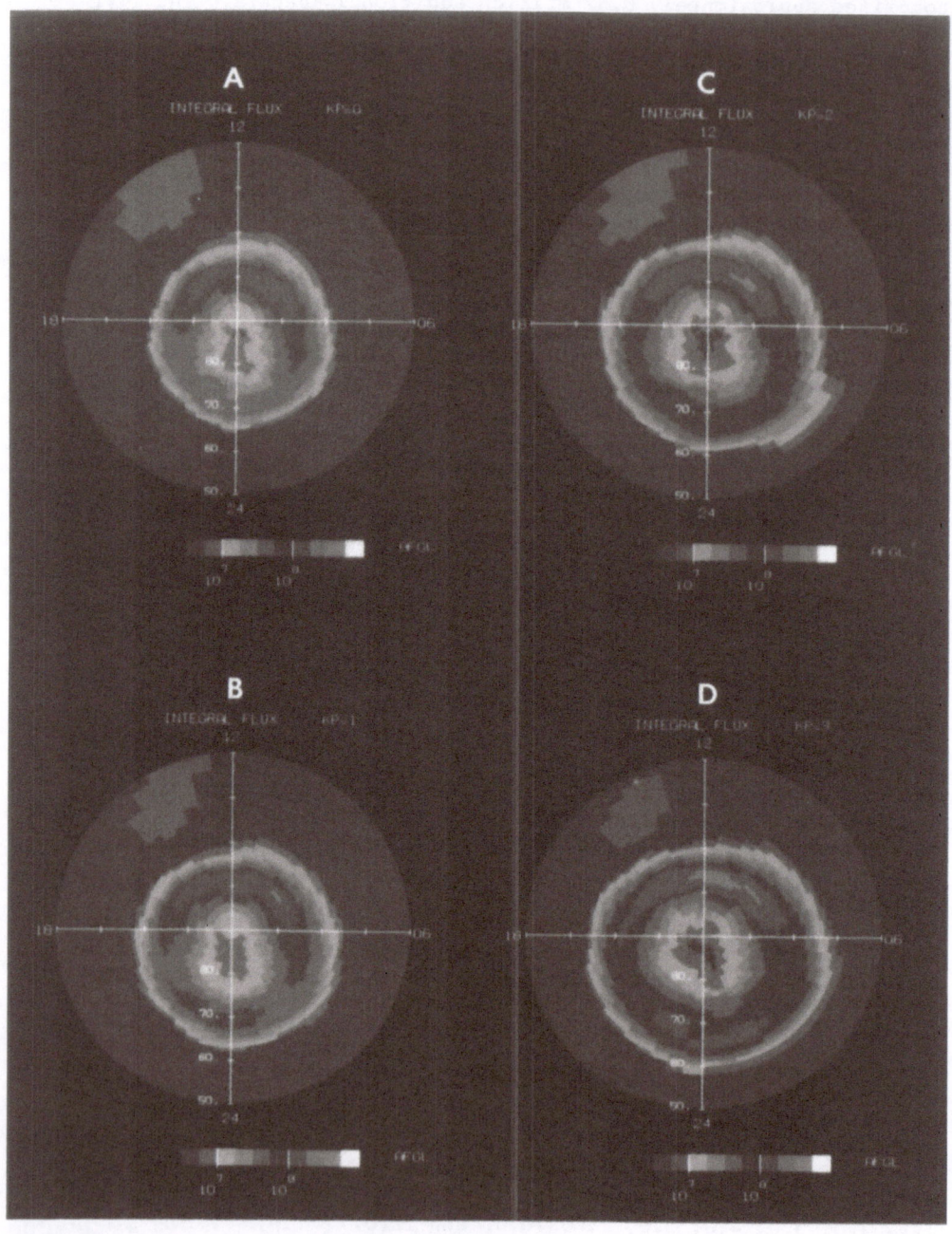

Figure 1a-d

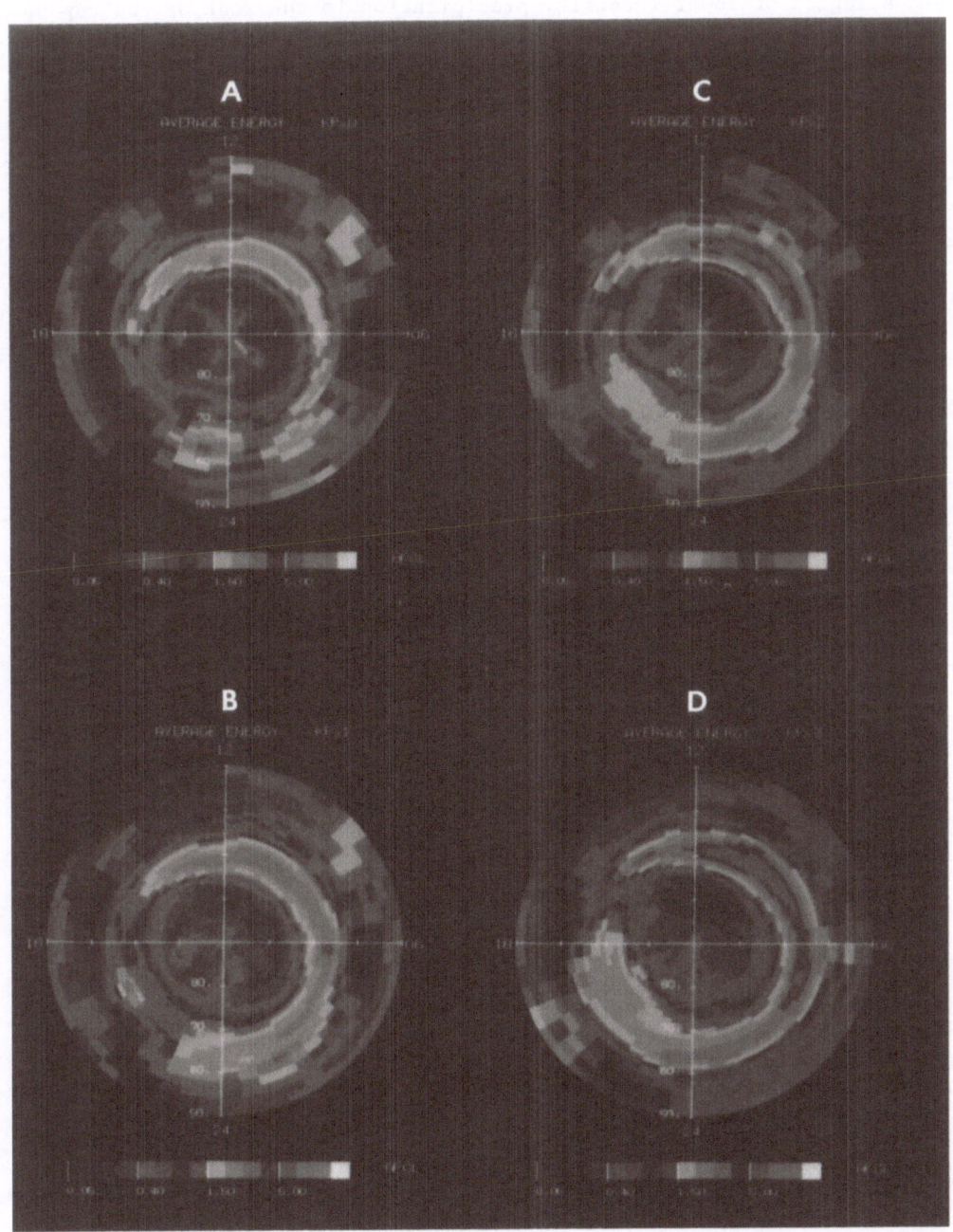

Figure 2a-d

roughly annular region whose low latitude edge is the equatorward edge of the auroral zone, while the colder plasma fills the remaining area between the poleward edge of the annulus and the geomagnetic pole.   The colder electron region is composed of a band of relatively intense precipitation (Figure 1) bounding the poleward edge of the hot plasma and a region of lower intensity precipitation in the rest of the cap up to the pole.   In the latter population of precipitating electrons, the dayside portion with energy less than 600 eV has the following properties:

1.   The highest number flux of cold electrons is found within the dayside portion of the electron precipitation region.   At the first five activity levels (only four are shown) there is a clear crescent-shaped region of cold electrons roughly symmetric about noon or slightly skewed towards prenoon.   The crescent is most clearly evident for Kp = 0 and Kp = 1.   Here the flux exceeds 5×10^7 el/(cm^2 -sec-ster) and extends in MLT over the entire dayside region and one to several hours into the nightside region.   The region extends closer to midnight on the morning-side than the evening-side.   The same general behavior is found for the next three activity levels but is obscured in the color plots by the increasing integral flux on the nightside from hot electrons.   For the two highest activity cases there is still an extended region of low energy precipitation on the dayside, but it is not as well organized as for the lower activity cases.

2.   The intensity of the integral number flux within the dayside region shows little if any increase with increasing activity.   The integral number flux is typically between 5×10^7 and 2×10^8 el/(cm^2-sec-ster).   Although the level of flux within the region is relatively constant, the total flux of electrons in the region increases with increasing activity because the precipitation occurs over a larger spatial region.   For the seven levels of activity the total downcoming flux over the entire dayside with energies between 50 eV and 660 eV are 7.65×10^{24}, 8.89×10^{24}, 1.00×10^{25}, 1.21×10^{25}, 1.97×10^{25} and 3.78×10^{25} el/ (sec-ster).   These numbers were obtained by determining the integral flux for electrons with energies between 50 eV and 660 eV in each spatial element on the dayside, multiplying each by the area of the spatial element, and summing.   Except for Kp > 6-, this trend is fit well by the equation:

$$I = 7.8 \times 10^2 \exp (0.2 \text{ Kp}) \text{ el/(sec-ster)}$$

3.   Within the dayside region of cold electron precipitation, there is a clear prenoon maximum in the number flux.   In Table I the parameters for these maxima are listed.   The maximum is relatively constant in MLT while the latitude decreases with increasing activity as the oval expands.   The integral number flux increases only from 3.05 to 4.1×10^8 el/(cm^2-sec-ster) from Kp = 0 to Kp > 6-.   Both the integral energy flux and average energy of the maximum similarly occur within a narrow range except at the two highest Kp activities.   The increases at higher Kp are attributable to an increase in the spatial variability of

the oval that result in inclusion of some hot electron spectra in the
determination of the averages.

Table I. Prenoon maximum in integral number flux. The values given
for the integral number flux are in units of 10^8 electrons/(cm^2-sec
-ster). The integral energy flux is in units of 10^7 keV/(cm^2-sec
-ster).

Kp	Latitude	MLT	Integral Number Flux	Integral Energy Flux	Average Energy (eV)
0	79°	0800-0830	3.05	7.04	232
1	78°	0800-0830	3.22	8.45	272
2	78°	0930-1000	3.32	8.43	256
3	77°	1100-1130	3.80	7.03	187
4	74°	0830-0900	3.61	19.5	552
5	73°	0830-0900	4.10	16.6	465

4. In the dayside region of cold electrons there is a minimum in
average energy. The location and electron characteristics at the
minimum are listed in Table II. The location in latitude shows a total
variation of 5° with activity. The average energy varies by less than
about 20 percent and decreases slightly with increasing activity. The
integral energy flux similarly does not vary much. The average energy
minima are located near the poleward edge of the crescent-shaped region
of cold electron precipitation.

Table II. Minimum in average energy. The integral energy flux is in
units of 10^7 keV/(cm^2-sec-ster) and the average energy is in eV.

Kp	Latitude	MLT	Integral Energy Flux	Average Energy (eV)
0	81°	1100-1130	4.57	199
1	81°	1130-1200	4.88	183
2	81°	1130-1200	3.83	168
3	79°	1100-1130	4.44	165
4	78°	1130-1200	3.96	162
5	78°	1230-1300	3.18	147
>6-	76°	1200-1230	7.29	184

These results are more clearly illustrated in Figure 3, where
selected isocontours of the average energy and integral number flux are
plotted between 75° and 90° latitude for the cases Kp = 1 and Kp = 3.
The contour closest to the minimum is approximately elliptical in shape
being more extended in MLT than in latitude. The gradient in average

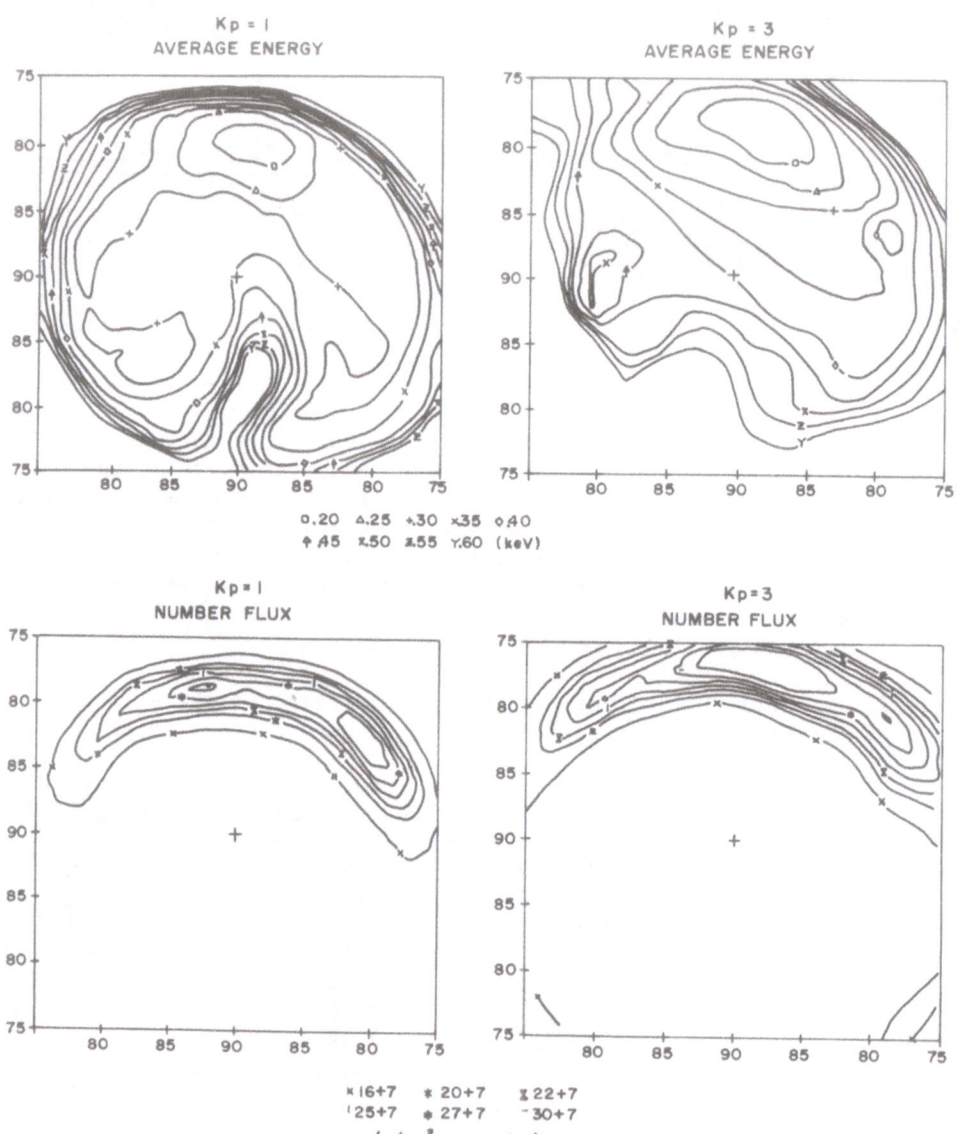

Figure 3.   Selected isocontour plots of the integral number flux and average energy for electrons with energies between 50 eV and 20 keV as a function of magnetic local time and latitudes between 75° and 90° for Kp = 1 and Kp = 3.

energy varies with direction away from the minimum. Equatorward of the minimum the gradient is steepest. Above the minimum, in the polar cap, the gradients are weaker and are similar to that found in the polar rain (See Figure 6 and discussion).

Based upon the above we identify the region near the average energy minimum as the cusp, and the remaining portions of the low energy crescent as the cleft.

5. POLAR RAIN

The polar cap lies at magnetic latitudes above the cleft. Figures 1a-d and 2a-d show the cap to be a region of low average energy and relatively large flux, more than 10^7 el/(cm^2-sec-ster), which decreases only slightly with distance from the cleft. The polar cap is least populated at highest latitudes on the nightside.

Individual passes through the cap show the two different types of characteristic precipitation that occur, called polar rain and polar showers by Winningham and Heikkila (1974). Polar rain is weak, structureless, near-isotropic electron flux with an average energy of about 100 eV. Polar showers are high density bursts of electrons of about the same average energy distributed throughout large portions of the cap. Polar rain is most common when magnetic activity is high, and Bz is negative. Polar showers occur for Bz north situations. Since the occurrence frequencies for Bz south and north are approximately the same, statistical maps that include both types of precipitation are dominated by polar showers.

Figures 4 and 5 show nearly identical noon-midnight satellite paths across the southern polar cap to illustrate this difference. Figure 4 is a Bz negative case (polar rain) and Figure 5 is a Bz positive case (polar showers). In each case the DMSP/F4 satellite passes through the energetic dayside diffuse aurora, the polar cleft, the polar cap up to latitudes of 85° CGM, and through the nightside auroral oval. In Figure 4 the cap is very clearly delineated by the polar rain. In Figure 5 the satellite encounters highly structured low energy fluxes throughout the cap until it passes through the pre-midnight diffuse aurora. Polar cap arcs occur when the cap is agitated in this manner (Hardy et al., 1982).

Results are given in Figure 6 from the second gridding technique where only clear cases of polar rain were included. The main selection criterion for polar rain was persistent uniformity for more than 120 consequtive spectra (or over 500 km). The color-coded maps in Figure 6 show average values of the number of spectra used, the integral flux and average energy when the only separation other than by MLT and CGM made is by Kp. In Figure 6 the maps in the left hand column are for Kp 0 to 2, and the right hand column for Kp 3 to 5.

The following conclusions can be drawn from the plots:

1. Polar rain falls in a roughly circular area offset toward pre-midnight and increasing with increasing magnetic activity. This

circle "fits" behind the crescent regions of high flux, low average
energy seen in Figures 1a-d.

2. The axis of symmetry of polar rain variations is pre-noon to
pre-midnight. Along this axis the integral flux decreases by about a
factor of 20, from 2×10^7 to 10^6 el/(cm^2-sec-ster), and the average
energy increases by a factor of 4, from 150 eV to 600 eV.

3. The polar rain flux increases and the average energy decreases with
increasing magnetic activity. The behavior is shown in Table III.

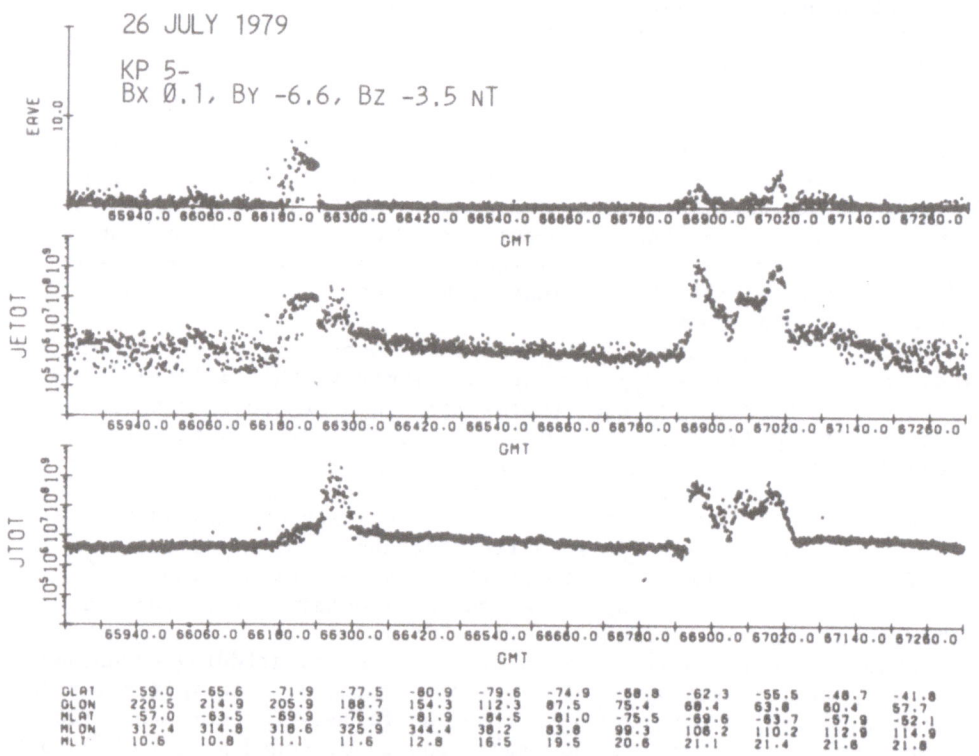

Figure 4. A south pole pass of the DMSP F4 satellite on 26 July 1979.
The integral flux (JTOT) in el/(cm^2-sec-ster), energy flux (JETOT) in
keV/(cm^2-sec-ster), and average energy (EAVE) in keV are plotted as
functions of universal time, geomagnetic and corrected magnetic
latitude and longitude, and magnetic local time.

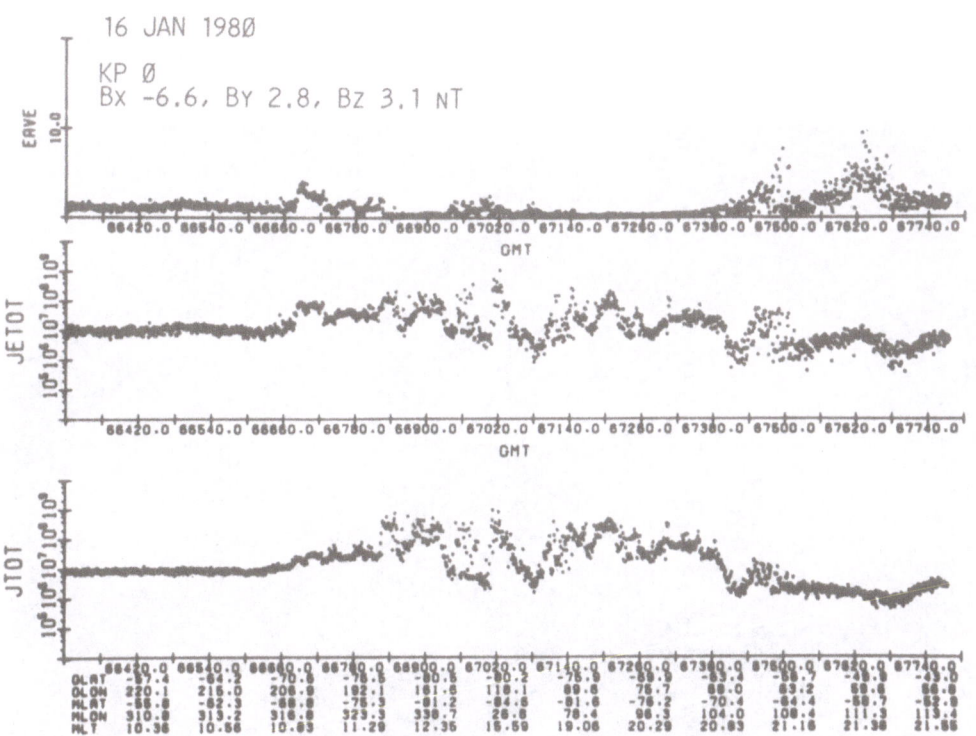

Figure 5. Same as Figure 4, but for 16 January 1980.

Table III. Integrated values of the polar cap area, the total number flux (NF) and the total energy flux (EF) for two ranges of Kp. The precentage difference between the two Kp ranges is also given.

	Area($cm^2 \times 10^{17}$)	NF(10^{23}/sec-ster)	EF(10^{20}eV/sec-ster)
Kp 0 to 2	0.86	3.3	0.94
KP 3 to 5	1.09	6.7	1.73
% Diff	21%	51%	46%

Figure 6. The average value of the polar rain integral number flux in el/(cm²-sec-ster) for Kp 0 to 2 and Kp 3 to 5 in equal logarithmic steps (top row); average energy in keV for Kp 0 to 2 and Kp 3 to 5, color-coded in equal linear steps (bottom row). The axes are labeled by degrees along the noon-midnight (top to bottom) and dusk-dawn (left to right) meridians.

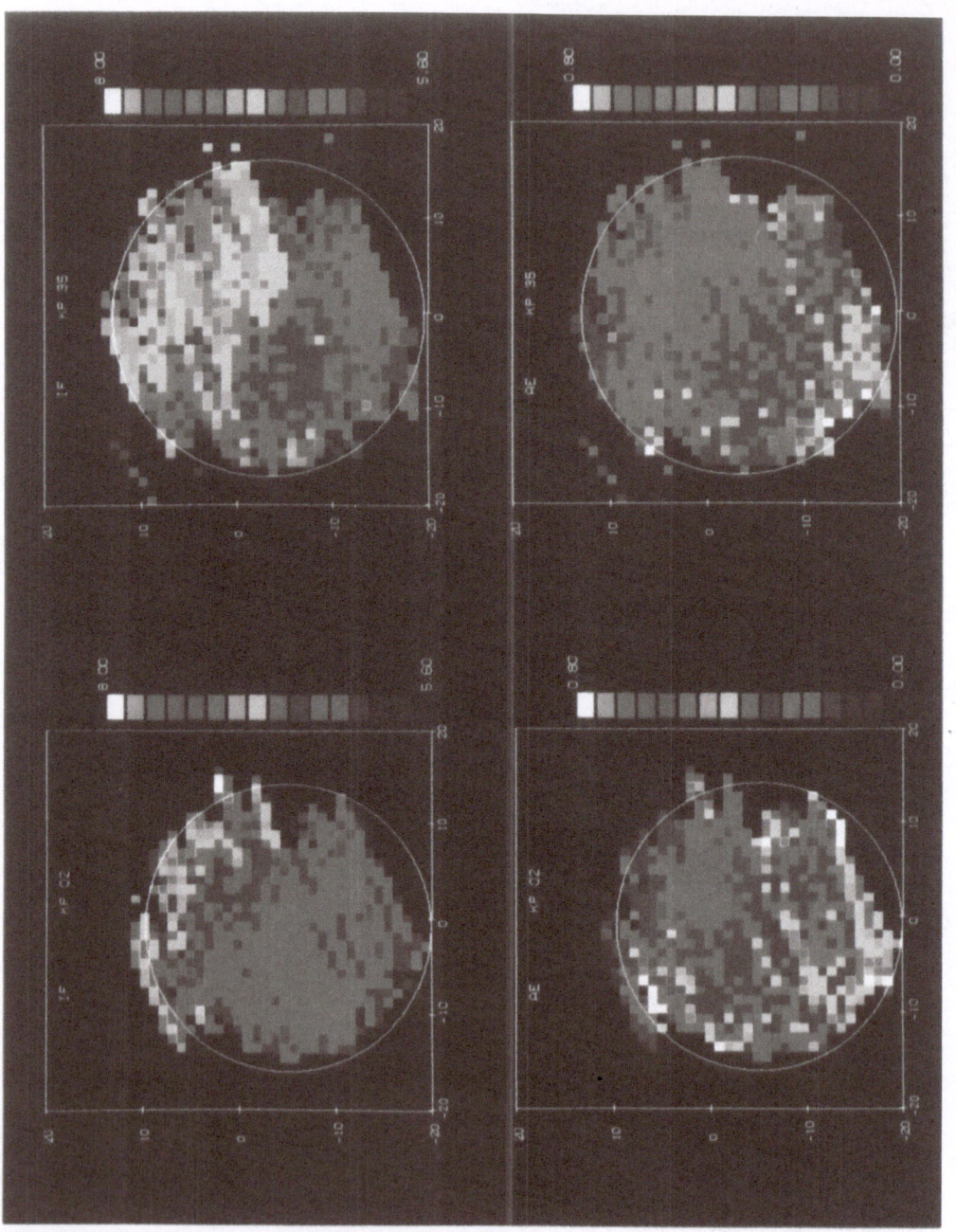

Figure 6

6. SUMMARY

In the above statistical studies we have shown the existence of three separate regions of low energy electron precipitation on the dayside polar cap and their statistical variations in magnetic latitude, local time and magnetic activity. The cusp is seen to be a spatially-confined elliptical region of minimum electron average energy near local noon. The cleft is a crescent-shaped region of high number flux extending from dawn to dusk that expands and moves to lower latitudes with increasing magnetic activity. The average energy in the cleft increases away from the cusp. The polar cap is only well-defined when the precipitation is polar rain. The polar rain number flux is maximum on the dayside, and has lowest average energies there as well.

7. ACKNOWLEDGEMENTS

The authors gratefully acknowledge N. Heinemann, R. Burkhardt, and E. Holeman for assistance in data selection and statistical analysis.
This research was supported by the Air Force Geophysics Laboratory under Contracts F19628-81-K-0032, F19628-82-K-0039, and F19628-82-K-0011 and the Office of Naval Research - London Branch.

8. REFERENCES

Hardy, D.A., W.J. Burke, and M.S. Gussenhoven, J. Geophys. Res., 87, 2413 (1982).
Winningham, D., and W. Heikkila, J. Geophys. Res., 79, 949 (1974).

THE CHARACTERISTICS OF A PERSISTENT AURORAL ARC AT HIGH LATITUDE IN THE 1400 MLT SECTOR

D. S. Evans
NOAA/ERL/Space Environment Laboratory
325 Broadway
Boulder, CO   80303   USA

ABSTRACT.   An analysis of precipitating particle energy flux observations show a local intensity maximum at ~ 77 invariant latitude and 14 MLT, a location where a number of other geophysical processes show intensity maximum. Detailed examination show these energy fluxes to be composed entirely of 0.3 - 3 keV electrons precipitating in latitudinally thin structures. The source populations from whence these electrons originated had densities of order 2 cm^{-3} and temperatures of order 150 eV. Parallel electric field acceleration of ~ 1000 V (but no more than 3000 V) was responsible for the electron rich energy flux enhancements. The source of EMF for this process probably originates in a fairly direct solar-wind magnetosphere interaction.

1.   INTRODUCTION

Potemra (1983) has surveyed a number of magnetospheric and ionospheric phenomena which appear to exhibit intensity maxima at high latitudes in the 14-16 MLT sector. Among these are intense upward flowing field-aligned currents, a maximum in the downward fluxes of low energy electrons, a maximum in the intensity of auroral emissions, and a relative maximum in the occurrence of upstreaming ion fluxes and ion cyclotron waves (Iijima and Potemra, 1978; McDiarmid et al., 1975; Shepherd, 1979; Cogger, 1977; Kintner et al., 1979). The incoherent scatter radar at Sondre Stromfjord also shows a persistent enhancement in F-region electron densities during this same magnetic local time sector (Robinson, et al., 1984). A statistical analysis of energy fluxes carried into the atmosphere by precipitating particles using observations made by instruments on the TIROS/NOAA satellites show that during quiet times there is a local maximum in the 13-15 MLT sector at invariant latitudes of 78°. Figure 1 shows the pattern, in invariant latitude and MLT, of average energy fluxes assembled from measurements made during periods of unusually low activity. The enhancement in the 13-15 MLT sector represents about a factor of 2 increase in energy fluxes over adjacent regions. At moderate levels of activity the global pattern of energy flux to the atmosphere

J. A. Holtet and A. Egeland (eds.), The Polar Cusp, 99–109.
© 1985 by D. Reidel Publishing Company.

increases so as to engulf this time sector so that a significant local
maximum is no longer apparent.

Because of the mass of evidence that the region λ ~ 75°, MLT 13-
15 is the focal point of unusual magnetospheric plasma processes, a
more extensive study of charged particle observations taken when the
TIROS-N satellite passed through that sector was done. The purpose of
this study was to specify the characteristics (species, energies,
locations, intensities, etc.) of the charged particle precipitation
encountered in that region of space in the hope that insight could be
gained into the underlying reason for the uniqueness of this
phenomena.

2.   OBSERVATIONS MADE BY THE TIROS/NOAA SATELLITES

The TIROS-N satellite is one of a series of meteorological spacecraft
which have sun-synchronous, near circular orbits (98° inclination, 850
km altitude).   TIROS-N is in a local time orbit which crosses the
equator northbound at 1530 local time so that the satellite passes
over the northern hemisphere during the early afternoon and pre-noon
daylight hours.

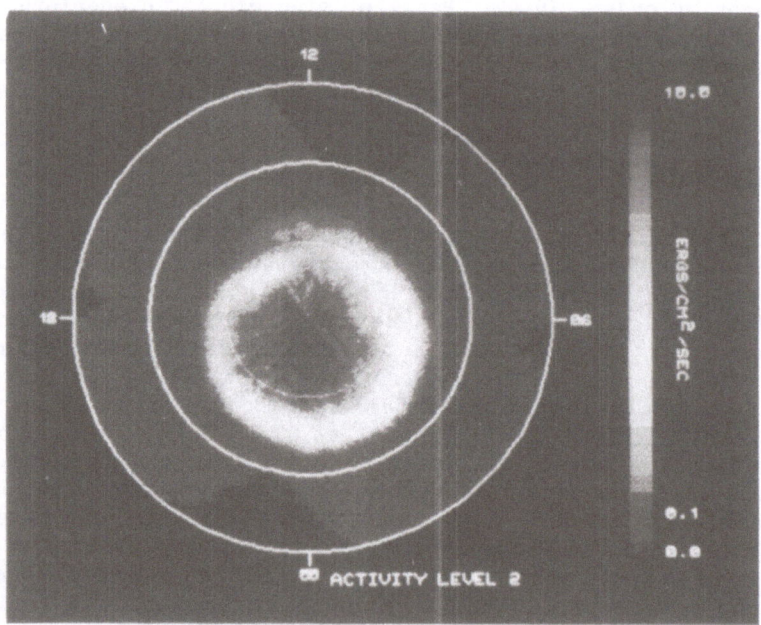

Figure 1.   A statistical pattern of energy flux to the atmosphere by
precipitating particles constructed from TIROS/NOAA observations made
during periods of low geomagnetic activity.   In the color version of
this plot the region 75-80° invariant latitude, 13-15 MLT is clearly a
well defined local maximum in the energy flux.

Instruments on board this satellite measure precipitating particles over the energy range 0.3 keV to > 100 keV. Above 30 keV solid state detectors measure the intensities of electrons (3 integral energy channels) and protons (5 differential channels) at small pitch angles, well within the loss cone, every two seconds. Between 0.3 keV and 20 keV, measurements of electrons and positive ions are made at two pitch angles (both within the loss cone) by electrostatic analyzer systems. These analyzers are swept in energy with a 1 second period and alternate between measuring electrons and ions. On board processing coverts the measurements made during a single sweep to the directional energy flux carried by those particles. A transformation of the local pitch angles along B to 120 km altitude followed by a 2 point angular integration yields the omni-directional energy flux into the atmosphere separately for ions and electrons over the energy range 0.3 to 20 keV. During each analyzer sweep the energy at the peak in the energy flux spectrum is identified and that energy together with the particle intensity at that energy are telemetered.

These primary observations of the 0.3 – 20 keV particles are supplemented, at a low duty, by a 4 point energy spectrum. For those cases, the 4 point spectrum may be combined with the 0.3 to 20 keV energy integral, knowledge of the energy and particle intensity at the maximum in the spectrum, and the > 30 keV particle observations to reconstruct a more detailed particle intensity vs. energy spectrum over the entire energy range from .3 keV to > 100 keV.

In the course of routine data procesesing observations of the 0.3 to 20 keV particle energy flux are averaged over $1°$ intervals in geomagnetic latitude (based upon the location where the particles actually enter the atmosphere). The contribution of both electrons and ions are included in this averaged energy flux although experience has shown that it is rare for ions to carry more than 20% of the total energy flux and even rarer for the total energy flux to exceed 1.0 ergs $cm^{-2}sec^{-1}$ in such instances. This condensed data is in a form convenient for survey studies and was used to construct the low activity pattern of energy input to the atmosphere shown in Fig. 1.

3.   PARTICLE PRECIPITATION IN THE 12–16 MLT, HIGH LATITUDE SECTOR BASED UPON ONE DEGREE AVERAGED MEASUREMENTS

The initial survey of charged particle precipitation in the magnetic latitude range $69°$ to $81°$ and 12–16 MLT sector was done using condensed ($1°$ averaged) energy flux data gathered by TIROS-N during 1980. The local time of the TIROS-N orbit insures favorable sampling of this magnetic local time sector and in the course of one year about 2500 northern hemisphere passes through this region of space will be made.

Of this number of passes, about 45% encountered a precipitating energy flux averaging more than 1 erg cm^{-2} sec^{-1} over $1°$ in latitude. Individual 2 second energy flux measurements within that average may exceed 1 erg $cm^{-2}sec^{-1}$ by a large amount. Fig. 2 shows the percentage of passes through each 15 minute MLT and $1°$ magnetic latitude box over the 12–16 MLT, $69°$ – $81°$ latitude region which recorded an averaged energy flux greater than 1 erg cm^{-2}sec^{-1} The percentages range from a few percent just past noon to more than 20% later in the afternoon. The highest probability of encountering such fluxes is in the latitude range 75° to 80° magnetic.

The 'island' of enhanced energy flux that appears in Fig. 1 is not replicated here probably because all levels of magnetic activity (1980 was a year of high activity) were included in the construction of Fig. 2. As mentioned above, the 13-15 MLT sector loses its unique identity with higher activity. At any rate, the analysis of 1° averaged observations show that significant, although not large, precipitating particle energy fluxes are a persistent feature of the post-noon high latitude polar region.

4.   DETAILED CHARACTERIZATION OF THE PARTICLE PRECIPITATION IN THE 14-15 MLT SECTOR

While some gross features of the high latitude, post-noon precipitation can be extracted from the 1° averaged data, a quantitative analysis requires the complete data set which has 2

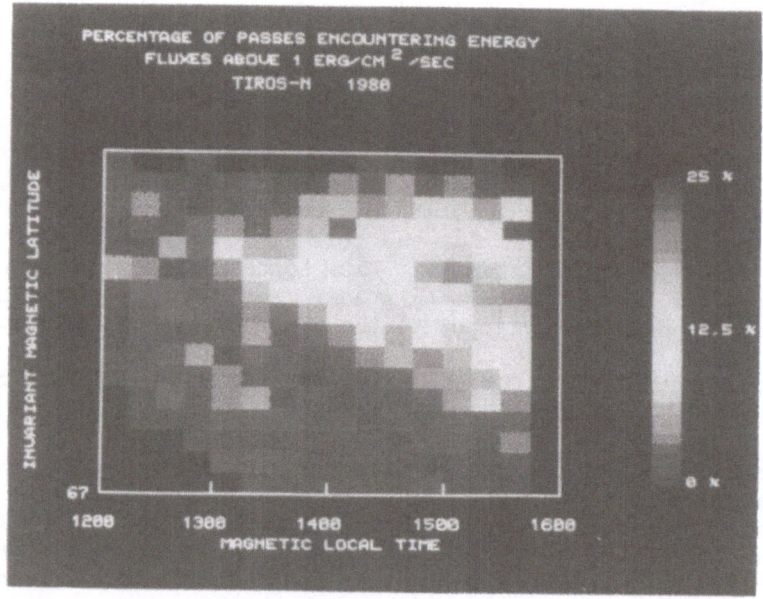

Figure 2.   A plot coding the probability of encountering energy fluxes of greater than 1 erg cm^{-2}sec^{-1} (averaged over 1° latitude) within the invariant latitude interval 69° to 81° (at the top) and between 12 and 16 MLT. The bin size is 1° by 15 min. The maximum occurrence probability within a single bin is .2.

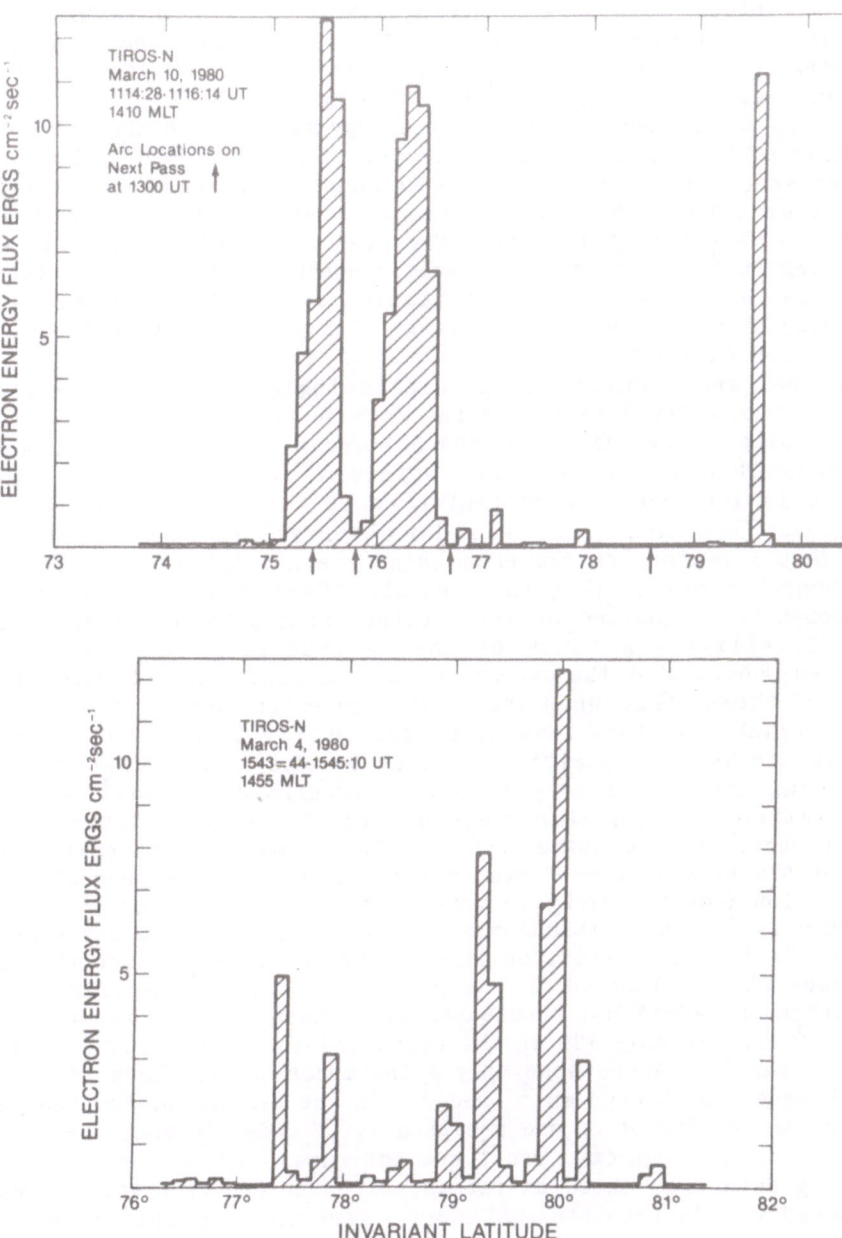

Figure 3. Two examples of the energy flux profiles observed during passes through the high latitude, 14-15 MLT sector. The spatial resolution is about 10 km. The arrows in the top panel show the locations of energy flux structures encountered on the next pass 100 minutes later. Multiple structures in a single pass are most common.

second resolution. Six periods during 1980 (1–10 Jan, 1–10 Mar, 1–10 May, 1–10 July, 1–10 Sept, and 1–10 Nov) were selected. Detailed data were extracted from 84 individual passes through the 14–15 MLT sector which had encountered 1 erg cm^{-2} sec^{-1} energy fluxes.

Fig. 3a and 3b show the electron energy flux to the atmosphere as a function of invariant latitude for two of the 84 passes. Only the electron energy flux is plotted as in none of the 84 passes did the energy flux carried by ions exceed 0.25 erg cm^{-2} sec^{-1} and was generally a factor of 5–10 less. Whenever the total energy flux was above 1 erg cm^{-2} sec^{-1}, the ions seldom contributed more than 1% and often less than 0.1%. Ion precipitation in the 0.3 – 20 keV range effectively plays no role in the energy flux enhancements observed in the post-noon high latitude region.

Several enhancements in precipitating electron energy flux, spaced apart by a few tenths to a few degrees in latitude, encountered during a single pass (Fig. 3a and 3b) are a somewhat more common occurrence than a single well defined flux enhancement. This structure is interpreted a spatial (10's of km) rather than temporal (1–10 sec). Incoherent scatter radar observations in the post-noon sector show F-region enhancements which are spatially structured but persistent temporally (Robinson, et al., 1984) which are presumably the ionospheric signature of the electron precipitation being sensed by the satellite. A study of the latitudinal widths of these structures showed that the average width was about 0.2° (20 km). Less than 4% of these structures had widths exceeding 0.6°. The minimum width resolvable in this study was about 0.1°. This distribution of structure widths is very similar to that found by Evans (1981) for the latitudinal extents of very intense (>60 ergs cm^{-2} sec^{-1}) auroral precipitation observed almost exclusively in the late afternoon to midnight sector of the magnetosphere. This similarity extends to the fact that the evening sector events were also virtually pure electron precipitation (corresponding to very bright auroral arcs).

The only apparent dissimilarity between the post-noon and evening arc-like electron precipitation (apart from the location) are the very much lower fluxes observed in the post-noon sector. The peak energy flux within the 14–15 MLT structures had a mean value of about 3 ergs cm^{-2} sec^{-1} and in only 12% of the cases did the peak value exceed 10 ergs cm^{-2} sec^{-1}. There were only 3 instances out of about 240 where the peak exceeded 30 ergs cm^{-2} sec^{-1}. In the evening sector the peak fluxes at the center of an arc are usually of order 10 ergs cm^{-2} sec^{-1} and values of 100 ergs cm^{-2} sec^{-1} are not rare.

Using data from these 84 passes, electron energy spectrums were constructed for all possible instances where the directional energy flux exceeded 0.1 ergs cm^{-2} sec^{-1} $ster^{-1}$. Three hundred sixty five such spectrums were constructed, each using data accumulated over one second. Figure 4 shows 4 representative spectral types observed. The spectrum in the lower right, (type 1) which show electron intensities decreasing monotonically above the lowest energy sampled, is a very common type accounting for 40% of all examples. Spectrums that exhibit a knee (type 2, upper right) or a distinct peak (type 3, lower left) account for 50% of all spectrums constructed. The fourth type

of spectrum shown in Fig. 4 (upper left) is a Maxwellian like spectrum
which was the rarest type amongst the 365.

- Electron energy spectrums with peaks or knees are commonly
observed in the evening to midnight sector of the auroral belt in
association with discrete auroral arcs. In such cases, the energy at
which the peak or knee is located in the spectrum is often taken to be
the magnitude of a magnetic field-aligned potential difference which
accelerated the electrons downwards into the atmosphere. Because of
the similarity in spectral shapes observed in the two regions of the
auroral belt and because in both regions the precipitation is

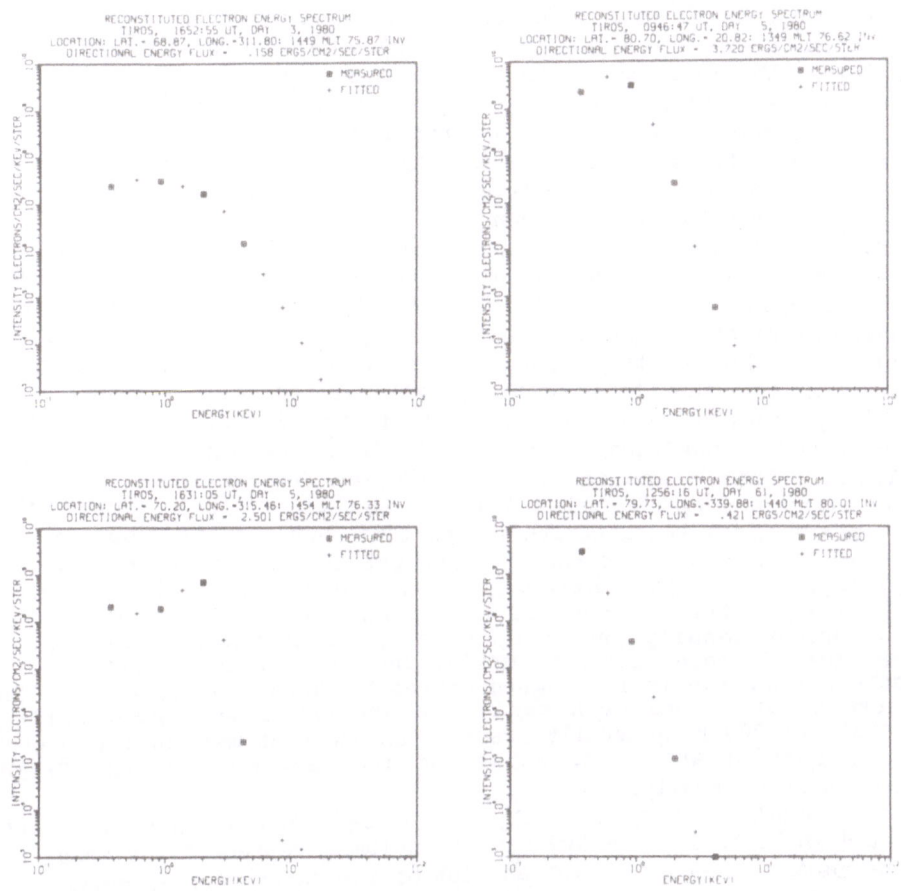

Figure 4.   Examples of the 4 different types of electron energy
spectrums encountered in the regions of energy flux enhancements. The
spectrums in the upper right and lower left show evidence of
acceleration through a potential difference. Spectrums of this type
occurred in about 50% of the examples. The type shown in the upper
left is near Maxwellian and the rarest type found.

completely dominated by electrons, suggesting preferential
energization, an identification of a field-aligned potential
difference with the peak or knee feature is made for these post-noon
high latitude events.   It should be pointed out that of the ~ 170
constructed spectrums that exhibited a peak or knee, the highest
energy for this feature was 3000 eV.   Similar spectrums taken in the
evening to midnight sector often show such features located at much
higher energy.

Knight (1973), Evans (1974), Lyons (1980), and Fridman and
Lemaire (1980), have modelled the electron distribution, originating
from a source specified by a density and temperature that has been
accelerated by a field aligned potential difference.   Their work
indicates that features in the accelerated distribution, such as the
slope of the energy spectrum at high energies, the particle
intensities, the location of discontinuities in the spectrum, and the
integrated energy flux carried by the accelerated plasma, may be used
to infer the number density and temperature of the source plasma as
well as the magnitude of the accelerating potential and an estimate of
the field-aligned current density carried by electrons transiting from
the magnetosphere to the ionosphere.

The set of constructed electron energy spectrums were subjected
to such an analysis to obtain estimates of these parameters (the
Maxwellian-like, type 4 spectrums were discarded from this study).
The temperature of the parent electron distribution was obtained from
the slope of the electron spectrum at energies above the spectral
feature that we identify with the parallel potential difference.   If
the energy spectrum was monotonically decreasing above the lowest
energy measured, the parallel potential difference was assumed to be
300 Volts.   This potential difference was further assumed to be
located at about 1 R_E altitude along the magnetic field line where the
scaler field is 0.1 the strength in the ionosphere.   With knowledge of
the source temperature and accelerating potential, and the assumption
of the location of the acceleration, the measurement of the energy
flux carried by the accelerated electrons may be used to infer the
source number density and, from these parameters, the integrated
number flux of those electrons moving from the magnetosphere to the
ionosphere (i.e. the field aligned current).   The assumption that, in
the absence of a peak or knee in the spectrum, the accelerating
potential was 300 V, generally leads to an underestimate of the source
electron density and a corresponding underestimate of the field
aligned current density.

The results of this analysis of 345 energy spectrums are
tabulated in Table I.   The 90% and 10% columns represent the value of
that parameter exceeded by 90% and 10% of the cases respectively.

The mean electron source temperature of 135 eV is in reasonable
agreement with a previous study by Burch et al (1976) of the plasma
properties in the dayside, high latitude magnetosphere although 135 eV
is significantly higher than the electron temperatures reported by
Zanetti et al (1981).   This latter disagreement may stem from the
inability of the TIROS-N instrument to produce a useful response for
electron distributions of less than about 60 eV temperature. The
temperatures inferred for the electron source population responsible

Table I

Parameter	90%	Mean	Average	10%
Directional Energy Flux ergs cm^{-2} sec^{-1} $ster^{-1}$	.15	.50	.93	2.5
Temperature eV	75	135	191	390
Density electrons cm^{-3}	.6	1.8	2.5	5.0
Field Aligned Current μA m^{-2}	.7	2.4	3.2	7.0

Results of the analysis of 345 electron energy spectrums observed in the 14-15 MLT, high latitude auroral belt.

for the high latitude, 14-15 MLT electron precipitation are similar to those ion temperatures measured in the magnetosheath but very much less than the keV electron temperatures characteristic of the evening and night time magnetospheric particle population.

The number densities inferred for the source electron population are of the same order as those found in the solar wind although generally less than the magnetosheath densities of 10 cm^{-3} or more. These densities are larger than the 0.1 - 1.0 cm^{-3} found in the nightime magnetosphere (DeForest and McIlwain, 1971).

Both the electron number densities and temperatures inferred in this study are not dissimilar to those inferred by Lundin (1985), from measurements in that part of the magnetospheric boundary layer connected to the polar cusp.

The field aligned electron currents that are inferred to be associated with the electron precipitation are of order 2μA m^{-2}. This current density is in good agreement with those currents observed in the post noon (14-16 MLT) at high latitude regions by the TRIAD satellites (Potemra, private communication).

5.   SUMMARY AND DISCUSSION

A region of persistent energy influx to the atmosphere, particularly well defined during low geomagnetic activity, is located at invariant latitudes between 75° and 80° and magnetic local time between 14 and 16 hours. This same location has previously been identified as a site of enhanced field aligned current flow, auroral luminosity, plasma wave turbulance, and F-region ionization.

The energy input is in the form of low energy (< 3 keV) electron precipitation and positive ions or more energetic particles are essentially absent. The precipitation appears in one or, generally, more thin (~ 10 km latitudinal extent) probably arc like structures. The peak energy flux within a structure averages about 3 ergs cm^{-2} sec^{-1} and only rarely exceeds 30 ergs cm^{-2} sec^{-1}.

The electron energy spectrums within these structures often show evidence of the electrons having been accelerated through a parallel potential difference. The magnitude of this potential is generally 1000 V or less and, in 345 examples, never exceeded 3000 V.

The number densities and temperatures of the source population of these precipitating electrons averages about 2 cm^{-3} and 150 eV respectively. These number densities and temperatures are different than those found either in the solar wind or within the magnetosphere proper. They are similar to those found in the magnetosheath or boundary layer plasma populations.

The field-aligned current carried by these precipitating electrons within the thin structures averages 2-3 μAm^{-2}. This current density agrees well with the field aligned currents inferred from TRIAD magnetic field measurements made in the region of the current density maximum around 14 MLT and 77° latitude. It seems reasonable to conclude that the currents inferred by TRIAD for this locale are primarily due to electrons originating from some boundary layer population that have been accelerated downwards by a parallel potential difference.

The combination of field aligned current flow, parallel potential difference and thin regions of almost pure energized electron precipitation bears a strong resemblance to the discrete auroral arcs occurring over the evening and night side of the earth. The only qualitative differences between the precipitation at these two locations appears to be that the post-noon precipitation involves lower parallel potentials, higher source number densities and lower source temperatures than do the nightime precipitation.

While the basic processes that lead to the creation of an EMF at high altitude and a current system which links this EMF with the ionosphere are almost certainly similar for the post-noon and nightime electron arc systems, it is unlikely that the precipitation in the 14-16 MLT sector is a simple extension of the evening phenomena. At low geomagnetic activity the energy flux observed at high latitude and 14 MLT is completely detached from the nightime precipitation. This points toward the source of EMF for the post-noon electron, arc-like, precipitation to be a result of a more direct solar wind-magnetosphere interaction while the nightside EMFs result from processes less directly linked to a solar wind interaction.

REFERENCES

Burch, J. L., Fields, S. A., Hanson, W. B., Heelis, R. A., Hoffman, R. A. and Janetzke, R. W., J. Geophys. Res., 81, 2223, 1976.
Cogger, L. L., Murphree, J. S., Ismail, S., and Anger, C. D., Geophys. Res. Lett., 4, 413, 1977.
DeForest, S. E. and McIlwain, C. E., J. Geophys. Res., 76, 3587, 1971.
Evans, D. S., J. Geophys. Res., 79, 2853, 1974.
Evans, D. S., EOS, 62, 664, 1981.
Fridman, M., and Lemaire, J., J. Geophys. Res., 85, 664, 1980.
Iijima, T., and Potemra, T. A., J. Geophys. Res., 83, 599, 1978.
Kintner, D. M., Kelley, M. C., Sharp, R. D., Ghielmetti, A. G., Temerin, M., Cattell, C., Mizera, P. F., and Fennell, J. F. J. Geophys. Res., 84, 7201, 1979.
Knight, L., Planet. Space Sci., 21, 741, 1973.
Lyons, L. R., J. Geophys. Res., 85, 17, 1980.
Lundin, R., this volume, 1985.
McDiarmid, I. B., Burrows, J. R., and Budzinski, E. E., J. Geophys. Res., 80, 73, 1975.
Potemra, T. A., pa. 335 in High Latitude Space Plasma Physics (Eds. B. Hultqvist and T. Hagfors) Plenum, 1983.
Robinson, R. M., Evans, D. S., Potemra, T. A., and Kelly, J. D., Geophys. Res. Lett., (in press) 1984.
Shepherd, G. G., Rev. Geophys. Space Phys., 17, 2017, 1979.
Zanetti, L. J., Potemra, T. A., Doering, J. P., Lee, J. S. and Hoffman, R. A., J. Geophys. Res., 86, 8957, 1981.

THE EFFECTS OF MAGNETOSHEATH ELECTRONS ON CHARGE EXCHANGE, RADIATION TRAPPING AND OTHER ATOMIC AND MOLECULAR PROCESSES IN THE MID-DAY POLAR CUSP THERMOSPHERE

G. G. Sivjee
Departament of Physics, University of Alaska
Fairbanks, Alaska, U.S.A.

ABSTRACT. Magnetosheath electrons precipitating in the mid-day cusp region dissipate most of their energy in the thermosphere above 200 km where atomic species are more abundant than at lower altitudes. Part of the atmosphere where these electrons excite mid-day auroras is also sunlit. The high abundance of atomic species, their longer mean free path and the presence of sunlight in the atmospheric region where mid-day auroras are formed, permit studies of certain inter-actions among atmospheric constituents, as well as radiation trans-port process, which cannot be observed in laboratories or in lower altitude night time auroras. Examples of such upper atmospheric processes amenable to investigations in the mid-day cusp region are (i) charge exchange reactions between $O^+(^2D)$ and N_2, (ii) radiative entrapment of extreme ultraviolet radiation and its effect on auroral emissions, (iii) resonant scattering of sunlight by atmospheric spe-cies, (iv) buildup of vibrationally excited molecules above the level expected from the kinetic temperatures in the thermosphere, (v) variations in the abundance of thermospheric N and He, and (vi) the propagation of atmospheric disturbances from the thermosphere to the mesosphere.

1. INTRODUCTION

Cusp auroras are formed at relatively high altitudes, in the daytime polar thermosphere, where atomic species are abundant (Sivjee, 1983a,b). Consequently, the intensity of atomic lines in these auroras is enhanced over the molecular bands (compare Figures 1 and 2 and 3a and 3b) leading to a major simplification of the aurora spec-trum (Gault, 1981; Sivjee, 1983a). Atomic lines which have not been observed in night-time auroras, partly because their wavelengths coincide with those of the more intense molecular bands, are easily detected in mid-day auroras. The relative weakness of the auroral molecular bands in mid-day cusp auroras also permits daytime observa-tions of the chemiluminescent airglow OH emissions from the polar mesosphere. Changes in intensity and rotational temperature of OH emissions can therefore be monitored in the midday cusp region to

J. A. Holtet and A. Egeland (eds.), The Polar Cusp, 111–126.

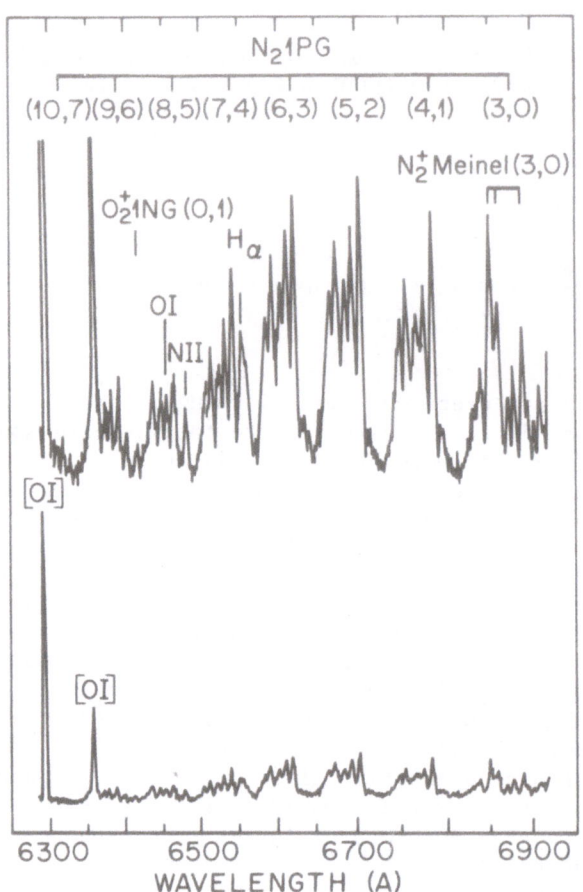

Figure 1.   Near infrared spectrum of the nightside aurora at two
            gain settings to show the intensity of N_2 1PG relative
            to 6300/6364A [OI].

infer the propagation of atmospheric disturbances, due to particle
precipitation, joule heating and other high latitude phenomena, to
lower altitudes in the mesosphere. The relatively low density of
(and hence longer mean free path in) the atmosphere, at altitudes
where mid-day auroras are formed, permits more metastable species to
radiate their excess energy, before they are collisionally quenched,
than in lower altitude night-time auroras. Hence, the abundance of
the metastable species in cusp auroras can be determined with relatively higher precision through measurements of their optical emissions. In turn, the effects of these long-lived particles, with
excess energy, on neutral and ion chemistry of the auroral atmosphere

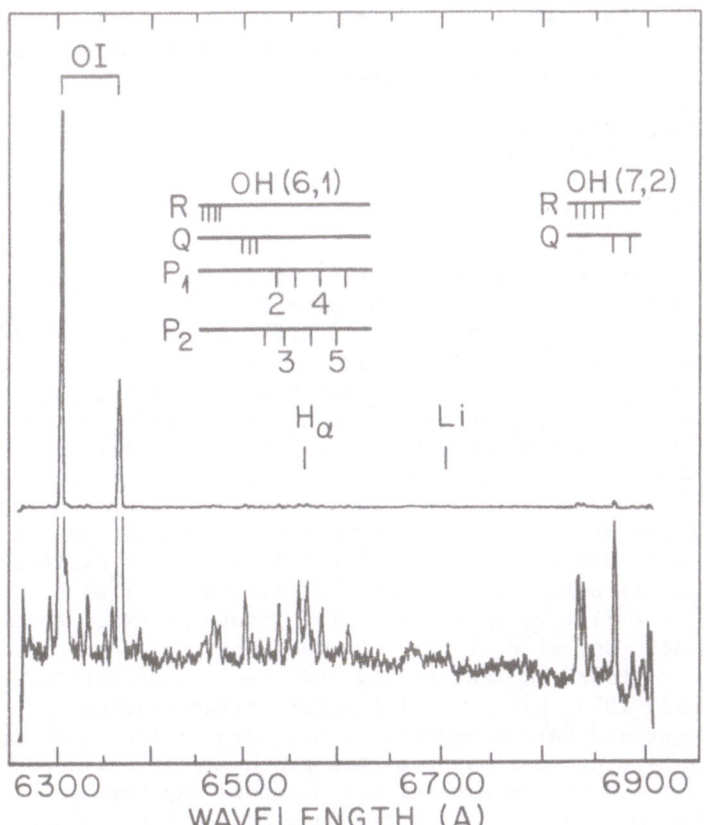

Figure 2. Near infrared spectrum of the dayside aurora.

may be investigated with greater confidence. Progress made in our understanding of these atmospheric processes, and the outstanding problems requiring more detailed studies, will be reviewed in the following sections.

2. $O^+ + N_2$ CHARGE EXCHANGE PROCESS IN THE THERMOSPHERE

The major source of N_2^+ ions in night-time auroras (most of which peak around 100 km) is electron impact on N_2 (Rees et al., 1975), i.e.,

$$N_2 + e \rightarrow N_2^+ + 2e. \tag{1}$$

On the other hand, in mid-day cusp auroras, the N_2^+ population, derived from N_2^+ 1NG observations (Sivjee, 1983a) is at least twice

the value expected from (1) (Deehr et al., 1980). Similar results
have been reported from twilight airglow observations (Broadfoot and
Hunten, 1966). The additional N_2^+ ions in the high altitude cusp
auroras and twilight airglow may be produced by the asymmetric (or
accidental) resonance charge transfer process:

$$O^+(^2D_J, ^2P_J) + N_2(X^1\Sigma_g^+, v' = 0,N) \rightarrow$$

$$O(^3P_J') + N_2^+(A^2\pi_g, v' = 1, N') + \Delta E \tag{2}$$

with $\Delta E \ll kT$ for favorable combination of (J, N, J', N'). According
to Omholt (1957), "With O^+ (^2P), there is 1.92 eV excess energy which
must be absorbed as vibrational or kinetic energy. With O^+ (^2D), there
is almost exact resonance for $v' = 1$." Initial auroral results
(Hunten, 1958; Reed, 1958; Harrison and Vallance-Jones, 1959), based
on the vibrational distribution of the N_2^+ Meinel ($A^2\pi_g - X^2\Sigma_g^+$)
emissions, indicated that emissions from the $v' = 1$ level might
indeed be enhanced. From his auroral spectroscopic observations,
Hunten (1958) noted that, "There are obvious variations in the relative intensities of the Meinel (1,0) and (2,1) bands. This may be
considered good evidence that the $v' = 1$ level of the excited state
(of N_2^+ ($A^2\pi_g$)) is being supplied by the reaction O^+ (^2D) +
N_2 ($X^2\Sigma_g^+$) $\rightarrow$ O (^3P) + N_2^+ ($A^2\pi_g$, $v' = 1$). There is exact resonance to an accuracy of 0.01 eV." On the other hand, more recent
IR studies with improved sensitivity and resolution (Gattinger and
Vallance-Jones, 1973; Baker et al., 1977; Vallance-Jones and Gattinger,
1978; Gattinger and Vallance-Jones, 1981) have failed to establish
unequivocally a contribution from this process. It now appears that
the interpretation of the earlier measurements may have been impacted
by an underestimate of water vapor absorption, which tends to reduce
the apparent emission rate in the N_2^+ M (0,1) band relative to that in
the (1,2) band and to similarly depress the apparent emission rate
of the (2,1) band relative to that of the (1,0) band.

The very extensive N_2^+ 1NG data set from cusp auroral measurements requires an additional source of N_2^+ (other than reaction (1))
to explain the observed intensities. Since O^+ emissions are relatively intense (100R to 1000R) in the high altitude mid-day cusp
auroras, reaction (2) is the most likely source of N_2^+ ions not ac-
counted for by (1). To verify the contributions of (2) to cusp
auroral N_2^+ abundance, one must search for some identifiable signatures of this reaction. Fortunately, reaction (2) produces two
distinct effects in the rotational and vibrational disturbances of
N_2^+ emissions. First, relative to the (0,0) band of the N_2^+ Meinel
system the (1,0) and (1,1) bands from reaction (2) should be enhanced
compared to the same emissions from (1). Secondly, according to
Dalgarno and McElroy (1966), the charge transfer reaction (2) "may
proceed via the formation of a collision complex and the apparent rotational temperature of N_2^+ may be large. The (charge-exchange)
reaction should lead to a nearly uniform distribution of population
among the rotational sub-states of the product ion." Some indication

of such anomalous rotational distribution in twilight observation of N_2^+ emissions has been reported (Broadfoot and Hunten, 1966).

Much of the auroral spectroscopic data relating to reactions (1) and (2) has been taken in bright and very bright auroras, corresponding to $I(3914) \simeq 100kR$. In such cases, the mean emission height is expected to be near 100 km where $[0]/[N_2] \sim 0.1$. Hence, measurements in very intense auroral events are biased toward the direct electron-impact process for producing N_2 ($A^2\Pi_g$, v' = 1), i.e.,

$$e + N_2(X^1\Sigma_g^+, v' = 0) \rightarrow 2e + N_2^+(A^2\Pi_g, v' = 1) \qquad (3)$$

The rapid increase in the ratio $[0]/[N_2]$ with altitude above 110 km suggests an increase in the importance of process (2) relative to process (3) in "normal" and higher altitude events. Recent determinations of rate coefficients for the $0^+(^2D)+N_2$ and $0^+(^2D)+0$ charge-exchange reactions have yielded $(8\pm2) \times 10^{-10} cm^3 s^{-1}$ (Johnson and Biondi, 1980) and $(2\pm1) \times 10^{-11} cm^3 s^{-1}$ (Orsini et al., 1977), respectively. Hence, the charge exchange of $0^+(^2D)$ with N_2 remains the dominant loss mechanism for this ion to rather high altitudes, while the ratio of the volume production rates for $0^+(^2D)$ to that for N_2^+ ($A^2\Pi_g$, v' = 1), through direct electron-impact ionization, is expected to vary as the ratio $[0]/[N_2]$. As a result, the importance of process (2) in the production of N_2^+ ($A^2\Pi_g$, v' = 1) should increase fairly dramatically with auroral height in the region above 100 km. The thermal 0^+ (^2D) + N_2 ($X^1\Sigma_g^+$, v' = 1), charge transfer process is expected to preferentially produce N_2^+ ($A^2\Pi_g$, v' = 1), with no final state channels corresponding to v' > 1.[8] Other less direct evidence for the importance of process (2) in N_2^+ production at high altitudes is found in the literature. For example, recent satellite-based investigations for the thermosphere have resulted in models which include a major role for process (2) in F-region N_2^+ production (Torr and Torr, 1979), in agreement with earlier suggestions. (See Banks and Kockarts, 1973 for a discussion of this idea and for pertinent references).

From the above summary, it appears that a better understanding of the role of reaction (2) in the ion chemistry of the thermosphere would require underlined simultaneous observations from the same aurora of: (1) The vibrational distribution of N_2^+ M bands in the IR region, with emphasis on emissions from v' = 0, 1, 2; (2) The 0^+ (^2D, ^2P) emissions in the near UV and IR regions; and (3) High resolution data on the rotational distribution of N_2^+ 1NG and M emissions.

3. THE VARIATIONS IN N AND 0 ABUNDANCES IN THE POLAR THERMOSPHERE

Various observational and theoretical (numerical model) investigations of the polar thermosphere show enhancement of certain species in the winter hemisphere compared to their values in the summer hemisphere (Keating et al., 1970; Sivjee, 1976; Strauss and Christopher, 1979; Sivjee et al., 1980). These enhancements have been explained mostly as a consequence of meridional transport. One of the alternative mechanisms for locally increasing certain minor constituents, such as N and O, particularly in the polar thermosphere, is the dissociation of the molecular constituents by particle impact follow- ed by diffusion. Such a build up is difficult to observe spectrosco- pically in the spatial region of molecular dissociation during any one auroral event because the atomic emissions are mostly indicative of the dissociation by-products and do not provide information about similar atoms already present in the ambient atmosphere prior to the onset of the particular aurora under investigation. Most of the past auroral spectroscopic observations were made in the more frequently occurring bright night-time auroras, and the few permitted atomic

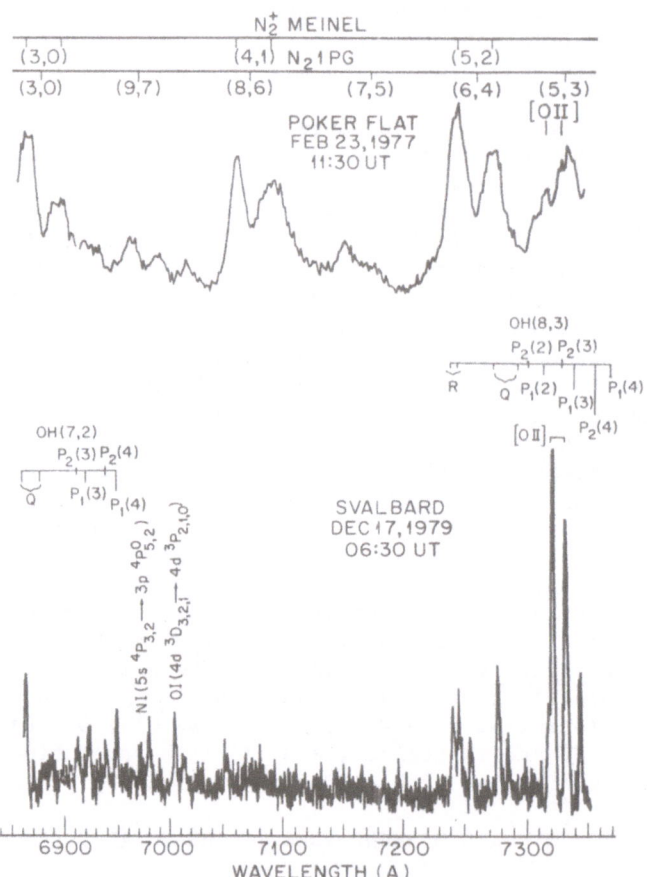

Figure 3.   (a) Near infrared spectrum of the nightside aurora from Poker Flat, Alaska. (b) Showing the same wavelength region of the day side aurora from Svalbard.

line emissions detected in these auroras pertain to molecular dissociation, rather than to direct particle impact on ambient upper
atmospheric atoms. Additionally, several atomic lines, from permitted transitions, were not detected because of the stronger molecular emissions whose wavelengths are close to or coincident with the
wavelengths of these atomic emissions. Figure 3 shows night-time and
mid-day auroral near-infrared spectra. While the $N_2 1P$ bands dominate
the former, weak N and O emissions, not previously observed in bright
auroras, are evident in the latter.

The difficulty of observing auroral atomic emissions has contributed to a lack of reliable information about variations in N and O
column densities in auroras. Consequently, almost all atmospheric
models used in studies of auroral processes completely neglect ambient N abundance (i.e., atmospheric N density prior to N_2 dissociation
in that particular aurora being modelled). Such models also arbitrarily adjust O density to reproduce the observed brightness of certain
forbidden O emissions. The neglect, in auroral atmospheric models,
of thermospheric N atoms which may be present prior to the onset of
any one particular auroral event is a serious defect of these models
especially since some of the optical measurements from high altitude,
such as those shown in Figure 3, point to relatively high N density
($> 10^7$ cm^{-3}) in the thermosphere above 200 km (Sivjee, 1983b). Almost
complete absence of N_2 and O_2 emissions from these high altitude mid-
day auroras clearly show that the observed atomic emissions could
only be produced by electron impact on N.

The relatively high N density implied by the cusp auroral optical emissions raises the possibility of higher O density in these
auroras than is currently assumed. Additionally, changes in the flux
of particles precipitating in the cusp region, as well as in meridional transport and other solar related geophysical disturbances, may
lead to significant variations in daytime polar thermospheric atomic
densities. Questions relating to absolute abundances of atomic
species in the thermosphere, and their variations under various disturbances, can be clarified through observations of NIR emissions
from N and O in high altitude auroras wherein molecular emissions are
relatively weak or absent. The atomic lines are much more intense
(and do not overlap with other atmospheric emissions) in the NIR
($0.6\mu - 1.6\mu$) region. Since changes in column densities of thermospheric N and O are directly reflected in their permitted-line-emis-
sion intensities when these emissions result from direct electron
impact on the atoms, absolute intensity measurements of some of these
atomic NIR line emissions can be used to derive upper atmospheric
abundance of N and O in the polar region.

The OI multiplet at about 1.32μ may prove to be a useful index
for column density of O. It has been established in laboratory studies
of $e + O_2$ collisions that the cross-section for producing this multiplet is very small (5×10^{-20} cm^2 @ 100 eV), whereas the theoretical
cross-section associated with the e+O process (Smith, 1976) is relatively large [$(\sigma_{max} (4s^3 s^0) = 2 \times 10^{-18}$ cm^2]. Since the 1.32-μ
m multiplet is a persistent feature in auroral spectra and can be detected in weak aurora, it can be used for monitoring thermospheric O.

4. HELIUM ABUNDANCE IN POLAR THERMOSPHERE

Observations of He 3889A line in the cusp and the polar cap regions, shown in Figure 4, indicate occasional enhancements of this
emission, by a factor ranging from 3 to 20, above the level expected
from resonant scattering of sunlight by, as well as from particle
excitation of, the ambient thermospheric He in the winter polar
region. The frequency of these enhancements may be much greater than
these recent observations suggest, since the relatively strong
N_2^+ 1NG (0,0) and (1,1) bands invariably mask the weak He 3889A emission. Additionally, no information is presently available concerning
the source of the observed occasional enhancements in He emission.
In this regard it will be helpful to observe any relation that may
exist between He emission and proton precipitation levels, as well as
with geomagnetic and solar disturbances. Simultaneous measurements
of He (1.083μ (which is relatively free of contamination, from atmospheric band emissions, compared to He 3889A line) and H_α (6563A)
emissions from various magnetospheric regions on an almost continuous
(24 hour) basis in mid-day auroras would provide the data required
for such studies.

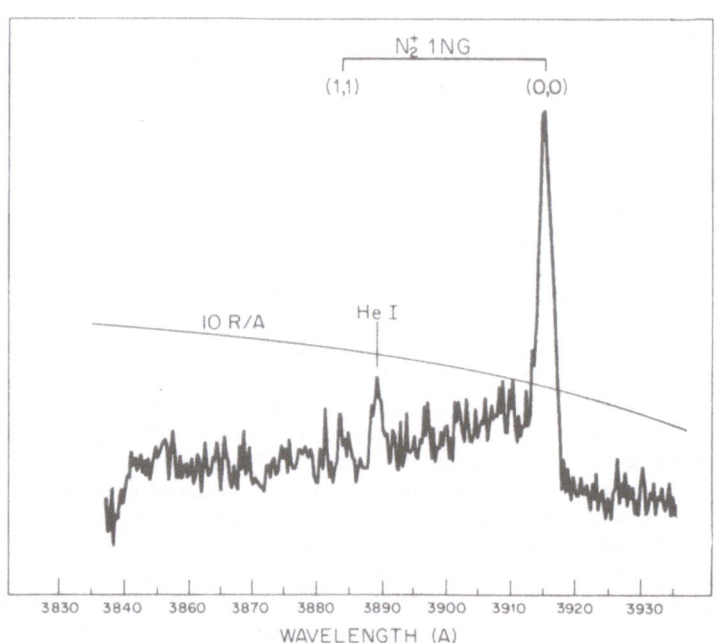

Figure 4. Emission spectrum featuring He 2889A from the polar
cap and cusp aurora.

5. NON-LTE VIBRATIONAL DISTRIBUTION OF THERMOSPHERIC MOLECULES

Through interactions with the atmospheric constituents, low energy ($\sim$ 2-4 eV) auroral electrons lose much of their energy in exciting the vibrational levels of N_2 and O_2. These homonuclear molecules cannot readily radiate their excess vibrational energy as (LTE) NIR emissions, instead, they tend to relax vibrationally (v-v transfer) before exchanging energy with other species. Hence, significant levels of non-local thermal equilibrium (LTE) vibrational population distribution can develop, particularly in the cusp region where low energy electrons excite $N_2(v')$ for almost eight hours everyday around local magnetic noon.

Current models predict only modest departures of the N_2 vibrational distribution from LTE in the lower thermosphere under all but extreme conditions, principally as a result of the severe quenching of vibrationally excited N_2 by O. Brieg et al. (1973) and Jamshidi and Kummler (1973) have used experimental data for this quenching process (McNeal et al., 1972; McNeal et al., 1974) to establish it as the dominant N_2^* loss process in the upper atmosphere between about 100 and 200 km. The effect of quenching is to hold the N_2^* distribution close to LTE unless unusually strong sources of N_2^* are involved. Diffusion is expected to dominate the volume loss rate of N_2^* above about 200 km. Hence, a substantial departure of the N_2^* distribution from LTE is a strong possibility in the cleft aurora, where sustained precipitation of soft magnetosheath electrons into the region above 200 km should 'vibrationally heat' the N_2 to temperatures significantly in excess of the ambient kinetic temperature.

Non-LTE vibrational distribution of N_2 has been extensively (but unsuccessfully) searched for, in non-sunlit night-time auroras, through measurements of the auroral N_2^+ 1NG emissions. Lack of success may be attributed to the rarity of high altitude auroral events on the nightside, and severe Rayleigh scattering problems in the near UV leading to an overlap of N_2^+ 1NG emissions from different heights of auroral arcs and from different auroras concurrently formed in the sky over any one observing station. Scattering problems can be minimized by monitoring N_2 and O_2 emissions from high altitude auroras in the NIR (between 0.6μ and 1.6μ) where Rayleigh scattering is about two orders of magnitude less severe than in the near UV region. Recent success in observing changes in O_2 vibrational distribution (Henriksen et al., 1983) suggests that observations of N_2 1P NIR emissions from the high altitude cusp auroras may help resolve the question relating to the extent of non-LTE N_2 vibrational distribution in the polar thermosphere.

6. RADIATIVE ENTRAPMENT OF AURORAL EXTREME ULTRAVIOLET RADIATION

The importance of radiative entrapment of extreme ultraviolet (EUV) radiation in the thermosphere is illustrated by the intensity of OI 7990A multiplet (Figures 5 and 6) and the intensity ratio of OI (8446A)/OI (7774A) (Figure 7) observed in auroras. Typical values of the I (7990A) fall within the range of 100-200 R (Christensen et al.,

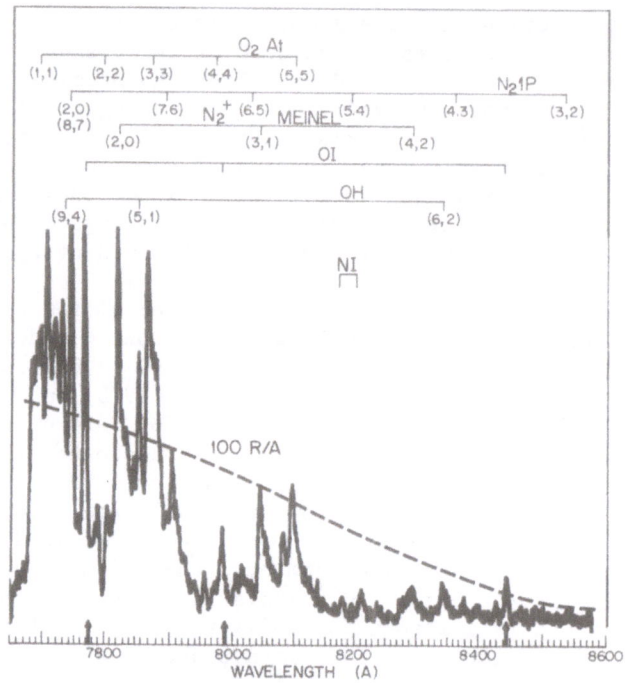

Figure 5.   Auroral spectrum featuring 7774A, 7990A and 8446A (OI)
            emissions.

1983) while I (8446A)/I (7774A) varies between 1 and 6 for auroras
peaked at altitudes between 100 and 400 km (Sivjee et al., 1984).
     Figure 8 shows a partial term diagram for atomic oxygen.   The
upper state ^{3}D 3s'; can either radiate XUV 989A photon or the NIR
7990A photon. Since the observed intensity of 989A is about 100 R
(Christensen et al., 1977; Feldman and Gentieu, 1982) and the branching ratio for ^{3}D 3s' state is ≤ 10^{-4} (Christensen et al., 1978; Pradhan
and Saraph, 1977; Christensen and Cunningham, 1978; Zipf et al.,
1979), the intensity of OI (7990A), from electron impact excitation
of O should be 0.1R, i.e., three orders of magnitude smaller than the
observed values. Currently, the only plausible explanation of this
discrepancy between the observed and the expected intensities of OI
(7990A) emission is the radiative entrapment of the resonant 989A
radiation by oxygen in the thermosphere. The line-center optical
depth for 989A radiation in the atmosphere is about 15,000 at 120 km
and 710 at 200 km (Christensen et al., 1977). Consequently, there is
almost total radiative entrapment of 989A at auroral heights. The
recycling of the 989A photons in auroras increases the population of
^{3}D 3s' by several orders of magnitude above the level expected from

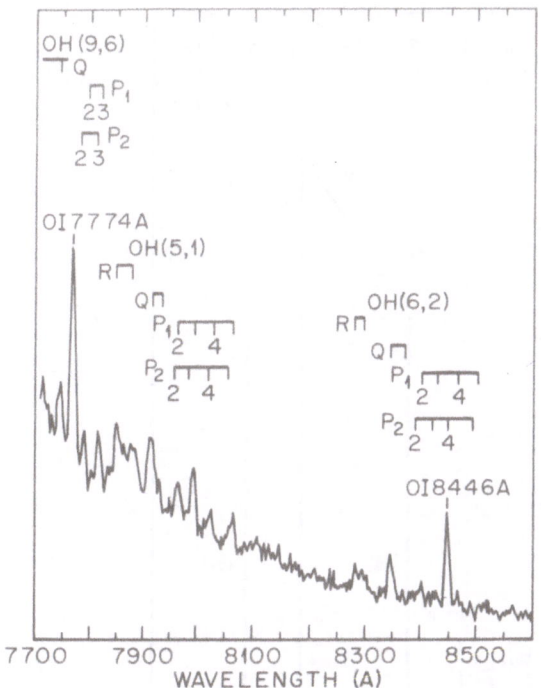

Figure 6. Auroral spectrum featuring 7774A, 7990A and 8446A (OI) emissions.

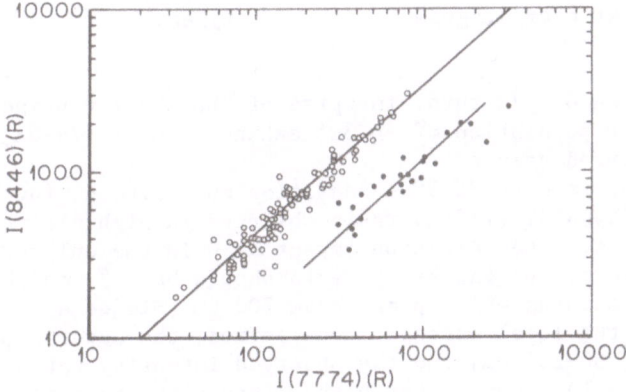

Figure 7. Showing the relationship between 7774A (OI) and 8446A (OI) emission from various auroras.

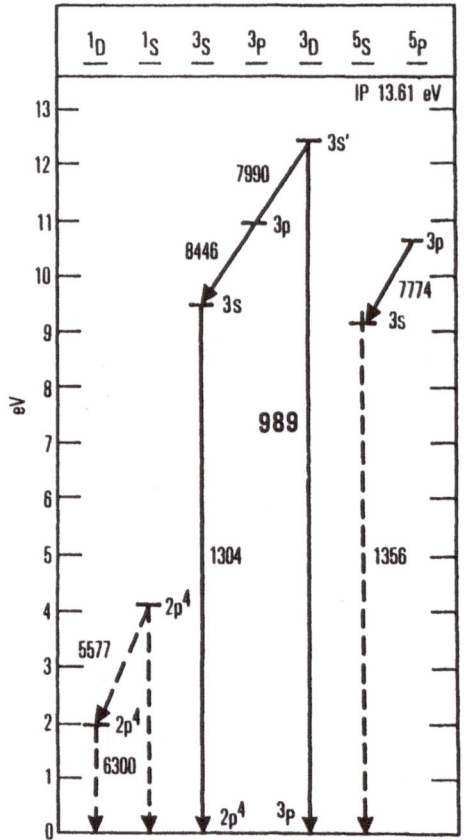

Figure 8. Partial term diagram for atomic oxygen.

electron impact on O. In turn, in spite of the smaller branching
ratio, the higher population of ^{3}D 3s' enhances the I (7990A) emission to the observed values.

Radiative entrapment of 989A radiation must also be invoked to
explain the I (8446A)/I (7774A) ratio observed in high altitude
(> 150 km) auroras. When electron impact on O is the only source of
these NIR emissions, the intensity ratio should be ≃ 3, as observed
in mid-day cusp auroras which peak above 200 km (Sivjee et al.,
1984). Outside the local magnetic noon period, the auroras peak at
lower altitude (150-200 km) and the observed intensity ratio is ≃ 6
(Gault et al., 1981). The latter higher intensity ratio in intermediate altitude auroras implies an additional source 8446A radiation
(beside electron impact on O). Again, in Figure 8 it is seen that
the ^{3}P3p state can be populated both e + O as well as through cascade
from ^{3}D 3s'. Since radiative entrapment of 989A radiation enhances

the 7990A emission, it also increases the ^{3}P3p level to which the ^{3}D 3s' state decays to produce the 7990A emission. Because of the much larger optical depth of 989A radiation in the 150-200 km, compared to higher altitudes, the 8446A intensity would also be higher in the intermediate altitude range.

The extent of the redistribution of auroral emissions through radiative entrapment described above depends on various atmospheric parameters, in particular the atomic oxygen abundance in the thermosphere. Meier's (1982) Monte Carlo calculations of OI (989A) radiative transport in the thermosphere illustrate the relation between singlet to doublet ratio of 7990A emissions and the O and O_2 densities. This relation derives from the feeding of singlet, doublet, and triplet lines of the 7990A multiplet by branching of the corresponding transitions in OI (989A) multiplet which in turn depends on the optical depth, and hence the density of O. Thus radiative entrapment of auroral XUV radiation in the thermosphere provides a convenient method of remote sensing O density and its variations, from ground-based observations of the OI (7990A) multiplets.

7. WAVE-INDUCED PERTURBATIONS OF NIR OH AND O$^+$ EMISSIONS

The suggestion that regular and irregular fluctuations in the intensities and 'temperatures' of the terrestrial OH Meinel night airglow features are associated with atmospheric waves was apparently first made by Krassovsky (1957). Subsequent airglow studies by Krassovsky, Shekov, and co-workers provided indirect evidence in general agreement with this suggestion. More recently, direct confirmation of the general validity of this suggestion has dramatically emerged from the photographic observations of Peterson and Kieffaber (1973) and Moreels and Herse (1977) as well as from the low-light-level video recordings of Crawford et al. (1975). Wave phenomena in the night airglow emissions from O_2^* have also been inferred from observations of fluctuations in intensities and temperatures. However, in this case, there are very few reported measurements on which to draw. Noxon (1978) has reported fluctuations in the intensity and rotational temperature associated with the (0,1) band of the atmospheric system of O_2. Simultaneous observations of fluctuations in the OH Meinel emissions provided complementary data which supported the idea that the observed variations were manifestations of gravity-wave-induced perturbations of the 80-100 km region.

As summarized by von Zahn (1983), a growing body of data points to a very dynamic mesosphere at high latitudes during the winter months. Results obtained to date for the high latitude 'mesopause temperatures' during the winter season exhibit periodic and erratic fluctuations of up to about ±30% from the mean. Both gravity waves, with periods ranging from 10 minutes to several hours, and planetary waves, with time scales of days to weeks, are expected to contribute to this variability. Winter-time soundings (acoustic-grenade and pitot-probe techniques) above Point Barrow, Alaska (71°N) (Heath et al., 1974 and references therein) revealed changes of up to 80K in the kinetic temperature near 85 km in 3 hours.

Recently, the OH (5,1) and (6,2) bands have been monitored almost continuously in the cusp, even during the period of magnetosheath particle precipitation. Figures 5 and 6 show atmospheric emission spectra in the wavelength region between 7600A and 8500A from night and day time auroras. While N_2 1P, O_2 At and N_2^+ M bands clutter the nocturnal auroral spectrum, the mid-day cusp spectrum is devoid of these intense band emissions and the weak OH (5,1) and (6,2) band emissions from the mesosphere can be easily detected. Preliminary analyses indicate large amplitude harmonic variations in OH intensity and rotational temperature (Myrabø et al., 1983). Detailed studies of the propagation of upper atmospheric wave disturbances, generated through solar terrestrial interactions in the high altitude thermosphere, to the mesosphere requires better temporal and spatial resolution in observing OH intensity and temperature variations than is possible with one spectrometer. At least three high throughput spectrometers will be needed to monitor simultaneously the same OH band from different regions of the sky.

Finally, in studying the effects of mesospheric gravity-wave disturbances on the OH Meinel emissions, it is necessary to supplement the OH intensity/ temperature measurements with adequate characterizations of the wave activity, i.e., horizontal wavelength (λ_h), and horizontal phase velocity (c), the spectrum of $\underline{k}$ and ω_k, the characteristic vertical distance H_g for the growth of gravity waves as well as the saturation of the amplitude growth of gravity waves (Weinstock, 1978). A large number of simultaneous observations of this type at different geographical locations would aid in the development of better 'picture' of the sources and dominant characteristics of such wave phenomena. These observations would also guide modeling efforts directed toward an evaluation of the consequences of the 'breaking of gravity waves' for the energy budget and overall dynamics of the middle atmosphere.

ACKNOWLEDGEMENTS

Financial support for this work was provided by NATO (Research Grant #513-82) and by the National Science Foundation (Grant #ATM82-00114).

REFERENCES

Baker, D. J., W. Pendleton, A. Steed, R. Huppi, and A. T. Stair, J. Geophys. Res., 82, 1601, 1977.
Banks, P. M. and G. Kockarts, Aeronomy, Academic Press, N.Y., 1973.
Breig, E. L., M. E. Brennan and R. J. McNeal, J. Geophys. Res., 78 1225, 1973.
Broadfoot, A. L. and D. M. Hunten, Planet. Sp. Sci., 14, 1303, 1966.
Christensen, A. B., G. J. Romick and G. G. Sivjee, J. Geophys. Res., 82, 4997, 1977.
Christensen, A. B. and A. J. Cunningham, J. Geophys. Res., 83, 4393, 1978

Christensen, A. B., G. G. Sivjee and J. H. Hecht, J. Geophys. Res., 88, 4911, 1983.
Crawford, J., P. Rothwell and N. Wells, Nature, 257, 650, 1975.
Dalgarno, A. and M.B. McElroy, Planet. Space Sci., 14, 1321, 1966.
Deehr, C.S., G. G. Sivjee, A. Egeland, K. Henriksen, P.E. Sandholt, R.W. Smith, P. Sweeney, C. Duncan and W. J. Gilmer, J. Geophys. Res., 85, 2185, 1980.
Feldman, P.D. and E.P. Gentieu, J. Geophys. Res., 87, 2453-2458, 1982.
Gattinger, R.L. and A. Vallance-Jones, Can, J. Phys., 51, 287, 1973.
Gattinger, R.L. and A. Vallance-Jones, Can. J. Phys., 59, 480, 1981.
Gault, W. A., R. A. Koehler, R. Link and G. G. Shepherd, Planet. Space Sci., 29, 321, 1981.
Harrison, A.W. and A. Vallance-Jones, J. Atm. Terr. Phys., 13, 291, 1959.
Heath, D. F., E. Hilsenrath, A. J. Kruger, W. Nordberg, C. Prabhahara and J. S. Theon, in Structures and Dynamics of the Upper Atmosphere, (ed. F. Verniani), Elsevier Scientific Publishing Co., N.Y., 1974.
Henriksen, K., G. G. Sivjee, C. S. Deehr and H. K. Myrabø, Submitted to J. Geophys. Res., 1984.
Hunten, D. M., Ann. Geophys., 14, 167, 1958
Jamshihidi, E. and R. H. Kummer, (abstract), EOS Trans, AGU, 54, 705, 1973.
Johnson, R. and M. A. Biondi, J. Chem. Phys., 73, 190, 1980.
Keating, G. M., J. A. Mullings and E. J. Prior, Space Res., 11, 987, 1972.
Krassovsky, V. I. and V. T. Lulashenia, Doklady-Akad. Nauk S.S.S.R., 80, 735, 1951.
Krassovsky, V. I., Des Memoires in -8° de la Societe des Sciences de Liege, Quatrieme Serie, 18, 58-67, 1957.
Krassovsky, V. I. Ann. Geophys., 28, 739, 1972.
McNeal, R. J., M. E. Whitson, Jr. and G. R. Cook, Chem. Phys. Lett., 16, 507, 1973.
McNeal, R. J., M. E. Whitson, Jr. and G. R. Cook, J. Geophys. Res., 79, 1527, 1974.
Meier, R. R., J. Geophys. Res., 87, 6307, 1982.
Moreels, G. and M. Herse, Planet. Space Sci., 25, 265, 1977.
Myrabø, H. K., C. S. Deehr and G. G. Sivjee, J. Geophys. Res., 88, 9255, 1983.
Noxon, J. F., Geophys. Res. Lett., 6, 25, 1978.
Omholt, A., J. Atmos. Terr. Phys., 10, 320, 1957.
Orsini, W., D. G. Torr, M. R. Torr, H. C. Brinton, L. H. Brace, A. O. Nier and J. C. G. Walker, J. Geophys. Res., 82, 4829, 1977.
Peterson, A. W. and L. N. Kieffaber, Nature (London) 242, 321, 1973.
Pradhan, A. K. and H. E. Saraph, J. Phys. B., 10, 3365, 1977.
Reed, J. M., M.Sc. Thesis, Univ. of Saskatchewan, Saskatoon, Saskatchewan, 1958.
Rees, M. H., G. G. Sivjee and K. A. Dick, J. Geophys. Res., 81, 6046, 1976.

Sivjee, G. G., K. Henriksen and C. S. Deehr, J. Geophys. Res., 85, 6043, 1980.

Sivjee, G. G., J. Geophys. Res., 88, 1983a.

Sivjee, G. G., Geophys. Res. Lett., 10, 349, 1983b.

Sivjee, G. G., A. B. Christensen, K. Henriksen and A. E. Belon, Ann. Geoph., (in press), 1984.

Smith, E.R., Phys. Rev., 13, 65, 1976.

Strauss, J. M. and L. A. Christopher, J. Geophys. Res., 84, 1241, 1979.

Torr, D. G. and M. R. Torr, J. Atm. Terr. Phys., 41, 797, 1979.

Vallance-Jones, A. and R. L. Gattinger, Can. J. Phys., 5, 1933, 1972.

Vallance-Jones, A. and R. L. Gattinger, J. Geophys. Res., 83, 3255, 1978.

von Zahn, U., Project MAP/WINE Campaign Handbook, 1983.

Weinstock, J., J. Geophys. Res., 83, 5175, 1978.

Zipf, E. C., R. W. McLaughlin and M. R. Gorman, Planet. Space. Sci. 27. 719, 1979.

THE HeI 3889Å LINE IN POLAR CLEFT SPECTRA

K. Henriksen and K. Stamnes
The Auroral Observatory, University of Tromsø
N-9000 Tromsø, Norway

C.S. Deehr and G.G. Sivjee
Geophysical Institute, University of Alaska
Fairbanks, Alaska 99701, USA

ABSTRACT. Using ground-based spectrometers, we find that the HeI 3889Å line is an inherent feature of dayside spectra on the latitude of the polar cleft, and its average intensity is well correlated with the solar depression angle. Superposed on this smoothly varying intensity level, there intermittently occur intensity enhancements on the day- and nightside. These enhancements are most likely caused by precipitating hot α-particles from the solar wind, whereas the smoothly varying average intensity level is due to resonance scattered solar light by orthohelium atoms.

INTRODUCTION

Four He emissions from the upper atmosphere have been observed. They are the orthohelium emissions $\lambda 10830$Å ($2\,^3P - 2\,^3S$), $\lambda 3889$Å ($3\,^3P - 2\,^3S$), $\lambda 5876$Å ($3\,^3D - 2\,^3P$), and the parahelium emission $\lambda 584$Å ($2\,^1P - 1\,^1S$), and these transitions are indicated in the term diagram shown in Fig. 1. The resonance lines at 10830Å and 3889Å have been observed both in twilight and sunlit aurora (Mirnov et al. 1959; Federova 1961; Shefov 1961a,b; 1963; 1968; Schecheglov 1962; Brandt et al. 1965; Harrison and Cairns 1969; Christensen et al. 1971; Sivjee et al. 1980), whereas the 5876Å emission is associated with aurora (Ivanchuk and Sukhoivanenko 1958; Stoffregen 1968; Henriksen 1978; Sivjee et al. 1980). The resonance line at 584Å is clearly seen in spectra obtained by rockets (Feldman 1985).

The 3889Å line is easily measured, using ground-based spectrometric equipment (Sivjee et al. 1980), and here we will present some new observations.

It is agreed that in twilight photoelectron impact on He($1\,^1S$) atoms generates the He($2\,^3S$) metastable atoms, and that solar resonance scattering is the main mechanism for producing the He($3\,^3P$) atoms (McElroy 1965; Brandt et al. 1965). Due to increased low-energy electron flux in aurora the He($2\,^3S$) excitation rate is increased propor-

127

J. A. Holtet and A. Egeland (eds.), The Polar Cusp, 127–135.
© 1985 by D. Reidel Publishing Company.

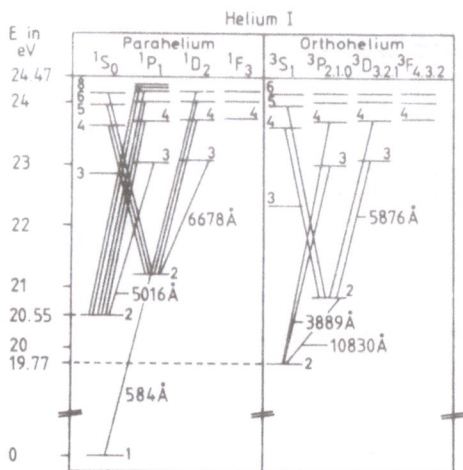

Figure 1.   Energy levels and transitions in helium atoms.

tionally.   In aurora precipitating helium ions from the solar wind
cause significantly intensity enhancements, whereas direct excitation
of the 3 ^{3}P level by electron impact on ambient helium gives an insignificant contribution.

RECENT SPECTROMETRIC SEARCH FOR THE HeI 3889Å LINE

In December 1979 and January 1980 the 3889Å emission was observed for
five days on the latitude of the polar cleft from Longyearbyen, Svalbard, using the same experimental set-up as Sivjee et al. (1980).
An integrated spectrum is shown in Fig. 2.   The HeI 3889Å line is
stronger than the N_2^+1NG(1,1) band at 3885Å, and the intensity of the
He line exceeds 60R.   This spectrum is taken in early twilight, but
the nearby Fraunhofer lines have already become significant.   At
3889.05Å and 3886.29Å the H_n and FeI Fraunhofer lines appear, and their
equivalent widths are 2.346Å and 0.920Å (Moore et al. 1966).   Therefore
in this spectrum about 10R and 4R are required to compensate for the
Fraunhofer depression of the continuum level at the HeI and N_2^+ emissions.

    Our spectrometric measurements indicate that the HeI 3889Å emission
is an inherent feature of the dayside spectrum, as shown by Fig. 2.
During the night this emission only intermittently appears and then
the intensity normally is below 10R.   The diurnal intensity variation
for two days is shown in Fig. 3, and there is an obvious correlation
between the intensity and the solar depression angle, when SDA is less
than 25°.   On the other hand, the measured dayside intensities during
the three days shown in Fig. 4 do not have any correlation with SDA.

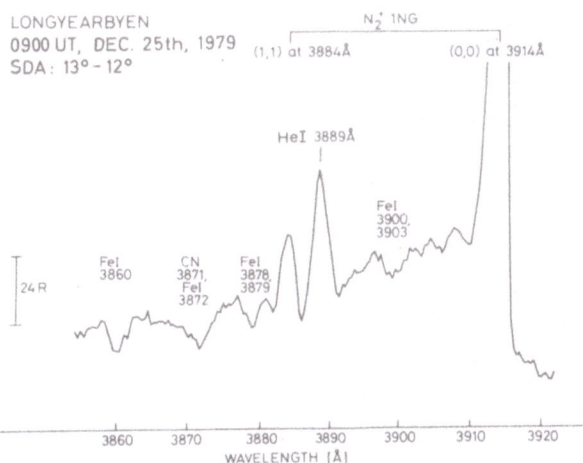

Figure 2. Polar cusp spectrum comprising $N_2^+1NG(1,1)$ and $(0,0)$ bands and the HeI 3889Å emissions, integrated for one hour. Fraunhofer lines are distinct in the spectrum, but the strongest Fraunhofer line H_η with an equivalent width of 2.346Å can not be seen due to the coincidence with the HeI 3889Å emission.

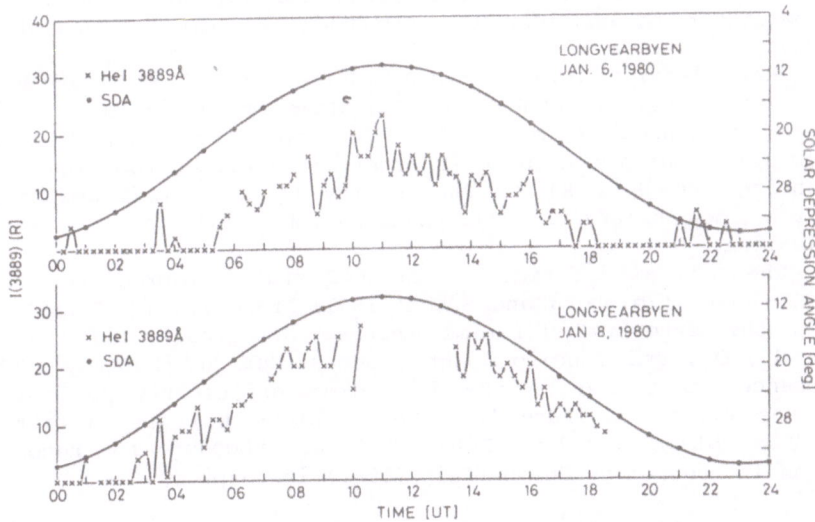

Figure 3. Two days of 3889Å intensity measurements, showing an overall correlation with the solar depression angle, SDA. The emission occurs only intermittently when SDA exceeds 25°.

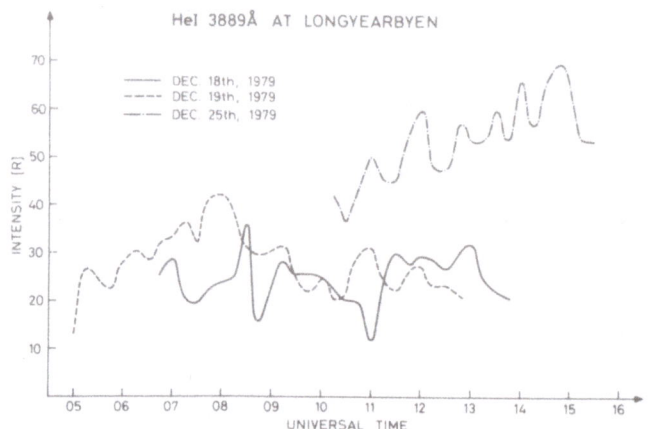

Figure 4. Comparison of HeI 3889Å intensities during three days.
No significant correlation with SDA can be seen.

RESONANCE EXCITATION

Our measurements in Svalbard indicate that resonance excitation can
play an important part in generating the HeI 3889Å line, and we find
it pertinent to consider the resonance excitation mechanism in some
detail.

Photoelectron impact on the ambient He(^{1}S) atoms produces column
densities of the metastable He($2\,^3$S) atoms exceeding 10^7 $\#$/cm^2 (McElroy
1965; Christensen et al. 1971). In aurora the integral electron flux
below 1 keV can increase by an order of magnitude (Doering et al. 1975)
enhancing the He($2\,^3$S) column density by a similar amount, when the
solar shadow height and the penetration depth of the electrons are
equivalent.

The natural lifetime of the $2\,^3$S state is predicted to be close
to 2.5 hours (Feinberg and Sucker 1971; Drake 1971). From 500 km downwards the effective lifetime reduces drastically due to collisions
with N$_2$, O$_2$, and O being about 1 sec at 200 km (McElroy 1965). Solar
resonance scattering on the $2\,^3$S atoms efficiently produce ^{3}P atoms,
and the excitation rates of the ^{3}P atoms will be proportionally to
the $2\,^3$S column density n($2\,^3$P) when the procedure for resonance excitation introduced by Chamberlain (1961) is adapted.

$$E(^3P) = n(2\,^3S)g_{SP}$$

The g-value equals

$$g_{SP} = \pi\ F_\nu\ \frac{\pi e^2}{mc}\ f_{SP}$$

and

$$F_\nu = F'_\nu \, r_\nu \; .$$

These quantities are specified as:

$\pi \, F_\nu$ = monochromatic radiant flux at one astronomical unit from the sun in quanta per unit area normal to the beam per second per unit frequency at frequency ν matching the appropriate S-P transitions.

$\dfrac{\pi e^2}{mc}$ = $2.646 \cdot 10^{-2}$ cm^2/s = integrated absorption coefficient per atom for unit f-value in c.g.s. units.

f_{SP} = absorption oscillator strength for the S-P transitions.

r_ν = fractional depression of the solar continuum F'_ν.

The above quantities are independently obtained, and g-values for the two S-P transitions calculated:

$2^3S \rightarrow 3^3P$		$2^3S \rightarrow 2^3P$	
$r^{1)}$	= 0.30	$r^{2)}$	= 0.95
$\pi \, F_\nu^{\;3)}$	= $1.15 \cdot 10^2$	$\pi \, F_\nu^{'\;3)}$	= $1.53 \cdot 10^3$
$f_{SP}^{\;4)}$	= 0.06446	$f_{SP}^{\;4)}$	= 0.5391
g_{SP}	= 2.20 $(\dfrac{\text{photon}}{\text{atom s}})$	g_{SP}	= 780 $(\dfrac{\text{photon}}{\text{atom s}})$

1) Delbouille et al. (1973)
2) Brandt et al. (1965)
3) Allen (1964)
4) Wiese et al. (1966).

Cascading into the ^3P states is of minor importance, because resonance excitation into the actual higher energy states involves forbidden transitions with relatively low transition probabilities. Since the ^3P-^3S transitions are allowed, collisional deactivations are negligible compared with radiation. Then the radiation can be expressed in terms of the excitation rates, and the branching ratios are specified by the transition probability factors A for resonance emissions

$$3^3P - 2^3S \rightarrow 3889\text{Å}$$

$$I(3889) = n(2^3S) \, g_{SP}(3889) \cdot \frac{A_1}{A_1 + A_2} \; ,$$

where A_1 is the transition probability to the 2^3S state and A_2 to the 3^3S state.

The HeI 3889Å intensity of ~ 1R measured in twilight (Brandt et al. 1965) is most likely caused by solar resonance scattering and requires a column density of He(2^3S) atoms of ~ $\underline{5 \cdot 10^6 \; \#/cm^2}$. The Svalbard measurements showing as much as 70R can partly be explained by this mechanism when taking into account enhancement of the electron density due to auroral activity, enhancing the HeI(2^3S) population. However, the lack of correlation of I(3889) with solar depression angle in Fig. 4 can be mainly caused by changing auroral activity, increasing the HeI(2^3S) population by electron impact and the HeI(3^3P) population by precipitation of solar wind α-particles, and these mechanisms are treated in the next chapters.

It can be added that resonance excitation of the 2^3S state by the process

$$He(1^1S) + h\nu \rightarrow H(2^3S)$$

is very slow mainly because the absorption oscillator strength for this forbidden transition is decreased by a factor of ~ 10^{-12} relative to the ^{3}S-^{3}P transitions.

ELECTRON IMPACT ON AMBIENT He

The main reason for the aurora is electron impact on atmospheric constituents, and this process has to be considered for the He emissions observed in aurora

$$He(1^1S) + e \rightarrow He(k) + e$$

He(k) denotes an excited atom. When an emission between two states k and j are observed, ignoring cascading into the k state, the column emission rate can be calculated by the relation

$$I(kj) = \int Q(k,j,h)n(He,k,h) \int I_e(E,h)\sigma_k(E) \; dE \; dh \; ,$$

where $Q(k,j,h)$ is the Stern-Volmer factor, $n(He,k,h)$ number density of excited He atoms, $I_e(E,h)$ electron intensity at energy E and height h, and $\sigma_k(E)$ cross-section for electron impact excitation into the excited state.

The Stern-Volmer factor is defined as

$$Q(k,j,h) = \frac{A_{kj}}{\sum_i A_{ki} + \sum_k d_k(X)n(X,h)n(He,k,h)}$$

where A_{kj} is the transition probability between k and j states, and $d_k(X)$ the rate coefficients for quenching of an excited He atom in collision with an atmospheric constituent X, having $n(X,h)$ number density at height h.

The cross-section for electron impact into the 2^3S and 2^3P are experimentally determined (see review article by Jusick et al. (1967)), and these cross-sections are quite similar and the maximum values occur close to 50 eV. Using these cross-sections, ambient He densities from U.S. Standard Atmosphere model (1966) for high-latitude winter with $T_\infty = 1100^\circ K$, and auroral electron flux in an IBCII aurora having an integral number flux of 10^{10} electrons/cm^2 s below 1 keV as measured by McEwen and Sivjee (1972), the resulting 3889Å intensity will be about 0.5R.

The intermittently observed enhancements of 10-50R of the HeI 3889Å line might be generated by electron impact on ambient helium in IBCIII aurora, but such bright aurora was not observed. During these observations the dayside aurora was within class IBCI, and the electron precipitation must have been much too low to account for the observed intensities. Therefore it seems likely that electron impact on ambient He atoms should be responsible for only minor parts of the irregular intensity fluctuations of the intensity measurements shown in Figs. 3 and 4. The most reasonable cause of these fluctuations is precipitating α-particles from the solar wind.

SOLAR WIND α-PARTICLE INTERACTION WITH AIR

It is well established that the solar wind contains moderate fluxes of He^{2+} ions, and that they penetrate along with electrons and protons into the auroral zone (Shelley et al. 1976). The neutralization process starts above 500 km, and singly ionized helium generates emissions as auroral protons do

$$He^+(E) + M \xrightarrow{\sigma_k(E)} He(2^3P, 3^3P, 3^3D, E) + M_i^+ ,$$

where M_i denotes ambient atmospheric constituents. The resulting column intensities can be estimated by the equation

$$I(kj) = \sum_i \int Q(k,j,h) n_i(h) F(He^+,E,h) \sigma_k(E) \, dE \, dh ,$$

where $n_i(h)$ is the number density of atmospheric constituent i at height h, $F(He^+,E,h)$ the intensity of singly ionized helium with energy E at height h, and $\sigma_k(E)$ the cross-section for electron capture by the helium ion that leaves the resulting helium atom in an excited state k.

Cross-sections for the 3^3P excitations in He^+ impact on He, N_2, and O_2 are experimentally obtained by Head and Hughes (1965). In collisions with N_2 the 3^3D cross-section is determined down to 10 keV, and the value exceeds 10^{-17} cm^2 around 50 keV. Below 10 keV extrapolated values has to be used, and we estimate the 3^3D and 3^3P cross-section to be ~ 10^{-17} cm^2 for the first decade of kinetic energy. Using these estimates and the He^{2+} intensity measured in the low-altitude polar cleft (Shelley et al. 1976) with its peak at 6 keV and an integrated intensity of ~ (2.4±1) × 10^7 #/cm s, the corresponding He emissions at 3889Å are found to be about 30R. The penetration depth of the 6 keV helium atoms is ~ 140 km interaction with all the atmospheric particles above this altitude and this number is ~ 10^{18} per cm^2. Therefore precipitating α-particle can be considered as the most likely source for enhanced 3889Å emission observed in aurora at high latitude.

CONCLUDING REMARKS

HeI 3889Å line intensities of the order of 10-100R can be expected when integral fluxes of 10^7-10^8 cm^{-2} s^{-1} of He^{2+} ions with average energy about 10 keV penetrate into the upper atmosphere. This source is the most reasonable explanation of the night time observations.

     As helium emissions sporadically occur in auroral spectra, the relation to solar wind α-particles must be further studied to unravel how these particles find their ways through the magnetosphere and into the upper atmosphere. The sporadic 3889Å enhancements are not confined to the polar cleft of the nightside auroral oval, but occur over much wider areas which we consider as a result of scattering of the precipitating particles in the upper atmosphere. More information about excitation processes can be obtained when measurements of the HeI emissions are correlated with solar depression angle, electron and proton aurora, and incoming hot α-particles (Lundin 1985).

     The order of magnitude more intense resonance scattered HeI 3889Å intensity observed on the latitude of the polar cleft than at mid-latitude (Brandt et al. 1965) may be due to increased He density in the polar regions. We can always speculate about increased density due to solar wind α-particles precipitation or meridional transport from lower latitudes, but firm conclusions can not be reached before sufficient measurements are carried out both at mid-latitude and in polar regions.

ACKNOWLEDGEMENTS

The support from NATO Research Grant No. 513/82 is acknowledged.

REFERENCES

Allen, C.W. 1963, Astrophysical Quantities, University of London, The Athlone Press, Second edition.
Brandt, J.C., Broadfoot, A.L., and McElroy, M.B. 1965, Astrophys. J., 141, 1884.
Chamberlain, J.W. 1958, Astrophys. J., 127, 54.
Christensen, A.B., Patterson, T.N.L., and Tinsley, B.A. 1971, J. Geophys. Res., 76, 1764.
Delbouille, L., Roland, G., and Neven, L. 1973, in Photometric atlas of the solar spectrum from λ3000 to λ10000. (Edited by Institut d'Astrophysique de l'Université de Liege.)
Doering, J.P., Peterson, W.K., Bostrom, C.O., and Armstrong, J.C., 1975, J. Geophys. Res., 80, 3925.
Drake, G.W.F. 1971, Phys. Rev. A, 3, 908.
Federova, N.I. 1961, Planet. Space Sci., 5, 75.
Feinberg, G. and Sucker, J. 1971, Phys. Rev. Lett., 26, 681.
Feldman, P. 1985, This volume.
Harrison, A.W. and Cairns, C.D. 1969, Planet. Space. Sci., 17, 1213.
Head, C.E. and Hughes, R.H. 1965, Phys. Rev., 139, A1392.

Henriksen, K. 1978, in Proceedings of Esrange Symposium, Ajaccio, 24-29 April 1978, ESA SP-135, European Space Agency.
Ivanchuk, V.I. and Sukhoivanenko, P.Ya. 1958, Astronomical Circular, Press Bureau of Astronomy, Reports of the USSR Academy of Sciences, No. 196.
Jusick, A.T., Watson, C.E., Peterson, L.R., and Green, A.E.S. 1967, J. Geophys. Res., 72, 3943.
Lundin, R. 1985, This volume.
McElroy, M.B. 1965, Planet. Space Sci., 13, 403.
McEwen, D.J. and Sivjee, G.G. 1972, J. Geophys. Res., 77, 5523.
Mirnov, A.V., Prokudina, V.S., and Shefov, N.N. 1959, Spectral Electro-photometrical and Radar Researches of Aurora and Airglow, 1, 20.
Moore, C.E., Minneart, M.G.J., and Houtgast, J. 1966, Nat. Bur. of Stand. Monogr., Washington, D.C. 2040.
Schecheglov, P.V. 1962, Soviet Astronomy - AJ, 6, 118.
Shefov, N.N. 1961a, Planet. Space Sci., 5, 75.
Shefov, N.N. 1961b, Ann. Geophys., 17, 395.
Shefov, N.N. 1963, Planet. Space Sci., 10, 73.
Shefov, N.N. 1968, Planet. Space Sci., 16, 1103.
Shelley, E.G., Sharp, R.D., and Johnson, R.G. 1976, J. Geophys. Res., 81, 2363.
Sivjee, G.G., Henriksen, K., and Deehr, C.S. 1980, J. Geophys. Res., 85, 6043.
Stoffregen, W. 1968, Planet. Space Sci., 17, 1927.

MODELLING OF CUSP AURORAS: THE RELATIVE IMPACT OF SOLAR EUV RADIATION AND SOFT ELECTRON PRECIPITATION

K. Stamnes[1], M.H. Rees[2], B.A. Emery[3] and R.G. Roble[3]
[1]Auroral Observatory, Institute of Mathematical and Physical
Sciences, University of Tromsø, 9000 Tromsø, Norway
[2]Geophysical Institute and Department of Space Physics and
Atmospheric Sciences, University of Alaska, Fairbanks,
AK 99701, U.S.A.
[3]National Center for Atmospheric Research, Boulder, CO 80303,
U.S.A.

ABSTRACT. A quantitative evaluation of the relative impact of solar
EUV radiation and auroral electron precipitation in cusp auroras shows
that the solar EUV input may contribute substantially to the electron
density and temperature in the cusp. The relative solar EUV contribution to atomic line emissions changes from being unimportant for solar
zenith angles (SZA) of 104° to dominant for SZA = 92° for typical cusp
auroras. The solar EUV contribution to the red line (OI 6300Å)
emission is about 70% for SZA = 92°, 35% for SZA = 98°, and 5% for SZA
= 104°. The direct electron impact contribution to the N_2^+ 1NG band
system is negligible compared to the resonant and fluorescent scattering contributions. However, the cusp auroral precipitation is very
important for this band system since it may enhance the efficiency of
the resonant (or fluorescent) scattering mechanism by an order of
magnitude by substantially enhancing the N_2^+ density.

1. INTRODUCTION

During the last five years a multinational program of ground-based
optical observations of the dayside aurora has been carried out at
Longyearbyen and Ny-Ålesund, Svalbard (Deehr et al. 1980). This island
group at the edge of the polar sea north of Norway is the only
accessible site in the northern hemisphere that is located both at a
high enough geographic latitude (near 79^ON, 15^OE) to minimize the
effect of scattered sunlight, and at an appropriate geomagnetic latitude to be under the geomagnetic cusp.

Around winter solstice the sun is never less than 9° below the
horizon in Adventdalen close to Longyearbyen on the island West
Spitsbergen (Deehr et al. 1980) where the main observatory building was
permanently established in November 1983.

It is generally appreciated that resonant and fluorescent
scattering of sunlight may contribute to the optical midday aurora (cf.

137

J. A. Holtet and A. Egeland (eds.), The Polar Cusp, 137–147.
© *1985 by D. Reidel Publishing Company.*

e.g. Sivjee 1976). In particular, this mechanism leads to the observed enhancement of the 4278-Å N_2^+ 1NG emission (Deehr et al. 1980). However, no quantitative evaluation of this effect for cusp auroral situations or of the relative contribution of solar EUV photons as compared to particle precipitation to the midday aurora has yet appeared in the literature. The purpose of the present paper is to evaluate this effect quantitatively and to compare the relative impact of solar EUV radiation and soft electron precipitation in cusp auroras. To facilitate the use of these results we focus our attention on computations that will be directly useful for the interpretation of optical observations made from Svalbard. However, our results are also relevant to other locations (such as e.g. Søndre Strømfjord, Greenland) since the most important parameter determining the solar EUV contribution is the local solar zenith angle.

2. DESCRIPTION OF THE MODEL

To compare the relative importance of solar EUV photons and soft electron precipitation in cusp auroras we use the time-dependent auroral model described by Roble and Rees (1977), Stamnes and Rees (1983) and Rees et al. (1983). Starting with a neutral atmosphere, we use this auroral model to simulate the ionospheric response to solar EUV fluxes and precipitating auroral electron fluxes typically observed in the cusp (Heikkila and Winningham 1971; Potemra et al. 1978). As explained by Stamnes and Rees (1983) this model provides a simultaneous solution of a coupled system of equations consisting of the energy equations for ions and electrons, the ion continuity equation and the electron transport equation.

We use the mass spectrometer/incoherent scatter (MSIS) model atmosphere (Hedin et al. 1977a,b) modified to high latitude by multiplying the molecular nitrogen density by 1.4, molecular oxygen by 2.0 and atomic oxygen by 0.35. The solar EUV flux and the photoionization and photoabsorption cross sections are taken from Tables 2 and 3 of Torr et al. (1979). These cross sections are based on a comprehensive compilation by Kirby-Docken et al. (1979) and the solar EUV fluxes on measurements by the EUV spectrometers onboard the Atmosphere Explorer satellites. Torr et al. (1979) present a reduced set of cross sections and solar EUV flux data comprising 37 wavelength intervals covering the wavelength range 50-1050Å. This set, which is obtained by averaging fluxes and cross sections over broader bands, is suitable for computing primary photoelectron spectra.

Cusp auroras are characterized by soft particle precipitation. Thus, to simulate the auroral input we adopted a maxwellian electron spectrum between 0 and 100 eV with a characteristic energy of 50 eV. Since there is usually a small amount of background precipitation present in the polar atmosphere we added another maxwellian spectrum with a characteristic energy of 200 eV and a total energy input of only 0.05 ergs cm^{-2} sec^{-1} to describe this background aurora which we shall hereafter refer to as the "polar drizzle".

One of the standard instruments used to monitor the aurora at Svalbard is a meridian-scanning-photometer (MSP) which consists of several photometers that scan the sky in the magnetic meridian to measure the intensity of the aurora at a fixed set of wavelengths. When the MSP points to the zenith it measures the intensity integrated along the line of sight for the ionosphere above the observing site. However, when it points just above the horizon to the south most of the contribution to the line of sight integrated intensity comes from a region of the sky that is located about 10° south of the observing site for typical cusp auroras.

This situation is schematically illustrated in Figure 1 where we have assumed that the peak of the emitting layer is situated at about 250 km. Since our auroral model is designed to describe the ionosphere along a geomagnetic field line at any given geographic location, we may simulate the situation depicted in Figure 1 by computing the iono-spheric response to solar EUV and auroral electron precipitation at three different locations corresponding to pointing the MSP in the zenith, at 15° and 5° above the horizon in the magnetic meridian. Table 1 gives the geographic coordinates for these three locations, the local solar time at which they are under the cusp and the corresponding solar zenith angle and solar EUV energy input inferred from our model. Typical auroral electron energy inputs in the cusp are a few tenths ergs cm^{-2} sec^{-1} (J.D. Winningham, private communication, 1984). Thus we have chosen a cusp auroral energy input spectrum yielding 0.32 ergs cm^{-2} sec^{-1} (which is the same as the solar EUV input at the southern-most location) to represent a "typical" cusp aurora and we use an energy input a factor 10 lower than this value (i.e. 0.032 ergs cm^{-2} sec^{-1}) to represent a "weak" cusp aurora.

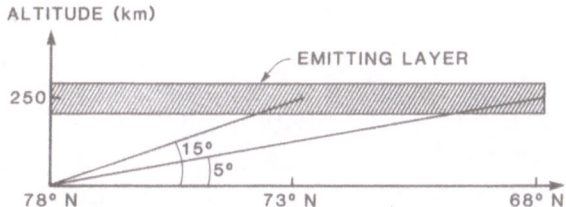

Figure 1.   Schematic illustration of a cusp auroral situation.

TABLE I.   Coordinates - Solar EUV Input

Local Time at Geomag. Noon	Geographic Latitude	Geographic Longitude	Solar Zenith Angle	Solar EUV Energy Input ergs cm^{-2} sec^{-1}
11:00 A.M.	68°N	38°E	92.2°	0.32
10:30 A.M.	73°N	30°E	97.7°	0.074
9:30 A.M.	78.2°N	15.6°E	104°	0.003

3. RESULTS

In Figures 2-5 we show height profiles of the relevant ionospheric
parameters at 78 N (corresponding to the MSP pointing overhead) and
68°N (corresponding to the MSP pointing just above the southern
horizon). The dot-dash curve corresponds to solar EUV photons and
background "polar drizzle" only (SEUV + PD), the broken line to a
"weak" aurora (0.032 ergs cm^{-2} sec^{-1}) superimposed on SEUV + PD and
the solid line to a "typical" aurora (0.32 ergs cm^{-2} sec^{-1}) superimposed on SEUV + PD. We should point out that the contribution to
the ionization rate (Fig. 2) below 200 km is mainly due to scattered
geocoronal radiation (Ly-β) and the "polar drizzle". Figures 2 and 3

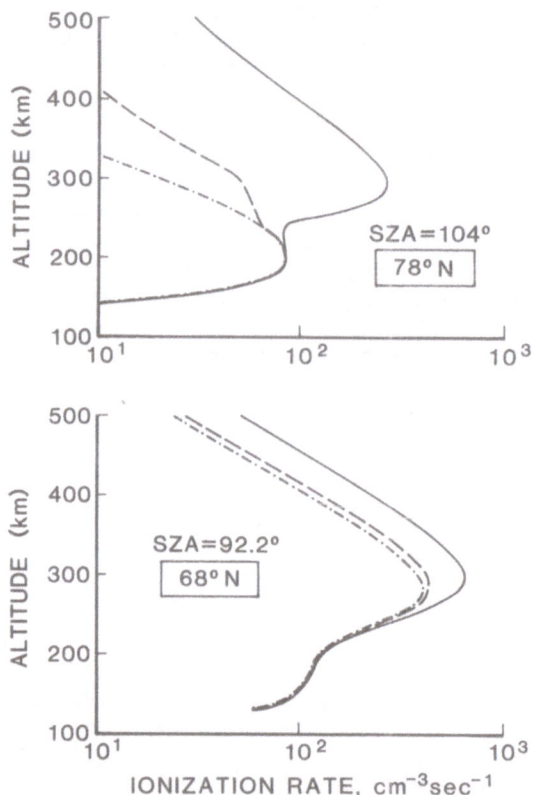

Figure 2. Computed ionization rate in the cusp at 68°N and 78°N for
           solar and auroral input.
           —·—·— Solar EUV and background polar drizzle only (SEUV+PD)
           — — — Weak aurora (0.032 ergs cm^{-2} sec^{-1}) superimposed
           ———— Typical aurora (0.32 ergs cm^{-2} sec^{-1}) superimposed on
           SEUV + PD.

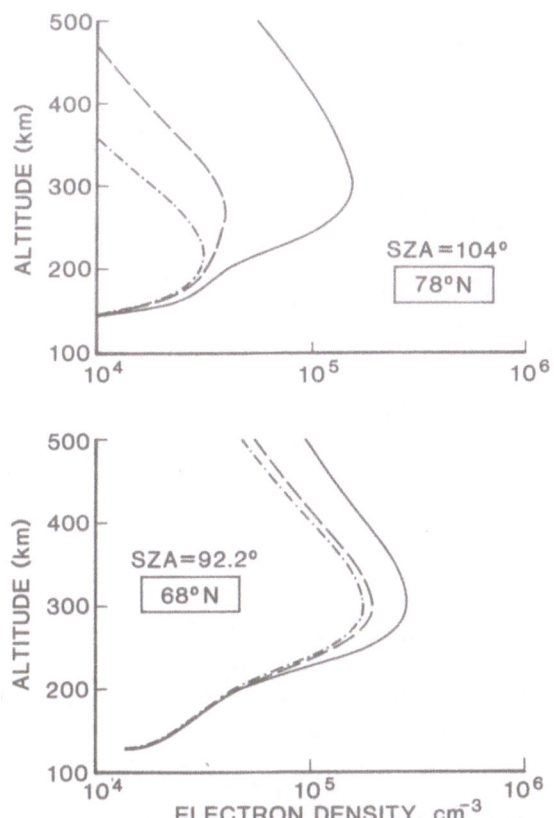

Figure 3. Computed electron density in the cusp for solar and auroral
           input. See Fig. 2 for symbol explanation.

show that the solar EUV input contributed substantially to the ionization rate and electron density at 68°N but provides only a minor
contribution at 78°N, as one would expect. A similar conclusion may
be drawn for the electron temperature shown in Figure 4.
     The height-integrated emission rates computed by the auroral model
are shown in Table 2 for several auroral radiations. While the solar
EUV input contributes negligibly to all the atomic lines at 78°N, this
source is the dominant contributor to the atomic oxygen red (OI 6300Å)
and green (OI 5577Å) lines at 68°N for "typical" cusp auroras. For the
OII 7320Å line the solar EUV and the "typical" cusp auroral input are
of equal importance, and the solar EUV source contributes about twice
as much as the "typical" cusp aurora to the NI 5200Å line at 68°N.
Figure 5 shows the corresponding altitude profile for the red line
emission and Figure 6 the percentage solar EUV contribution to the red

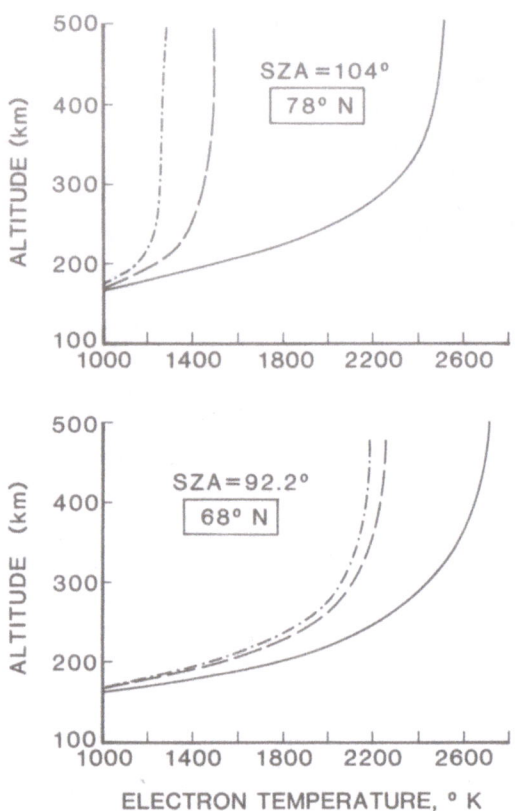

Figure 4. Computed electron temperature in the cusp for solar and
           auroral input. See Fig. 2 for symbol explanation.

TABLE II. Emission Rates (Rayleighs)

Energy Input ergs cm^{-2} sec^{-1}		6300Å	5577Å	7320Å	5200Å	4278Å Electron Impact	Flour Scatt.
68°N							
Solar EUV	0.32	2000	273	192	89	6	1055
Cusp	0.032	100	19	19	4	2	1107
Aurora	0.32	920	181	182	42	24	1468
73°N							
Solar EUV	0.074	614	45	73	20	2	579
Cusp	0.032	123	23	19	9	2	623
Aurora	0.32	1076	209	102	67	25	1143
78°N							
Solar EUV	0.003	57	4	3	5	1	144
Cusp	0.032	121	26	19	8	3	281
Aurora	0.32	1143	230	180	77	27	992

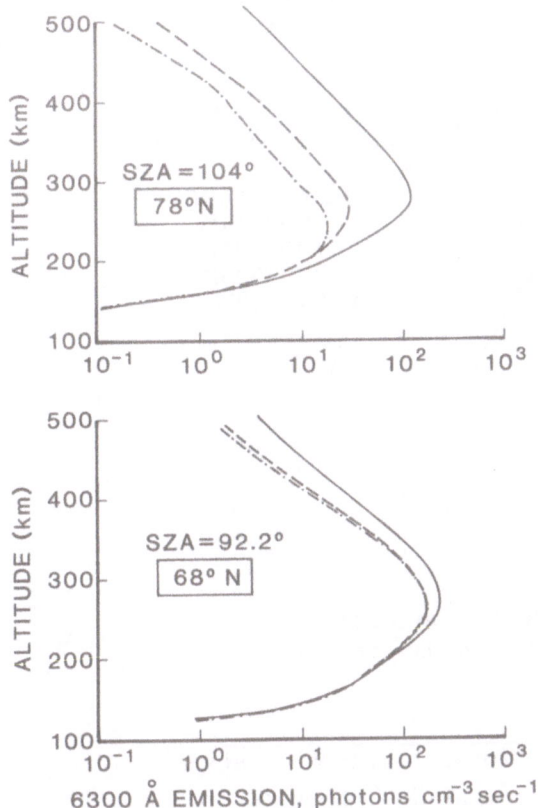

Figure 5. Red line emission rate due to solar and auroral input. See
Fig. 2 for symbol explanation.

line emission at the different locations. Since the solar zenith angle
is also given in Figure 6, this information may be used by investigators working at locations other than Svalbard to infer the relative
importance of solar EUV input to the red line emission.

The auroral model includes photoionization and electron impact
ionization: both processes are sources of N_2^+ production. The
resulting ion density height profiles, computed with the time dependent
model, are shown in Figure 7 for the SEUV and drizzle case only, and
for the auroral situations adopted as examples in this work. We note
that at 78°N the typical cusp aurora leads to a substantial increase
in the N_2^+ density, about a factor of seven in the column integrated
value. The fluorescent scattering contribution to the N_2^+ 1NG band
emission is given by

$$4\pi I = gN \text{ photons cm}^{-2} \text{ sec}^{-1} \quad (10^{-6} \text{ rayleighs})$$

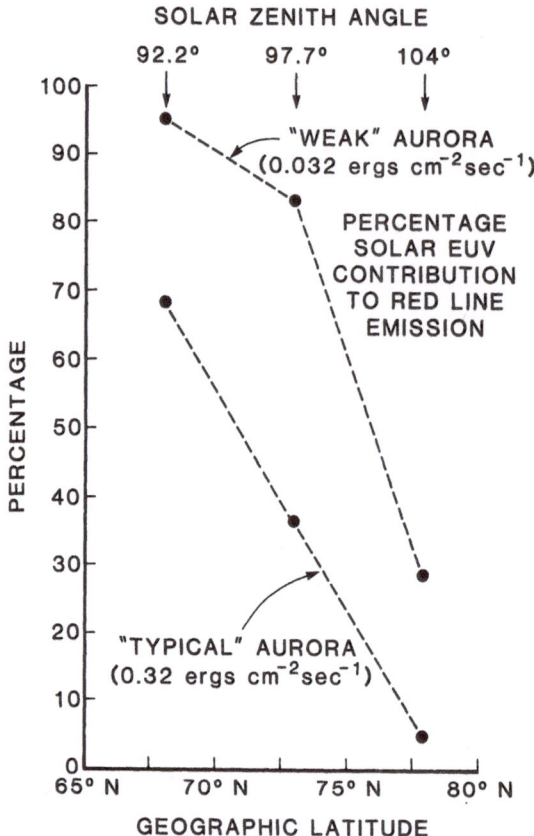

Figure 6.   Percentage solar EUV contribution to the atomic oxygen (OI 6300Å) emission rate for different solar zenith angles.

where N is the column density and the g-factor gives the number of photons scattered per second per molecule. Isotropic scattering is assumed. The value adopted for $g\left[N_2^+ \, 1NG(0,1)\right]$ is 0.015 (Chamberlain 1978). Contributions to the 4278Å emission from direct electron impact and from fluorescent scattering are summarized in Table 2. The latter source dominates in the cusp under all conditions.

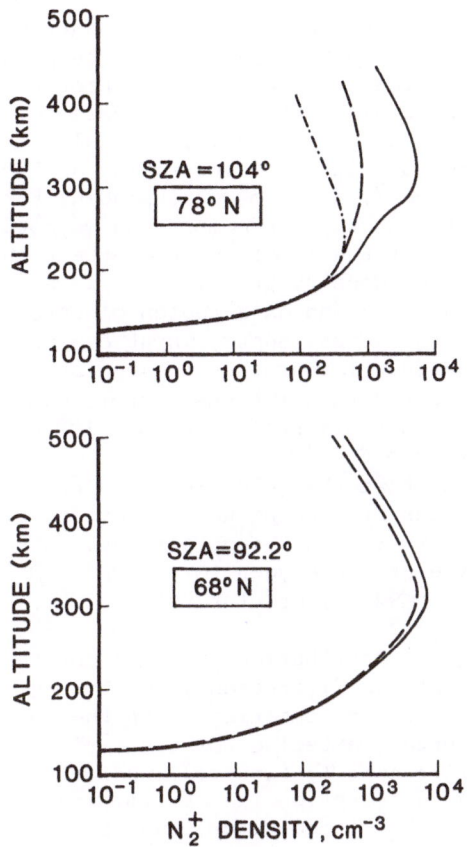

Figure 7.   Computed N_2^+ density for solar and auroral input.   See
            Fig. 2 for symbol explanation.

4.   DISCUSSION AND CONCLUSIONS

Our auroral model simulates the behaviour of the ionospheric plasma
inside a geomagnetic flux tube in the collision-dominated part of the
atmosphere.   It accounts for physical and chemical processes (in a
weakly ionized medium) that are the consequence of auroral particle
precipitation and solar EUV radiation.   Since the model is one-
dimensional horizontal convection caused by orthogonal electric fields
and the wind system cannot be computed self consistently, but may be
arbitrarily imposed.   However, this limitation has little consequence
for our present effort aimed at comparing the relative impact of solar
EUV radiation and cusp auroral particle precipitation on computed
ionospheric parameters and spectroscopic auroral emission features.
However, in order to compare the predicted electron density with

measurements, one would have to account for ionization convected in and out of the field of view.

We find that the solar EUV input may in general contribute substantially to the electron density and temperature in cusp auroras. This finding has important implications for the interpretation of incoherent scatter radar data obtained at Søndre Strømfjord, Greenland. Thus, for SZA = 92.2° (SZA = solar zenith angle) the solar EUV radiation leads to an electron density profile peaking at about 300 km with a value of about 2×10^5 cm^{-3}, and the corresponding F-region electron temperature is about 2100°K. For comparison we note that at 78°N (SZA = 104°) where the solar EUV input is unimportant our typical cusp aurora leads to an electron density profile that is similar to the one for SZA = 92.2° (no aurora) and an F-region electron temperature of about 2400°K. Since the actual energy input can occasionally be more than an order of magnitude larger than the typical value (0.32 ergs cm^{-2} sec^{-1}) adopted in this study, the cusp aurora can conceivably give rise to F-region electron densities of more than 10^6 cm^{-3} and electron temperatures larger than 3000°K.

Our computations indicate that the relative solar EUV contribution to atomic line emissions changes from being unimportant for SZA = 104° to dominant for SZA = 92.2° for "typical" cusp auroras. Likewise the solar EUV contribution to the atomic oxygen (OI 6300Å) red line emission is about 70% for SZA = 92.2°, 35% for SZA = 97.7°, and 5% for SZA = 104°.

The direct electron impact contribution to emissions that are also efficiently excited by resonant (or fluorescent) scattering (such as the N$_2^+$ 1NG system) is negligible in comparison with the contribution due to resonant (or fluorescent) scattering for transitions having large photon scattering coefficients (g-factors). However, for the N$_2^+$ 1NG system the cusp auroral precipitation is very important since it may enhance the efficiency of the resonant (or fluorescent) scattering mechanism by an order of magnitude by substantially enhancing the N$_2^+$ density. The N$_2^+$ 1NG band emission which has been very useful to evaluate auroral energy deposition in night time aurora, cannot be used for this purpose in the cusp, or daytime aurora for that matter. However, the OII (7320Å) emission might be a suitable transition for optical diagnostics because it provides a measure of the ionization rate and the energy deposition rate by energetic electrons and it has a negligible resonant/fluorescent scattering contribution.

ACKNOWLEDGEMENTS

This work was supported by the National Science Foundation under grant ATM 83-12883.

REFERENCES

Chamberlain, J.W., 1978, Theory of Planetary Atmospheres, Academic Press, New York.

Deehr, C.S., Sivjee, G.G., Egeland, A., Henriksen, K., Sandholt, P.E., Smith, R., Sweeney, P., Duncan, C., and Gilmer, J., 1980, J. Geophys. Res., 85, 2185.

Hedin, A.E. et al., 1977a, J. Geophys. Res., 82, 2139.

Hedin, A.E. et al., 1977b, J. Geophys. Res., 82, 2148.

Heikkila, W.J. and Winningham, J.D., 1971, J. Geophys. Res., 76, 883.

Kirby-Docken, K., Constantinides, E.R., Babeu, S., Oppenheimer, M., and Victor, G.A., 1979, At. Data Nucl. Data Tables, 23, 63.

Potemra, T.A., Doering, J.P., Peterson, W.K., Bostrom, C.O., Hoffman, R.A., and Brace, L.A., 1978, J. Geophys Res., 83, 3877.

Rees, M.H., Emery, B.A., Roble, R.G., and Stamnes, K., 1983, J. Geophys. Res., 88, 6289.

Roble, R.G. and Rees, M.H., 1977, Planet. Space Sci., 25, 991.

Sivjee, G.G., 1976, J. Atmos. Terr. Phys., 38, 533.

Stamnes, K. and Rees, M.H., 1983, J. Geophys. Res., 88, 6301.

Torr, M.R., Torr, D.G., Ong, R.A., and Hinteregger, H.E., 1979, Geophys. Res. Lett., 6, 771.

POLAR CUSP DYNAMICS

R.H. Eather
Department of Physics
Boston College
Chestnut Hill, MA 02167

ABSTRACT. The position of dayside aurora (as measured from South Pole station in 1981) is compared to the interplanetary magnetic field B_z component and to the AE index. The results support our earlier work showing a close relationship with AE and little correlation with B_z. Two recent papers have presented data that purport to contradict this interpretation, claiming a dominant B_z correlation. A reanalysis of the data sets used in those papers does not support B_z dependence, and in fact reaffirms a close dependence on AE. We conclude that the position of the dayside cusp is largely controlled by substorm processes internal to the magnetosphere rather than by direct merging and erosion processes with the interplanetary field.

1. INTRODUCTION:

The dynamics of the polar cusp and dayside aurora have continued to be of prime interest in recent years. Many studies have been concerned with the factors that influence the location of the polar cusp and midday aurora, and there is still disagreement on the relative importance of a number of possible mechanisms. Besides B_z and AE, cusp and midday auroral location are also expected to be affected by the solar wind pressure and, during magnetic storms, by the strength of the ring current (as represented by D_{st}; Siscoe, 1979).

In an earlier paper (Eather et. al., 1979) we presented dayside auroral data from the South Pole that led to our conclusion that there was a detailed and predictable 1:1 correlation between auroral (or cusp) position and substorm timing and intensity (as represented by the AE index). In that paper we asserted that previous work that claimed direct correlation with B_z was misleading; we suggested that B_z changes were not the causative driving mechanism, but that sustained negative B_z (>1 hour) increased the probability of substorms. However two recent papers (Sandholt et. al., 1983; Meng, 1983) have challenged our claims, and maintain there is a far better correlation with B_z. We feel that the data shown do not support such conclusions.

In this paper, which is a shortened version of Eather (1984b), we first present further South Pole auroral data that illustrate good

149

J. A. Holtet and A. Egeland (eds.), The Polar Cusp, 149–162.

correlation of cusp position on AE, and a lack of correlation with B_z. We then re-examine the data shown by Sandholt et. al. (1982) and Meng (1983) that purports to show good correlations between cusp position and B_z, and show that in fact AE is a much better correlative index.

2. SOUTH POLE DATA

An image-intensified monochromatic (6300Å) slit camera was operated at South Pole during the 1980 austral winter, allowing the position of the cusp to be monitored in the range ~ 81° → ~ 69° (Eather, 1984a). Figure 1 shows representative data within ± 4 1/2 hours of local noon for 21 days during the 1980 austral winter (about half the available data set for clear moonless nights). Plotted on the same time scale are the AE index and interplanetary B_z (ISEE-3). Not shown are most quiet days showing no auroral or B_z activity, and most days of weak activity that show small but seeemingly random variations in AE, B_z and auroral position.

Comparison of auroral position and B_z in Figure 1 shows that there is no correlation between B_z and auroral position on < 10s of minutes time scale. When B_z stays negative for long periods (> 1 hour), then a gross correlation sometimes becomes apparent (e.g. days 144, 145) but there is rarely correlation in detailed structure with auroral position. In many more cases one sees well-defined longer-term B_z changes with no response of auroral position (e.g. days 187, 193, 197, 226). On all days a much better correlation with AE is immediately apparent. [Note that we have not offset the auroral and B_z data by ~ 1 hour to allow for the solar wind transit time from the position of ISEE 3 to the Earth. However, examination of the data of Figure 1 shows that such a time offset would not change the conclusion of lack of correlation.]

None of these results is particularly surprising in view of our earlier work (Eather et. al., 1979) where we presented a statistical analysis that gave a correlation coefficient of 0.13 between auroral position and B_z. On the other hand, the correlation coefficient between auroral position and AE for the ± 3 hr. time period around local noon was 0.89.

Examination of the 1980 South Pole data (as well as 1981 data not shown here) confirmed our earlier conclusions concerning a lack of association of cusp auroral positions and B_z, except during sustained periods of southward B_z that lead to high substorm probability. In view of the high correlation we found between cusp auroral position and AE, (Eather et. al, 1979) we concluded that the cusp position correlation with B_z at such times was misleading. Although the cusp aurora moved equatorward during (−)ve B_z periods, the detailed correlation was still with substorm activity as represented by AE.

This conclusion is important as it casts doubts on the importance of magnetic field merging and subsequent erosion of the dayside magnetosphere as a mechanism for driving substorms. Thus we feel it important to look in detail at the two recent papers that challenge our conclusions.

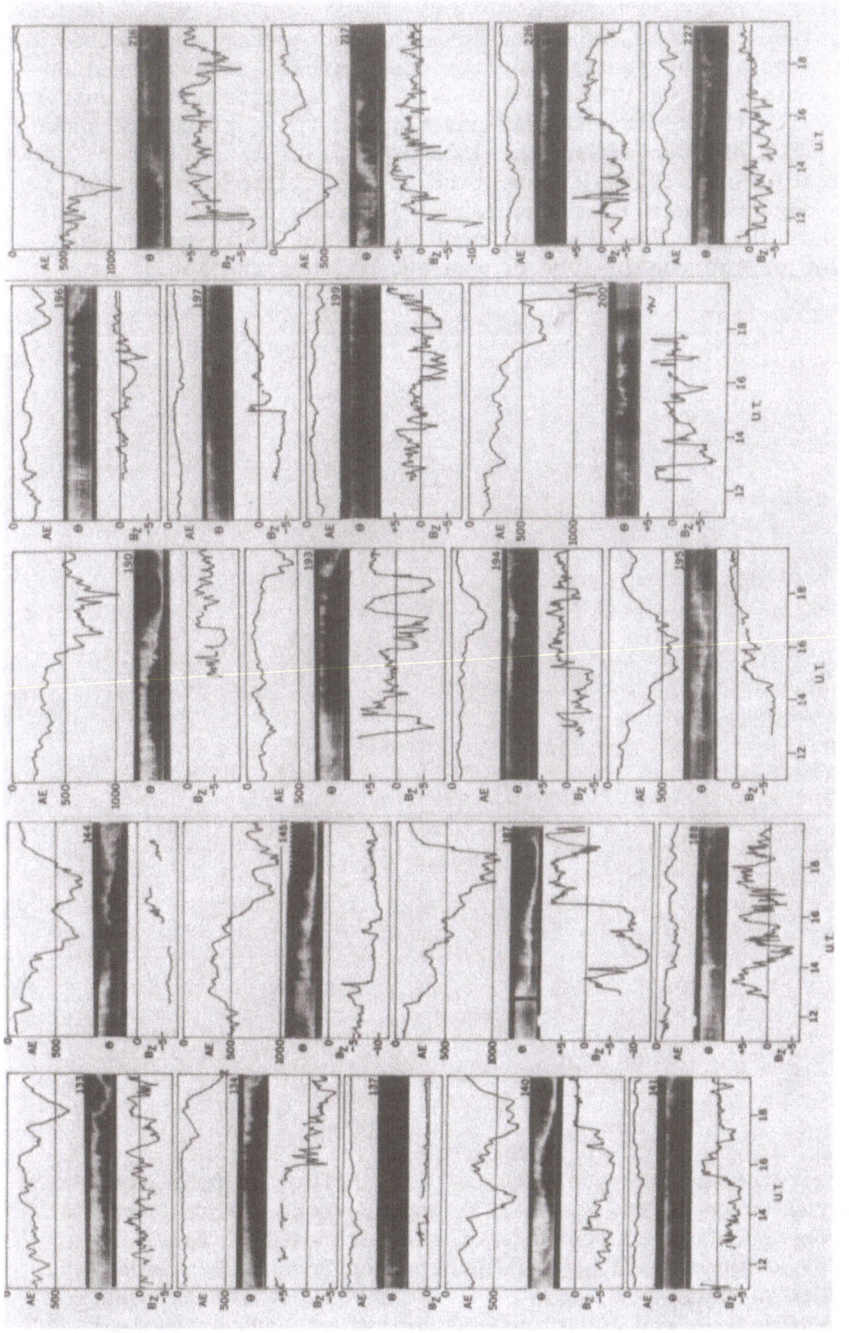

Figure 1: Representative data from South Pole, 1980, showing movements of dayside aurora (± 4 1/2 hours from local noon), together with AE and B_z. The gray scale keograms of 6300 OI emission cover a latitude range of approximately 69° to 81°. Note that the B_z time base has not been offset to allow for solar wind transit time between the satellite and the magnetopause (~ 1 hour).

3. SANDHOLT ET. AL., 1983:

In this investigation, meridian scanning photometers on Svalbad
were used to measure the position of the cusp aurora, as evidenced by
the 6300 OI emmision. We rescaled the data and replotted it on uniform
time scales, with the addition of AE and B_z, and these plots are shown
in Figure 2. [We did not offset the ISEE-3 Bz plots to allow for
the ~ 1 hour solar wind transit time from the satellite to near the
front of the magnetosphere (and the ISEE-1 location; see Sandholt et.
al., 1983, their Figure 7). However such an offset would not in any
way change any of the conclusions at the end of this section.]

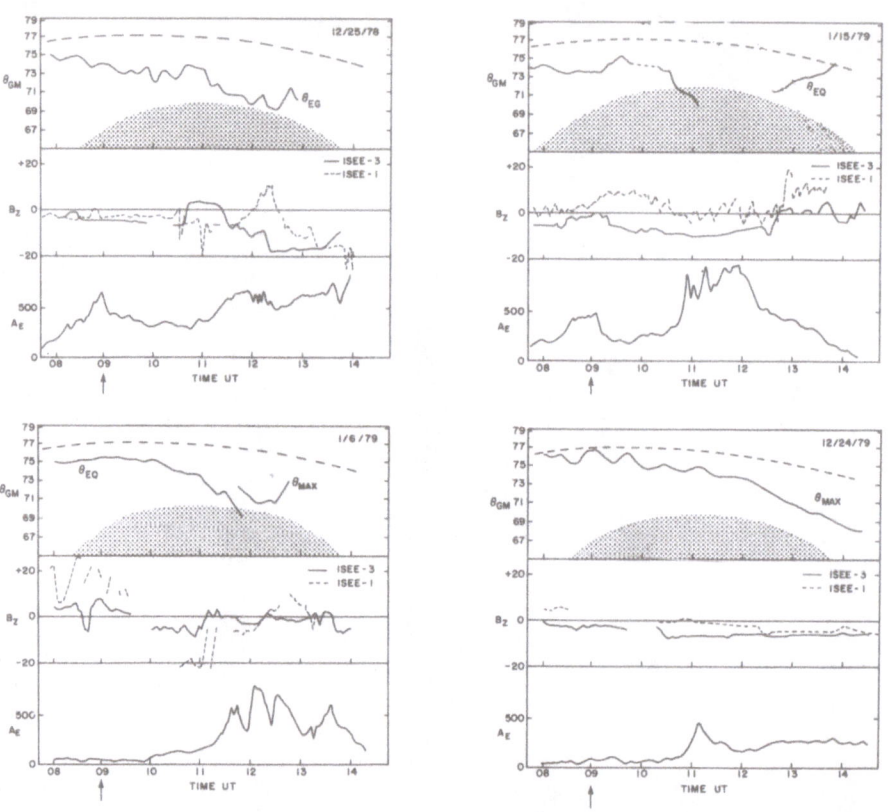

Figure 2: Plots of auroral position, B_z and AE for the four days
analyzed in the Sandholt et. al. (1983) paper. Most auroral position
data is for the equatorward boundary of the aurora (Θ_E), but the
position of peak intensity (Θ_{max}) is plotted on December 26, 1979 and
the latter part of January 6, 1979. Also shown is Θ_E for the Starkov
(1969) quiet-time boundary location (dashed curve), and an estimate of
the region of unreliable detection of cusp-associated 6300 OI (hatched
area) because of solar contamination. The arrow indicates local
magnetic noon (0900 UT).

Sandholt et. al. made the following comments concerning B_z correlations based on the data shown in Figure 2:
1. Dec. 25, 1978 "Notice the close correlations between variations of the IMF B_z and dayside auroral location"
2. Jan. 6, 1979 ".. a shift of the IMF vector from the northeast to the southwest direction at 09.40 UT coincident with the beginning of activity on the day-night sides of the oval"
3. Jan. 15, 1979 [No comments re auroral position and B_z. No figure for B_z.]
4. Dec. 24, 1979 [interval 1217 to 1630 UT] "The response of geomagnetic activity and the dayside aurora to the sharp IMF B_z variations was nearly instantaneous in this case."

Then in the summary section, it is stated: "Close correlation is observed between changes in the direction of the interplanetary magnetic field vector, dayside auroral shifts, and circum-oval geomagnetic variations... The equatorward shift of dayside auroral forms and the growth of associated geomagnetic activity are directly related to southward movement of the IMF vector... The poleward return of auroral forms accompanied by decreasing amplitude of the associated magnetic activity corresponds to a northward shift of the interplanetary magnetic-field vector (Dec. 25, 1978 is a good example).... The response of dayside auroras and the associated geomagnetic disturbance to the IMF variation impinging at the bow shock is observed to occur typically within 10-20 minutes. On Dec. 24, 1979 the response was nearly instantaneous."

We cannot agree that any of these statements are supported by the data in Figure 2 and suggest that the data in fact better supports our own conclusions that there is little if any correlation between midday aurora position and B_z on these time scales.

Concerning AE correlations, examination of Figure 2 shows a reasonably good association, with our previously measured relationship of $1°$ of latitudinal movement per 150 γ of AE applying fairly well. The exceptions are:
(1) 0900 UT Dec. 25, 1978 where the AE excursion for 20 minutes above 400 γ does not have the expected $\sim 3°$ auroral shift counterpart.
(2) 1110 UT Dec. 24, 1979 where the AE excursion for 20 minutes above 250 γ does not have the expected $\sim 1 1/2°$ auroral shift counterpart. Note however that only Θ_{max} is plotted, so it is not clear how Θ_E behaved.

Without trying to suggest ad hoc explanations for these two events, we conclude this discussion by stating we still believe these data offer much better support for AE being the major correlative index, and in fact confirm a lack of correlation with B_z.

4. MENG, 1983:

In this paper, Meng examined the latitudinal variation of the polar cusp during three intense geomagnetic storms. He compared these variations with the IMF B_z, D_{st} and a preliminary North American AE index to determine which correlated best with cusp position. One variable Meng did not consider in his analysis was the solar wind pressure; variations in ram pressure could contribute ~ 3° to the equatorward shift of the cusp (see Eather et. al. 1979).

We rescaled the data and replotted it on uniform time scales, together with hourly averages of
1.   actual ISEE 3 IMF B_z (rather than the preliminary "approximate" hourly averages used by Meng), time shifted by 1 hour to make an approximate allowance for the position of ISEE 3 upward from the magnetosphere
2.   correct AE values, rather than the limited preliminary North American AE index used by Meng
3.   solar wind pressure measured by ISEE 3 (not considered by Meng). These data are shown in Figure 3, where we plotted just the equatorward boundary (Θ_E) the cusp as published by Meng.

Before commenting further on the data set, some general comments on the expected nature of some correlations are in order.
1.   This data set is from large magnetic storms. Such storms are associated with long periods of southward B_z, and so from our previous work (Eather et. al. 1979) we would expect a gross correlation between cusp position and B_z, as the probability of substorms is high.
2.   The ring current intensity (as represented by D_{st}) becomes appreciable during these storms, so the ring current effect on cusp position (Siscoe, 1979) might be expected to be significant.
3.   AE is probably not a good measure of substorm effects during large magnetic storms. Because of limited ground station distribution at high latitudes, it is well known that AE may not be a reliable substorm indicator for small substorms occurring during very quiet conditions (Akasofu et. al., 1983; Kamide and Akasofu, 1983). A similar problem occurs during magnetic storms, when the auroral oval widens and moves to lower latitudes not well monitored by the AE contributing ground stations.

Figure 3 (over): Top panel: Hourly average of solar wind ram pressure (x 10^{-8} dyne cm^{-2}) from ISEE-3.
         2nd panel: Hourly average D_{st} (γ)
         3rd panel: Hourly average of interplanetary B_z (γ) component. Note that the B_z time base has not been offset to allow for solar wind transit time between the satellite and the magnetopause (~ 1 hour).
         4th panel: Hourly average of AE index (γ)
         Lower panel: Position of equatorial boundary of dayside cusp precipitation as measured by DMSP satellites in the northern (x) and southern (o) hemispheres, from Meng (1984). The histogram represents calculated latitude from the multiple linear regression equation (see Table 1). Θ_E = .008 D_{st} − 0.54P − 0.010 AE + 79.1, and the heavily shaded area is the difference between the measured and calculated values.

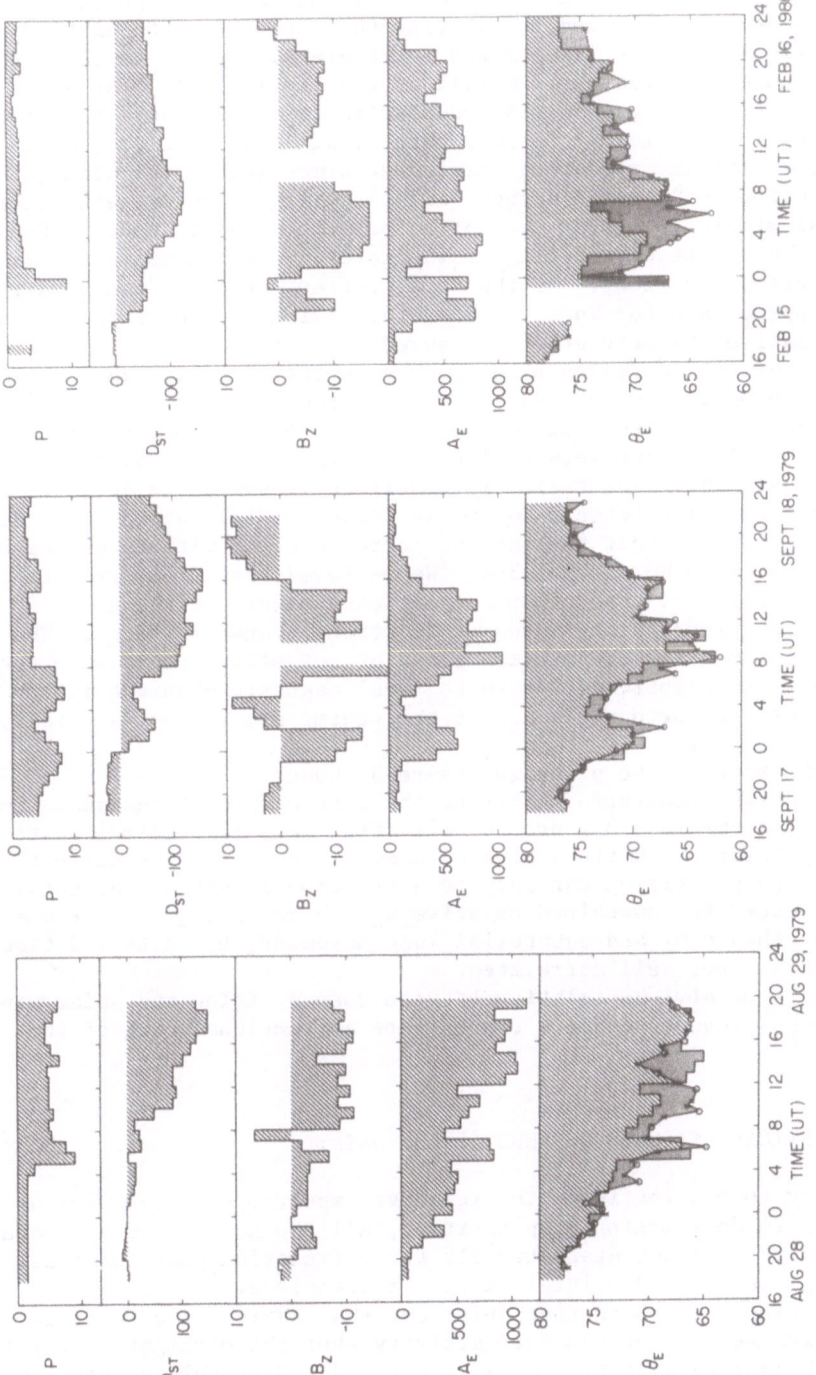

Figure 3.

Both the widening of the auroral oval and the shift to lower
latitudes results in AE underestimating the total strength of the
westward electrojet, as illustrated in Figure 4. Figure 4a shows the
distribution of AE stations, as well as the statistical location of the
poleward and equatorward auroral boundaries (midnight) as a function of
Kp (Whalen, 1970). It may be seen that AE will not be a good measure
of substorm activity a) during very quiet times when local midnight is
over the European-Siberian sector, and b) during large magnetic storms
when local midnight is over the North American-Atlantic sector. Figure
4b shows the effect of widening of the auroral oval on AE for a given
total electrojet current. As the total current spreads from a narrow
region (appropriate for Kp < 1) to a wide region (appropriate for Kp >
4) the relative magnetic effect measured by a ground station located
under the current decreases by as much as a factor of 3. Thus during
magnetic storms when Kp exceeds about 4 we would expect that the
measured AE for a given electrojet current to be considerably less than
if that same electrojet were confined to a narrow region above one of
the stations. This expectation is confirmed if one looks at the
statistical relation between Kp and AE (Allen, 1983), shown in Figure
4c. It may be seen that the rate of increase of AE with Kp decreases
significantly when Kp exceeds 3-4. We believe this effect results from
a wider auroral oval that often locates equatorward of the AE
monitoring stations during moderate to strong magnetic storms. The
situation is worst if the maximum phase of the storm occurs when the
North American-Atlantic sector is at local magnetic midnight i.e.
2300→0900 UT. Unfortunately all of the storms studied by Meng fall
into this category.
        Examination of the plots in Figure 3 shows:
1. Although the equatorial shift of the cusp occurs at the same time
D_{st} decreases, minimum D_{st} occurs well after peak latitudinal shift and
while cusp position is recovering poleward. Thus any ring current
effects on cusp position can only be a fractional part of the total.
2. As expected for sustained negative B_z, there is a gross correlation
between southward B_z and equatorial cusp movement, but detailed time
structures are not well correlated.
3. There is no obvious relation between cusp position and solar wind
pressure, so pressure effects can only be a fractional part of the
total.

5. STATISTICAL ANALYSIS OF MENG (1983) DATA:

        To try to best estimate the relative importance of the various
parameters in determining cusp position, all the hourly average data of
Figure 3 were analyzed statistically for correlation coefficients,
linear regressions and multiple linear regressions. Because of the
arguments presented above that point out that AE is not a reliable
quantitative measure of substorm activity when the midnight sector is
over North America-Atlantic, we also created a data subset that
excluded data points if the cusp position fell below 69° (approximately
10° from its quiet-time location) between 2300 and 0900 UT.

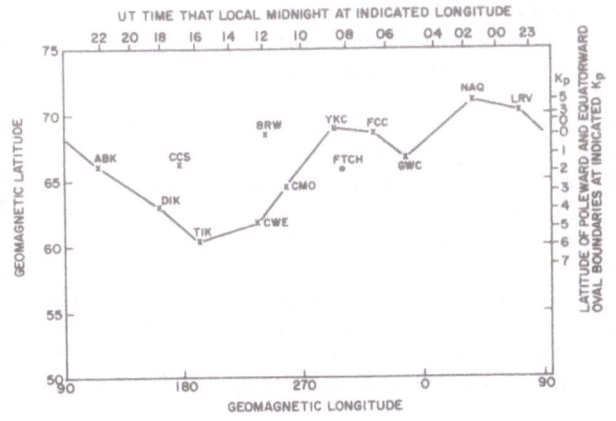

a) Locations of stations used to derive the AE index. Also indicated is the U.T. time when local midnight is at the indicated longitude, and the approximate width of the midnight auroral oval for various Kp values (from Whalen, 1970).

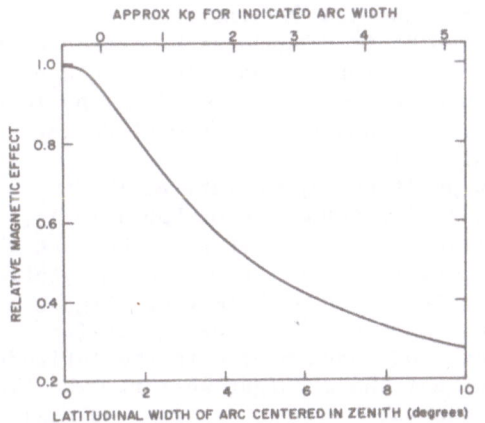

b) Calculated north-south magnetic effect (relative) for a given total electrojet current centered overhead, as a function of latitudinal width of the current. Also indicated is the expected Kp associated with the indicated arc width (from Whalen, 1970)

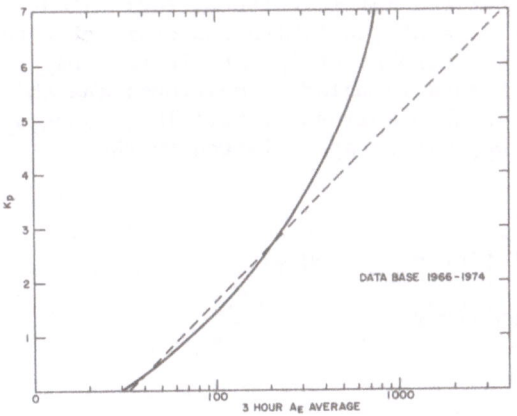

c) Statistical relation between Kp and 3 hr. averaged AE for the 1966-1974 data base (Allen, 1979), compared to a linear relationship (dashed line).

Figure 4.

Correlation coefficients, simple linear regression slopes and
intercepts, and multiple regression coefficients and residuals, are
shown in Table 1. Units used were degrees (for latitude), gammas (for
AU, AL, AE, D_{st} and B_z), and 10^{-8} dynes cm^{-2} (for pressure P = nmv^2).
     An examination of Table 1 shows:
1.   The best correlation is between Θ_E and AE, with correlation
coefficients of 0.75 (all data) and 0.83 (data subset). Correlation
coefficients between Θ_E and B_z are 0.61 and 0.64, respectively.
2.   a) The linear regressions show the least data scatter with
AE, with a probable error in the regression slope of 9%, compared
to 16% for B_z, 15% for D_{st} and 32% for P.
     b) The intercept is also most physically realistic with AE as the
variable; the equatorial edge of the cusp is known to locate at > 79°
under quiet conditions, and so this is the expected intercept. The AE
intercept is 76.6°, which is closest to the expected results; B_z gives
the worst result of 72.7°.
3.   The multiple linear regression analysis gives the best
results (in terms of approaching the expected 79° intercept and
minimizing the average residual) when B_z is dropped as an independent
variable. This multiple regression equation was used to calculate the
expected cusp position, and these predictions are plotted in the lower
panels of Figure 3, where the heavily shaded area is the difference
between measured and calculated values.
     In general the predicted cusp positions agree very well with the
measured position. The most serious discrepancy is on February 16,
1980 between 04-07 UT, and we believe this to result from the fact that
the nightside aurora was well equatorward of the AE monitoring stations
at this time (see Figure 4) so that AE was underestimated. Figure 5
plots the poleward and equatorward limits of nightside aurora (from
Meng, 1984) for the days considered, and superimposed is the latitude
of the most equatorward AE station near the midnight sector. The time
period 02-08 UT on February 16 is the worst case of a broad auroral
region located well equatorward of the AE stations.
     The coefficients of the linear and multiple linear regressions
cannot be taken too seriously because of the limitation of forcing to a
linear fit. The expected rates of decrease of Θ_E with increasing
values of Dst (Siscoe, 1979), P (pressure balance equations) and AE
(see Section 4 above) are all less than a linear rate. The expected
theoretical dependence of $\Delta\Theta$ on D_{st} may be approximated by the
empirical function

$$\Delta\Theta = 0.4 \times |D_{st}| = D* \quad (for\ D_{st} < 200\ \gamma)$$

The ram pressure balance equation gives

$$P/P_0 = \cos^{12}\Theta_0/\cos^{12}\Theta$$

where P_0 is a representative quiet-time pressure ($\sim$ 1.0 x 10^{-8} dyne
cm^{-2}) and Θ_0 is a representative quiet-time cusp latitude ($\sim$ 79°).

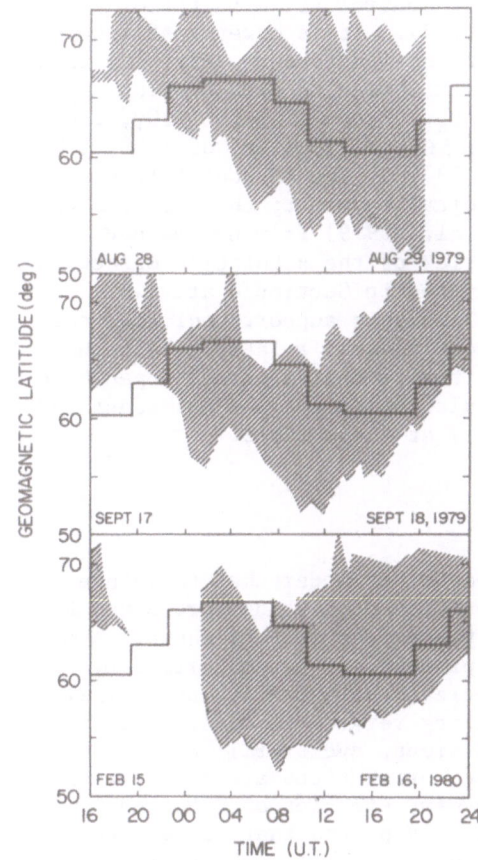

Figure 5. Poleward and equatorward boundaries of near midnight auroral precipitation as measured from DMSP satellites (from Meng, 1984). Superimposed is the latitude of the most equatorward AE monitoring station during the indicated 3-hour UT intervals.

This may be approximated by the empirical function

$$\Delta\Theta = 5.5 \times 10^3 \ (P-1) = P^* \ (\text{for } P \leqslant 20 \times 10^{-8} \text{ dyne cm}^{-2})$$

We used these functions to replace D_{st} and P in the multiple linear regression analysis, with the rationale that this should be more realistic than a forced linear fit. The resultant multiple regression fit gave

$$\Delta\Theta = 0.33 \ D_{st}^* - .008 \ AE - 2.03 \ P^* + 79.2$$

Average Residual = 1.17

As this is a limited data set, not a lot of confidence should be
placed on the derived regression coefficients, or the inference that
the theoretical models overestimate D_{st} control (by a factor of 3) and
underestimate P control (by a factor of 2). It is however of interest
to substitute representative values of the independent variables in the
above equation to see the relative contribution of each to cusp
movement. Typical maximum values for D_{st}, P and AE were -160γ, 8.0 x
10^{-8} dyne cm^{-2} and 1100γ, resulting in $\Delta\theta$ contributions of 1.7°, 3.0°,
and 8.8° respectively, for a total of 13.5°. The AE coefficient
corresponds to 125 γ per 1° of latitudinal movement; this is less than
the 150γ per 1° derived by Eather et. al. (1979) from non-magnetic
storm data, and is not surprising in view of the saturation effect on
AE at higher activity levels, as discussed in Section 4 above.

We believe that this statistical analysis supports our view that
the position of the dayside cusp is most closely related to AE, and
shows little dependence on B_z. We reiterate that apparent dependence
on B_z in longer time average data results only from the AE associated
with the increased substorm probability at these times.

7. CONCLUSIONS:

The analysis presented in this paper was undertaken to refute
recent claims of B_z control of the position of dayside aurora and the
polar cusp. It is demonstrated that these data in fact support a much
closer association with the substorm process, as paramaterized by the
AE index. Given the limitations associated with the AE index (viz.
unreliability in some time sectors during very quiet conditions, in
other time sectors during active conditions, and overall for
longitudinally isolated substorms) the correlations are surprisingly
detailed. In fact we have studied so many examples of detailed
correlations (Eather et. al. 1979 and this paper) that we tend to
suspect cases of poor correlation as resulting from one of the
above-mentioned limitations inherent in AE.

It seems difficult to lay to rest the various claims over the
years of an intimate dependence of dayside cusp position on B_z. We
reiterate once more our belief that there is essentially zero
correlation on short time scales, and that increasing apparent
correlation with longer-time averaged data results simply from the
increased probability of substorm occurrence. We believe that the good
AE correlation on all time scales indicates that cusp position is being
largely controlled by internal magnetospheric processes, probably
involving reconfigurations of the large-scale substorm current system
(Eather et. al., 1979). Recent theoretical calculations (Goldstein,
1983) indicate that the associated magnetic fields are only 10% of
those required, casting doubt on our suggested model, but it is not
clear that these current systems are well enough understood to allow
reliable quantitative modelling. However, we do acknowledge that
these data in no way preclude the possibility of a different mechanism
that drives both dayside auroral position and substorm current
intensity in a coherent manner.

Table 1

STATISTICAL ANALYSIS SUMMARY

	All Data									Data Subset					

A. Correlation Coefficients:

	Θ	Dst	P	B_z	AU	AL	AE		Θ	Dst	P	B_z	AU	AL	AE
Θ	1.00	.57	−.31	.61	−.57	.70	−.75	Θ	1.00	.58	−.46	.64	−.69	.76	−.83
Dst		1.00	.05	.30	−.26	.61	−.56	Dst		1.00	−.04	.29	−.31	.64	−.59
P			1.00	.05	.38	−.14	.25	P			1.00	.04	.38	−.19	.29
B_z				1.00	−.36	.58	−.58	B_z				1.00	−.48	.65	−.67
AU					1.00	−.46	.73	AU					1.00	−.52	.76
AL						1.00	−.94	AL						1.00	−.95
AE							1.00	AE							1.00

B. Linear Regressions:

$$\Theta_E = (.047 \pm .007)\, Dst + 74.5$$
$$= (-0.719 \pm .210)\, P + 73.8$$
$$= (.318 \pm .052)\, B_z + 72.7$$
$$= (-.020 \pm .003)\, AU + 74.8$$
$$= (.012 \pm .001)\, AL + 75.3$$
$$= (-.011 \pm .001)\, AE + 76.7$$

C. Multiple Linear Regressions:

$\Theta_E = .016 D_{st} + .173 B_z - 0.71P - .0049AE + 78.5$
Average Residual = 1.51

$\Theta_E = .032 D_{st} - .315 B_z - 0.77P + 78.2$
Average Residual = 1.59

$\Theta_E = .008 D_{st} - 0.54P - 0.010AE + 79.1$
Average Residual = 1.45

$\Theta_E = .005 D_{st} - 0.010AE + 77.4$
Average Residual = 1.77

$\Theta_E = .014 D_{st} + .120 B_z - 0.76P - .0058AE + 78.9$
Average Residual = 1.08

$\Theta_E = .029 D_{st} + .298 B_z - 0.76P + 78.1$
Average Residual = 1.21

$\Theta_E = .005 D_{st} - 0.44P - 0.008AE + 79.0$
Average Residual = 1.04

$\Theta_E = .010 D_{st} - 0.010AE + 77.7$
Average Residual = 1.51

ACKNOWLEDGEMENTS:

    AE indices were provided by World Data Center C2 for geomagnetism,
Kyoto University. ISEE-3 magnetic data was provided by Dr. C. Russell.
We would like to thank Martha Kane (1980) and Bill Gail (1981) for
operation of the keogram camera at South Pole station during the long
austral winters. This research was supported by NSF grants DPP-8215312
and ATM-8207974.

REFERENCES

Akasofu, S.-I., Y. Kamide and J.H. Allen, A note on the accuracy of
    auroral electrojet indices, J. Geophys. Res., 88. 5769, 1983
Allen, J.H., Comparison of geomagnetic indices, paper presented at
    Mayaud Symposium XVIIth General Assembly IUGG, Canberra, Australia,
    December, 1979.
Eather, R.H., Dayside auroral dynamics, J. Geophys. Res., 89, 1695,
    1984a.
Eather, R.H., Polar cusp dynamics, presented to J. Geophys. Res.,
    1984b.
Eather, R.H., S.B. Mende and E.J. Weber, Dayside aurora and relevance
    to substorm current systems and dayside merging. J.Geophys. Res.,
    84, 3339, 1979.
Goldstein, B.E. and D.B. Beard, Configuration of the dayside
    magnetopause; effects due to the plasma sheet and Birkeland
    currents and relevance to dayside erosion during substorms,
    submitted to J. Geophys. Res., 1983.
Kamide, Y. and S.-I. Akasofu, Notes on the auroral electrojet indices,
    J. Geophys. Res., 88, 1647, 1983.
Meng, C.-I., Case studies of the storm time variation of the polar
    cusp, J. Geophys. Res., 88, 137, 1983.
Meng, C.-I., Dynamic variation of the auroral oval during intense
    magnetic storms, J. Geophys. Res., 89, 227, 1984.
Sandholt, P.E., A. Egeland, C.S. Deehr, G.G. Sivjee and G.J. Romick,
    Effects of interplanetary magnetic field and magnetospheric
    substorm variations on the dayside aurora, Planet. Space Sci., 31,
    1345, 1983.
Siscoe, G.L., A D_{st} contribution to the equatorward shift of the
    aurora, Planet. Space Sci., 27, 997, 1979.
Starkov, G.V., Analytical representation of the equatorial boundary of
    the oval auroral zone, Geomag. Aeron., 9, 614, 1969.
Whalen, J.A., Auroral oval plotter and nomograph, Report No. AFGRL
    70-0422, Air Force Geophys. Res. Labs., Bedford, MA., 1970

LARGE - AND SMALL - SCALE DYNAMICS OF THE POLAR CUSP REGION

P.E. Sandholt, A. Egeland, J.A. Holtet,
B. Lybekk, K. Svenes, and S. Åsheim
Institute of Physics
University of Oslo
0316 Oslo 3, Norway

ABSTRACT. Photometric observations from two stations on Svalbard, Norway, have been used to map the location and dynamics of polar cusp auroras. Cases showing the behaviour of cusp auroras and the local magnetic field related to changes in the interplanetary magnetic field (IMF) and irregularities in the solar wind plasma are presented. Dynamical phenomena with different time scales are studied. South- and northward expansions of the midday sector of the auroral oval are connected with south- and northward turnings of the IMF, measured just outside the bow-shock (ISEE-1 data). No direct relationship to substorm activity was observed. Intensifications and rapid motions of discrete auroral structures in the cusp region are shown to be associated with local Pi-type magnetic pulsations, each event lasting a few minutes. These small-scale dynamical phenomena are discussed in relation to the concept of impulsive penetration of plasma irregularities across the dayside magnetopause, from the magnetosheath to the polar cusp region of the magnetosphere.

1.   INTRODUCTION

During the last decade new information has been collected on the electrodynamic coupling between the solar wind and the dayside polar ionosphere. Of particular relevance to this study are the ionospheric effects of variations in the interplanetary magnetic field (IMF) and plasma irregularities in the solar wind. Friis-Christensen and Wilhjelm (1975) studied the relationship between geomagnetic disturbances on the ground and the IMF. The DPY mode of the disturbance field, related to IMF B_y, was recognized. Satellite-borne magnetometers have been used to map the overall field-aligned current system of the cusp region (Sugiura and Potemra 1976; Iijima and Potemra 1976). Iijima and Potemra reported field-aligned current sheets poleward of the region 1 currents, called cusp region field-aligned currents. McDiarmid et al. (1978) found a close connection between IMF B_y and the tangential magnetic deviation when the satellite crossed the oval in the midday sector. Wilhjelm et al. (1978) investigated the relation-

163

ship between ionospheric DPY current and field-aligned current above the cusp ionosphere (cf. also Bythrow et al. 1982; Rich and Kamide 1983).

D'Angelo (1980) considered the cusp field-aligned currents as extensions of Chapman-Ferraro magnetopause currents flowing parallel to the earth's magnetic field lines. Based on this assumption Primdahl and Spangslev (1981) derived a quantitative expression for the DPY component of the ground magnetic perturbation in terms of IMF B_Y and ionospheric conductivities.

The relationship between expansions and contractions of the dayside part of the auroral oval and interplanetary magnetic field as well as storm and substorm variations have been the topic of several studies (e.g. Burch 1973; Kamide et al. 1976; Horwitz and Akasofu 1977; Eather et al. 1979; Meng 1983; Sandholt et al. 1983). According to Eather et al. all substorms result in equatorward shifts of the dayside aurora. No equatorward motion without simultaneous substorm activity was observed. In Sandholt et al. counter-examples to these statements were presented, indicating the importance of the IMF B_Z component.

The transport of plasma across the dayside magnetopause, from the magnetosheath to the distant polar cusp, has been investigated by means of in situ measurements (e.g. Paschmann et al. 1976; Eastman and Hones 1979; Russell and Elphic 1979; Paschmann et al. 1979; Lundin 1984 and 1985) and theoretical modelling (e.g. Lemaire et al. 1979; Sonnerup et al. 1981; Heikkila 1982; Lundin 1984 and 1985).

This paper presents two cases of midday cusp auroral dynamics in relation to changes in the solar wind. Attention is focused on small-scale dynamics of individual auroral forms in the midday cusp as well as more large-scale behaviour of the oval in the magnetic midday sector. The case discussed in Sect. 3.1 illustrates the response of the optical aurora and the geomagnetic field in the cusp when a compression in the solar wind impinges on the earth's magnetosphere. The effect of the IMF B_Y-component on the geomagnetic signature at different stations close to the cusp aurora is shown. In Sect. 3.2 we describe the response of the midday part of the oval to distinct south- and northward turnings of the interplanetary magnetic field. Section 5 includes a discussion of certain small-scale auroral dynamics as possible effects of impulsive plasma transport across the dayside magnetopause.

2.   OPTICAL AND MAGNETIC OBSERVATIONS

The auroral observations were made by two sets of photometers operated in the meridian-scanning mode from Ny Ålesund (NYÅ) and Longyearbyen (LYR), Svalbard, Norway (geomagnetic coordinates: 75.4, 131.4 (NYÅ); 74.4, 130,9 (LYR), geographic coordinates: 79.0 N, 12.0 E (NYÅ); 78.2 N, 15.7 E (LYR)). The scan period (horizon to horizon) was 12 s. The midday aurora was therefore observed within the range $69-80^\circ$ geom. latitude. Local magnetic noon at the recording sites corresponds to ~ 0830 UT.

ALL-sky cameras operated at both stations provided supplementary information relative to the meridian profiles recorded by the photo-meters. Geomagnetic disturbances were detected by normal and pulsa-tion magnetometers at Ny Ålesund and normal magnetometer at Hornsund (geomagnetic coordinates: 73.5, 127.8) The interplanetary magnetic field data were obtained from the satellites ISEE-1 and 3. The orbit of ISEE-1 had an apogee of 23 earth radii and a perigee of 280 km with a period of ~ 58 hours. ISEE 3 was moving in a halo orbit about the libration point ~ 250 earth radii upstream of the earth. Since the ISEE-3 data cannot fill the need for accurate representation of the IMF immediately upstream of the earth - a key parameter to the present study of the dynamics of the polar cusp aurora - the ISEE-1 data are of primary importance (cf. Russell et al. 1980; Crooker et al. 1982).

3.   DATA PRESENTATION

3.1 Auroral oval dynamics following the sudden commence-ment on Nov. 30, 1979.

A.   Magnetic signatures

A discontinuity in the interplanetary medium characterized by a factor 3 increase of the magnetic field, was detected from ISEE-3 at ~ 0645 UT and reached spacecraft ISEE-1, just outside the bow-shock, at ~ 0735 UT (cf. Fig. 1). The IMF compression was followed by a geomagnetic sudden commencement (SC) at ~ 0739 UT. Figure 1 shows the signature of the SC at the ground stations Ny Ålesund, Hornsund and Tromsö (67.1° geom. lat.).

This SC was detected by spacecraft MAGSAT on its southbound orbit above the South Atlantic Ocean. The global ionospheric current system for the preliminary impulse of the SC is discussed in Araki et al. (1983) based on combined MAGSAT and ground data. Magnetograms from auroral stations near the midnight meridian show that the SC occurred just after the beginning of the expansion phase of a sub-storm (Araki et al.). Another phenomenon to notice in Fig. 1 is the clear modulation of the magnetic field of the cusp region by the IMF B_Y-component.

The relative amplitude of disturbance in the H-component at the two stations Ny Ålesund and Hornsund changes as the aurora moves in latitude between these stations. Between 08 and 10 UT the largest response ($\Delta H/IMF$-B_Y) is seen at Hornsund, when the center of the aurora is close to that latitude. Between 10 and 12 UT the belt of luminosity extends to higher latitudes. This change in auroral exten-sion is accompanied by an increase in the ratio $\Delta H/IMF$-B_Y at Ny Ålesund. A typical value of this ratio when the aurora is nearly overhead is 5.

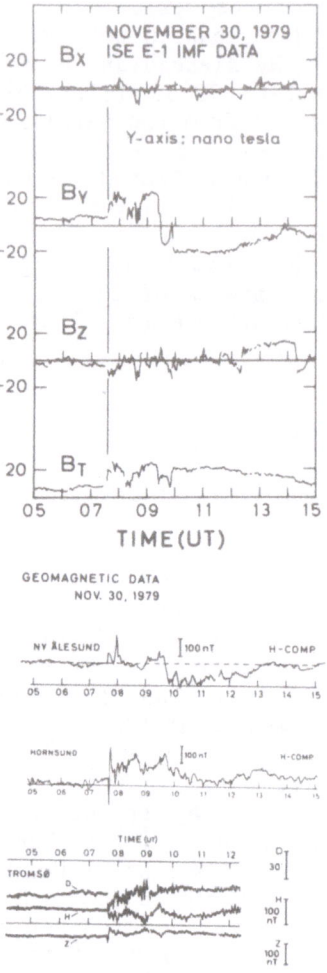

Fig. 1   Upper panel: ISEE-1 magnetic field data from outside the
                            bow-shock.
           Lower panel: Geomagnetic data from ground stations on
                            Svalbard (Ny Ålesund, Hornsund) and from
                            Tromsö (Northern Norway).

B.   Optical signatures
      The first signature in the midday cusp aurora following this
sudden commencement was enhanced intensity of the emissions at
4278 Å, 5577 Å and 6300 Å at 0740 UT (cf. Fig. 2). The maximum
intensity of the dominating emission at 6300 Å increased from below 1
to ~ 5 kR. During the next 15 min, from 0741 to 0756 UT, the aurora
was very stable in intensity, form and location.
      The main intensification during the whole midday period occured
at 0757 UT, when a bright red corona of visible intensity was switch-

ed on (cf. Fig. 2). During the following 5 min. (0757 - 0802 UT) the bright form moved poleward with an average drift speed of ~ 750 m/s. A maximum intensity of 40 kR in the red line was reached at 0802 UT while the green line (not shown) was generally decreasing in intensity after the initial intensification at 0757 UT, with a maximum of 2 kR. This trend was interrupted by two minor transient enhancements at 0759 and 0802 UT. Typical intensities of the midday cusp aurora at 6300 Å are a few kilo Rayleighs (kR), with the ratio I6300Å/I5577Å ~ 2 (cf. Deehr et al. 1980).

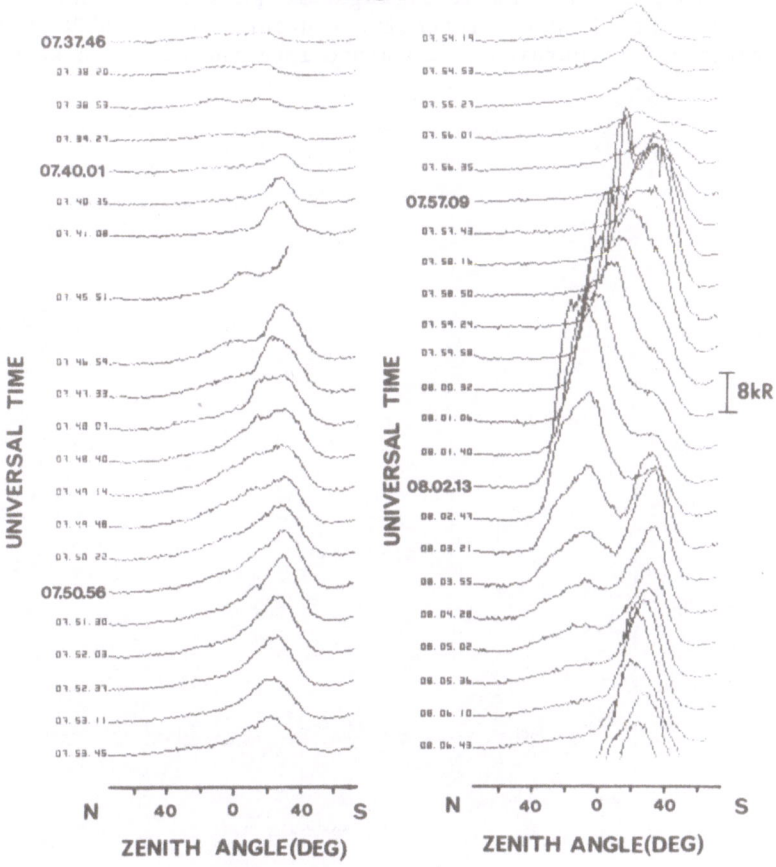

Fig. 2   Meridian scanning photometer recordings of the red oxygen line at 6300 Å, illustrating changes in intensity and latitudinal location of the midday cusp aurora from 0737 to 0806 UT on Nov. 30, 1979.

Inspection of all sky camera pictures from Longyearbyen shows that the auroral intensification moved through the arc from east to west - appearing in the south-east and ending in the north-west. The north-westward motion is consistent with the convection flow pattern in the throat region for positive IMF B_y (cf. Heelis 1984). Similar intensifications and associated poleward expansions as that reported between 0757 and 0802 UT, although not that bright, were observed during the following intervals: 0812-0817 UT, 0820-0824 UT, 0859-0904 UT, 0910-0916 UT, 0921-0926 UT, 0928-0934 UT. Notice that five minutes is a typical time for each luminosity burst.

C.   Relationship between small-scale auroral
     dynamics and magnetic pulsations
     Figure 3 illustrates the connection between dynamical behaviour of the cusp aurora and local magnetic pulsations during this event. The upper panel shows intensity contours of the 6300 Å emission in elevation-time coordinates obtained from the meridian scanning photo-

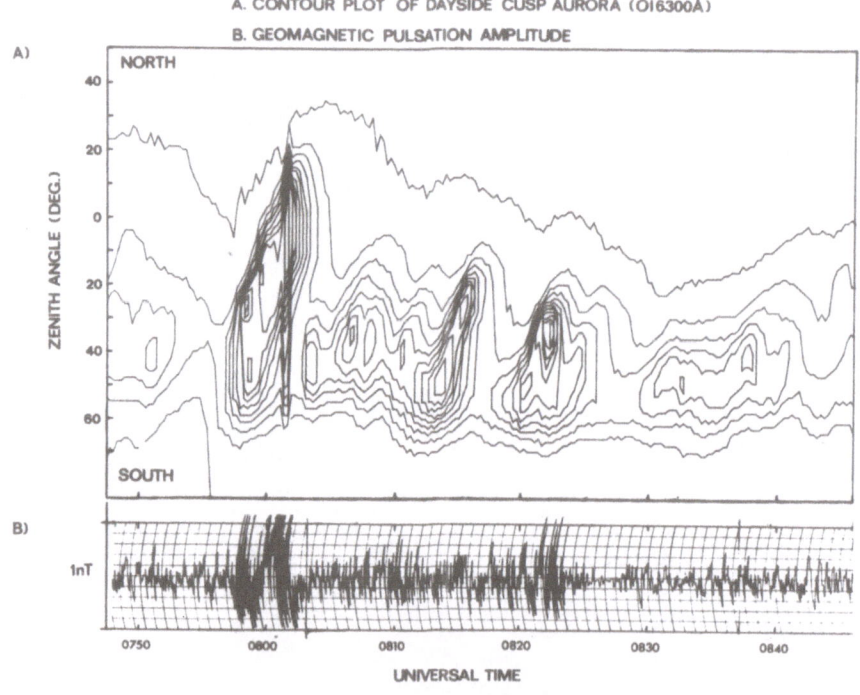

Fig. 3   Panel A: Midday cusp auroral luminosity (red oxygen line)
          in zenith angle versus time coordinates. Intensity levels
          are in steps of 2 kR starting at 1 kR.
          Panel B: Magnetic Y-component pulsation amplitude. Full
          scale is 1 nanotesla.

meter (MSP). We recall the strong intensification and the associated latitudinal expansion starting at 0757 UT (cf. the preceding section). Two other brightenings/expansions, in the intervals 0812-0817 and 0820-0824 UT, are also shown. In the lower panel are plotted the recorded pulsations in the Y-component of the local magnetic field (the X-component is missing). The pulsation activity during this period started with a transient impulse at the time of the SC (0739 UT). From 0740 UT the average level of pulsation activity was well above the pre-existing activity for several hours.

A close correlation between the auroral intensity and the amplitude of the local Pi-type pulsations is recognized. These short periodic pulsations are superposed on more slow variations. Pronounced events of associated auroral brightening/latitudinal expansions and geomagnetic pulsation activity were also observed during the time intervals 0910-0916 UT and 0928-0934 UT (not shown). They were local events associated with bursts of auroral luminosity (particle precipitation) in the cusp.

3.2 Large-scale auroral oval dynamics

The contour-plot in the lower panel of Fig. 4 is a three-dimensional representation of the red oxygen emission (6300 Å) above Svalbard on Jan 16, 1981. From ~ 0710 UT we observe a significant thinning and equatorward shift of the oval. This is identified as the response to a significant transition of the IMF from northward to southward orientation occurring slightly before 07 UT (see ISEE-1 IMF data in upper frame). A sharp transition back to northward IMF orientation was recorded at ~ 0750 UT. This was followed by a northward contraction of the auroral belt near 08 UT. The location of the aurora was mainly unchanged between 0830 and 1010 UT. Then it again shifted southward. This was preceded by a distinct southward change of IMF direction at ~ 0945 UT. At 1020 UT IMF B_z was almost back to zero, before another southward transition took place. The oval location was unchanged between 1015 and 1035 UT, then it moved steadily southward again. This southward expansion was associated with brightening and thinning of the oval aurora.

In Fig. 4 we have marked the time of onset of two magnetic substorms recorded on the nightside. According to Boulder substorm log onsets occurred at 0745 and at 1050 UT. At these times we do not observe any distinct response in the dayside aurora.

JAN.16,1981

1. INTERPLANETARY MAGNETIC FIELD (Z-COMP.)

2. CONTOUR PLOT OF DAYSIDE CUSP AURORA

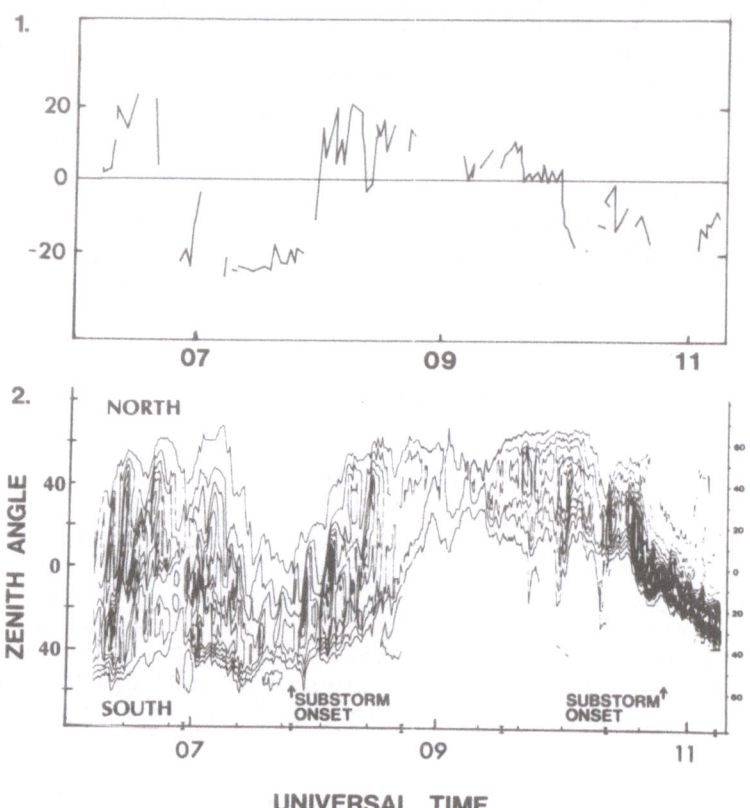

UNIVERSAL TIME

Fig. 4:   Upper panel: IMF B_z component measured from spacecraft
                          ISEE-1. The vertical scale is in nanotesla.
           Lower panel: Contour plot of dayside aurora (6300 Å)
                          observed at Longyearbyen, Svalbard, on Jan.
                          16, 1981. Intensity levels are in steps of 1
                          kR starting at 3 kR. Onsets of magnetic
                          substorms are indicated.

4.   SUMMARY

i)   The Nov. 30, 1979 event
     An amplification of the interplanetary magnetic field by a
factor 3 was followed by Pi-type magnetic pulsations and enhanced
auroral luminosity in the midday cusp region. These enhanced levels

of activity persisted for some hours. Short-lived ($\sim$ 5 min) bursts of auroral brightening/latitudinal expansions were accompanied by local intensifications of the Pi-activity. Magnetograms from local ground stations in the cusp region show a clear DPY component of the geomagnetic disturbance field, which is often reported in the summer hemisphere (cf. Troshichev 1982). The source of such a signature is reported to be a zonal Hall current in the cusp ionosphere (cf. Friis-Christensen and Wilhjelm 1975; Primdahl et al. 1979).

The observed optical spectral ratio (I6300Å/I5577Å) during the events of luminosity burst and enhanced Pi-activity - ranging between 10 and 40 - indicates a soft particle spectrum. Thus, the main part of the particle flux does not affect the state of E-layer ionization. However, the increase of the ratio $\Delta H/IMF-B_Y$ when the aurora comes close to the observing station indicates that the high-energy tail of the particle spectrum has some effect on the E-layer conductivity. Notice the positive spike in the Ny Ålesund magnetogram at $\sim$ 0800 UT when the bright red aurora expanded up to the station. Increased response in the local magnetic disturbance at Ny Ålesund was observed from $\sim$ 10 UT onwards (Fig. 1), when the luminosity covered a larger latitudinal extension (not shown), including both Ny Ålesund and Hornsund latitudes.

Precipitating particles were measured on spacecraft NOAA-6 which crossed the Svalbard meridian (photometer scanning direction) at 0820 UT on Nov. 30, within the auroral belt (1 degree south of luminosity maximum). The energy flux of larger than 300 eV particles, precipitating into the atmosphere, was $\sim$ 1 erg cm^{-2} sec^{-1} with the characteristic energy (E_0) of the whole spectrum less than 300 eV (D.Evans, personal communication 1983).

ii) The Jan. 16, 1981 event

Northward (southward) transitions of IMF was followed within 10 to 25 min (ref. ISEE-1 recordings outside the bow-shock) by poleward (equatorward) shifts of the auroral oval in the midday sector (cf. Fig. 4). The onset of geomagnetic substorms was not observed to have any clear effect on the midday aurora.

5.   DISCUSSION

In this report we have presented two events illustrating the ground observations of magnetic and optical dynamical phenomena in the polar cusp region. From these data and those presented in Sandholt et al. (1983) we conclude that IMF B_z is a major influence on the large-scale dynamic motion of the midday aurora (cf. also Meng 1983). We note that the ISEE-3 IMF data should be used with caution in this type of study, because the precise timing of the arrival at the earth of the observed IMF structure is a serious problem. The time delay between the ISEE-3 location and the earth is shown to be highly variable (Russell et al. 1980; Crooker et al. 1982). Thus our event study is restricted to periods when spacecraft ISEE-1 was outside the bow-shock.

Several events showing no relationship between substorm activity and dayside auroral dynamics are observed (Sandholt et al. 1983). Thus, not all substorms have an effect on the dayside aurora. This does not mean that IMF B_z necessarily is the only influence. In fact, cases showing a close relationship between substorms and latitudinal movements of the dayside aurora are reported by Eather et al. (1979). According to one model auroral oval expansions and contractions are considered to be the net result of dayside merging and nightside re-connection (cf. Akasofu 1977 p. 210).

Our observations seem to be in favour of the view that the energy supply from the solar wind gives rise to two different modes of energy dissipation in the magnetosphere and upper atmosphere, one directly driven by solar wind-magnetosphere dynamo mechanisms and another more explosive release of energy stored in the magnetotail (unloading process), constituting the expansion phase of the substorm (cf. Nishida 1983). Manifestations of the driven process are coherent modulations of region 1 field-aligned current, the DP2 component of the geomagnetic disturbance field and expansion/contraction of the auroral oval (Sandholt and Åsheim 1984).

In the following we will discuss the more small-scale auroral intensifications in the cusp region (Figs. 2,3) in relation to the phenomenon of impulsive penetration of magnetosheath plasma irregu-larities into the magnetosphere, reported most recently by Lundin (1984 and 1985) cf. also Lemaire et al. 1979). Particle observations in the low-altitude cusp region (satellite- and rocket-data) consistent with the idea of impulsive entry from the magnetosheath are reported by Maynard and Johnstone (1974) and Torbert and Carlson (1976). Hoffman (1972) was able to identify considerable temporal-spatial structures within the soft zone of precipitation at dayside high latitude. He called it the zone of bursts. The auroral bursts repor-ted here might be the optical signature of the same phenomenon.

Lundin proposed an MHD model for the observed plasma penetra-tion, generating a local E-field in the magnetospheric boundary layer, powering an electric current system transferring energy to the ionosphere. An estimate of the range of lifetimes for a penetrating plasma element, representing the order of magnitude of the decay time of the energy transfer process, as observed in the dayside high-latitude ionosphere, was given. Typical values were found to be a few minutes, corresponding to an ionospheric conductivity of the order of 1 mho, which is a representative value for the dayside high latitude winter hemisphere (Spiro et al. 1982). The inferred lifetimes compare well with the duration of the observed optical intensifications in the cusp region.

According to Lemaire et al. the "penetration and capture of solar wind irregularities in the cleft regions is possible for almost any orientation of the interplanetary magnetic field direction". The same lack of IMF dependence also holds for our observations of cusp auroral bursts.

Phenomena with some similarity to the impulsive plasma penetrations discussed above are the so-called flux transfer events reported by Russell and Elphic (1979) (cf. also Rijnbeek et al. 1984). As the impulsive injection model was first suggested for interactions of the solar wind with a closed magnetosphere, the flux transfer events, which are most frequently observed when IMF-B_Z is negative, are explained in terms of a transient reconnection process. Heikkila (1982) has discussed impulsive injections in relation to an open magnetic field geometry, indicating that the magnetopause reconnection (flux transfer events) and impulsive plasma injections are phenomena which may coexist and interact (Burch 1983).

Finally, we give some more comments on the signatures which were seen in the local magnetic field on Nov. 30, 1979 (Fig. 1).

The reported relationship between DPY ionospheric current and field-aligned current in the dayside cusp (e.g. Wilhjelm et al., 1978) suggests that the DPY disturbance (cf. Fig. 1) is the ground signature of a three-dimensional current system related to the magnetospheric cusp, like that proposed by D'Angelo (1980).

Heacock and Chao (1980) could explain their observations of Pi activity in the cusp region by fluctuations in a three-dimensional current system driven by the magnetospheric electric field at times of enhanced plasma convection. They noticed larger Pi-amplitudes at locations where auroral precipitation increased the E-region conductivity. This was explained in terms of enhanced response of the current loop with lowered ionospheric resistance.

For the case studied we can estimate the upper limits of the conductivity ratio Σ_H/Σ_P and the response function $\Delta B_{DPY}/IMF\text{-}B_Y$ from the information on the spectrum of precipitating particles (cf. Primdahl and Spangslev 1981). According to Spiro et al. (1982) $\Sigma_H/\Sigma_P = (E_0/1\text{ keV})^{5/8}$. Thus, in our case $\Sigma_H/\Sigma_P < 0.5$. The relatively weak E-layer ionization may give rise to some Hall current which can explain the DPY signature we observe on the ground. The model proposed by Primdahl and Spangslev, applied to our geometry (location and width of Hall current sheet in the cusp ionosphere), gives

$$\Delta B_{DPY}/IMF\text{-}B_Y \sim 15 \cdot (\Sigma_H/\Sigma_P) < 8.$$

On Nov 30, 1979, a typical value for this ratio was 5 when the cusp aurora was overhead (cf. Fig. 1).

ACKNOWLEDGEMENT

We express our thanks to Drs. V.A. Troitskaya (Institute of Physics of the Earth, Moscow, USSR) and C.T. Russel (University of California, Los Angeles, USA) for providing magnetometer data from ground-station Ny-Ålesund (pulsation magnetometer) and spacecraft ISEE-1, respectively. NOAA-6 data from Dr. D. Evans (National Oceanic and Atmospheric Administration, Boulder, USA) and Longyearbyen all-sky pictures provided by Dr. C.S. Deehr (University of Alaska) have been very useful.

References
Akasofu, S.-I. 1977, Physics of magnetospheric substorms.
        D. Reidel Publishing Company.
Araki, T., Iyemore, T., Tsunomura, S., Kamei, T., and Maeda, H. 1983,
        Geophys. Res. Lett., 9, 341.
Burch, J.L. 1973, Radio Sci., 8, 955.
Burch, J.L. 1983, In U.S.National report to International Union of
        Geodesy and Geophysics 1979-1982, p. 463. American Geo-
        physical Union.
Bythrow, P.F., Potemra, T.A., and Hoffman, R.A. 1982, J. Geophys.
        Res., 87, 5131.
Crooker, N.U., Siscoe, G.L., Russell, C.T., and Smith, E.J. 1982,
        J. Geophys. Res. 87, 2224.
D'Angelo, N. 1980, Ann. Geophys. 36, 31.
Deehr, C.S., Sivjee, G.G., Egeland, A., Henriksen, K., Sandholt,P.E.,
        Smith, R., Sweeney, P., Duncan, C., and Gilmer, J. 1980,
        J. Geophys. Res. 85, 2185.
Eastman, T.E. and Hones, E.W. 1979, J. Geophys. Res. 84, 2019.
Eather, R.H., Mende, S.B., and Weber, E.J. 1979, J.Geophys. Res.,
        84, 3339.
Friis-Christensen, E., and Wilhjelm, J. 1975, J. Geophys. Res.,
        80, 1248.
Heacock, R.R., and Chao, J.K. 1980, J. Geophys. Res., 85, 1203.
Heelis, R.A. 1984, J. Geophys. Res. 89, 2873.
Heikkila, W. 1982, Geophys. Res. Lett., 9, 159.
Hoffman, R.A. 1972, In Magnetosphere - Ionosphere Interactions
        (Ed. K. Folkestad), p. 117, Universitetsforlaget, Oslo.
Horwitz, J.L. and Akasofu, S.-I. 1977, J. Geophys. Res., 82, 2722.
Iijima, T. and Potemra, T.A. 1976, J. Geophys. Res., 81, 5971.
Kamide, Y., Burch, J.L., Winningham, J.D., and Akasofu, S.-I. 1976,
        J. Geophys. Res. 81, 698.
Lemaire, J., Rycroft, M.J., and Roth, M. 1979, Planet. Space Sci.,
        27, 47.
Lundin, R. 1984, Planet. Space. Sci., 32, 757.
Lundin, R. 1985, This volume.
Maynard, N.C. and Johnstone, A.D. 1974, J. Geophys. Res., 79, 3111.
Mc.Dairmid, I.B., Burrows, J.R., and Wilson, M.D. 1978, J. Geophys.
        Res. 83, 5753.
Meng, C.I. 1983, J. Geophys. Res. 88, 137.
Nishida, A. 1983, Space Sci. Rev., 34, 185
Paschmann, G., Haerendel, G., Sckopke, N., and Rosenbauer, H. 1976,
        J. Geophys. Res. 81, 2883.
Paschmann, G. Sonnerup, B.U.O., Papamastorakis, I., Sckopke, N.,
        Haerendel, G., Bame, S.J.., Asbridge, J.R., Gosling, J.T.,
        Russell, C.T., and Elphic, R.C. 1979, Nature, 282, 243.
Primdahl, F. and Spangslev, F. 1981, Ann. Geophys., 37, 529.
Primdahl, F., Walker, J.H., Spangslev, F., Olesen, J.K., Fahleson,
        U., and Ungstrup, E. 1979, J. Geophys. Res., 84, 6458.
Rich, F.J., and Kamide, Y. 1983, J. Geophys. Res. 88, 271.
Rijnbeek, R.P., Cowley, S.W.H., Southwood, D.J., and Russel, C.T.
        1984, J. Geophys. Res., 89, 786.

Russel, C.T., and Elphic, R.C. 1979, Geophys. Res. Lett. 6, 33.
Russel, C.T., Siscoe, G.L., and Smith, E.J. 1980, Geophys. Res.
         Lett., 7, 381.
Sandholt. P.E., Egeland, A., Lybekk, B., Deehr, C.S., Sivjee, G.G.,
         and Romick, G.J. 1983, Planet. Space Sci., 31, 1345.
Sandholt, P.E. and Åsheim, S. 1984, Ann. Geophys. (in press).
Sonnerup, B.U.O., Paschmann, G., Papamastorakis, I., Sckopke, N.,
         Haerendel, G., Bame, S.J., Asbridge, J.R., Gosling. J.T.
         and Russel, C.T. 1981, J. Geophys. Res. 86, 10049.
Spiro, R.W., Reiff, P.H., and Maher, L.J. Jr. 1982, J. Geophys. Res.,
         87, 8215.
Sugiura, M., and Potemra, T.A. 1976, J. Geophys. Res., 81, 2155.
Torbert, R.B., and Carlson, C.W. 1976, In Magnetospheric Particles
         and Fields (Ed. B.M. McCormac), p. 47.
Troshichev, O.A. 1982, Space Sci. Rev., 32, 275.
Wilhjelm, J., Friis-Christensen, E., and Potemra, T.A. 1978, J. Geo-
         phys. Res. 83, 5586.

POLAR CUSP FEATURES OBSERVED BY DMSP SATELLITES

C.-I. Meng
The Johns Hopkins University
Applied Physics Laboratory
Laurel, Maryland 20707
USA

M. Candidi
IFSI/CNR
CP 27
00044 Frascati (Roma)
Italy

1. INTRODUCTION

Among various techniques to monitor the polar cusp region, the best available way is by using several low-altitude satellites all on the noon-midnight polar orbit to perform particle precipitation measurements. Fortunately, such an opportunity occurred in the USAF Defense Meteorological Satellite Program (DMSP). Two identical satellites had circular orbits in the noon-midnight meridian at ~ 840 km altitude. Thus the polar cusp regions over both the northern and southern polar regions could be monitored with a temporal resolution of at least once every hour per hemisphere. In this paper, we will summarize recent findings of polar cusp characteristics and dynamics based on DMSP observations.

In the following, we will first establish the identification of the polar cusp region by using low energy (50 eV to 20 keV) electron measurement. Then we will investigate (1) the relation of the solar wind density and the polar cusp electron intensity, (2) the variation of the polar cusp morphology with the interplanetary magnetic field orientation, (3) the dynamics of the polar cusp during intense magnetic storms, and (4) the conjugacy of the polar cusp region.

2. POLAR CUSP IDENTIFICATION

The identification of the polar cusp region is rather subjective since the identification must be based on particular types of particle or field measurements used by individual investigators. The particle signature of the polar cusp is similar to that of the magnetosheath plasma, namely, very soft electron or proton spectra. However, during the entry of particles into the cusp, the ions suffer energy dispersion while the electrons do not, an effect which results in dependences of cusp size and location on particle species and energy. The very soft electrons are more suitable for the identification of the polar cusp region attributed to the fast entry of electrons (Burch, 1972). Indeed in the DMSP auroral electron precipitation data the polar cusp region is rather easy to identify. An example of a polar

177

J. A. Holtet and A. Egeland (eds.), The Polar Cusp, 177–192.
© *1985 by D. Reidel Publishing Company.*

cusp crossing is shown in Figure 1 to illustrate the identification of the polar cusp region. Three electron precipitation parameters are calculated once every second from DMSP measurements of 16 channels covering 50 eV to 20 keV. These parameters are total precipitating electron flux, energy flux, and average electron energy. In the example shown, the cusp region is defined by the extremely intense electron fluxes of $\sim 10^9$ electrons/cm^2-s-sr and the low average energy of < 200 eV (i.e., very soft but intense precipitation region). These criteria are based on the electron characteristics of the polar cusp region observed by ISIS-1 satellite (Heikkila and Winningham, 1971). The polar cusp region is identified from 0206:25 (-78.7° gm lat) to 0207:23 (-75.8° gm lat) for the February 15, 1980 polar crossing approximately along 1030 magnetic local time meridian.

3.   SOLAR WIND AND POLAR CUSP ELECTRON INTENSITY

In this section, we study temporal variations of the intensity of the precipitating low energy electron flux in the polar cusp region and the relations of these variations to changes of the solar wind density and the IMF. Regardless of the particle entry mechanism, the intensity of the precipitating electron flux in the polar cusp has to be related to the solar wind density, if the solar wind is indeed the direct source of the cusp particles. However, their quantitative relation is expected to be influenced by the intensity and the polarity of the IMF components. The following analysis reveals that the intensity of the precipitating electron flux in the polar cusp is related to the solar wind density by a positive dependence; that is, the cusp electron flux increases with increasing solar wind density, but not linearly. Furthermore, the polarity of the IMF B_z component affects the intensity of the precipitating cusp electron flux as well; a southward interplanetary magnetic field orientation appears to be the favorable condition for the transport of electrons from the solar wind into the polar cusp region. There is a factor of ~ 2 increase over the positive B_z condition.

Figure 2 is a time sequence plot of the hourly averaged IMF B_z component from IMP 8, the 5 min solar wind ion density from ISEE 3, and the 47 eV cusp electron flux from the DMSP observations. There is a positive relation between the intensity of the precipitating electron flux in the cusp region and the density of the solar wind ions. The extreme minimum of the solar wind ion density on day 212 is reflected by a depression of the 47 eV cusp electron flux; its recovery early on day 213 corresponds to the recovery of the cusp electron flux. Conversely, a relation between temporal variations of the cusp electron flux and the solar wind density also exists, such as minima on days 212 when the extremely low solar wind density was observed, and maxima on days 208 to 210, when the solar wind density had corresponding extremely high values.

The influence of the IMF B_z component, shown in the top panel of Figure 2 is evident at several times. For instance, the solar wind density exhibits three relative maxima in the late hours of each day from day 207 to day 209, with corresponding cusp electron flux

TABLE 1

MLT eV	6-7	7-8	8-9	9-10	10-11	11-12	12-13	13-14	14-15	15-16	16-17	17-18
50	.75	.86	.56	.52	.42	1.2	12	17	14	17	2.9	3.2
77	.68	.73	.58	.46	.46	1.6	17	22	16	22	4.7	4.3
110	.50	.59	.39	.43	.47	2.3	18	26	22	24	4.1	6.6
183	.49	.41	.34	.34	.46	2.6	15	29	27	27	5.6	12
434	1.3	.25	.37	.18	.47	3.5	17	45	95	32	18	54
1045	2.2	.20	.40	.18	.99	2.7	10	120	77	32	18	206
3790	.61	.98	.52	.40	1.1	2.6	2.3	2.5	5.0	2.8	6.8	29
8990	.86	.81	.71	.70	.91	1.3	1.3	1.5	1.6	1.4	1.4	1.6

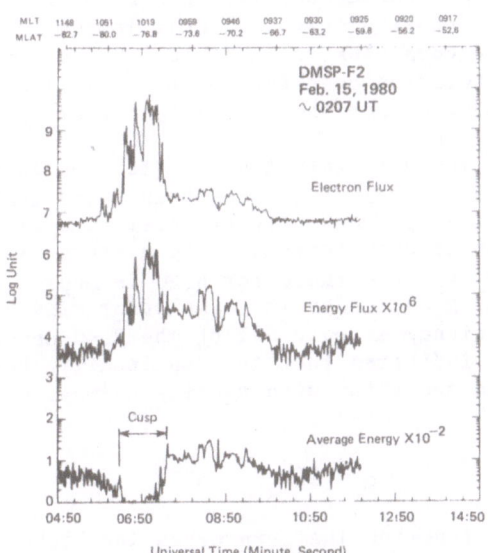

Fig. 1 Typical example of the DMSP precipitating electron observation over the dayside auroral oval. Trajectory in geomagnetic latitude (MLAT) and local time (MLT) is shown on the top.

variations. The three solar wind density maxima are increasingly
higher and longer in duration in these three days; however, the maxima
of the respective cusp fluxes decrease in intensity. This may be
attributed to the IMF effect. The IMF data are available only through
06 UT on day 209; the IMF B_z component is extremely intense during the
second half of day 207 (< -12 nT between 21 and 24 UT) and generally
negative but > -2.1 nT during the second half of day 208. This
slightly negative B_z is possibly responsible for the lower intensity
of the second maximum of the cusp electron flux. Similarly, the cusp
electron fluxes around 0000 UT of day 206 (IMF B_z is generally
negative) are larger than those around 0000 UT of day 207 (IMF B_z
generally positive) even though the solar wind density is not very
different.

The aforementioned qualitative correlations between cusp electrons and solar wind parameters are examined quantitatively as
expressed by the linear regression in Figure 3. Figure 3 illustrates
the correlation between the logarithm of the 47 eV cusp average electron flux of each crossing and the logarithm of the corresponding (1
hour lagged) ISEE 3 solar wind hourly average density. The two
quantities are correlated with large scattering. The 0.60 correlation
coefficient of the regression is not very high, indicating that other·
parameters also influence the observed cusp electron flux and give
rise to the scattering of the data points with respect to the best fit
line. The slope of this best fit line is approximately 0.4, which
indicates a nonlinear dependence; this was already implicit in the
time sequence plots of Figure 2. A linear regression of the actual
values of the 47 eV cusp flux to the actual values of the solar wind
density yields correlation coefficients of 0.46, i.e. significantly
lower than 0.60 obtained for the logarithmic regression, indicating a
nonlinear relationship.

By separating the data into two subsets, one including only B_z>0
cases and the other B_z<0 cases, it has been found that negative B_z is
a favorable condition for the entry of solar wind electrons down to
the low-altitude polar cusp regions. The average ratio of the cusp
electron fluxes for B_z<0 to those for B_z>0 is approximately 1.6,
indicating that for B_z>0, under the same solar wind density conditions
the transport efficiency is only 60% of the B_z<0 conditions.

This analysis indicates that the nonlinearity between the cusp
electron flux ϕ and the solar wind density n implies a quantitative
relation

$$\phi = A(B_z) \quad n^{0.4} \tag{1}$$

where $A(B_z)$ is the function that expresses the higher fluxes associated with the negative B_z. Due to the large scatter of the data the
value of 0.4 for the slope of the best fit curve is to be only taken
as an indication that the dependence of ϕ on n is nonlinear and close
to a $\sqrt{n}$ dependence.

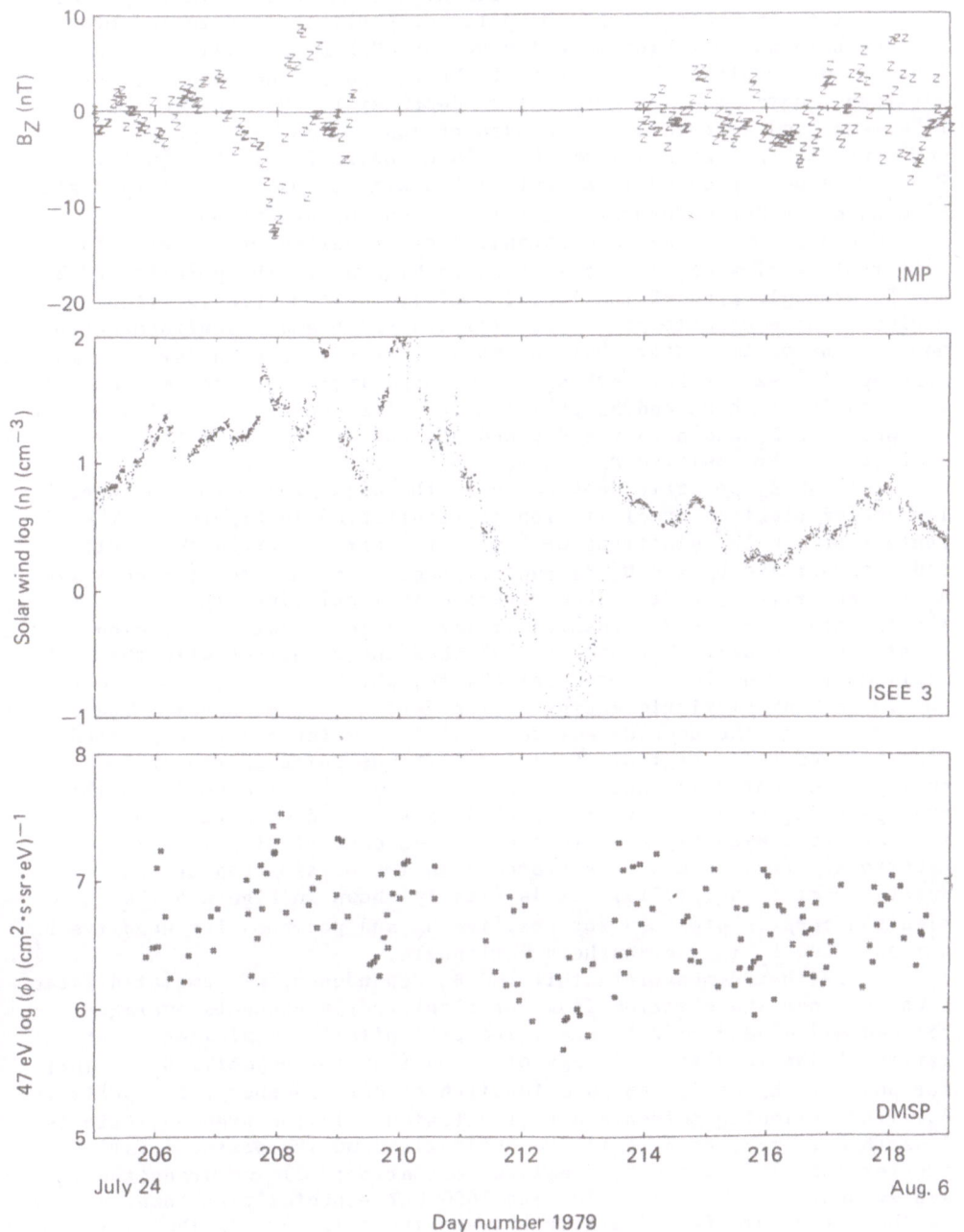

Fig. 2 Temporal variation of the IMF B_z, solar wind density, and the DMSP 47 eV cusp electron flux from July 24, 1979, to August 6, 1979.

4. CUSP ELECTRON DISTRIBUTION AND THE IMF

The purpose of this section is to illustrate the statistical spatial
distribution in the intensity of polar cusp electrons detected over
the southern polar region detected by the DMSP-F2 satellite in two
local summer months. The effect of IMF B_y and B_z components is also
examined. Data from 874 orbits from December 1, 1977 to January 31,
1978 were used to create a data base of four second averages between
-60° and -90° gm lat and from 06 to 18 magnetic local time (MLT).
These data were grouped to a spatial bin size of 0.4° in latitude and
30 minutes in MLT under different IMF B_y and B_z conditions.

       Contours of the average intensity distribution of the electron
differential flux of 50 eV are shown in Figure 4. The polarity of B_y
and B_z strongly affects the location of the maximum precipitation
region. These effects are an equatorward shift and a confinement to
noon of the maximum distribution, when B_z is negative; a larger dawnside spatial extent for both B_y and B_z negative; a prenoon maximum of
the cusp for both B_y and B_z positive; an afternoon maximum of the cusp
for positive B_z and negative B_y; and postnoon equatorward boundary
shift of 3° for positive B_y and positive B_z.

       The IMF B_y polarity dependence in the distribution of intense,
low energy electron precipitation is highlighted in Figure 5. The
contour of 3×10^9 electrons cm^{-2} s^{-1} sr^{-1} keV^{-1}, for 50 eV electrons
and for positive B_y and B_z is superimposed over that for the negative
B_y and positive B_z case. When reversed in local time, the shape of
the two contours are remarkably similar. A lower latitude region
exists between 08-16 MLT and 75°-79° MLAT and coincides with the
maximum electron flux recorded at 434 eV, which can be assumed to be
the lower characteristic energy of the dayside plasma sheet. Thus,
this region is the dayside auroral oval into which are precipitated
electrons of all energies. A higher latitude bulge exists of low-
energy electrons only, and this region above 79° corresponds to the
precipitating cusp electrons, which is associated with both an
increase of low-energy electrons and a decrease of higher-energy
electrons. This is also associated with the dayside gap of the
auroral oval (Meng, 1981). It is clearly shown in Figure 5 that the
electron cusp is pre-noon for positive B_y and postnoon for negative B_y
for positive B_z in the Southern Hemisphere.

       To further demonstrate this IMF B_y dependence, we generated Table
1 which shows the electron flux for eight energy channels averaged
between -77.4° and -83.3°, where the precipitation maximizes. The
ratios of the resulting average electron flux for negative B_y to that
for positive B_y are given as a function of MLT and energy for positive
B_z. The following comments are illustrated: 1) The prenoon ratio is
less than 1 implying B_y positive domination and the postnoon ratio is
greater than 1 implying B_y negative domination; 2) two transitions
are evident at 0800, 1100-1200 and 1600 MLT especially at lower energy
as the ratios are fairly stable between these times, 3) the ratios
decrease (prenoon) and increase (afternoon) with increasing energy up
to 434 eV.

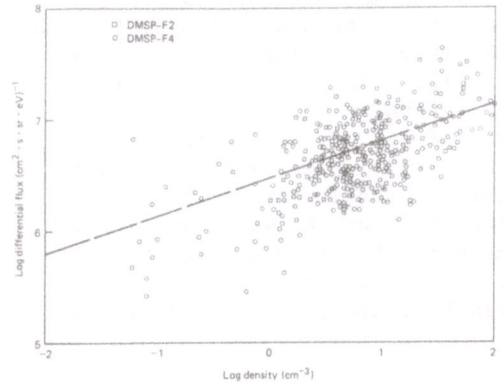

Fig. 3   Scatter plot of the cusp electron
flux logarithm versus the logarithm
of corresponding solar wind density.
The slope (0.4) indicates a non-
linear dependence of the cusp
electron flux on the solar wind
density.

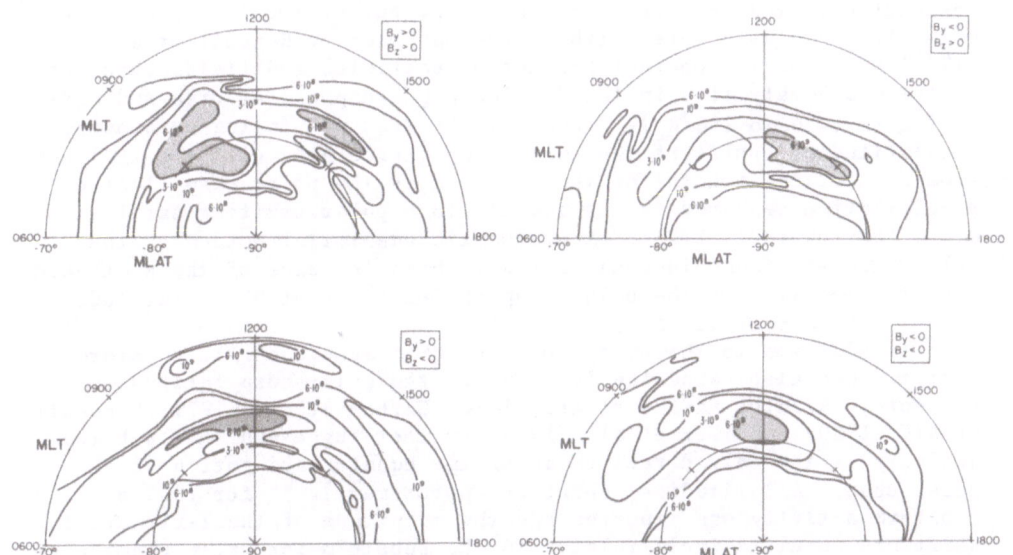

Fig. 4   Contour plots of 50 eV electron flux (electron/cm^2-s-sr-keV) sorted by IMF polarity.

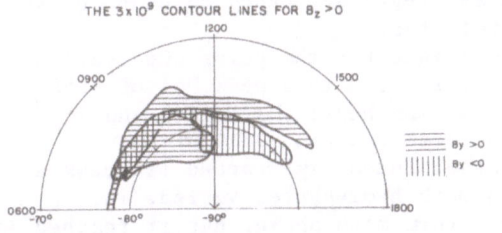

Fig. 5   Regions of maximum precipitating
50 eV electron flux for IMF $B_z > 0$
under different B_y orientations.

The field-aligned current structure of the northern dayside polar regions has been extensively studied. It has been established (Iijima et al., 1978 and Saflekos and Potemra, 1980) that there is an asymmetry between morning and afternoon sectors, in the sense that the upward and downward current systems are unbalanced. The net current in the afternoon sector is upward and the net current in the prenoon sector is downward. Iijima and Potemra (1976) place the peak upward current at 1500 MLT and 75° with a current density of 6×10^{12} electron $m^{-2} s^{-1}$. From the data shown in Figure 4, we derive a peak current density of 8×10^{12} electrons $m^{-2} s^{-1}$ for 50 eV electrons at 10-11 MLT and -79°. Thus it is fair to conclude that the precipitating low-energy cusp electrons observed here are responsible for the upward, cusp field-aligned current.

5. POLAR CUSP DYNAMICS DURING STORMS

The first report of the possible polar cusp latitudinal variations responding to the changes of the B_z component of the interplanetary magnetic field associated with storms was made by Russell et al. (1971) based on the characteristics of particles and fields observed by the OGO-5 satellite in the dayside magnetosphere at high and mid-altitudes (~ 6 to $2.6 R_e$). However, the most convincing evidence of the latitudinal shift of the polar cusp (associated with the IMF) came from statistical studies of the location of the polar cusp electron precipitation measured by the low altitude polar orbiting satellite OGO-4 (Burch, 1972, 1973). A systematic equatorial motion of the polar cusp was found in conjunction with an increase of the southward IMF; the location of the polar cusp varied by about 5° in latitude from IMF $B_z \approx 6$ nT to -6 nT.

In addition to the effect of the IMF, magnetospheric substorm activity can also cause the lowering of the polar cusp latitude (Akasofu, 1972a,b; Kamide et al., 1976; Eather et al., 1979; Sandholt et al., 1980). Eather et al. (1979) further suggested that substorm activity is the only direct cause of the equatorward motion of the polar cusp. A latitudinal shift of approximately 5° for 1000 nT substorm activity was reported and the magnitude of the latitudinal shift was quantitatively related to the substorm intensity (Kamide et al., 1976). Without doubt, the polar cusp location is related to both the magnitude of the IMF north-south component and the intensity of the magnetospheric substorm (i.e., to both external and internal dynamic processes of the magnetosphere).

In this section, the latitudinal variations of the polar cusp during an intense geomagnetic storm are reported, based on consecutive observations from DMSP-F2 and F4 satellites.

The largest geomagnetic storm examined for the polar cusp variation was the storm of September 18-19, 1979, with a peak Dst of -156 nT. Figure 6 illustrates variations of the hourly Dst index and the polar cusp location. The main phase commenced at $\sim$ 0000 UT on September 18, 1979, and the ring current intensity reached its peak at 1600 UT. The polar cusp latitude in both hemispheres varied, in general, with Dst during most of the storm main phase, but it reached

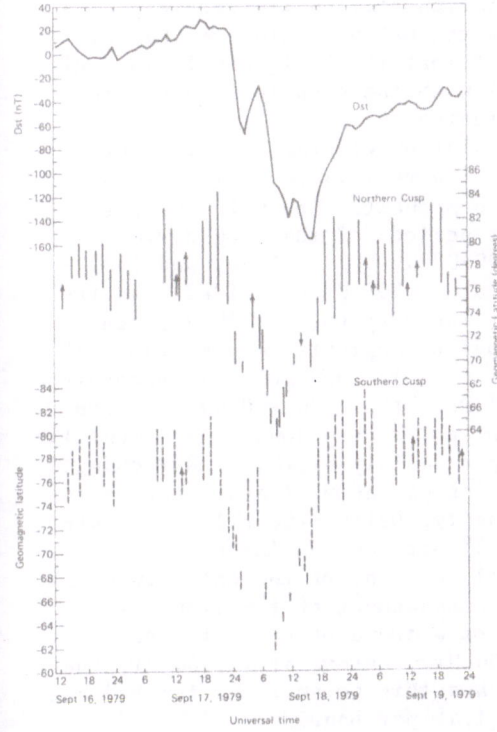

Fig. 6   Temporal variations of polar cusp region during an intense storm detected over both hemispheres.

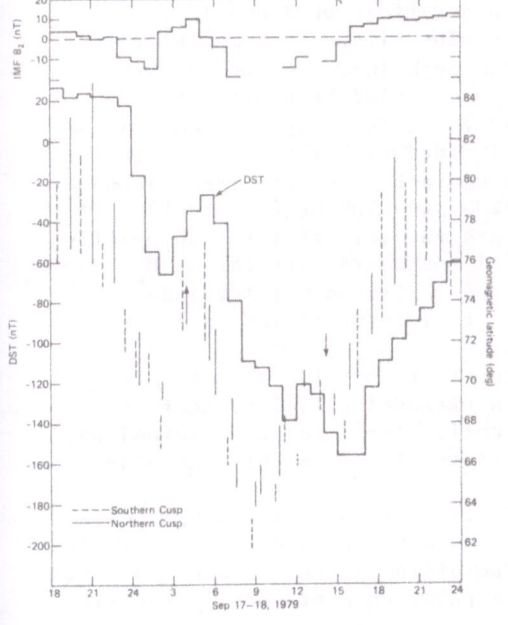

Fig. 7   The high-resolution plot of the temporal variations of the IMF B_z, the DST index and the polar cusp.

the equatormost latitude many hours before the time of the peak ring
current intensity. The polar cusp location started its recovery
toward the normal latitude long before the Dst variation reached its
peak. It is also obvious in this event that the latitudinal width of
the cusp was rather narrow ($\sim 1°$ - $2°$) when the cusp shifted equatorward, similar to the other examples studied.

Figure 7 is the high temporal resolution diagram illustrating
details of the cusp movement together with variations of the ground-
based Dst index and the ISEE 3 IMF B_z component. The main phase of
this storm consisted of three distinct periods of the ring current
growth (i.e., dDst/dt < 0), namely 0000-0300, 0600 - 1100, and 1300 -
1600 UT. The onset of the main phase at $\sim$ 0000 UT coincided with the
initiation of the equatorward shift of the cusp (at $\sim$ 2330 UT), and
within 3 hours the cusp latitude dropped by approximately $8°$ to $\sim 67°$
while Dst value decreased by $\sim$ 82 nT. A rapid cusp poleward recovery
to $73°$ corresponded to the recovery of Dst at $\sim$ 0300 - 0400 UT. The
second equatorward movement brought the polar cusp down to the lowest
latitude position ever observed at approximately $61.7°$, as Dst changed
by $\sim$ 82 nT to -110 nT at $\sim$ 0900 UT (which was about 7 hours before the
ring current reached its maximum intensity, Dst = -156 nT). The rate
of cusp equatorial movement was about $3°$ per hour during both
events. A third significant equatorial movement of the polar cusp was
from 1200 to 1600 UT during the third enhancement of the ring current. This movement was superimposed on a trend of general cusp
poleward recovery which began at $\sim$ 0900 UT. Later, at $\sim$ 1900 UT, the
cusp recovered to its normal latitude and this poleward motion had a
nearly constant rate of approximately $1.4°$ per hour.

The interplanetary magnetic field measurement during this storm
was made also by ISEE 3 at $\sim$ 235 R_e upstream. There were two periods
of extended southward orientation with durations of 3 and 10 hours,
respectively. The Dst index was decreasing (i.e., the main phase of
this storm) during negative B_z, and the peak intensity of the ring
current occurred at the end of the second period of southward IMF
orientation. Comparing the IMF B_z and the polar cusp latitude, it is
noticed that the equatorward shift coincided with the increase of the
$-B_z$ magnitude and the lowest position was achieved when IMF reached
the largest $-B_z$ value, such as near 0200 and 0900 UT ($\sim$ -19 nT). It
is also obvious that during the poleward recovery of the cusp position
(i.e., after 0900 UT) the IMF was still southward with large $-B_z$
component. The cusp completely returned to its normal latitude
of $\sim 76°$ just 3 hours after the northward turning of the interplanetary magnetic field.

The combined DMSP F2 and F4 measurements provide the first continuous polar cusp observation over an extended period so that the
dynamics of the polar cusp during magnetic storms can be examined and
reveal the following morphological features based on this and other
examples.

1. The dynamic latitudinal variations of the polar cusp region
during geomagnetic storms are seen to be very consistent over both the
northern and southern hemispheres. Therefore it can be inferred that
the spatial extent and location of the polar cusp are nearly identical

in the opposite polar region, even though the instantaneous cusp
regions in conjugate hemispheres were not monitored simultaneously.

2. The latitudinal movement of the polar cusp is very large
during intense storms, and the cusp equatorial edge can shift down to
about $62°$ from its normal latitude of $\sim 76°$. Such a large latitudinal
shift of the cusp during storms with a Dst of -150 nT has not been
observed previously. It is also noticed that the equatorward shift of
the polar cusp takes place during the main phase of intense storms
when the value of Dst decreased sharply (i.e., dDst/dt < 0 and large)
in association with the rapid growth of the ring current. But there
is no apparent relationship between the lowest latitude of the cusp
position and the peak of the ring current intensity.

3. The latitudinal extent of the polar cusp region defined by
the present method varies drastically during a storm. It becomes very
narrow, about $1° - 2°$ wide, during the equatorward shift of the cusp
and widens significantly ($> 5°$) after the cusp recovers to its normal
location. A rapid change of the cusp latitudinal extent is also seen;
its width can vary by a factor of 2 - 3 in about 1 hour.

4. The large amplitude equatorward shift of the polar cusp
region invariantly follows the rapid large decrease of the north-south
component of the interplanetary magnetic field, namely, increasing
southward IMF magnitude. However, the poleward recovery takes place
under any IMF B_z conditions (northward turning, constant negative B_z,
or decreasing $-B_z$ mgnitude), and the rate of recovery is not related
to variations of B_z.

5. The lowest latitude of the polar cusp is observed to occur at
(or near) the time of the largest southward component of the interplanetary magnetic field. There is some indication of a relationship
between the equatormost cusp position and the maximum magnitude of $-B_z$
during each storm.

6. The large-amplitude variation of the position of the polar
cusp associated with the geomagnetic storms does not have a close
relation with the variation of magnetospheric substorm activity.

These are the general features of the polar cusp dynamics, but
the detailed variations are different for each storm. The detailed
relationship of the cusp movement with the interplanetary magnetic
field north-south component is perplexing. The equatorward motion of
the cusp is indeed associated with the southward turning of IMF (i.e.,
decreasing the north-south B_z magnitude) during all the storms
examined, and the occurrence of the large displacement equatorward of
the polar cusp region takes place during the period of extended
intense southward interplanetary magnetic field. However, the
poleward movement of the cusp during the recovery does not have a
simple relation with the southward IMF component variation.

6. CONJUGACY OF THE POLAR CUSP

The observation of the cusp regions has heretofore been limited to
only one of the opposing hemispheres at a time and the conjugacy of
the cusp has not been investigated due to the difficulty of making
simultaneous conjugate observations. Simultaneous measurements at the

same geomagnetic local times, such as those reported here, are essential for the investigation of this subject because the local time variation of the cusp characteristics is likely to be a strong one.

On rare occasions the DMSP-F2 and F4 satellites were favorably located to perform simultaneous polar cusp observations at conjugate positions, and this section reports two such extremely rare coincidences. The trajectories for these two cases are illustrated in Figure 8. The conjugate cusp regions were observed less than 6 min apart, within about 1-2 h geomagnetic local time. The precipitating electron parameters for the two conjugate observations on 7 October 1979 and 6 October 1979 are shown in Figures 9 and 10, respectively. The regions marked as "cusp" are characterized by a very high electron flux ($\gtrsim 10^9$ el cm^{-2}s^{-1}sr^{-1}) and very low average energy (< 200 eV).

a. 7 October 1979 Event ($\sim$ 1350 UT)

The conjugate polar cusp detection shown in Figure 9 was made at the time of a sharp rise of AE from a quiet to an enhanced level of 1000 nT. Both satellites were travelling from lower to higher latitudes, on the opposite hemispheres, in the 0800-0900 MLT sector. DMSP-F2 on the North was leading DMSP-F4 on the South by 4 min. They first encountered a region of low intensity (< 10^8 el cm^{-2} s^{-1}sr^{-1}) precipitating electrons, with average energy between 1 and 4 keV, from 69 to 71° corrected geomagnetic latitude (MLAT). This flux is associated with the dayside extension of the plasma sheet. The low energy high intensity polar cusp electrons were detected before the satellites got out of this region; they persisted well after the plasma sheet electrons disappeared at 1350:11 UT on F2 and 1353:50 UT on F4, at 71° MLAT in both hemispheres. The Northern Hemisphere cusp electrons were observed by F2 between 1349:48 UT (70.0° MLAT) and 1351:20 UT (74.3° MLAT), and the Southern ones by F4 between 1353:40 UT (-70.7° MLAT) and 1354:50 UT (-74.3° MLAT). At higher latitudes on the Southern Hemisphere, the cusp electrons dropped sharply for a short time and reappeared shortly afterward, persisting until 1355:20 UT (-75.7° MLAT). After leaving the cusp regions, both satellites detected the low intensity "polar rain" over the polar cap regions. The equatorward and poleward boundaries of the observed cusp in both hemispheres were located at latitudes lower than the usual cusp latitude of $\sim$ 77° MLAT, due to the high level of geomagnetic activity and to the negative IMF B_z component.

b. 6 October 1979 Event ($\sim$ 1510 UT)

This conjugate cusp observation was made also under disturbed conditions while K_p was 6 between 1200 and 1600 UT. The two satellites were moving to the noon sector lower latitudes from higher latitudes near the 1500 MLT meridian, with DMSP-F2 on the Southern Hemisphere leading the F4 northern cusp observation by 6 min. We will limit the discussion of this pass (Figure 10) to the MLT sector between 1500 and noon. The Southern Hemisphere pass shows a clear observation of the polar cap until 1506:10 UT (-72.8° MLAT), followed by the polar cusp

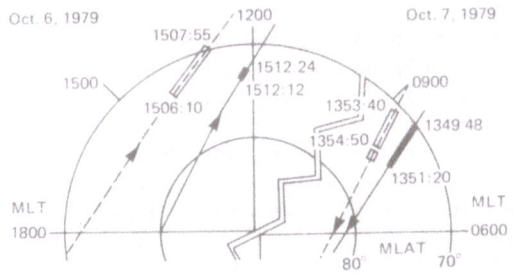

Fig. 8 Orbit plots of two simultaneous
        conjugate cusp observations.
        Dashed lines are southern and
        solid lines are northern polar region
        passes. The cusp extents are shown
        by boxes.

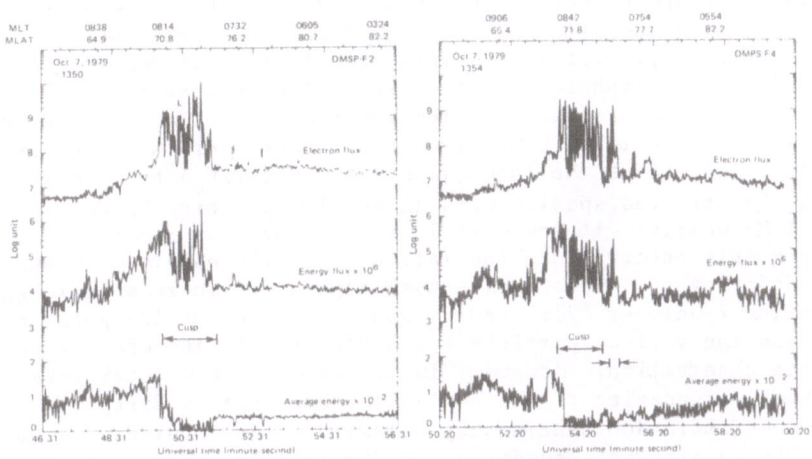

Fig. 9 Precipitating electron parameters for conjugate cusp crossing event on
        October 7, 1979.

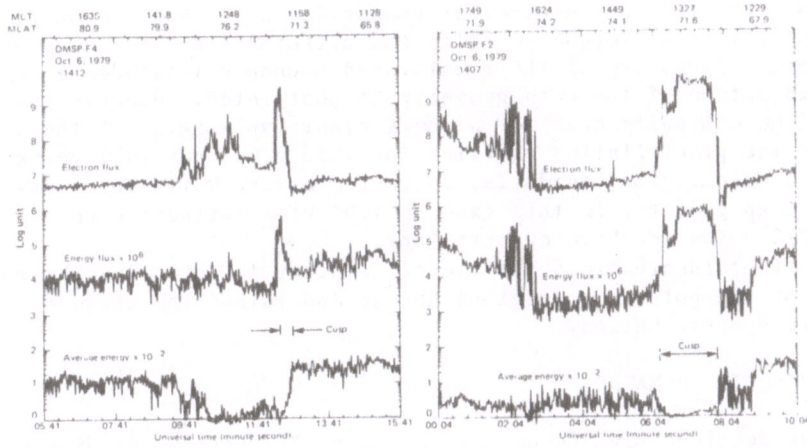

Fig. 10 Precipitating electron parameters for conjugate cusp crossing event on
         October 6, 1979.

high flux and low energy electrons between 1506:10 UT (-72.8° MLAT) and 1507:55 UT (-70.0° MLAT). The same sequence is apparent in the Northern Hemisphere observation, with the polar cap extended to 1509:32 UT (79.1° MLAT) and cusp-like precipitation observed between 1509:32 UT and 1512:34 UT (79.1-72.3° MLAT). In this Northern Hemisphere observation, we only identify the period 1512:12 UT (73.3° MLAT) to 1512:24 UT (72.3° MLAT) as the polar cusp region. The difference of nearly two orders of magnitude in intensity between the electron flux observed in this region and the flux observed between 15.09:32 and 15.12:12 UT at higher latitude (79.1-73.3° MLAT) does not allow a straightforward identification of the whole region as the cusp. The latitudes of the observed cusp equatorward and poleward boundaries are again lower than expected, presumably because of the high geomagnetic activity and the negative IMF B_z component.

The average precipitating electron differential energy spectra over conjugate hemispheres for the event of 6 October 1979 are shown in Figure 11. The spectra over the two hemispheres match very well. A similarly well matched comparison can be made for the other event. An exponential fit of the form $dj/dE = A \exp(-E/E_o)$ best characterizes the observed spectra with an e-folding energy E_o equal to 67 $\pm$ 3 eV. A Maxwellian fit would yield $kT \sim 50$ eV.

From this observation, the latitudes of the equatorward and poleward boundaries of the conjugate cusp region are in reasonable agreement in the 7 October 1979 event. Good conjugacy of the polar cusp region and the various particle boundaries can be therefore inferred from this observation. However, in the event of 6 October 1979 the equatorward boundaries of the conjugate cusp regions differ by 2.3° with the Northern Hemisphere cusp poleward of the Southern Hemisphere cusp. The equatorward boundaries of the cusp on opposite hemispheres are observed more than 1 h apart in geomagnetic local time. This longitudinal separation may explain the gross mismatch of the cusp regions equatorward boundary latitudes, if a severe distortion in the local time configuration of the cusp equatorward boundary is invoked. The two cusp equatorial boundaries were observed only with a 4.5 min UT time difference. This time difference may be invoked to explain the asymmetry of the equatorward boundary latitudes if a fast poleward motion of the cusp geometry is postulated. However the space/time ambiguity does not allow a clear explanation of the difference of the precipitation profiles for this 6 October 1979 event. The observed cusp region boundaries suggest a strong North-South asymmetry of the cusp geometry in this case (a 1.0° wide northern cusp region and a 2.8° wide southern cusp region).

Further observations are needed in order to determine the conjugacy of the polar cusp regions and to understand the cause of asymmetric distribution.

7. CONCLUDING REMARKS

Previous sections have demonstrated the usefulness of the USAF DMSP satellite observations in studying the polar cusp region. The results summarized here are just part of the DMSP's contribution. Detailed

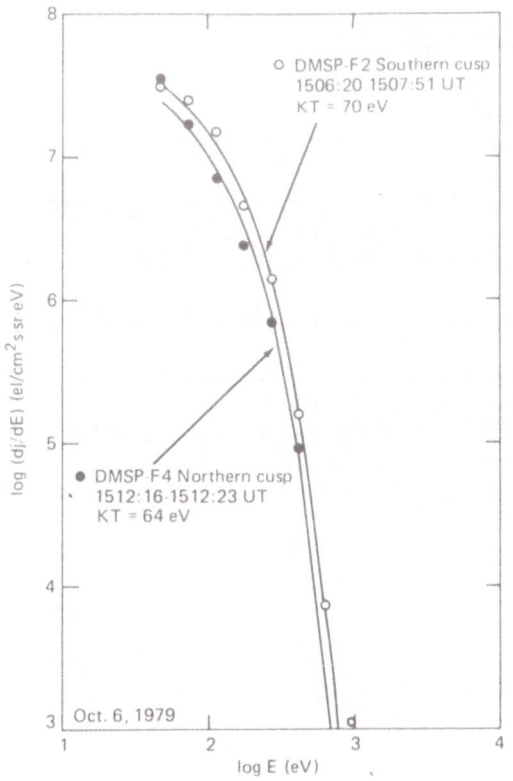

Fig. 11   Average electron energy differential spectra for the conjugate cusp observation of October 6, 1979.

observations and analyses have been published in scientific journals. The following is a partial list of our DMSP investigation concerning the polar cusp region and the details of the results reported here can be found in these publications.

Electron Precipitations in the Midday Auroral Oval by C.-I. Meng, published in J. Geophys. Res., 86, 2149, 1981.

Latitudinal Variation of the Polar Cusp During a Geomagnetic Storm by C.-I. Meng, published in Geophys. Res. Lett., 9, 60, 1982.

Case Studies of the Storm Time Variation of the Polar Cusp by C.-I. Meng, published in J. Geophys. Res., 88, 137, 1983.

Spatial Intensity of Dayside Polar Soft Electron Precipitation and the IMF by M. Candidi, H. W. Kroehl and C.-I. Meng, published in Planet. Space Sci., 31, 889, 1983.

Nearly Simultaneous Observations of the Conjugate Polar Cusp Region by M. Candidi and C.-I. Meng, published in Planet. Space Sci., 32, 41, 1984.

The Relation of the Cusp Precipitating Electron Flux to the Solar Wind and Interplanetary Magnetic Field by M. Candidi and C.-I. Meng, published in J. Geophys. Res. (in press), 1984.

REFERENCES

Akasofu, S.-I., 1972a, J. Geophys. Res., 77, 244.
Akasofu, S.-I. 1972b, J. Geophys. Res., 77, 2303.
Burch, J. L. 1972, J. Geophys. Res., 77, 6696.
Burch, J. L. 1973, Radio Sci., 8, 955.
Eather, R. H., Mende, S. B., and Weber, E. J. 1979, J. Geophys. Res.,
    84, 3339.
Heikkila, W. J. and Winningham, J. D. 1971, J. Geophys. Res., 76, 883.
Iijima, T. and Potemra, T. A. 1976, J. Geophys. Res., 81, 2165.
Iijima, T., Fujii, R., Potemra, T. A., and Saflekos, N. A. 1978, J.
    Geophys. Res., 83, 5595.
Kamide, Y., Burch, J. L., Winningham, J. D., and Akasofu, S.-I. 1976,
    J. Geophys. Res., 81, 698.
Russell, C. T., Chappell, C. R., Montgomery, M. D., Neugebauer, M.,
    and Scarf, F. L. 1971, J. Geophys. Res., 76, 6743.
Saflekos, N. A. and Potemra, T. A. 1980, J. Geophys. Res., 85, 1987.
Sandholt, P. E., Henriksen, K., Deehr, C. S., Sivjee, G. G., and
    Romick, G. J. 1980, J. Geophys. Res., 85, 4132.

OPTICAL-PARTICLE CHARACTERISTICS OF THE POLAR CUSP

D.J. McEwen
Institute of Space and Atmospheric Studies
University of Saskatchewan
Saskatoon, Saskatchewan, Canada S7N OWO

ABSTRACT. The use of extended ground based optical measurements of the dayside cusp has yielded much information on the general character- istics and dynamics of this region of the auroral oval. They show a region of stable (mostly 6300A) emission during the noon period but which becomes more active and marked by frequent short lived enhance- ments of 5577A emission beginning about 1 hour after local magnetic noon. It can be inferred that the enhancements are due to more ener- getic electron influx. Two rocket flights across the cusp region, one into post-noon conditions (VB-03 at 0004:58 UT on Dec. 6, 1977) and a second into conditions more noon-like (IVB-38 at 23:14:58 UT on Dec. 6, 1981), have yielded electron data which are compared with simultaneous ground-based photometric measurements at 6300A and 5577A.

1. INTRODUCTION

Detailed studies of dayside aurora have been carried out in recent years using photometric techniques at selected sites in the polar regions. Three such sites have been at South Pole (Eather et al., 1979), Sachs Harbour (Duncan and McEwen, 1979) and Longyearbyen (Deehr et al., 1980). These stations are all at or above 75° magnetic invariant latitude which places them beneath, or at least within view of the dayside cusp emissions. With the continuous operation of meridian scanning photometers near or through the winter solstice it has been possible to observe both the morphology and many of the dynam- ical characteristics of dayside auroras as well as to draw inferences on the nature of the particle precipitation giving rise to these emissions. The establishment of a rocket launching base at Cape Parry, N.W.T. (λ=75°) has permitted instrumented rocket flights poleward across the cusp in conjunction with simultaneous ground measurements from Sachs Harbour (λ=77°) beneath the flight paths. Some results from two such integrated investigations conducted in 1977 and 1981 will be reviewed here.

J. A. Holtet and A. Egeland (eds.), The Polar Cusp, 193–202.

2 INSTRUMENTATION

A 6-channel meridian scanning photometer with 5-inch diameter optics
and a 0.9° field of view was used for ground based optical studies of
dayside aurora. Filters of bandwidths of approximately 35A were
mounted on a filter wheel to sequentially measure OI 6300 and 5577A,
N_2 First Positive Bands around 6700A, the N_2^+ First Negative Band at
4278A and Hβ at 4861A. A sixth filter was used to monitor background
continuum at 4800A due to either twilight or moon and to allow a background correction. A 45° mirror was used to scan across the meridian
in 1° steps at a nominal rate of either 5 or 10 steps per second.
Complete meridian scans were thus obtained in 36 or 18 seconds. A
fuller description of the instrument is given by Duncan and McEwen
(1979).

The photometer in this configuration was used at Sachs Harbour in
1977 and Longyearbyen in 1978-79. New filters were installed in the
instrument prior to use at Sachs Harbour in 1981 in order to get better
twilight discrimination. Filters of about 2.5A bandpass were used for
the 6300 and 5577A emissions and similar filters at adjacent wavelengths were used to allow reliable background subtraction. The
filters were in each case rotated slightly from normal position to
adjust the wavelength of peak transmission to the desired value. In
field operation the photometer compartment containing the filters was
maintained at 20°C while the remainder of the photometer remained at
near ambient temperature. With these narrow band filters it was
possible to observe dayside aurora through the noon period at Sachs
Harbour (72°N latitude) for about 10 days on either side of winter
solstice.

The rocket borne spectrometers used for measuring precipitating
electron fluxes contained electrostatic analyzers of cylindrical plate
design. Energy scans from 18 keV to below 20 eV were obtained each
0.3 seconds. The spectrometers had a field of view of 10° by 11° and
an energy resolution $\Delta E/E$ of 0.15. A fuller description of spectrometer design and operation is given by McEwen (1977).

3. RESULTS

The strongest emission in dayside aurora is normally the 6300A OI line
which is easily detectable through the mid-day. The 5577A OI emission
is also continuous but at a lower intensity except for short-lived enhancements which frequently appear on either side of a normally quiet
period around local magnetic noon. A 2-month period of photometer
operation around the 1978 winter solstice at Longyearbyen reported by
Gilmer et al. (1984) has delineated many of the optical characteristics
of dayside aurora as depicted by these two emissions.

Illustrative data from that study are shown in Fig. 1. These are
meridian intensity profiles of 6300A and 5577A emissions recorded through

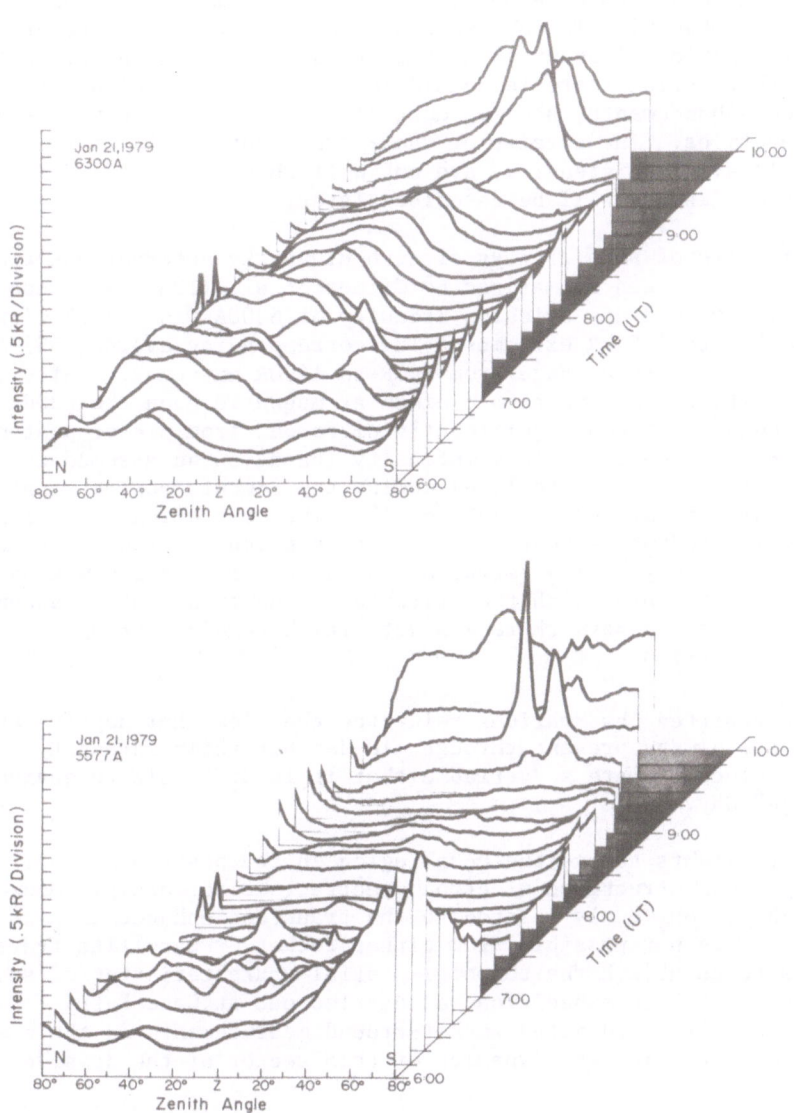

Fig. 1 Typical meridian profiles of 6300A and 5577A intensities
averaged over 10 minute intervals through magnetic noon (0840 UT) at
Longyearbyen (Λ=75°) on January 21, 1979. They illustrate the stable
conditions which normally exist for 1-2 hours at noon with a minimum
5577A intensity. The data loss along the southern edge of the scans
is due to morning twilight.

noon on January 21, 1979, They are 10 minute averages and extend from
0600 UT to 1000 UT. They were recorded at Longyearbyen (λ=75°) and extend from zenith angles of N 80° to S 80°. Local magnetic noon is
approximately 0840 UT. The oxygen emissions are seen to be relatively
stable for a period of about an hour centred on magnetic noon. Before
and after that period there is considerable variability with large but
short-lived enhancements, showing most clearly in the more prompt 5577A
emission profiles. The details of these transient events, often of
several kilorayleigh intensity, are not well shown by these 10 minute
averaged profiles but will be described later.

The results of some 23 days of such clear sky observations from
Longyearbyen have been summarized by Gilmer et al. (1984) who report
that the average mid-day emission itensity of 6300A was 1.7 kR with a
range from 0.9 to 4.5 kR extremes. The corresponding average 5577A intensity was 600R with a range from 240R to 1400R extremes. While the
5577A intensity was always at a minimum at magnetic noon the 6300A intensity tended to be not significantly different from pre or post-noon
times. The location of maximum intensity (based on an assumed emission
height of 225 km for the red line) during the 2 month period ranged
from an invariant latitude of 79° to 71°. The cusp emission region had
an average latitudinal extent of 2.5° with a slight narrowing on magnetically active days. The extent of the noon quiet period was on
average about 1½ hours (with the criterion of no transient enhancements)
while on 7 of the 23 days there was activity extending through noon
with no quiet period.

These reported observations reinforce the view that dayside auroral
emissions are always present through mid-day but their intensity and
latitudinal location are so variable that it is difficult to envisage
an "average" day.

Rocket flights across the cusp region in the post-noon sector have
invariably shown structured electron spectra (see for example McEwen,
1977) which are obviously related to the transient enhancements in 5577A
emission. It is not possible from either rocket or satellite traversals
themselves to establish the temporal-spatial characteristics of such
events and in 1977 an expedition was carried out at Cape Parry - Sachs
Harbour to obtain coordinated rocket-ground measurements in the post-
noon period to examine the dynamics of this sector of the dayside
auroral oval.

A Nike-Black Brant rocket (NVB-03) was launched from Cape Parry
NWT at 0004:48 UT on Dec. 6, 1977. It flew on a northward trajectory
which took it through the dayside auroral oval and into the polar cap.
The meridian scanning photometer was located approximately under rocket
apogee, at Sachs Harbour, and was oriented so that its meridian scans
were approximately along the plane of the rocket trajectory.

Fig. 2(a) shows a sequence of meridian scans of the OI 5577A and
6300A emissions beginning at the time of rocket lift-off and continuing

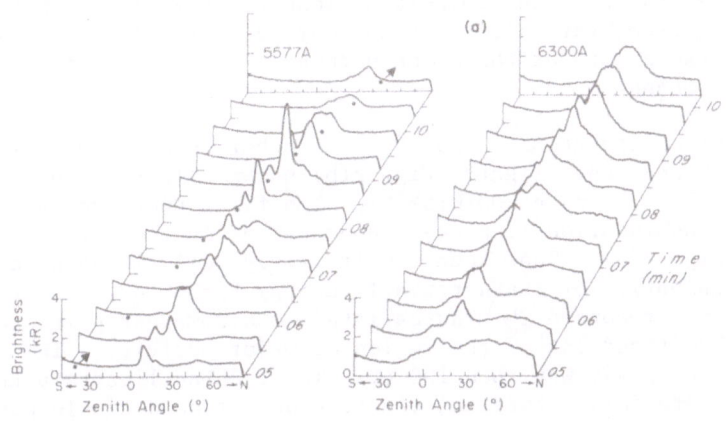

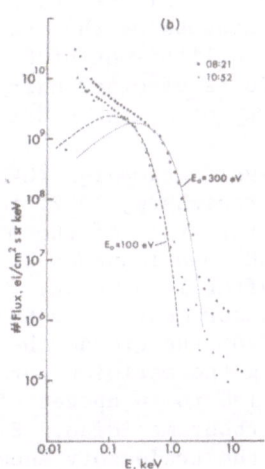

Fig. 2(a)   Photometer scans along the Cape Parry - Sachs Harbour meridian during the flight of rocket NVB-03.   (b)   Electron spectra recorded at 0008:21 (E_O=300 eV) and 0010:52 (E_O=100 eV) during the rocket flight at locations shown by the dots in Fig. 2(a).

until 0011 UT. The approximate location of the rocket as a function of time as it travelled northward with a speed of about 1 km/sec is shown by the row of dots on the 5577A meridian scans. The scans show the measured intensities of the emissions along the meridian from south to north. The extreme variability of the dayside auroral emissions, which run through the zenith at the Sachs Harbour site, is evident in these sequences of scans.

The elctron spectrometer aboard the rocket recorded electron spectra throughout the flight. Fig. 2(b) shows illustrative spectra recorded at 0008:21 and at 0010:52 UT. The first spectrum was recorded at the time and location of a major enhancement in the 5577A emission, as seen in Fig. 2(a). That spectrum is fairly well fitted by a Max-wellian distribution of characteristic energy $E_0=300$ eV. The second spectrum shown, recorded $2\frac{1}{2}$ minutes later, was somewhat poleward when the rocket had moved out of the region of major activity. The electron characteristic energy was then 100 eV. These data illustrate the need for extreme care in interpreting electron data from a single pass across the dayside auroral oval. During the few minutes taken by the rocket to traverse the cusp region there may be several electron bursts (inverted V events such as illustrated in Fig. 2(b)) with corresponding 5577A enhancements. The 110 second lifetime of the 6300A red line emission leaves that emission profile relatively unresponsive to some of these shorter-lived events. A preliminary study of such events in the afternoon sector of the auroral oval by Creutzberg and McEwen (1983) using TV all-sky cameras shows that they are mostly narrow arcs, frequently appearing and with lifetimes of only 1-2 minutes on average. The 6300/5577A optical ratio is of order unity indicating an average precipitating electron energy of a few hundred eV.

In November-December 1981 a campaign (CENTAUR) was conducted at Cape Parry-Sachs Harbour (Creutzberg, 1982) to investigate both the particle and optical characteristics of the mid-day region of the cusp. A Black Brant rocket (IVB-38) was launched at 23:14:58 UT on December 6, 1981 approximately 1 hour after local noon. The path of the rocket relative to ground based measurements from Sachs Harbour is shown in Fig. 3. Viewed optically from the ground the aurora was stable and largely 6300A emission along the meridian through Cape Parry and Sachs Harbour. The rocket travelled to an apogee of 603 km but a few degrees to the west of the Sachs Harbour meridian. Electron fluxes were measured along the part of the trajectory shown with an added dashed line.

The rocket was within the cusp region virtually its whole flight poleward, as evidenced from the electron data recorded. A typical spectrum recorded before rocket apogee at a height of 500 km is shown in Fig. 4. This is the electron flux measured at pitch angles of 40-70°. The spectrum was extremely soft and is fitted fairly well through the low energy region by a Maxwellian of $E_0=10$ eV. There were some indications that the rocket traversed some inverted-V type events. These can be identified in Fig. 5 which is a plot of the 170 eV elec-tron flux measured through the rocket flight. The solid portions of the graph are for 170 eV electrons of pitch angles in the range 50-70°.

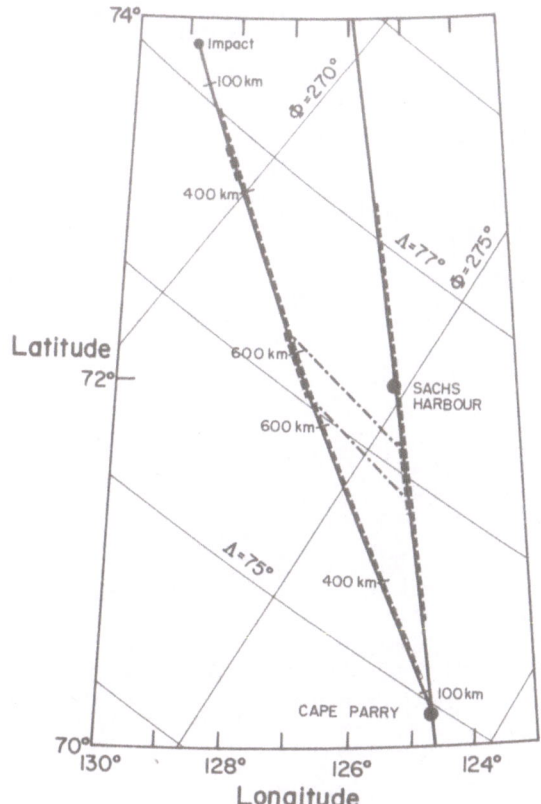

Fig. 3 Rocket IVB-38 path poleward and the coverage of the photometer scans from Sachs Harbour. Dashed lines show the cusp extent with electron enhancements just after apogee and near the poleward edge of the cusp.

The orientation of the spectrometer in the rocket was such that due to coning of the rocket about its velocity vector there were intervals when the spectrometer was viewing only electrons of pitch angles greater than 90°. During those periods the precipitating electron fluxes were estimated from the measured upward fluxes and they are shown with dashed lines. The flux of 170 eV electrons was about 3×10^8 el/cm^2 sec ster keV across most of the cusp region while the rocket was above the absorbing atmosphere. At three times in the flight it sharply increased to about 8×10^9 el/cm^2 sec ster keV, in the middle portion of the cusp region about 23:22 UT and at the poleward edge about 23:26 UT. The rapid decrease in flux beyond 23:26 UT indicated rocket exit into the polar cap. This was at an invariant latitude of 77°.

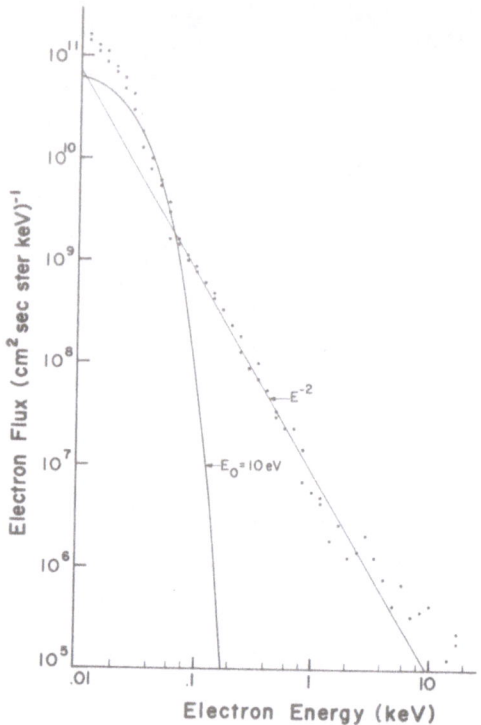

Fig. 4   A typical electron spectrum, recorded at a height of 500 km.
The solid lines are Maxwellian and power law energy distributions.

      The meridian scanning photometer was operated at Sachs Harbour
during the time of the rocket flight. Prior to launch and during the
early part of the flight there was 6300A emission observed extending
from 40° elevation in the north through the zenith to 140° elevation
angle. The 5577A emission was very low. There was then an increase in
5577A intensity reaching a peak about 23:23:01 UT. Fig. 6 shows the
meridian scan starting at that time with the enhancement at 110° elevation. This appears to be produced by the increased electron flux shown
in Fig. 5 just after rocket apogee. The relative locations of the simultaneous electron and optical enhancements are indicated in Fig. 3.
They suggest the auroral form was probably a narrow E-W transient arc.

4.   DISCUSSION

Ground based optical studies of cusp emissions have been most illumina-
ting in allowing a fuller interpretation of incoming particle fluxes

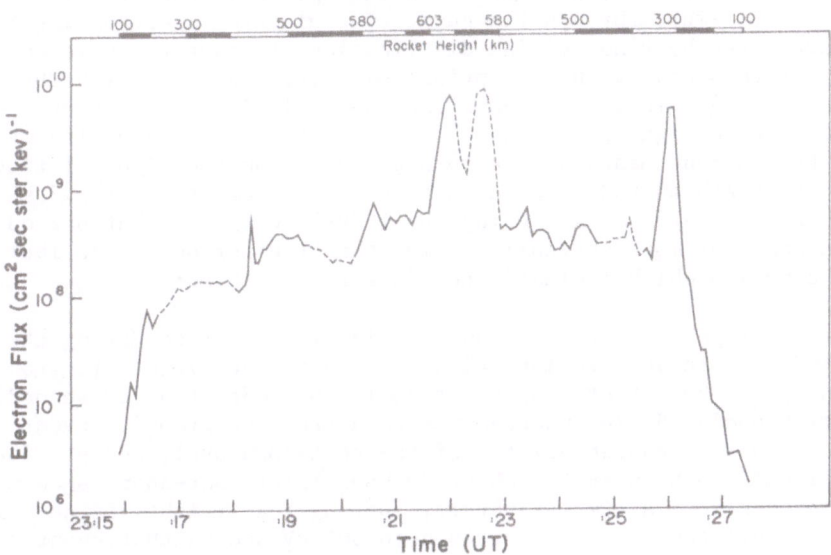

Fig. 5 A plot of the 170 eV electron flux measured through the IVB-38
rocket flight.   Several brief enhanced fluxes are seen.   The solid
lines are for electron pitch angles of 50-70° while the dotted lines
are estimated fluxes.

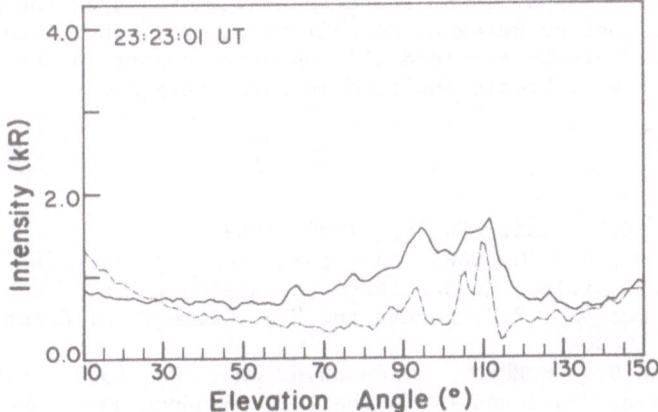

Fig. 6 Photometer scans of 6300A (solid) and 5577A (dashed) intensities
measured from Sachs Harbour. The 5577A enhancement (110° elevation
from the North) corresponds to the 170 eV electron increase in Fig. 5.

measured in rocket traversals of the dayside oval. The mid-day region
is characterized by very soft electron influx, as illustrated in the
rocket IVB-38 flight through the cusp. The electron spectrum was fairly
well represented by a Maxwellian distribution of characteristic energy
10 eV. For this condition the optical signature was largely 6300A emission. The 5577A emission was small and more difficult to measure re-
liably in the twilight background. The measured 6300A intensity was
about 1 kR. Ground measurements through noon from the higher latitude
site at Longyearbyen had given an average 6300A intensity of 1.7 kR and
a 5577A intensity of 0.6 kR during the 1978-79 winter. That period
appeared somewhat more optically active than the period of the 1981
CENTAUR campaign which included the above rocket flight.

There were events of electron energization recorded during the
rocket flight. These were most clearly seen in the 170 eV electron flux
which increased by a factor of about 30 for periods of a few seconds,
with corresponding 5577A enhancements seen from the ground. Events of
this kind are more characteristic of the post-noon oval, and previous
rocket flights such as VB-03, about an hour later post-noon, show more
energetic events of a similar nature. During that flight, as seen in
Fig. 2, the electron energy increased to 300 eV and in that event the
green line intensity rose to 5 kR. Some optical enhancements reported
by Gilmer et al. (1984) show 5577A intensities of 25 kR or more. Current studies of such electron-optical enhancements using sensitive TV
all-sky cameras show they are narrow forms but with considerable longitudinal extent. Their lifetimes are typically not more than a minute
or two.

ACKNOWLEDGEMENTS

This work has been supported by research grant A5904 from the Natural
Science and Engineering Research Council of Canada. The Canadian
Centre for Space Science provided all logistic support in the rocket
investigations. D.S. Steele assisted in data analysis.

REFERENCES

Creutzberg, F. EOS trans. AGU, 63, 1080, 1982.
Cretuzberg, F. and D.J. McEwen. Adv. Space Res., 2, 85, 1983.
Deehr, C.S., G.G. Sivjee, K. Henriksen, A. Egeland, P.E. Sandholt, R.
        Smith, P. Sweeney, C.N. Duncan and W.J. Gilmer. J. Geophys. Res.,
        85, 2185, 1980.
Duncan, C.N. and D.J. McEwen. J. Geophys. Res., 84, 6533, 1979.
Eather, R.H., S.B. Mende and E.J. Weber. J. Geophys. Res., 84, 3399, 1979.
Gilmer, W.J., J.A. Koehler, C.N. Duncan, D.J. McEwen, D.C. Agarwal and
        G.G. Sivjee. J. Geophys. Res., submitted 1984.

CHARACTERISTICS OF LARGE-SCALE BIRKELAND CURRENTS IN THE CUSP AND
POLAR REGIONS

T. A. Potemra and L. J. Zanetti
The Johns Hopkins University
Applied Physics Laboratory
Laurel, Maryland 20707

ABSTRACT. Particle observations from satellites acquired over a decade ago confirmed the presence of the dayside cusp regions (Heikkila
and Winningham, 1971; Frank, 1971). Later studies showed that a wide
variety of phenomena was associated with the cusps, including auroral
emissions (Meng 1981), kilometric radiation (Alexander and Kaiser,
1977), ULF-ELF noise (Gurnett and Frank, 1978), ionospheric disturbances (Greenwald 1982). and Birkeland currents (Iijima and Potemra,
1976a). The fact that large scale Birkeland currents flow into and
away from the cusps has been confirmed by several satellite experiments. Their locations, intensities, and flow directions are closely
related to the orientation and magnitude of the interplanetary magnetic field (IMF). The characteristics of cusp-region Birkeland
currents and their relationships to the IMF, polar convection, ionospheric currents, and ULF-VLF waves are reviewed here. The characteristics of a new large-scale Birkeland current system (referred to
as the "NBZ" system) in the dayside polar region which intensifies
during periods of strongly northward IMF are also reviewed.

1. INTRODUCTION

Satellite observations of disturbances transverse to the geomagnetic
field have been used in numerous studies to determine characteristics
of field-aligned "Birkeland" currents which flow into and away from
the polar regions (see, for example, the review by Potemra, 1979 and
Saflekos et al., 1982). The large-scale Birkeland currents are concentrated in two principal areas which encircle the geomagnetic pole
(Zmuda and Armstrong, 1974; Iijima and Potemra, 1976a,b). These are
shown in Figure 1 (from Iijima and Potemra, their Figure 6). The
Birkeland current flow regions have been arbitrarily designated by
Iijima and Potemra (1976a) as "Region 1" located at the poleward
side, and "Region 2" located at the equatorward side. The Region 1
currents flow into the ionosphere in the morning sector and away from
the ionosphere in the evening sector. The Region 2 currents flow in
the opposite direction at any given local time.

J. A. Holtet and A. Egeland (eds.), The Polar Cusp, 203–222.
© 1985 by D. Reidel Publishing Company.

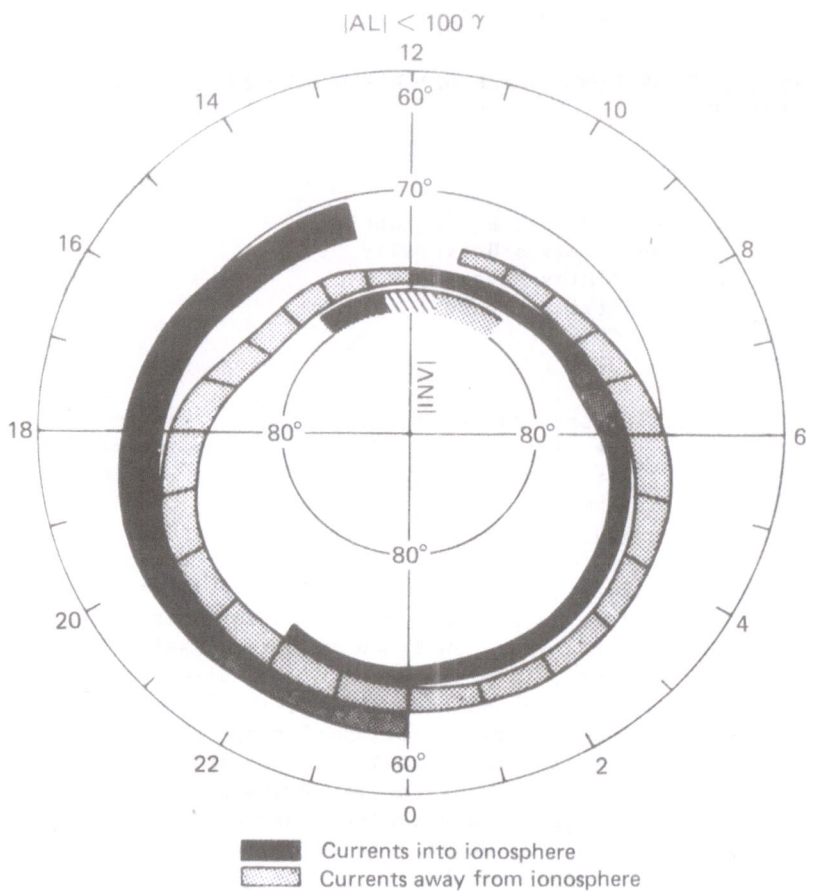

Fig. 1   A summary of the distribution and flow directions of large-scale
field-aligned currents determined from data obtained on **493**
passes of TRIAD during weakly disturbed conditions
($|AL| < 100\ \gamma$) (from Iijima and Potemra, their Fig. 6).

The densities of the Region 1 and Region 2 currents are not
necessarily equal at given local time (i.e., the upward and downward
flowing currents are not always completely coupled by ionospheric
Pedersen currents flowing along a geomagnetic meridian). The regions
of greatest imbalance between the densities of the Region 1 and
Region 2 systems occur near 1400 MLT (magnetic local time) and 0800
MLT (see Figure 14 of Iijima and Potemra, 1978). The largest Region
1 currents flow into the ionosphere near 0800 MLT and away from the
ionosphere near 1400 MLT producing the "net" or "unbalanced" field-
aligned currents discussed by Sugiura and Potemra (1976). The region
near 1400 MLT may have a special significance with regard to particle
precipitation, auroral emissions, and topology of the magnetosphere
(Potemra, 1983).

A three-region flow pattern of Birkeland currents exists in 2200 to 2400 MLT sector (sometimes referred to as the "Harang discontinuity region") and may be thought of as an "overlapping" of the two-region flow patterns usually observed in the surrounding MLT sectors; it is not possible to distinguish between Region 1 and Region 2 Birkeland currents in this MLT sector. The basic flow pattern of Birkeland currents is the same during geomagnetically inactive and active periods, although the regions widen and shift to lower latitudes during disturbed periods (similar to the variations in the statistical "auroral oval" determined by Feldstein, 1966). Complicated small-scale structures are superimposed upon the large-scale Birkeland current features, especially near the high-latitude noon region associated with the cusp and on the nightside during substorm events.

A region of Birkeland currents exists on the dayside, between 1000 and 1400 MLT and poleward of Region 1, between ~ 78° and 81° invariant latitude (Iijima and Potemra, 1976b). Because of their location and the close correlation of their intensities and increases of the southward interplanetary magnetic field component, these field-aligned currents are believed to be associated with the magnetospheric cusp. Their characteristics are described in detail in Section 2.

More recently, a stable pattern of Birkeland currents in the sunlit polar cap has been discovered to persist during periods of northward IMF (Iijima et al., 1984; Zanetti et al., 1984). This system, referred to as the "NBZ Birkeland current" system (because of its relationship to northward B_z flows downward on the afternoon side and upward on the morning side (opposite to the flow of the Region 1 system). Characteristics of the NBZ currents are also closely related to the IMF and these details will be discussed in Section 3.

2. CUSP-REGION BIRKELAND CURRENTS

Complicated patterns of magnetic disturbances have been observed with various low altitude satellites including the TRIAD and ISIS-2, S3-2, and HILAT spacecraft in the high latitude region near magnetic noon (Iijima and Potemra, 1976b; McDiarmid et al., 1978; Iijima et al., 1978; Doyle et al. 1981). The disturbances at the poleward edge of this region were believed to be due to the flow of Birkeland currents in the cusp because of their location (for example, between 78° and 80° invariant latitude and between 1000 MLT and 1400 MLT during quiet geomagnetic conditions) and because of their relationship to the IMF B_z component (larger intensities were usually observed during periods of southward B_z). These cusp-region Birkeland currents were determined by Iijima and Potemra (1976b) to be flowing away from the ionosphere in the pre-noon hours (MLT) and into the ionosphere in the post-noon hours (as shown in Figure 1), from a data base that included both sector orientations of the IMF. Subsequent analysis of transverse disturbances over the north and south polar regions acquired by the TRIAD, ISIS-2, and S3-2 spacecraft (Wilhjelm et al., 1978; McDiarmid et al., 1978b; Iijima et al., 1978; Doyle et al., 1981) have revealed that one flow direction is observed more often

than the other, depending upon the polarity of B_y. This dependence
had been predicted from theoretical considerations (e.g., Gizler et
al., 1977; Levitin et al., 1977; Reiff et al., 1978).

2.1 Relationship to the Interplanetary Magnetic Field

McDiarmid et al. (1978b) have demonstrated from the ISIS-2 magnetometer experiment that triangular-shaped magnetic disturbances (which
may be interpreted as caused by oppositely directed double-sheet
Birkeland currents) are frequently observed in the northern polar cap
region between 1000 to 1500 MLT. For any given B_y polarity, the
direction of the largest magnetic disturbance associated with the
triangular pattern observed by ISIS-2 exhibits a preferred sense of
orientation in the geomagnetic east-west direction over nearly the
entire sector between 1000 and 1500 MLT. For $B_y > 0$ this maximum
disturbance is directed eastward and reverses direction for $B_y < 0$
for a large number of cases acquired (with only a few exceptions).
This relationship is evident in the TRIAD magnetometer data shown in
Figure 2 (from Saflekos and Potemra, 1980, their Figure 2). These
are polar plots of the magnitude and direction of the magnetic disturbance vectors measured by TRIAD over the south polar region.
Although the relationship between the orientation of the transverse
magnetic disturbances and the orientation of the IMF may be the same
in both hemispheres, the flow directions of the associated Birkeland
current sheets is systematically reversed. During an "away" IMF
sector orientation ($B_y > 0$), the predominant flow of cusp Birkeland
currents is away from the north polar cap and into the south polar
cap. These flow directions appear to reverse during periods of
"toward" IMF orientation ($B_y < 0$) (although this may be interpreted
as being due to a "shifting" spatial pattern of Birkeland currents,
Potemra, 1979).

   Attempts to schematically diagram the complicated relationships
between Birkeland current flow patterns and the IMF orientation are
shown in Figure 3 (from Potemra, 1978), Figure 4 (from McDiarmid et
al., 1979) and Figure 5 (from Saflekos et al., 1982 which they
adopted from D'Angelo, 1980). The three types of Birkeland current
systems identified by Iijima and Potemra (1976a,b) are shown in the
two panels of Figure 3 (i.e., Region 1, Region 2 and cusp). The
locations of the Region 1 and 2 Birkeland currents do not change
significantly with different polarities of B_y, and are drawn the same
in both panels of Figure 3 (labeled "1" and "2"). These regions were
not extended across the noon meridian because the intensities of the
Birkeland currents here were found to diminish near noon (see, for
example, the diurnal distribution of current intensities in Figure 8
of Iijima and Potemra, 1978).

   The region of cusp Birkeland currents is shown in Figure 3 as
the crescent-shaped area poleward of Regions 1 and 2. The demarca-
tion line between upward and downward flowing Birkeland currents is
depicted to shift to the afternoon or morning side of the cusp,
depending upon the polarity of B_y. This interpretation retains the
general flow pattern of downward flowing Birkeland currents on the
afternoon side of the cusp and upward flowing Birkeland currents on

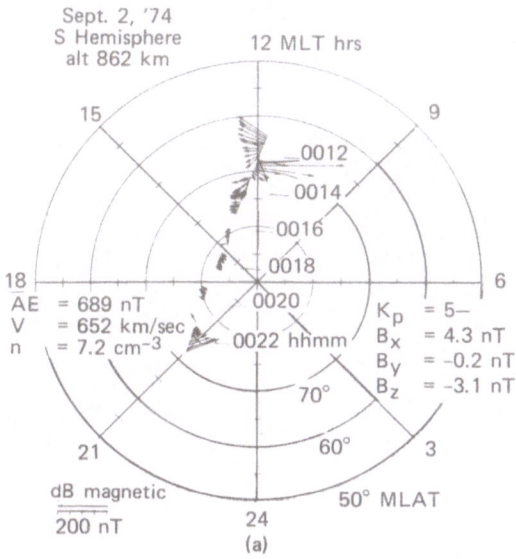

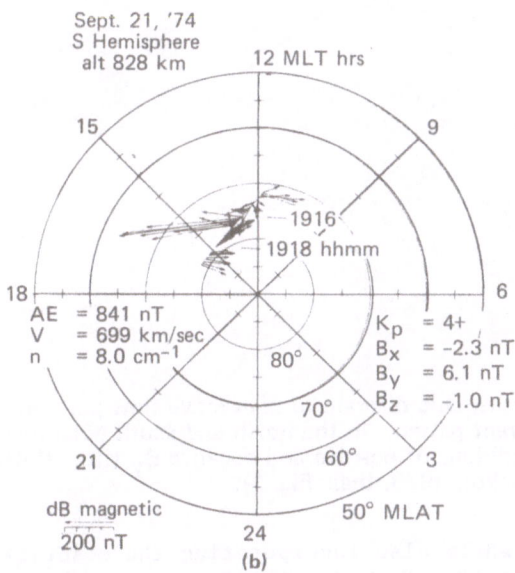

Fig. 2   Magnetic disturbances near the cusp region plotted on a MLT-
         INV dial. The view is inside the earth looking at the south pole
         ionosphere. (a) Data acquired on September 2, 1974, during a
         toward IMF sector. (b) Data acquired on September 21, 1974,
         during an away IMF sector (from Saflekos and Potemra, their
         Fig. 2).

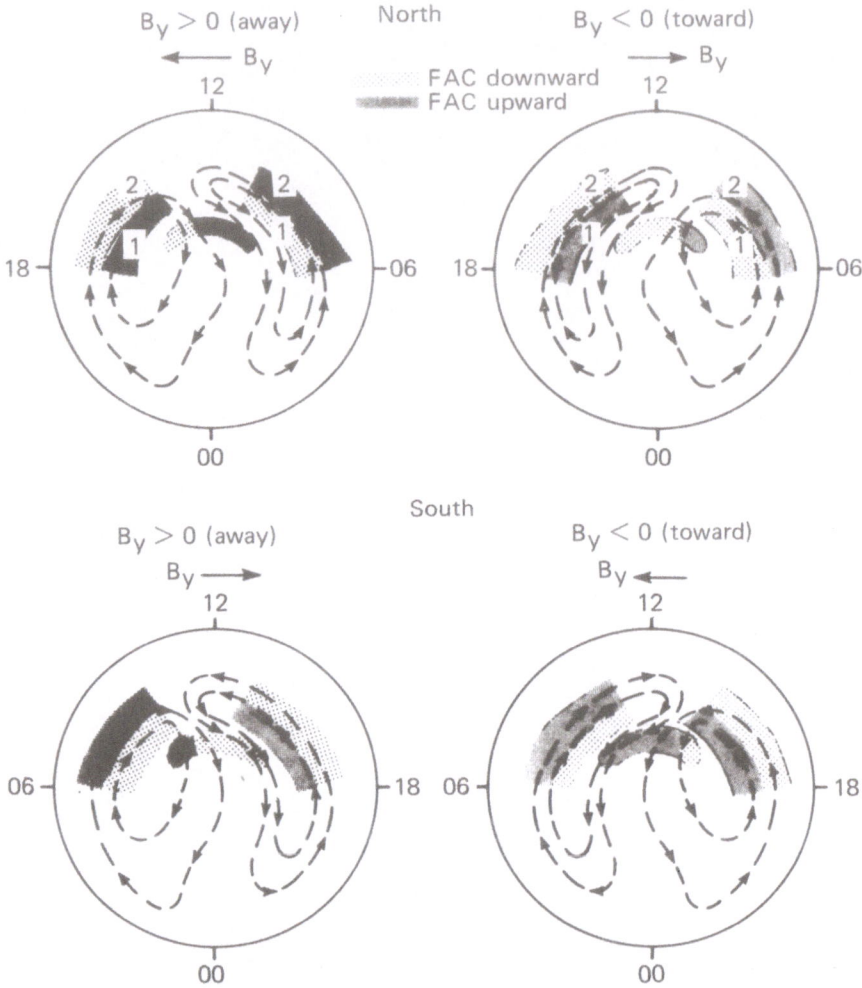

Fig. 3   A schematic diagram of convective flow patterns and Birkeland
current patterns in the north and south polar regions during
conditions of positive and negative B_y (from Potemra and
Saflekos, 1978, their Fig. 1).

the morning side, while also incorporating the observation of dominant upward current flow for positive B_y and dominant downward
current flow for negative B_y. The dashed arrows in this diagram are
intended to show the convection flow pattern for each B_y polarity,
and the DPY ionospheric current is believed to flow opposite to this
flow near noon between the cusp and Region 1.
      Figure 4 shows the distribution of Birkeland currents in the
northern polar region for different B_y polarities developed from the
ISIS-2 magnetic field and particle observations (from McDiarmid et

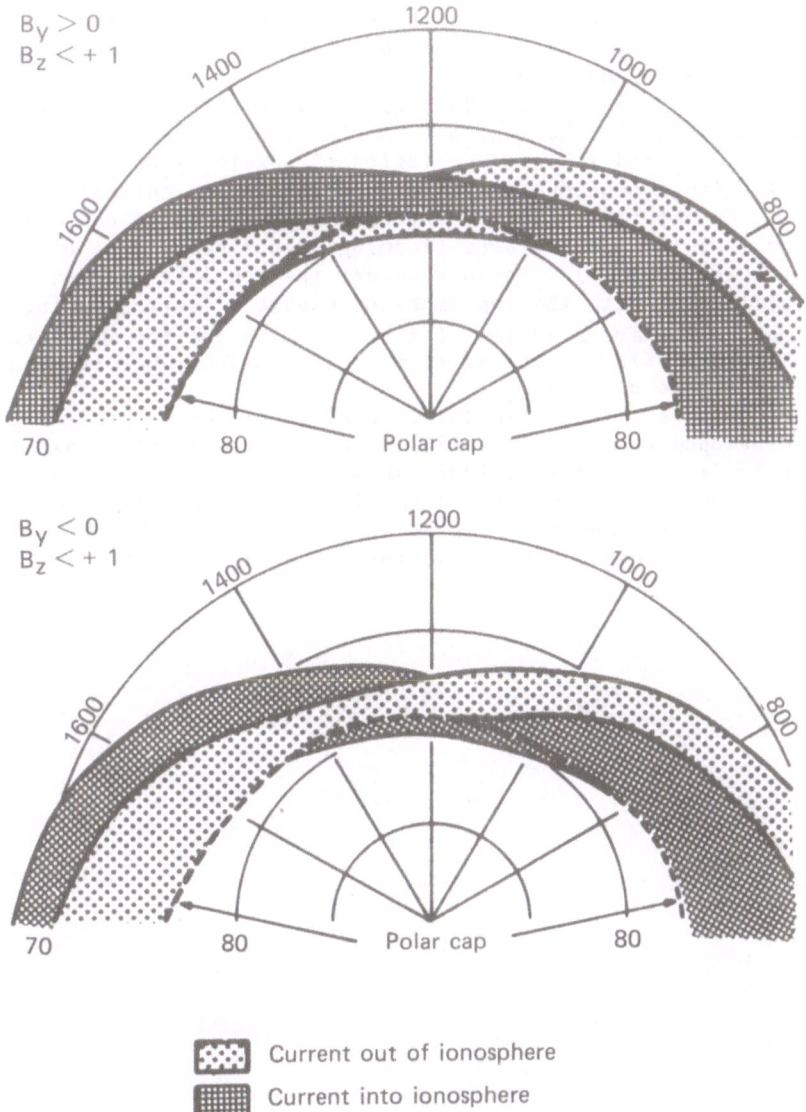

Fig. 4   Field-aligned current patterns inferred from the ISIS-2
         magnetic field observations (from McDiarmid et al. 1979,
         their Fig. 9)

al., 1979). The dashed line is interpreted as the polar cap boundary,
or the high latitude limit of closed field lines as determined from
the particle observations. The flow patterns in the dawn and dusk
sectors shown in Figure 4 are the same as those shown in Figure 3.
There are some apparent discrepancies between these two sets of
diagrams with regard to the distributies near noon, but these may be

due to interpretation. The diagram in Figure 4 shows that the flow
direction of Birkeland currents in the cusp (the small region poleward of the dashed line) is entirely upward for positive B_y and
entirely downward for negative B_y. The two sets of diagrams are
essentially the same if the set in Figure 3 is modified as follows;
in the left-hand panel, the Region 1 in the afternoon sector is moved
poleward and connected to the cusp region of upward flowing Birkeland
currents (forming a continuous area of upward flowing currents), and
Region 2 on the afternoon side is extended across noon to connect
with Region 1 on the morning side (forming a continuous area of
downward flowing current). In this manner the top left-hand panel of
Figure 3 will agree with the top panel of Figure 4 (corresponding to
$B_y > 0$), and a similar extension of the region in the top right-hand
panel of Figure 3 will make it agree with the bottom panel of Figure
4 (corresponding to $B_y < 0$).

Figure 5 shows a schematic interpretation of the cusp Birkeland
currents developed by D'Angelo (1980) (and extended to the southern
hemisphere by Saflekos et al., 1982, their Figure 7). These patterns
are almost identical to those shown in Figure 3 except that the
spatial extent of the Region 1 system also varies with the IMF. The
Region 1 system extends across noon from either the dawn or dusk

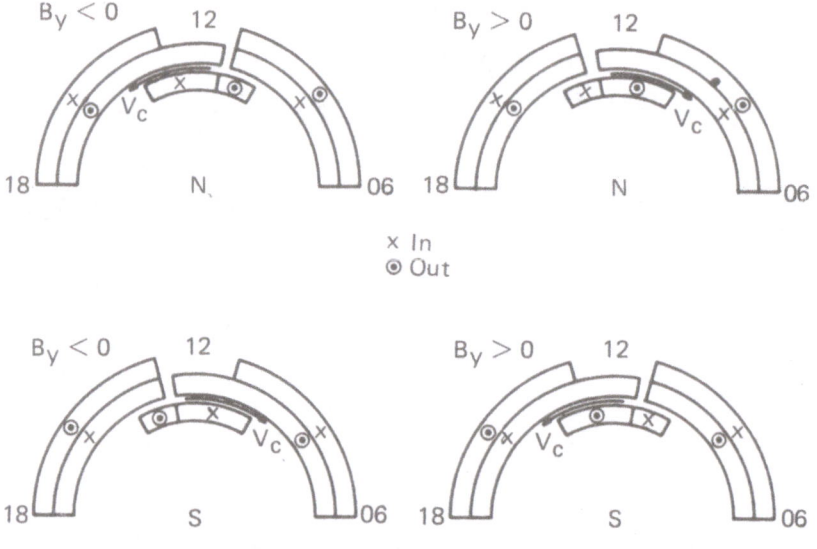

Fig. 5   Schematic of dayside, high-latitude field-aligned current systems
          for the north and south hemispheres (top and bottom panels,
          respectively). Left- and right-hand panels show the dominant
          polar cap convection direction and effects on the noon sector
          Region 1 and cusp systems caused by IMF $B_y < 0$ and IMF
          $B_y > 0$, respectively. (Figure was adapted from D'Angelo,
          1980, by Saflekos et al., 1982).

sides (depending upon the polarity of B_y) to match the spatial extent of the cusp Birkeland current with the opposite flow direction. The arrow labeled V_c shows the direction of convective flow for each B_y polarity. The direction of the ionospheric DPY current would be opposite to this.

2.2 Classification of Cusp-Region Birkeland Currents

Iijima et al. (1978) examined vector magnetometer data acquired from more than 230 TRIAD passes over the south polar region. They classified the transverse magnetic disturbances into the four categories drawn schematically in Table 1. These data were obtained from 121 TRIAD passes in the sector between 0900 and 1500 MLT. Simultaneous interplanetary magnetic field data were available for 102 of these passes. The occurrence number of each type is listed in this table, and the types are sorted according to MLT and polarity of the B_y component of the IMF. The hourly average value of B_y preceding the TRIAD observation was used (King, 1977; R. P. Lepping, personal communication, 1977).

The principal features evident from Table 1 can be summarized as follows. In the prenoon MLT sector the type 3 and 4 magnetic disturbances are observed most often. Type 4 can be interpreted as being due to a net or unbalanced Birkeland current flowing into the ionosphere, as studied by Sugiura and Potemra (1976), and corresponds

Table 1

Occurrence number of characteristic large-scale magnetic perturbations in geomagnetic east-west component observed in the South Polar Region between 0900 and 1500 MLT

MLT	Shape of Magnetic Perturbation 1 By>0	1 By<0	2 By>0	2 By<0	3 By>0	3 By<0	4 By>0	4 By<0
0900–1000	0	0	0	0	1	3	6	5.5
1000–1100	0	0	0	0	5.5	12	4.5	3.5
1100–1200	2	1	3	0	2.5	4.5	1.5	2.5
1200–1300	5	3	6	5	1	0.5	0	1.5
1300–1400	0	3	6.5	4	0	0	0	0
1400–1500	3	5	0.5	0	0	0	0	0
Total number	10	12	16	9	10	20	12	13

Note: A fraction number (like 5.5 etc) is due to the FAC events observed between MLT increments.

Only magnetic perturbations in geomagnetic east-west component larger than 100 nT were included.

to the Region 1 Birkeland current defined by Iijima and Potemra
(1976a,b) from their analysis of north pole data. The type 3 magnetic disturbance can be interpreted as being due to a pair of
oppositely flowing Birkeland currents. The downward flowing
Birkeland current on the equatorward side corresponds to the Region 1
system, and the upward flowing Birkeland current on the poleward side
is consistent with the cusp region Birkeland current as defined by
Iijima and Potemra (1976b) from TRIAD data obtained in the prenoon
MLT sector over the north polar region.

The type 1 and 2 magnetic disturbances are observed most often
in the postnoon MLT sector. The type 1 perturbation indicates the
presence of a large-scale net Birkeland current flowing away from the
postnoon sector similar to that determined by Sugiura and Potemra
(1976) from north polar TRIAD data, and they may be identified as
Region 1 Birkeland currents. The type 2 perturbation may be due to a
pair of oppositely directed Birkeland currents, the equatorward (upward flowing) Birkeland current being interpreted as a Region 1
Birkeland current and the poleward (downward flowing) Birkeland
current as a cusp region Birkeland current.

In summary, the cusp region Birkeland currents are observed
approximately one-half the time that the Region 1 Birkeland currents
are observed. Furthermore, the examples assembled in Table 1 suggest
that the associated Birkeland currents display systematic flow
directions in each MLT sector (i.e., irrespective of B_y the cusp
region Birkeland currents generally flow into the south polar
ionosphere in the postnoon sector and away in the prenoon sector),
similar to the pattern determined in the north polar region by Iijima
and Potemra (1976b) shown in Figure 1.

Figure 6 shows the relationship between the amplitude of the
magnetic disturbance associated with cusp region Birkeland currents
for 55 cases detected by TRIAD over the south polar regions (from
Iijima et al., 1978, their Figure 9). This figure indicates that the
more intense cusp region Birkeland currents are associated with large
negative values of B_z. However some high intensity cusp currents
also occur during periods of northward IMF. The cusp Birkeland
currents identified by Iijima et al. (1978) during northward IMF may
be the system of NBZ Birkeland currents more recently identified by
Iijima et al. (1984) and Zanetti et al. (1984) from the MAGSAT data
(see Section 3).

2.3 Relationship to Ionospheric Currents

Wilhjelm et al. (1978) compared ground-based geomagnetic variations
measured in Greenland with TRIAD magnetometer data. They determined
that the DPY equivalent ionospheric current (which flows along curves
of equal invariant latitude near noon and was so named by Friis-
Christensen and Wilhjelm, 1975 because of its relationship with B_y)
was located between two Birkeland current sheets. For $B_y > 0$ the DPY
current flows eastward and is contained between a downward flowing
Birkeland current sheet on the equatorward side and an upward flowing
current sheet on the poleward side. These flow directions in the

north polar cap reverse for B_y < 0. Wilhjelm et al. (1978) suggested
that the directions of the Birkeland current sheets and the DPY current were consistent, if it was assumed that the ionospheric current
was a Hall current. An important conclusion of this study is that
the flow of the poleward Birkeland currents (presumed to be the cusp-
region Birkeland currents) is predominantly away from the north polar
ionosphere for B_y > 0 and predominantly into the north ionosphere for
B_y < 0. These relationships suggest that the cusp Birkeland currents
are the important link in accounting for the close correlation between surface geomagnetic variations in the polar regions and the
sector direction of the interplanetary magnetic field (Svalgaard,
1973; Mansurov, 1969).

2.4 Birkeland Currents, Convection Electric Fields, and ULF-ELF
     Waves in the Cusp

Saflekos et al. (1979) used nearly simultaneous observations from the
TRIAD and Hawkeye spacecraft over the south polar cap to investigate
relationships between Birkeland currents, convection electric fields,
ULF-ELF magnetic noise, broad-band electrostatic noise and the interplanetary magnetic field. Figure 7 (from Saflekos et al., 1979,
their Figure 2) shows the magnetic power spectral density determined

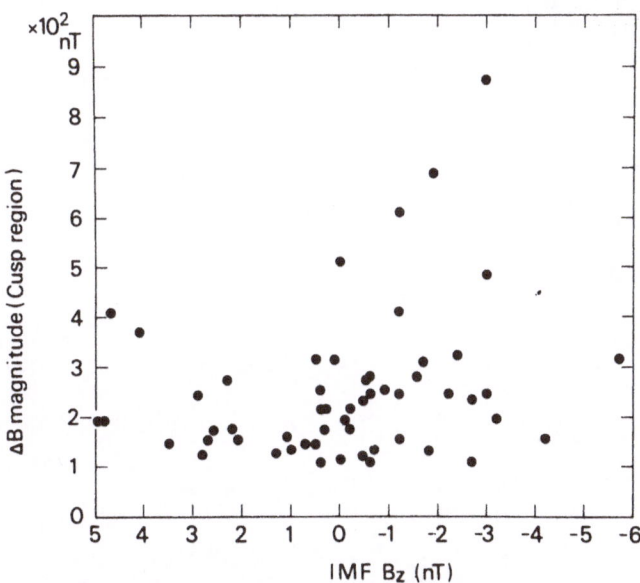

Fig. 6   Relationship between the ΔB (the magnitude of the magnetic
          perturbation in the geomagnetic east-west direction) of the cusp
          region Birkeland current for 55 events acquired in the 0900-1500
          MLT sector during the period from the end of March to August
          1974 in the south polar region and the hourly averaged B_z
          component (GSM) of the interplanetary magnetic field preceding
          the TRIAD observations (from Iijima et al., 1978, their Fig. 9).

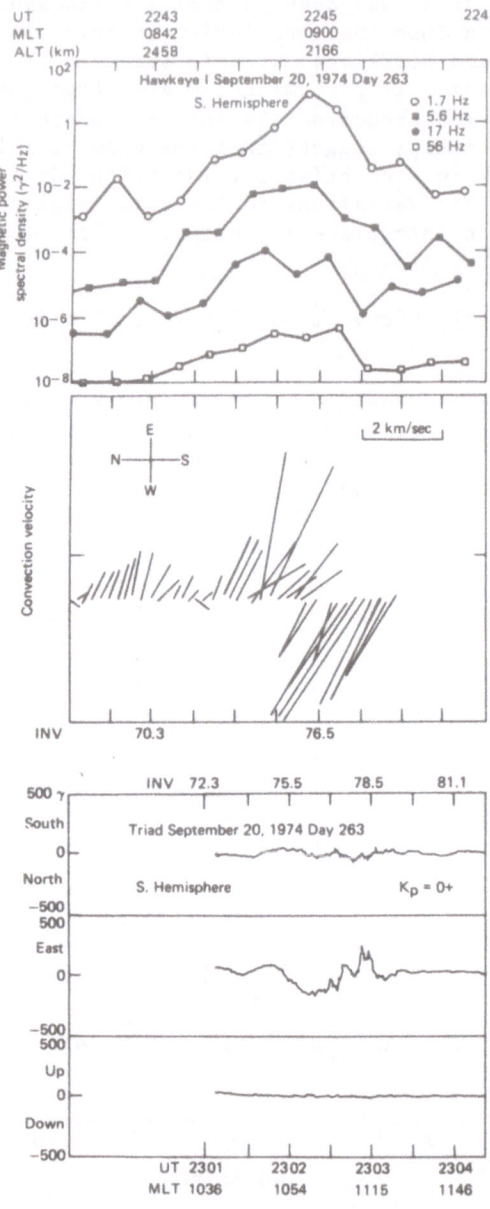

Fig. 7   Top panel represents the magnetic power spectral densities in the
frequency range 1.7-56 Hz from the Hawkeye satellite. Middle
panel represents the convection velocity vectors from the Hawkeye
satellite. Bottom panel shows the three components of magnetic
disturbances observed by TRIAD. The abscissa gives the UT, INV,
and MLT on day 263, 1974, for south hemisphere passes (from
Saflekos et al., 1979, their Fig. 2).

from the Hawkeye search coil experiment in the top panel, the
convection velocity vectors measured with the Hawkeye electric field
experiment in the middle, and the magnetic field disturbances
measured with the TRIAD magnetometer on September 20, 1974. The
three traces in the bottom panel represent the north-south, the east-
west, and radial magnetic disturbances due to Birkeland currents.

The observed intensity of Birkeland currents in this case ranges
from 0.77 to 1.68 $\mu A/m^2$. The reversal of the field-aligned current
sheets near $\Lambda = 76°$-77° coincides spatially with the velocity-shear
reversal $\Lambda = 76°$ and the position of maximum magnetic power spectral
density $\Lambda = 76°$. During this period of observation, the geomagnetic
conditions were very quiet, Kp = 0+. The interplanetary medium was
characterized by a fast solar wind velocity V = 580 km/s and an away
magnetic sector determined by satellite (King, 1977). The hourly
average of the interplanetary magnetic field in geocentric solar magnetospheric coordinates is (B = 6.6 nT, B_x = -5.1 nT, B_y = 2.5 nT and
B_z = 1.1 nT).

Figure 8 (Figure 3 from Saflekos et al., 1979) is an (MLT-INV)-
space plot of the transformed magnetic field disturbance from TRIAD
and the convection velocity pattern from Hawkeye. Both spacecraft
are moving toward the south pole. The view is from inside the earth
looking at the south polar ionosphere. The eastward direction is
counterclockwise from dawn- to noon- to dusk-side. A thick broken
line is drawn in as an aid to the reader and it represents to within

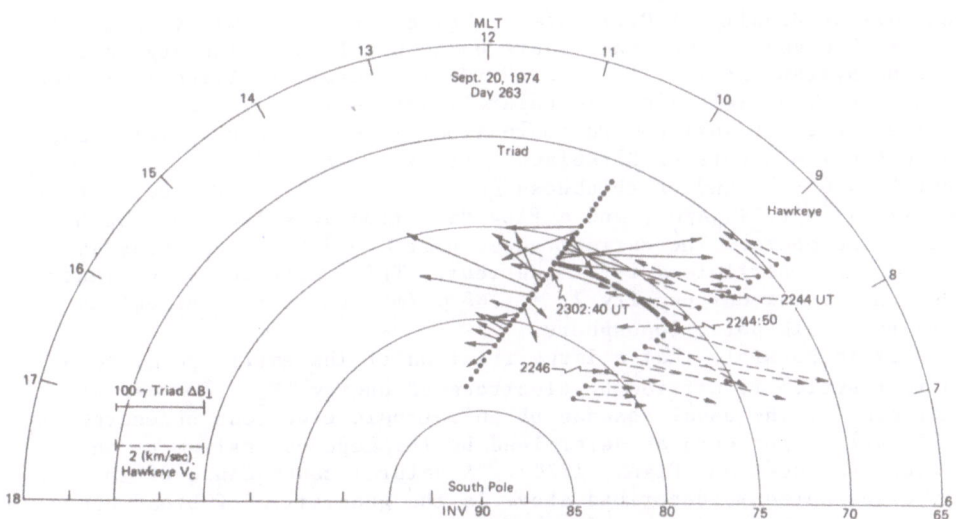

Fig. 8   A polar plot of the magnitude and direction of the magnetic
disturbance vector due to field-aligned currents and of the
convection velocity vector in the two-dimensional space of
magnetic local time and invariant latitude. Observations made
on September 20, 1974. Universal times of interest are marked
along the satellite tracks (from Saflekos et al., 1979, their
Fig. 3).

half a degree accuracy the invariant latitude line on which the
convection velocity or equivalently the convection electric field
reverses. Note that the convection speed is almost 5 km/s at Λ =
75°. Equatorward of the reversal, the electric field points northward (away from the south pole) and poleward of the reversal, it
points in the usual dawn-to-dusk direction as is usual for polar cap
convection electric fields for IMF conditions with dominant eastward
B_y (GSM component). At the electric field reversal, and poleward of
it, exists a Birkeland current sheet oriented in the east-west
direction and flowing away from the ionosphere. We identify it as a
polar cusp Birkeland current. This agrees well with the identification of this same region by the ULF-ELF magnetic noise, an established criterion for this purpose (Gurnett and Frank, 1978). Thus if
the DC electric field reversal and the Region 1 Birkeland currents
are good indicators of the equatorward boundary of the polar cusp, we
locate that boundary at an invariant latitude of 76°-77°.

Adjacent to, and equatorward of the electric field reversal,
there exists a second Birkeland current flowing into the south
ionosphere and having its sheet oriented in the east-west direction
parallel to the average direction of the convection velocity vectors. This is identified as a Region 1 current.

Figure 9 (Figure 6 of Saflekos et al., 1979) shows schematically
the electric field reversal, the general regions of eastward convection on the equatorward side and the general regions of westward convection on the poleward side of it. In the region of sunward convection there is embedded a Birkeland current sheet of thickness 280 km
and current density of 0.77 $\mu A/m^2$. The east-west termination points
of the observed current sheets are unknown. This is the Region 1
current system and it is placed inside the outer Van Allen radiation
zone on closed field lines immediately inside the trapping region.
In the region of antisunward convection close to the noon meridian,
there are two sheets of Birkeland currents. The wider of the two
near Λ = 76°-77° and of thickness ΔS = 280 km has an average current
density $J_\parallel$ = 1.19 $\mu A/m^2$, and a flow direction away from the south
polar ionosphere. The current sheet near Λ = 80° of thickness ΔS =
140 km may constitute a return current. This southernmost current
sheet has a current density $J_\parallel$ = 1.68 $\mu A/m^2$, and a flow direction
into the south polar ionosphere.

It is possible that a large fraction of the entire polar cusp
current system is carried by electrons of energy (E_e < 200 eV) as
indicated by the usual absence of anisotropic electrons of energy (E_e
> 200 eV) in the cusp as determined by the Lepedea instrument on
Hawkeye (Gurnett and Frank, 1978). A natural consequence of the
Birkeland currents described above is the generation of broad-band
VLF hiss and ULF-ELF waves (Shawhan, 1977; Stenzel, 1977).

This study confirmed that the Region 1 Birkeland current is
located equatorward of the region of antisunward convection (and
therefore with the region of closed geomagnetic field lines). The
cusp Birkeland current is located inside the region with strong
antisunward convection. The pair of currents poleward of the Region
1 current, shown in Figure 9, are believed to be cusp currents. The

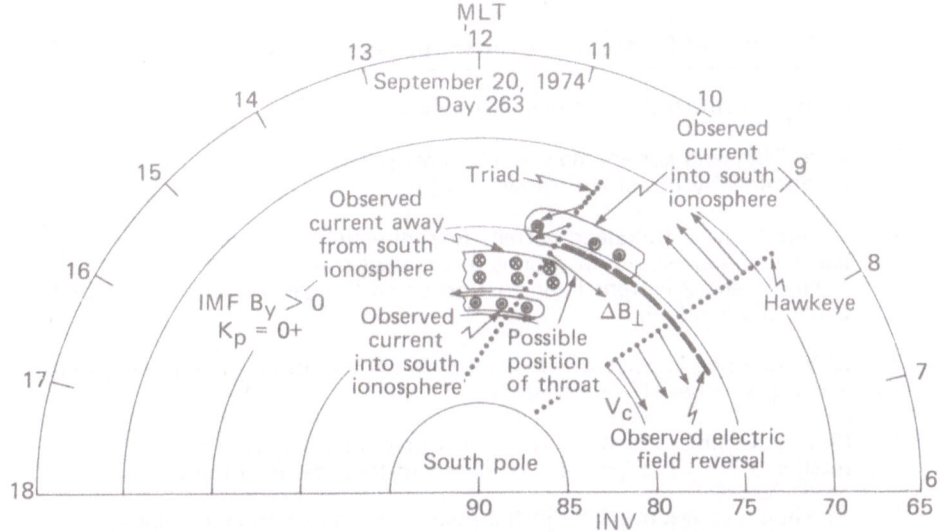

Fig. 9   Plots of the field-aligned current sheets in (MLT-INV)-space. Crosses represent current flow away from the south ionosphere. Conversely, dots represent current flow into the south ionosphere. Vectors along the TRIAD trajectory show the sense and direction of the magnetic disturbance vectors due to the current sheets. The vectors along the Hawkeye trajectory show the sense and direction of the convection velocities. On this day, September 20, 1974, the IMF B_y was positive and the planetary magnetic index Kp was 0+ (from Saflekos et al., 1979, their Fig. 6).

smaller and oppositely-directed current on the poleward side is interpreted as being due to the return flow of the large cusp Birkeland current to the equatorward side. These observations were also presented as evidence in favor of the idea that the fine structure Birkeland currents are associated with the generation of ULF-ELF noise in the cusp.

2.5 Summary of Current Characteristics

A summary of the patterns of Birkeland currents, ionospheric currents, and convection patterns is provided in Table 2 (from Potemra et al., 1979) and in Figure 3 shown earlier. All of these patterns apply to conditions of negative B_z only.

Table 2

Summary of B_y-dependent current patterns (for $B_z < 0$)

Positive B_y, Northern (Southern)[1] Hemisphere:

1. Cusp Birkeland current flow at the most poleward latitudes is predominantly upward[2,3] (downward)[4].

2. Region 1 and 2 Birkeland current flow directions are not affected by B_y and are contained within regions of sunward convection[5,6]. Larger Region 1 and 2 current intensities occur on the dawn side of the auroral zone[7] (dusk side)[8].

3. DPY ionospheric current is located between two Birkeland current sheets and flows eastward[2] (westward?).

4. Convective flow near noon is directed predominantly westward (eastward)[6,9]. The "throat" is shifted to the afternoon (morning).

5. Maximum convective flow in the polar cap (and associated E-fields) is located near dawn (dusk)[10].

All patterns and flow directions are systematically reversed for negative B_y.

[1] For each item the flow direction or spatial location in the Southern Hemisphere is given in parenthesis.
[2] Wilhjelm et al. (1978).
[3] McDiarmid et al. (1978b).
[4] Iijima et al. (1978).
[5] Iijima and Potemra (1976b).
[6] Saflekos et al. (1979).
[7] McDiarmid et al. (1978a).
[8] Saflekos and Potemra (1979).
[9] Heelis et al. (1976).
[10] Heppner (1972, 1973).

3. BIRKELAND CURRENTS DURING POSITIVE B_z

Iijima (1983) discovered the presence of a large-scale Birkeland current system in the northern summer polar region during a period of strongly northward IMF. He determined this from his analysis of a remarkably stable pattern of transverse magnetic disturbance which appeared in data acquired during 8 consecutive MAGSAT orbits over the northern hemisphere on May 10, 1980 (see his Figure 4). Araki et al. (1984) came to the same conclusion from their analysis of MAGSAT data acquired on the same day. This large-scale Birkeland current system occupies the entire region poleward of the dayside Region 1 Birkeland current system, and has the opposite flow direction. Namely, the polar region Birkeland current system flows into the ionosphere on the dusk side and away from the ionosphere on the dawn side.

Iijima et al. (1984) examined magnetic field data acquired during 146 orbits of MAGSAT over the south polar regions during

November, 1979 to January 1980. The characteristics of these polar disturbances include the following: (1) They occur at latitudes poleward of the Region 1 Birkeland current system at daytime magnetic local times (0600 MLT through noon to 1800 MLT). (2) The spatial distribution along the dawn-dusk direction resembles the "W" shaped distribution of electric fields observed in the polar cap during periods of positive B_z (Burke et al., 1979). (3) These patterns show remarkable stability showing little change from orbit to orbit, up to 7 orbits of MAGSAT (equivalent to a period of 10 hours). (4) As shown in Figure 10 (from Iijima et al. 1984), the magnitude of the peak disturbance, ΔB, correlates with a "complementary" magneto-spheric transmission function of the form: $\epsilon^* = (B_y^2 + B_z^2)^{1/2}$ $\cos \theta/2$, where θ is the angle between the positive z-axis and IMF. If the magnetic disturbances are interpreted in terms of Birkeland currents, they flow downward on the dusk side and flow away on the morning-side (identical to the cusp current flow reported by Iijima and Potemra, 1976b). During periods of negative B_y the region of morning-side (upward flowing) currents is much larger than the evening-side (downward flowing) current region in the southern hemisphere (as shown schematically in Figure 11 from Iijima et al., 1984). The density of the currents in the smaller spatial region is larger than the density of the currents in the bigger region. This pattern systematically reverses during periods of positive B_y. These observations are interpreted as evidence for a large-scale, stable Birkeland current system in the polar region which is associated with merging on field lines in the geomagnetic tail. This current system intensifies and is more stable as B_z becomes more northward (remini-scent of the behavior of the Region 1 current system with increasing southward values of B_z). The new stable polar cap current system is referred to as the "NBZ" Birkeland current system for "northward B_z" and is important because of its relationship to a variety of other "northward B_z" phenomena such as polar cap auroral arcs ("theta aurora") and multi-cell convective flow patterns (see, for example, Potemra et al., 1984).

Zanetti et al. (1984) used the radial component of magnetic field measurements from MAGSAT to deduce horizontal currents flowing in the ionosphere during periods of positive B_z. They compared the locations, intensities, and flow directions of the ionospheric cur-rents and Birkeland currents deduced from the same MAGSAT orbits. They discovered a B_y dependence in the characteristics of the ionospheric currents that matched a B_y dependence in the Birkeland currents noted by Iijima et al. (1984) above.

SUMMARY

Large-scale Birkeland currents flow into and away from the dayside high latitude areas identified as the dayside cusp regions. These "cusp-region" Birkeland currents are intimately associated with ionospheric currents, convection electric fields, magnetic turbu-lence, and electrostatic noise. The locations, intensities, and flow

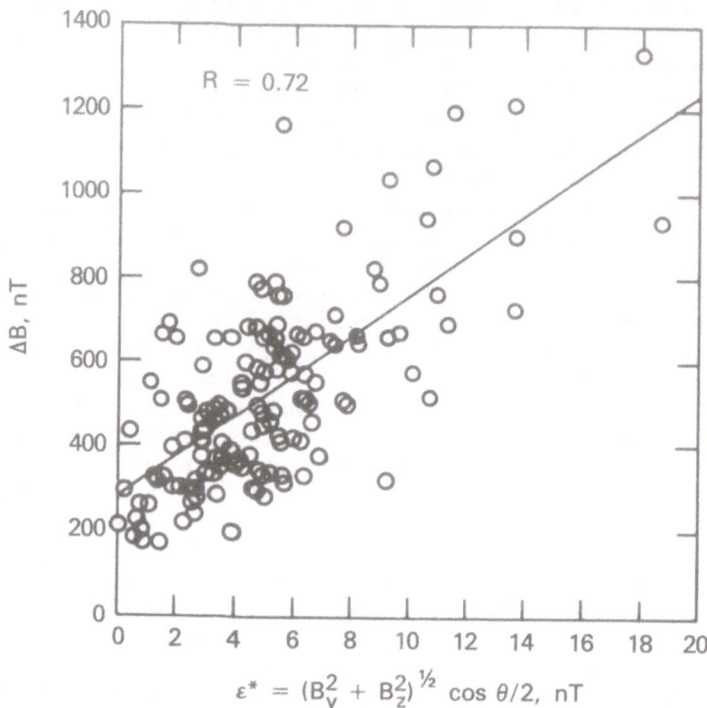

Fig. 10   The magnitude (ΔB) of peak triangular ΔS disturbance plotted
            versus the "complementary transmission function" $\epsilon^* = \cos \theta/2$
            for 146 MAGSAT events, analyzed by Iijima et al. (1984, their
            Fig. 10).

directions of the cusp-region Birkeland currents correlate with the
amplitude and orientation of the interplanetary magnetic field.
These relationships suggest that the cusp Birkeland currents are the
important link between the interplanetary medium and the dayside high
latitude ionosphere.   This Birkeland current system can account for
the close correlation between surface geomagnetic variation in the
polar regions and the sector direction of the interplanetary magnetic
field (Svalgaard, 1973; Mansurov, 1969).

     Although the Birkeland current systems in the auroral zone
diminish in intensity during periods of northward interplanetary
magnetic field (as do many of the standard geomagnetic indices), a
current system develops and intensifies in the dayside polar cap.
This "NBZ" system of Birkeland and ionospheric currents is important
because it demonstrates that a considerable amount of energy flows

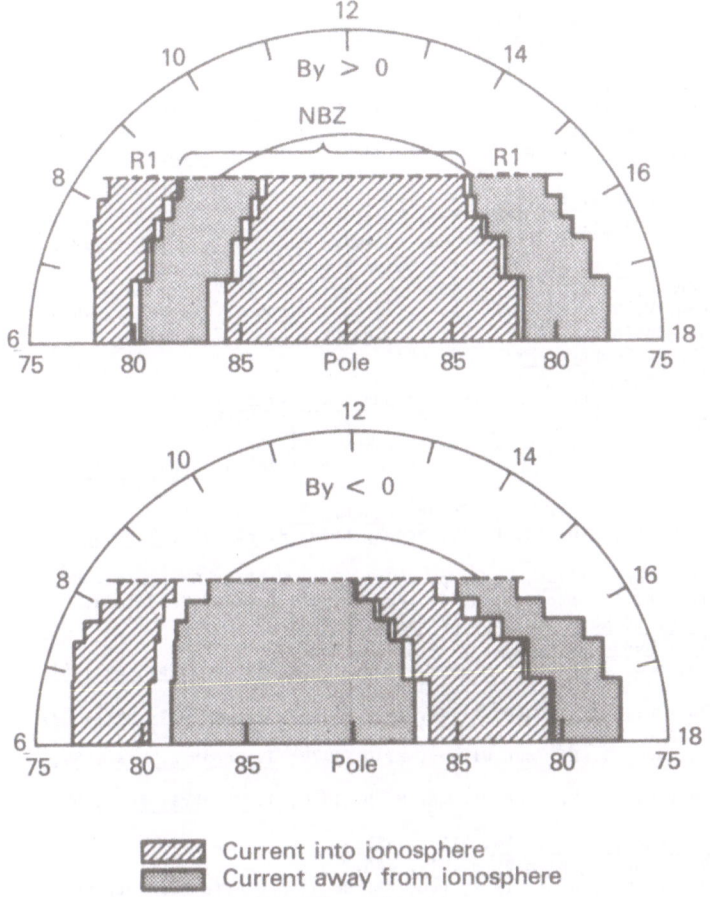

Current into ionosphere
Current away from ionosphere

Fig. 11   The spatial distribution of the flow directions of the large-scale NBZ and Region 1 Birkeland currents in the south polar region during $B_y > 0$ (the upper panel) and during $B_y < 0$ (the bottom panel). This statistical plot was determined from an analysis of 81 MAGSAT orbits during positive B_y and 63 orbits during negative B_y (from Iijima et al., their Fig. 11).

into the magnetosphere even during periods of northward IMF. These various systems of Birkeland currents comprise important elements in the interplanetary-magnetosphere-ionosphere network during almost all conditions.

ACKNOWLEDGEMENT

This work was supported by the Atmospheric Research Section of the National Science Foundation, the Office of Naval Research, and the Defense Nuclear Agency.

REFERENCES

Alexander, J. K., and Kaiser M. L. 1977, J. Geophys. Res., 82, 98.
Araki, T., Kamei, T., and Iyemori, T. 1984, Geophys. Res. Lett., 11, 23.
Burke, W. J., Kelley, m. C., Sagalyn, R. C., Smiddy, M., and Lai, S. T. 1979,
    Geophys. Res. Lett., 6, 21.
Friis-Christensen, E. and Wilhjelm, J. 1975, J. Geophys. Res., 80, 1248.
D'Angelo, N. 1980, Ann. Geophys., 36, 31.
Doyle, M. A., Rich, F. J., Burke, W. J., and Smiddy, M. 1981, J. Geophys. Res., 86,
    5656.
Feldstein, Y. I. 1966, Planet. & Space Sci., 14, 121.
Frank, L. A. 1971, J. Geophys. Res., 76, 5202.
Gizler, V. A., Semenov, V. S., Troshichev, O. A. 1977, EOS, 58 719, (Abstract).
Greenwald, R. A. 1982, Reviews of Geophys. & Space Phys., 20, 577.
Gurnett, D. A. and Frank, L. A. 1978, J. Geophys. Res., 83, 1447.
Heelis, R. A., Hanson, W. B., and Burch, J. L. 1976, J. Geophys. Res., 81, 3803.
Heikilla, W. J., and Winningham, J. D. 1971, J. Geophys. Res., 76, 883.
Heppner, J. P., 1972, J. Geophys. Res., 77, 4877.
Heppner, J. P. 1973, Radio Sci., 8, 933.
Iijima, T. 1983, in Magnetospheric Physics, Geophys. Monograph 28, 115.
Iijima, T., and Potemra, T. A. 1976a, J. Geophys. Res., 81, 2165.
Iijima, T. and Potemra, T. A. 1976b, J. Geophys. Res., 81, 5971.
Iijima, T., Fujii, R., Potemra, T. A., and Saflekos, N. 1978, J. Geophys. Res., 83,
    5595.
Iijima, T., and Potemra, T. A. 1978, J. Geophys. Res., 83, 599.
Iijima, T., Potemra, T. A., Zanetti, L. J., and Bythrow, P. F. 1984, J. Geophys.
    Res., in press.
King, J. H. 1977, NSSDC Rep. 77-04, Nat. Space Sci. Data Center, Greenbelt, Md.
Levitin, A. E., Belov, B. A., Afonina, R. G., Faermark, D. S. and Feldstein, Y. I.
    1977, USSR Academy of Sciences IZMIRAN Preprint N 17a.
Mansurov, S. M., 1969, Geomagn. Aeron., 9, 622.
McDiarmid, I. B., Burrows, J. R., and Wilson, M. D. 1978a, J. Geophys. Res., 83,
    5753.
McDiarmid, I. B., Burrows, J. R., and Wilson, M. D. 1978b, J. Geophys. Res., 83,
    681.
McDiarmid, I. B., Burrows, J. R., and Wilson, M. D. 1979, J. Geophys. Res., 84,
    1431.
McDiarmid, I. B., Burrows, J. R., and Wilson, M. D. 1980, J. Geophys. Res., 85,
    1163.
Meng, C.-I. 1981, J. Geophys. Res., 86, 2149, 1981.
Potemra, T. A. 1979, Geophys. Space Phys., 17, 640.
Potemra, T. A., and Saflekos, N. A. 1979, Proc. of Magnetospheric Boundary Layers
    Conference, Alpbach, 11-15 June 1979, 193.
Potemra, T. A., Saflekos, N. A., and Gustafsson, G. 1979, EoS Trans. AGU, 60, 348.
Potemra, T. A. 1983, Johns Hopkins APL Technical Digest, 4, 276.
Potemra, T. A., Zanetti, L. J., Bythrow, P. F., Lui, A. T. Y., and Iijima, T. 1984,
    J. Geophys. Res., in press.
Reiff, P. H., Burch, J. L., and Heelis, R. A. 1978, Geophys. Res. Lett., 5, 391.
Saflekos, N. A., Potemra, T. A., Kintner, Jr., P. M., and Green, J. Lauer 1979, J.
    Geophys. Res., 84, 1391.
Saflekos, N. A., and Potemra, T. A. 1980, J. Geophys. Res., 85, 1987.
Saflekos, N. A., Sheehan, R. E., and Carovillano, R. L. 1982, Rev. of Geophys. &
    Space Phys., 20, 709.
Shawhan, S. D. 1977, Res. Rep. 77-23, Univ. of Iowa, Iowa City.
Stenzel, R. L. 1977, J. Geophys. Res., 82, 4805.
Sugiura, M., and Potemra, T. A. 1976, J. Geophys. Res., 81, 2155.
Svalgaard, L. 1973, J. Geophys. Res., 78, 2064.
Wilhjelm, J. E., Christensen, E. Friis, and Potemra, T. A. 1978, J. Geophys. Res.,
    83, 5586.
Zanetti, L. J., Potemra, T. A., Iijima, T., Baumjohann, W., and Bythrow, P. F. 1984,
    J. Geophys. Res., in press.
Zmuda, A. J., and Armstrong, J. C. 1974, J. Geophys. Res., 79, 2501.

DAYSIDE HIGH-LATITUDE IONOSPHERIC CURRENT SYSTEMS

W. Baumjohann[*] and E. Friis-Christensen[**]
* MPI für extraterr. Physik, 8046 Garching, W-Germany
** Meteorological Institute, Copenhagen, Denmark

ABSTRACT.   This paper reviews the present understanding of the current flow in the dayside high-latitude ionosphere. We shall at first describe the morphology of this current circuit (on the basis of ground-based and low-altitude satellite magnetic measurements) and then point out its essential elements and their relation to the dynamics of the interplanetary magnetic field (IMF). Subsequently, we shall discuss possible generation mechanisms for these currents, especially merging between interplanetary and magnetospheric field lines.

1.   INTRODUCTION

The current flow in the high-latitude ionosphere is generated by different physical processes operating in quite different regions of the magnetosphere since the geomagnetic field lines intersecting this region map to various boundaries and plasma regimes at the magnetopause and inside the magnetosphere. This mapping is visualized in Figure 1 taken from Vasyliunas (1979).

The current flow in the nightside auroral oval is governed by processes taking place within the earth's magnetosphere, i.e. in the plasma sheet and the plasma sheet boundary (the latter is denoted "plasma boundary layer" in Fig. 1). The morphology of these currents, i.e. the convection and substorm electrojets, is well-known (cf. Kamide 1982; Baumjohann 1983) and also the chain of processes controlling their dynamics is rather well understood and has been even modelled numerically (Harel et al. 1981).

The remainder of the current flow, i.e. those currents flowing in and near to the polar cusp and in the dayside polar cap, are connected (along geomagnetic field lines) to an produced by processes taking place on the magnetopause. Since the latter are governed by the dynamics of plasma and magnetic field within the magnetosheath, and thus in the solar wind, the solar wind and especially the interplanetary magnetic field (IMF) exerts a controlling influence on these dayside high-latitude currents. In the following we will summarize the morphology of these currents and their dependence on the IMF and we will rehearse

J. A. Holtet and A. Egeland (eds.), The Polar Cusp, 223–234.
© 1985 by D. Reidel Publishing Company.

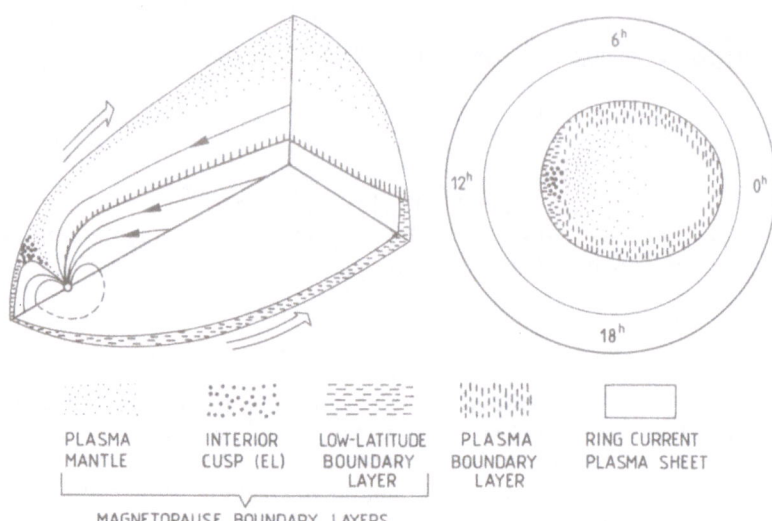

Fig. 1.   Left panel: Schematic diagram of various observed magnetospheric boundary layers. The great extent of the low-latitude boundary
layer into the nightside is speculative. Open arrows indicate the flow
of the magnetosheath plasma. Right panel: Schematic diagram of the
regions occupied by the various boundary layers, mapped down to the
ionosphere along magnetic field lines. Hours indicate local time. The
boundary between the low-latitude and the plasma sheet boundary layers
is highly uncertain (from Vasyliunas 1979).

some ideas on how these currents are generated. We should, however,
caution that these ideas are not at all so developed as those for the
nightside electrojet currents and that considerable work still has to
be done to understand and model the full chain of processes.

2.   MORPHOLOGY OF THE POLAR CAP CURRENT SYSTEMS

In the following we shall describe the morphology of the polar cap and
cusp currents and their dependence on the state of the IMF. Our description will mainly be based on two very recent studies by Friis-
Christensen et al. (1984) and Zanetti et al. (1984) who used ground-
based (Greenland chain) and low-altitude satellite (Magsat) magnetic
measurements, respectively. We will restrict our description to the
Hall current system since this come closest to the convection pattern.
The Pedersen current pattern, however, can easily be visualized if one
keeps in mind that these currents flow perpendicular to the Hall currents with a density of about half of the latter (cf. Friis-Christensen
et al. 1984). We will also omit a thorough discussion on the correspondence between the ionospheric currents and the electric field or
Birkeland current patterns but just note that this correspondence is

generally as expected. For a more thorough description of electric field and Birkeland current patterns the reader is referred to the companion reviews by Heelis and Potemra in this volume.

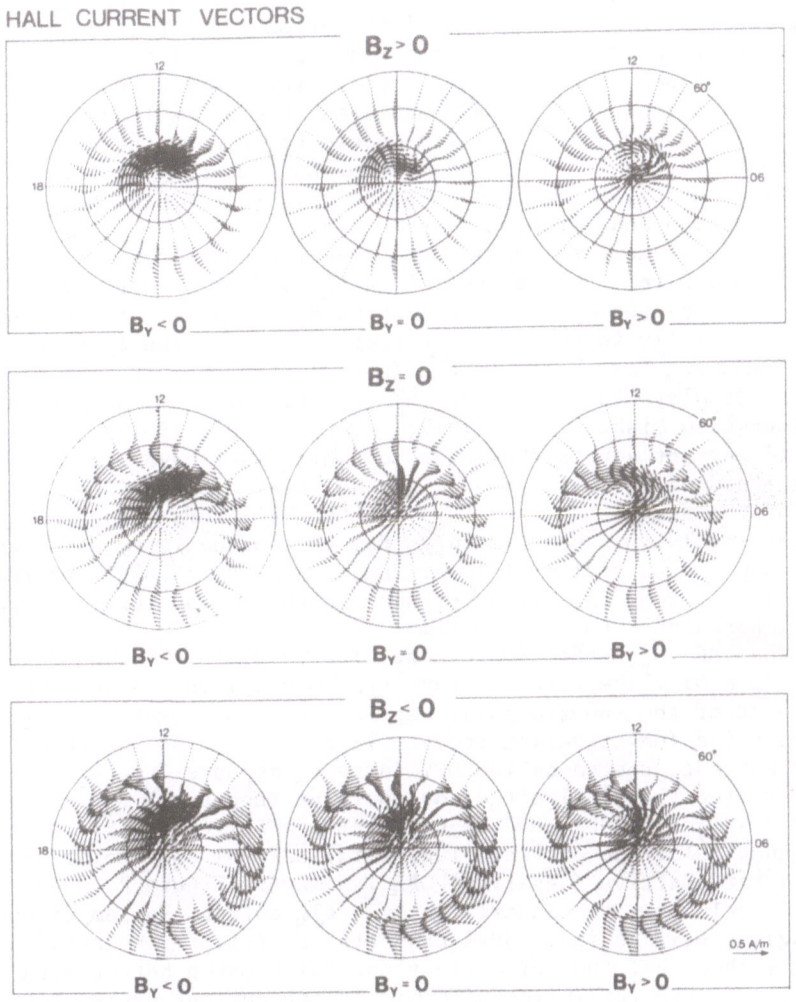

Fig. 2. Polar diagrams of Hall current vector distributions in the northern hemisphere calculated by means of an empirical model of magnetic variations (from Friis-Christensen et al. 1984). Nine different states of the IMF B_y and B_z components are considered (see text).

Friis-Christensen et al. (1984) used Greenland magnetometer data

together with high-resolution IMF data and the numerical model
described by Kamide et al. (1981) to obtain statistical relationship
between the IMF components and ionospheric and field-aligned currents
especially in the polar cap and cusp. As a first step a multiple linear
regression analysis was done to establish a linear relation between the
Greenland chain magnetic perturbation vectors and the IMF B_y and B_z
components as a function of local time and station. This empirical
model was then used to predict distributions of magnetic perturbation
vectors corresponding to selected and representative conditions of the
IMF. From these magnetic disturbance vectors corresponding equivalent
current functions were derived and used together with simple and analytical ionospheric conductivity models to compute the electric potential distribution in the polar region. Knowing conductivity and electric field distribution it was then simple to compute the pattern of
Hall, Pedersen and Birkeland currents.

Figure 2 shows the Hall current vectors in the northern hemisphere
for various directions of the IMF in the Y-Z plane. Around 70° geomagnetic latitude the eastward and westward auroral electrojets prevail,
which especially in the midnight sector depend strongly on the plasma
sheet dynamics, as stated in the Introduction. The $B_y = B_z = 0$ polar
diagram displays the so-called S_q^p system which is independent of the
IMF (see also Mishin et al. 1979; Levitin et al. 1982). It exhibits
sunward current across the polar cap while antisunward current flows
equatorward of the dawn- and dusk-side polar cap boundary. The rotation of the axis of symmetry of this two-cell current system towards
earlier local times stems mainly from the conductivity gradient between
polar cap and auroral oval (e.g. Yasuhara et al. 1983). The remaining
eight frames show the effect of non-vanishing transverse IMF components
superimposed onto the S_q^p system.

The IMF B_y effects are most clearly visible in the middle panel
(IMF $B_z = 0$). The total current system displays the well-known displacement of the sunward polar cap Hall current toward dawn (dusk) for
B_y positive (negative). Moreover, in the region of the polar cusp a
zonal current, the so-called DPY current, appears for $B_y \neq 0$ which is
eastward (westward) directed for positive (negative) B_y. Some authors
(e.g. Rostoker, 1980) have argued that the DPY current is merely a
continuation of the eastward (westward) auroral electrojet toward dawn
(dusk) for positive (negative) B_y. However, Figure 2 (especially
upper panel where the underlying two-cell current system is not so
strong) as well as individual case studies (cf. Friis-Christensen 1984)
clearly show that the DPY current does also match better with existing
theories on the generation of the polar cap current system (cf. next
section).

Also the effects of changes in the IMF B_z component are clearly
visible in Figure 2. For negative B_z (southward IMF) the two-cell
currents and especially the auroral electrojets are greatly enhanced.
This is due to the appearance of a new double-cell system, the so-
called DP2 currents, which have about the same geometry as the S_q^p
system. The DP2 system, however, is neither independent of the state
of the IMF nor is it unaffected by processes in the plasma sheet: it
typically accompanies the storage of solar wind energy in the geomag-

netic tail prior to a substorm - see, for example, Pellinen et al.
(1982) or Kamide and Baumjohann (1984). For $B_z = 0$ the DP2 system
vanishes and for $B_z > 0$ an antisunward directed Hall current component appears in the polar cap. If the S_q^p system is subtracted the
residual current system especially for $(B_z > 0, B_y = 0)$ corresponds
to a "reversed" two-cell current system earlier reported by, for
example, Maezawa (1976) or Horwitz and Akasofu (1979) and in agreement
with convection patterns for northward IMF derived from satellite electric field measurements (Burke et al. 1979).

In order to show the antisunward Hall current component more clearly and, furthermore, to exhibit the different directions of the DPY currents in the northern and southern hemisphere for the same IMF B_y component, we display in Figure 3 a schematic by Zanetti et al. (1984) which
comprises data from a couple of orbits of the Magsat satellite about
300 km above the southern polar cap. They used the vertical disturbance
component for a continuation towards the source in the ionosphere via a
Fourier analysis - for details of this method see, for example, Baumjohann et al. (1979) or Zanetti et al. (1983) - and then calculated the
Hall current density distribution for several passes during two days
with steady northward IMF.

Figure 3 shows three-cell patterns for positive and negative B_y
similar to Figure 2, but now, in the southern hemisphere, the DPY current is directed towards dawn (dusk) for positive (negative) B_y. This
hemispherical asymmetry is not restricted to positive B_z but prevails
for all states of the IMF B_z component (cf. Cowley 1981). For $B_y = 0$
a four-cell current patterns was clearly visible in Zanetti et al.'s
(1984) data with antisunward current (sunward convection) across the

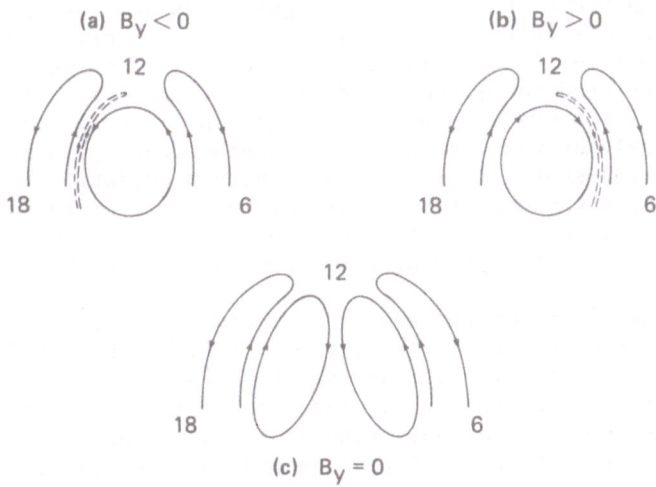

Fig. 3. Sketch of the ionospheric Hall current pattern in the southern
hemisphere for northward B_z as it appears to be influenced by the IMF
B_y component (from Zanetti et al. 1984).

polar cap. Moreover, Zanetti et al. speculated that also for $B_y \neq 0$ a fourth cell (shown by dashed lines) with a shear-like current reversal is present near the dawn- or dusk-side boundary of the polar cap, but if this cell exists it must be very thin and data with higher spatial resolution is needed to confirm it.

3. POSSIBLE GENERATION MECHANISMS FOR THE POLAR CAP CURRENTS

In this section we shall show that the polar cap current systems described in the last section are generated by two basic mechanisms: viscous interaction and magnetic merging. The way both mechanisms operate can be made clear best by discussing the generating processes for the S_q^p and DP2 current systems. The S_q^p system is independent of the state of the IMF, but correlated with the solar wind velocity (Mishin et al. 1979; Levitin et al. 1982) and has thus been attributed to viscous interaction. The DP2 system, having the same geometry but appearing for $B_z < 0$, on the other hand, has been ascribed to merging of northward-directed magnetospheric field lines with southward-directed ones in the dayside equatorial magnetosheath and subsequent reconnection in the geomagnetic tail.

Figure 4 (from Cowley 1982) nicely sketches these two processes and the resulting plasma flow in the magnetospheric equatorial plane and in the ionosphere: the hatched regions indicate the flow driven by a viscous process in the low-latitude boundary layer. Here the magnetic tubes should remain closed throughout the cycle, while the remainder of the flow is driven by reconnection with antisunward flow (sunward Hall current) on open polar cap field lines and sunward flow (eastward and westward electrojets) on closed field lines. We should, however, add that the data in the central frame of Figure 2 as well as similar S_q^p systems given by Mishin et al. (1979) and Levitin et al. (1982) indicate that the current flow ascribed to viscous interaction does not seem to be restricted to closed field lines but that part of the S_q^p current also flows sunward across the polar cap. This either means that the viscously driven ionospheric currents create polarization charges in the ionosphere (due to conductivity gradients) which are accompanied by a dawn-dusk cross polar cap potential difference or that viscous interaction does not only take place on closed low-latitude boundary layer field lines but also on open ones in the high-latitude lobe.

The reconnection geometry shown in Figure 4, and thus the DP2 current system, is that for a purely southward IMF. However, IMF B_y effects and thus the DPY currents can easily be incorporated if the dawn-dusk magnetic tension of the solar wind/magnetosheath end of the merged field line is taken into account. Figure 5 (taken from Cowley 1981) illustrates this situation. Beginning on the dayside, the tension on newly opened flux tubes has a net east-west component in the presence of a B_y field, which is oppositely directed in both hemispheres. In response, oppositely directed azimuthal plasma flows and hence Hall currents appear in the polar cusp region as discussed in the previous section (compare Figures 2 and 3).

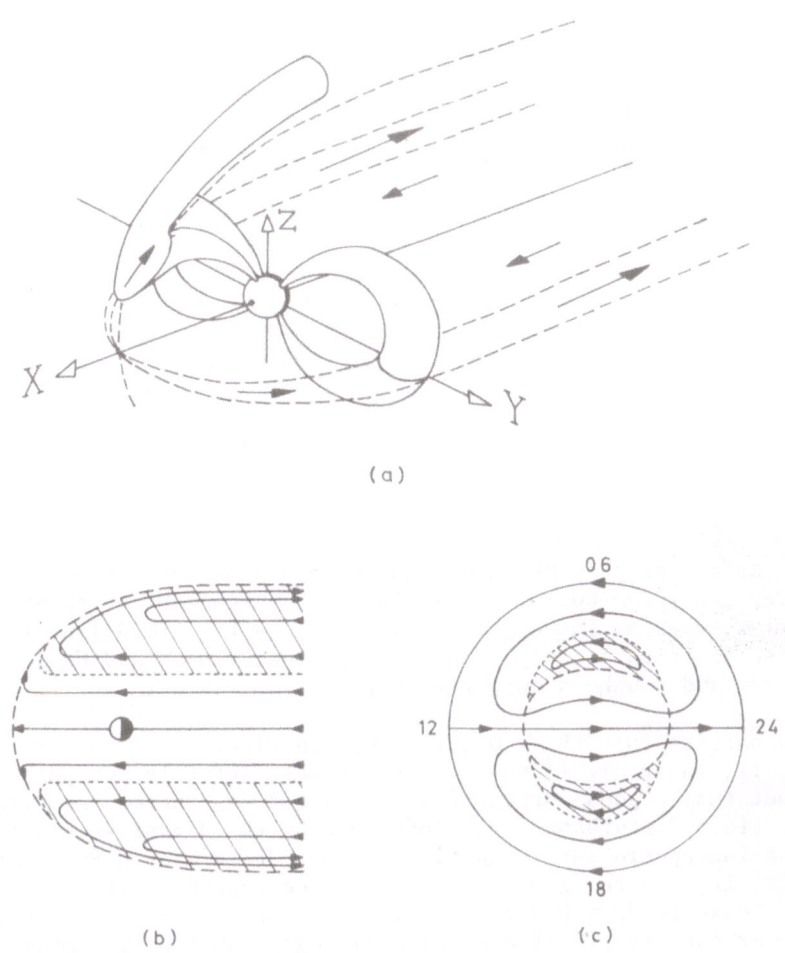

Fig. 4. Sketch (a) illustrates two basic processes contributing to solar wind-driven magnetospheric convection, reconnection and viscous interaction. The resulting plasma flow is shown in the lower diagrams, (b) in the magnetosphere and (c) in the ionosphere. In the latter the Hall current pattern would be given by just reversing the direction of the arrowheads. In (b) and (c) the hatched region corresponds to the flow driven by viscous interaction while the remainder is associated with reconnection (from Cowley 1982).

For purely northward IMF, merging cannot occur on closed dayside magnetopause field lines but rather on open lobe field lines north of the cusp. In this case also a two-cell convection and current pattern evolves (e.g. Maezawa 1976), but it is restricted to open polar cap field lines and the direction is reversed, in agreement with the

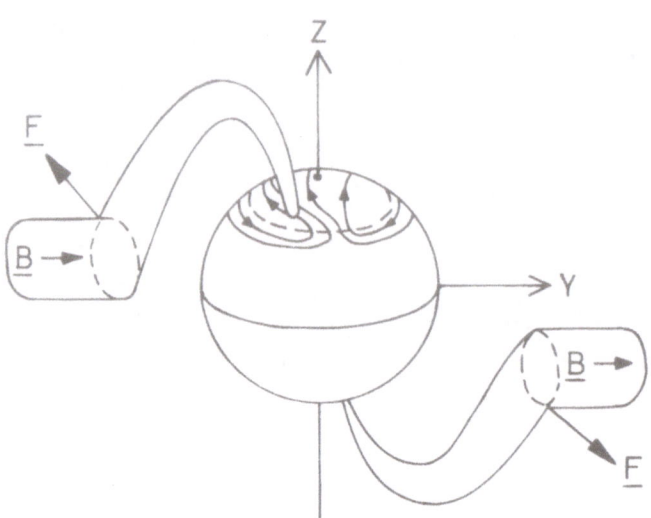

Fig. 5. Effect of the IMF B_y component on plasma convection in the dayside polar cap (from Cowley 1981). Shown is a schematic view of newly
opened flux tubes on the dayside, illustrating the field line tension $\underline{F}$
which results in oppositely directed azimuthal flows (and Hall currents)
in northern and southern cusp regions.

observed pattern for $(B_z > 0$, $B_y = 0)$ in Figure 3. A very recent
study by Iijima et al. (1984) even shows that not only the pattern
agrees but that a good correlation exists between the maximum transverse
magnetic field disturbance ΔB (observed by the Magsat satellite) just
above the ionosphere and a function of the IMF (given in Figure 6).
Since ΔB is, for the given geometry of the current system, proportional to the dusk-to-dawn Pedersen current and thus to the dusk-to-dawn
cross polar cap electric field while Iijima's function resembles quite
closely the merging electric field for the aforementioned geometry (cf.
Sonnerup 1974; Hill 1975; Kan and Lee 1979) this means that there is
not only qualitative but also some quantitative agreement between the
observed current system and the theoretically proposed merging with
northward interplanetary field lines. (Note that a good correspondence
between the cross polar cap potential and the dayside merging electric
field for southward IMF was already earlier established by Reiff et al.
(1981), Doyle and Burke (1983), and Wygant et al. (1983) on the basis
of low-altitude satellite electric field measurements.)
    We should, however, add that there is one aspect of merging with
northward IMF that is still unclear: since the interplanetary field
line merges with an open field line it is not known if the interplanetary field line merges with magnetically conjugate field lines in both
hemispheres (called "reclosure type") of if it merges with "different"
field lines in the northern and southern hemisphere (so-called "stirring
type"). Figure 7 (from Reiff 1982) illustrates these two different to-

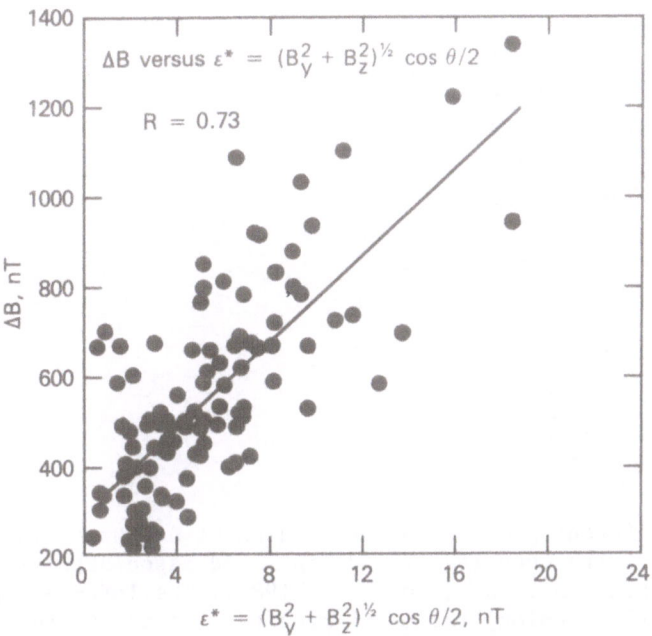

Fig. 6. Scatter plot of the peak transverse magnetic disturbance just above the polar cap ionosphere versus a function of the IMF for more than 100 transpolar Magsat satellite passes during northward IMF (from Iijima et al. 1984).

pologies (for IMF due northward) but also shows that the same four-cell convection and Hall current pattern (the two inner cells associated with merging and the two outer cells with viscous interaction) emerges for both merging geometries. Depending on the geometry of the geomagnetic tail (open or closed) one can differentiate between even more merging geometries for northward IMF (cf. Cowley 1981; Kivelson 1982) but they all result in the same basic polar cap convection pattern so that a distinction between them on the basis of polar cap current patterns is not possible.

While for southward IMF the basic two-cell convection due to merging prevails even for strong B_y components (only an asymmetry appears) the data presented in Figures 2 and 3 suggests that for northward IMF only a single convection cell remains in the polar cap if the IMF has a non-zero B_y component. Looking into Figure 8 (from Crooker 1979) it can be seen that also this data feature is reproduced by the pattern of field line and plasma convection caused by magnetic merging at the dayside magnetopause (Figure 8 was actually constructed by Crooker (1979) on the basis of merging between strictly antiparallel fields, but a qualitatively similar pattern would also emerge for component merging - see Cowley (1981)).

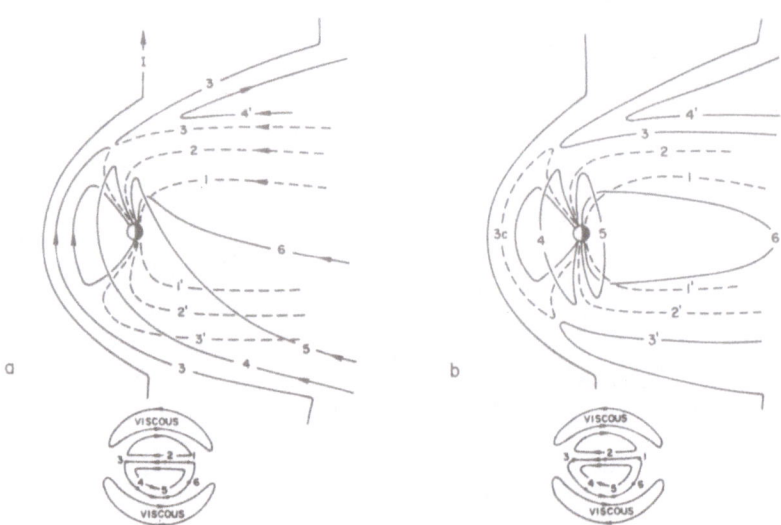

Fig. 7. Two different geometries for merging of northward interplanetary field lines with open lobe field lines and associated (identical)
polar cap convectionpatterns (from Reiff 1982). The left-hand panel
gives the magnetic topology for a stirring type of northward merging
while the right-hand panel gives the same for a reclosure type (which
is more likely to be quasi-conjugate).

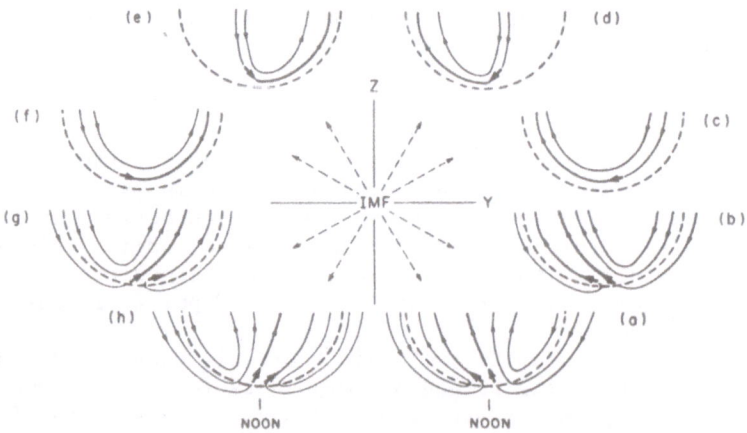

Fig. 8. Predicted polar cap convection patterns in the dayside polar
hemisphere due to (antiparallel) merging (from Cooker 1979). The short
heavy arrows represent the direction of flow initiated by dayside merging near the cusp. Note that convection due to merging is restricted to
open polar cap field lines for $B_z > 0$ while the convection crosses
the polar cap boundary (shown by the dahsed circle for southward IMF.

4. CONCLUSION

In this short review we have tried to show that most aspects and parts of the large-scale dayside high-latitude ionospheric current pattern can be explained by magnetic merging between interplanetary field lines of different orientations and closed or open dayside field lines of terrestrial origin intersecting the magnetopause. The remaining large-scale currents, i.e. the S_q^p system, are caused by viscous interaction between the tailward streaming magnetosheath plasma and the magnetospheric plasma, at least in the low-latitude boundary layer.

However, the good agreement between observed and expected patterns does not mean that everything is already known. The above description of convecting merged and reconnected field lines provides the pattern but not the strength of the polar cap currents. In order to estimate the latter, one must actually consider the MHD dynamos in the magnetospheric boundary layers which provide the energy dissipated by the currents in the resistive ionosphere. Since in order to do this all circuit elements (boundary layer generator, ionospheric load, and transmission line, i.e. (resistive) field lines) have to be taken into account simultaneously. This has been done for the nightside current circuit (cf. Harel et al. 1981) but for the dayside high-latitude currents only considerations restricted to one of the circuit elements have been made (cf. Vasyliunas, 1979).

ACKNOWLEDGEMENTS

One of us (W.B.) is grateful to the Deutsche Forschungsgemeinschaft for financial support through a Heisenberg-Fellowship.

REFERENCES

Baumjohann, W., 1983, *Adv. Space Res.*, 2(10), 55.
Baumjohann, W., et al. 1979, p. 49 in *Magnetospheric Study 1979* (ed. T. Obayashi). Japanese IMS Committee, Tokyo.
Burke, W.J., et al. 1979, *Geophys. Res. Lett.*, 6, 21.
Cowley, S.W.H. 1981, *NATO AGARD-CP*, 293, 4-1.
Cowley, S.W.H. 1982, *Rev. Geophys. Space Phys.*, 20, 531.
Crooker, N.W. 1979, *J. Geophys. Res.*, 84, 951.
Doyle, M.A., and Burke, W.J. 1983, *J. Geophys. Res.*, 88, 9125.
Friis-Christensen, E. 1979, p. 290 in *Magnetospheric Study 1979* (ed. T. Obayashi). Japanese IMS Committee, Tokyo.
Friis-Christensen, E. 1984, p. 86 in *Magnetospheric Currents* (ed. T.A. Potemra). AGU, Washington.
Friis-Christensen, E., et al. 1984, *J. Geophys. Res.*, 89, in press.
Harel, M., et al. 1981, *J. Geophys. Res.*, 86, 2217.
Hill, T.W. 1975, *J. Geophys. Res.*, 80, 4689.
Horwitz, J.L., and Akasofu, S.-I. 1979, *J. Geophys. Res.*, 84, 2567.
Iijima, T., et al. 1984, *J. Geophys. Res.*, 89, in press.
Kamide, Y. 1982, *Space Sci. Rev.*, 31, 127.

Kamide, Y., and Baumjohann, W. 1984, *J. Geophys. Res.*, 89, in press.
Kamide, Y., et al. 1981, *J. Geophys. Res.*, 86, 801.
Kan, J.R., and Lee, L.C. 1979, *Geophys. Res. Lett.*, 6, 577.
Kivelson, M.G. 1982, *J. Geophys. Res.*, 87, 5981.
Levitin, A.E., et al. 1982, *Phil. Trans. R. Soc. Lond. A*, 304, 253.
Maezawa, K. 1976, *J. Geophys. Res.*, 81, 2289.
Mishin, V.M., et al. 1979, p. 249 in *Dynamics of the Magnetosphere* (ed. S.-I. Akasofu). Reidel, Dordrecht.
Pellinen, R.J., et al. 1982, *Planet. Space Sci.*, 30, 371.
Reiff, P.H. 1982, *J. Geophys. Res.*, 87, 5976.
Reiff, P.H., et al. 1981, *J. Geophys. Res.*, 86, 7639.
Rostoker, G. 1980, *J. Geophys. Res.*, 85, 4167.
Sonnerup, B.U.Ö. 1974, *J. Geophys. Res.*, 79, 1546.
Vasyliunas, V.M. 1979, *ESA Spec. Publ.*, 148, 387.
Wygant, J.R., et al. 1983, *J. Geophys. Res.*, 88, 5727.
Yasuhara, F., et al. 1983, *J. Geophys. Res.*, 88, 5773.
Zanetti, L.J., et al. 1983, *J. Geophys. Res.*, 88, 4875.
Zanetti, L.J., et al. 1984, *J. Geophys. Res.*, 89, in press.

SMALL SCALE INTENSE FIELD ALIGNED CURRENT SHEETS IN THE NORTHERN POLAR CUSP.

Annick Berthelier and Christine Machard
CRPE - CNET - CNRS
4 avenue de Neptune - 94107 Saint-Maur - CEDEX
France

ABSTRACT. From the measurement of both AC and DC components of the magnetic field on board AUREOL-3 satellite, we have analysed the fine scale structure of Field Aligned Currents. Isolated pulses have been detected on the horizontal components of the AC magnetic field in regions where FAC are highly structured, and they have been identified as the signature of the traversal of Small Scale Intense upward Field Aligned Current sheets. Their characteristics have been derived from a modelisation of current sheet crossing.

Two examples of such structures observed in the northern polar cusp are presented here. They have typical width of 100 to 500 m, and intensity of 100–200 μA/m2, and in one case two successive crossings of a thin upward current sheet are observed imbedded in a mid-scale (∿10-20 km) downward return current of ∿10 μA/m2 intensity.

1. INTRODUCTION

During the past decade the main features of the large scale high latitude Field Aligned Currents have been derived from a number of satellite, rocket, radar or ground magnetometer measurements (see reviews by Potemra 1979, and in this volume).

It appeared also clearly from those measurements that FAC display in general a complex and highly variable spatial structure, and that the current density may vary from a few μA/m2 to some hundred μA/m2 while the spatial structures range from a few km to some tenths of km (e.g. Berko et al 1975 ; Saflekos et al 1978 ; Doyle et al 1981 ; Sugiura et al 1983 ; Burke et al 1983 ; Robert et al 1984).

The purpose of this paper is to present simultaneous measurements of the DC and AC magnetic field taken from instrumentation on board AUREOL-3 satellite, from which we have deduced the main characteristics of the fine scale structures in the Field Aligned Currents observed during crossing of the Northern polar cusp.

J. A. Holtet and A. Egeland (eds.), The Polar Cusp, 235–242.

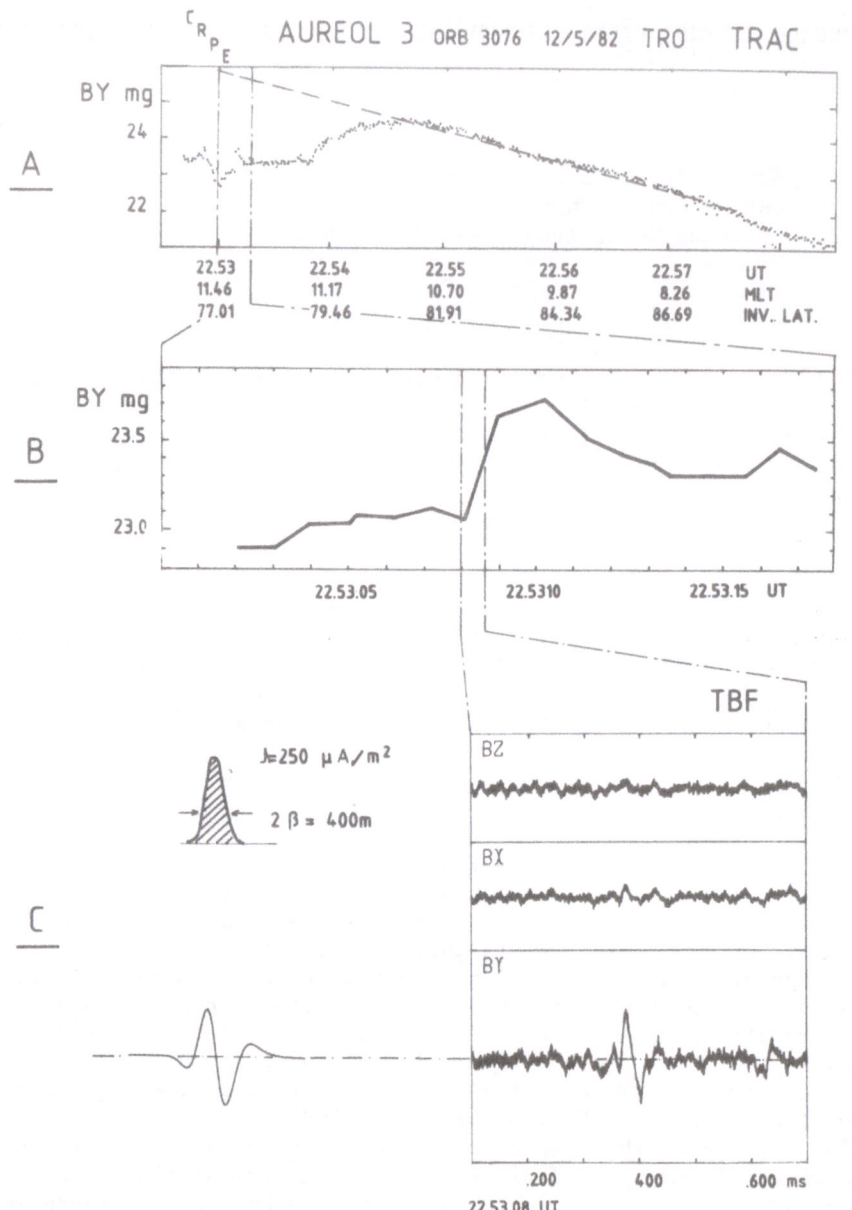

Figure 1-A- Variation of the BY component of the DC Magnetic field for the orbit 3076 ; the dashed line indicates the variation of BY as esti-mated in the absence of FAC - B- Same as A, on an extended time scale - C- Variation of the three components of the AC magnetic field (at right), and model of AC output signal on By corresponding to the characteristics of the observed pulses.

2. EXPERIMENTAL RESULTS

2.1. Instrumentation

AUREOL-3 is a soviet satellite with a joint French Soviet Scientific
payload ; it was launched on September 21st, 1981 into a low altitude
polar orbit with perigee ∿400 km, apogee ∿2000 km, inclination 82°30' ;
it is a three axis stabilized spacecraft with the X axis pointing in
the direction of the velocity vector, the Z axis along the local
ascending vertical and the Y axis completing the right hand frame of
reference. The TRAC experiment (Berthelier et al 1982a) processes the
data from the on-board soviet fluxgate magnetometer to perform measurements of the three axis of the DC magnetic field with a resolution
of ∿12 nT, and a sampling rate of 1 to 12 points per second, depending
on the telemetry mode. The TBF experiment (Berthelier et al 1982b)
measures the three components of the AC magnetic field in the frequency
range from ∿10 Hz to 1.5 KHz.

These experiments has worked satisfactorily since the launch, and
approximately 3-4 passes at high latitudes both in the Northern and
Southern hemispheres have been recorded daily during the first two years
of operation.

2.2. Observations

The transverse BY component of the DC magnetic field is represented as
a function of time in panel A of figures 1 and 2. Its variation gives
access to the average density $j_{//}$ of the crossed FAC at high latitudes,
using infinite current sheet approximation, namely :

$$j_{//} = \frac{1}{\mu_o} \frac{\Delta BY}{\Delta X}$$

$j_{//}$ is an averaged value over ∿7 km in the case of 1 point per second
sampling rate, and over ∿600 m in the case of 12 pts per s. sampling
rate.

During FAC crossings, it often happens that impulsive bursts are
detected on the wave form of the output AC channels. They commonly
appear as irregular series of pulses, sometimes isolated (typical
duration of 50-100 ms.), sometimes as "packets" extending over about 1
to a few seconds. They are generally seen only on BX and BY components ;
i.e. situated in the plane perpendicular to the main geomagnetic field,
and the more often on BY, i.e. transverse to the speed satellite direction X. Examples of such signatures on the AC signal are presented with
an extended time scale on panel C of figures 1 and 2. These pulses are
generally observed in regions where the BY component of the DC magnetic
field is rapidly varying which indicates that the satellite crosses a
structured FAC.

The AC signal may be due to either a temporal variation or a
spatial variation of the magnetic field. We shall restrict ourselves in
this paper to the cases of isolated pulses ; their typical shape (see
below) and their most common orientation roughly along lines of equal

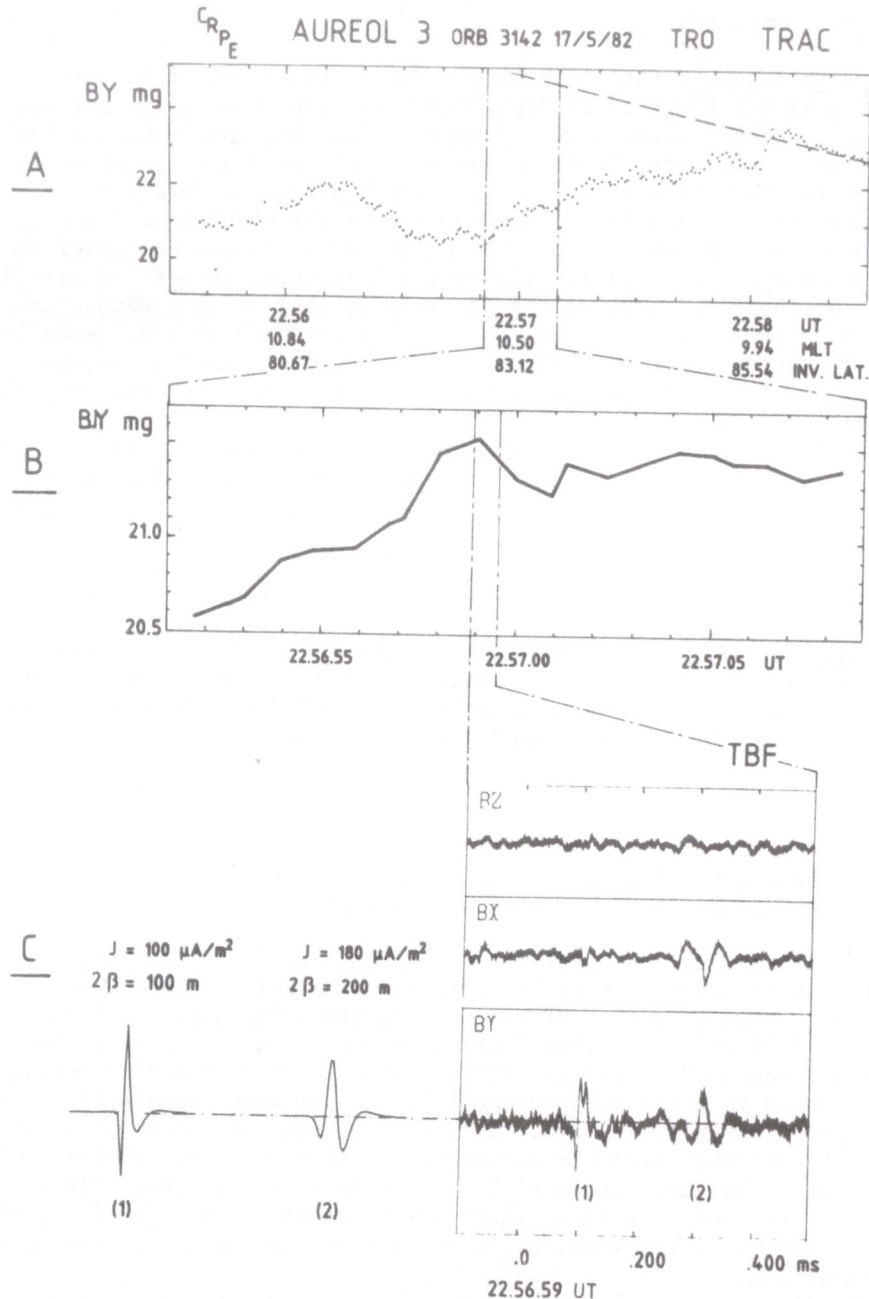

Figure 2 – Same as Figure 1 for orbit 3142

latitude are strong arguments in favour of a spatial variation and thus we have identified them as due to the crossing of very thin sheets of FAC.

3. MODELISATION

A rather simplified model is presented here, where the satellite is supposed to cross a field aligned current sheet infinite in Z and Y directions, with a current density profile along the crossing direction X represented a Gaussian function depending on two parameters :

J : maximum current density value
β : 1/e half width

Owing to the transfer function of the electronics, the output signal on the AC magnetic channels resulting from the crossing of such a current sheet appears as a pulse in BY, the shape and amplitude of which crutially depend upon β as shown in Figure 3. For a given β, the amplitude is

Figure 3 - Modelised output signal in BY for J = 100 μA/m2, and four different values of the half width β (The unit is in Volt/Telemetry).

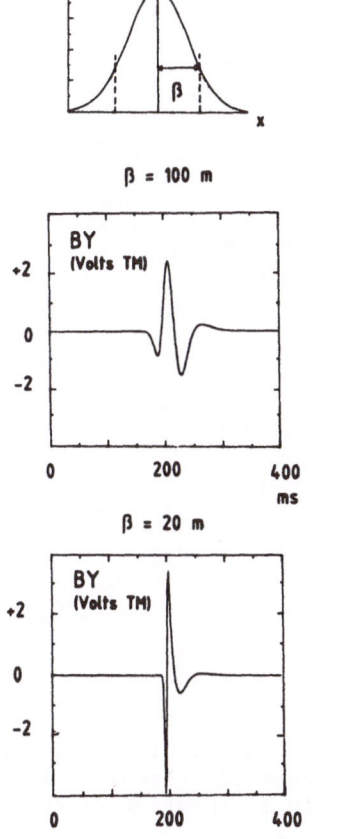

proportional to J, the sense of the current being obtained from the
shape of the signal. As mentionned above there is a striking coïnci-
dence between the asymmetric shape of the pulse as obtained from this
modelisation and the real signal (Panel C, Figures 1 and 2). This appears
to be a very strong argument in favour of this interpretation.

We have also examined a model of current tubes in place of current
sheets ; this model does not fit the observations which are coherent
with a longitudinal dimension of the structure much larger than the
latitudinal one.

The characteristics of the crossed current sheet: main orientation,
width, and intensity are thus determined from the direct comparison
between modelised and real pulses.

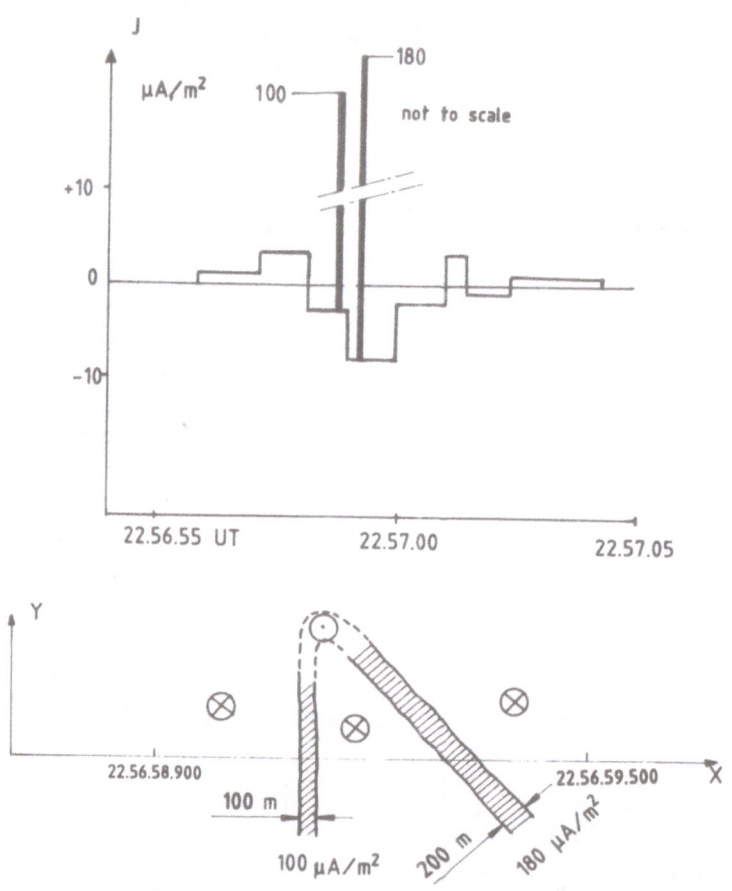

Figure 4 – Position of the two upward current sheets observed in the
midscale downward current in the case of the orbit 3142, and a
schematic interpretation of the folded arc structure.

4. INTERPRETATION AND DISCUSSION

In the two examples presented here Small Scale Intense FAC are observed inside the polar cusp region, identified by the typical precipitation of low energy protons and electrons (a few 100 eV up to 1 keV) as measured by a companion particle experiment (Bosqued et al 1982).

4.1. The pulse shown in the first example (Orb 3076, Fig. 1) corresponds to an upward current of $\sim$400 m width, and 250 μA/m2 maximum density ; it is entirely seen on the Y axis and thus, according to the orbit and attitude of the spacecraft, extends approximately along lines of constant latitude. (Panel C, Fig. 1). It occurs in a region where FAC are structured and predominently upward on a large scale (average density of 0.3 μA/m2, over $\sim$750 km, panel A). On a medium scale (panel B), one can see that it corresponds also to upward currents as averaged over $\sim$ 7 km (sampling rate = 1 pt/s. in this case).

4.2. In the second example (figure 2) one observes two pulses separated by $\sim$200 ms, that is $\sim$1500 m ; both correspond to upward currents. The first one has a maximum density J $\sim$100 μA/m2 and width 2 β $\sim$100 m, and is only detected on the Y component ; the second one has J $\sim$180 μA/m2, 2β $\sim$200 m, and is seen on both the X and Y channels.
      It can be seen from DC BY measurements that they occur in a region of large scale upward flowing currents of $\sim$0.6 μA/m2 on the average over 700 km (Panel A). The major difference compared to the first example is that the SSI FAC are here imbedded in a localized downward current (Panel B).
      This situation has been interpreted as resulting from a double crossing of a wavy thin arc, where upward SSI FAC are the signature of precipitating electrons in the arc, imbedded in a localised return downward current as it is schematically drawn on Figure 4.

5. SUMMARY AND CONCLUSION

Isolated pulses and more extended wave packets are commonly observed on the magnetic channels of the TBF experiment at high latitudes. The isolated pulses have been interpretated as originating from the crossing of thin sheet of very intense field aligned currents ; typical trans-verse dimensions are in the 100-500 m range, the current density may reach as high as 100-200 μA/m2, and the data so far analysed only reveal upward currents. These Smale Scale Intense FAC have been in particular observed in the polar cusp and two exemples are presented here ; one of them is extremely interesting since it may be explained by two succes-sive crossings of a distorted current sheet imbedded in a middle scale ($\sim$ 10-20 km) return current of significant intensity ($\sim$ 10 μA/m2). The geometry of the current sheet is close to what can be expected for fine scale instable auroral arcs (Webster and Hallinan 1973), and bears some similarity with auroral vortices described by Burke et al (1983).

6. ACKNOWLEDGEMENTS

We gratefully aknowledge the efforts of the people who participated in
the success of the TRAC and TBF experiments, in particular the principal
investigators JJ. Berthelier, Y. Galperin and O. Moltchanov. The
AUREOL-3 was realized in the frame of the French-Soviet cooperation and
on the French side the experiments have been financed by CNES under
grants 76.224 to 80.224. One of us (A.B.) is indebted to the NATO-ARW
organizing comittee for a financial support, and both of us thank
J.J. Berthelier and J.C. Cerisier for helpful discussions.

REFERENCES :

Berko, F.W., Hoffman, R.A., Burton, R.K., and Holzer, R.E. 1975,
    J. Geophys. Res., 80, 37.
Berthelier, J.J., Berthelier, A., Galperin, Y.I., Gladishev, V.A.,
    Gogly, G., Godefroy, M., Guérin, C., and Karczewski, J.F. 1982,
    Ann. Geophys., 38, 635.
Berthelier, J.J., Lefeuvre, F., Mogilevsky, M.M., Molchanov, O.A.,
    Galperin, Y.I., Karczewski, J.F., Ney, R., Gogly, G., Guérin, C.,
    Leveque, M., Moreau, J.M., and Séné, F.X. 1982, Ann. Geophys.,
    38, 643.
Bosqued, J.M., Barthe, H., Coutelier, J., Crasnier, J., Cuvilo, J.,
    Medale, J.L., Rème, H., Sauvaud, J.A., and Kovrazhkin, R.A. 1982,
    Ann. Geophys., 38, 567.
Burke, W.J., Silevitch, M., and Hardy, D.A. 1983, J. Geophys. Res.,
    88, 3127.
Doyle, M.A., Rich, F.J., Burke, W.J., and Smiddy, M. 1981, J. Geophys.
    Res., 86, 5656.
Potemra, T.A. 1979, Rev. Geophys. Space Phys., 17, 640.
Robert, P., Gendrin, R., Perraut, S., and Roux, A. 1984, J. Geophys.
    Res., 89, 819.
Saflekos, J.A., Potemra, T.A., and Iijima, T. 1978, J. Geophys. Res.,
    83, 1493.
Sugiura, M., Iyemori, T., Hoffman, R.A., Maynard, N.C., Burch, J.L.,
    and Winnigham, J.D. 1983, Technical memorandum 8505, NASA, Green-
    beld, Md.
Webster, H.F., and Hallinan, T.J. 1973, Radio Sci., 8, 475.

INTERFEROMETER OBSERVATIONS OF ION AND NEUTRAL DYNAMICS IN THE POLAR
CUSP

R.W. SMITH
School of Physical Science
Ulster Polytechnic
Newtownabbey, Co. Antrim, UK

ABSTRACT. Thermospheric winds, ion drifts, and temperatures, determined
by ground-based optical Doppler velocimetry in the dayside of the winter
polar cap, are presented and discussed in context with the systematic
errors of measurement and the possible influence of cusp-related
geophysical phenomena. The thermospheric wind observed from Spitsbergen
is predominantly westward on the dayside of the polar cap when plotted
in geographic coordinates. In geomagnetic coordinates, the flow is more
poleward but retains a westward zonal component on the average.
Separation of the data by cusp latitude shows that the westward
component is stronger when it is poleward of the station. The cusp
shows clearly as a temperature maximum from noon-midnight passes of the
DE satellite, but is not noticeable in the same way from the ground,
making a zonal pass underneath. Observations of vertical winds are
discussed along with the implications on the uncertainty of reduction of
observational data to horizontal components. Winds are found to be non-
uniform particularly near the minimum velocity seen in passage through
the evening vortex.

Optically measured ion drifts have the expected characteristic with
the change of sign of the IMF By component. For By positive, the zonal
component is westward near magnetic noon. For By negative, it is
eastward. The neutral wind is also a function of the By component, but
response is asymmetric. This result demonstrates the strong influence
of ion-neutral momentum coupling on the thermospheric wind and the
variation in its effectiveness to produce circulation cells depending on
the dawn and dusk ion convection cells. Results are quoted which
indicate that in winter conditions near the terminator, there is a 200-
400 m/s downward field aligned ion flow below the F peak. This is in
qualitative agreement with the predictions of the Sheffield ionospheric
model (Allen et al., 1984).

1. INTRODUCTION

The discussion has been limited to those types of measurement which are
made with high resolution optical interferometers. Specifically, these
will be assumed to include the measurement of the Doppler shift and
Doppler broadening of spectral lines in the visible or near infrared

J. A. Holtet and A. Egeland (eds.), The Polar Cusp, 243–260.

spectral regions which are emitted in the aurora and airglow in the
vicinity of the polar cusps. In principle, these observations may be
made by day and by night. Since the emission rates are low and the
timescales for change short, the instrument must have a large light
gathering capability. In addition, sufficient wavelength sensitivity is
required to enable Doppler shifts to be measured to a precision better
than the equivalent of a few tens of metres per second, and Doppler
broadening to the equivalent of a few tens of degrees K. Nighttime
measurements may be made with a single etalon Fabry–Perot Interferometer
(FPI). For daytime measurements, much higher spectral selectivity is
required because of the high daytime background illumination. In
effect, a scannable high resolution filter is required in order to
reduce the daylight background to the weak signal. For this, a multi-
etalon device is needed (Cocks and Jacka, 1979).

      Using a single etalon Fabry–Perot Interferometer many workers
(Armstrong, 1969; Hays and Roble, 1971; Nagy et al., 1974; Hernandez and
Roble, 1976; Jack et al., 1979; Smith and Sweeney,1980; Sipler et al.,
1981; Killeen et al., 1982; Rees et al., 1984a) have made nighttime
observations of neutral winds and temperatures. Also the Michelson
interferometer in WAMI form (Zwick and Shepherd, 1973; Smith, 1973;
Gault and Shepherd, 1983) is suitable for both wind and temperature
measurements. For daytime observations, multiple etalon Fabry–Perot
interferometers are now in use (Cocks et al., 1980, Hays, 1982; Rees et
al., 1982) in space-borne and ground-based observatories. The
difficulty in operation of multi-etalon devices increases as a very high
order function of the number of etalons used. However, modern computer
control techniques are making this task more tractable (Rees et al.,
1982). The technical details of these instruments differ and will not
be described here, but the principle of the observation is the same and
the interpretation of the data requires similar considerations.

      The interferometer receives light within a cone of angles centered
along the optical axis of the instrument. An interferogram is produced
containing spectral information on those spectral features admitted by
the prefilter. The line shape and central wavelength of interest must
be extracted by removal of the instrumental response and the unwanted
observational effects which also appear. Assuming this is done, then
standard techniques permit the determination of a mean Doppler shift and
mean Doppler temperature (Hays and Roble, 1971; Hernandez, 1972; Smith,
1980) representing the various volumes of emitting gas which lie in the
field of view. Any variation in bulk velocity or distributions of
random velocities along the line of sight distort the interferogram in a
way which is, in principle, unknown to the observer. Line of sight
observations of components of the three dimensional wind field are
normally "reduced to the horizontal" by a dividing by a cosine
correcting factor. Normally no account is taken of any component of
velocity in the vertical direction since it is often presumed small. We
shall see that this condition is by no means prevalent in the polar cap
in either ions or neutrals. However, neither are vertical velocities
steady in space or time, hence corrections are difficult to apply.
Further analysis readily yields vector plots of various types (eg.,
Smith, 1980; Tepley et al., 1984). These require additional

assumptions, either of uniformity of the wind field over quite wide
areas, or better (as in the case of Tepley et al., 1984) a prescribed
form of the wind gradient. These serve to give a pictorial
representation of the data but the only fully reliable measurements are
the original observed values which apply to the line of sight only.
Even those are averages along the line of sight which admit an
uncertainty of interpretation. There are a variety of possible auroral
and airglow emissions which could be observed. Without doubt, most work
has been done in the cusp region using the bright emission of 630 nm
from $O(1D)$ atomic oxygen. This is emitted at heights above 180 km with
a centroid at 230 km (Link et al., 1983). The upper boundary and
centroid of emission are dependent on the excitation mechanism for the
upper state of the transition. In the vicinity of the cusp this upper
limit could be above 400 km when hot thermal electrons from the high
energy Maxwellian tail or soft electrons directly entering through the
cusp or cleft are involved. The centroid of emission may then be over
300 km. This and other useful emissions are detailed in Table 1.

TABLE 1

USEFUL TRACER EMISSIONS IN CUSP			PARAMETERS
UPPER THERMOSPHERE:	(200-300KM)	630nm OI	TEMPERATURE AND WIND
UPPER THERMOSPHERE:	(200-300KM)	520nm NI	TEMPERATURE AND WIND
F-REGION IONOSPHERE:	(200-300KM)	732nm OII	ION TEMPERATURE, ION DRIFT
LOWER THERMOSPHERE:	(90-100KM)	558nm OI	TEMPERATURE AND WIND
AURORA:	(90-150KM)		(INTERPRETATION DIFFICULT)
UPPER MESOSPHERE:	(ABOUT 90KM)	589nm NaD	TEMPERATURE AND WIND
UPPER MESOSPHERE:	(89-90KM)	OH LINES	TEMPERATURE AND WIND

2. POLAR CUSP WIND AND TEMPERATURE DATA.

Bearing in mind the foregoing considerations, some observations in the
vicinity of the cusp will be described and discussed. A ground based
station can (weather permitting) monitor the evolution of the wind field
in its vicinity as a function of time. Although, using simplifying
assumptions, the wind vectors resulting from such monitoring can be
plotted in a time sequence, it must be recognised that this does not
represent the equivalent of a snapshot even if the geomagnetic
conditions have remained constant. There is a "UT" effect by which the
circulation pattern within the polar cap evolves with a periodicity of
24 hours. The plot of wind vectors shown in figure 1(a) illustrates the
time evolving pattern seen from the Longyearbyen station (LYR) on
Spitsbergen (78.2N, 15.6W, 75 Λ) typical of quiet conditions when the
cusp passes poleward of the station. Plotted in geographic coordinates

on a circle of 78 degrees geographic latitude, it emphasises a typical
dayside pattern in which a westward wind of 2-300 m/s exists on the
equatorward side of the cusp.

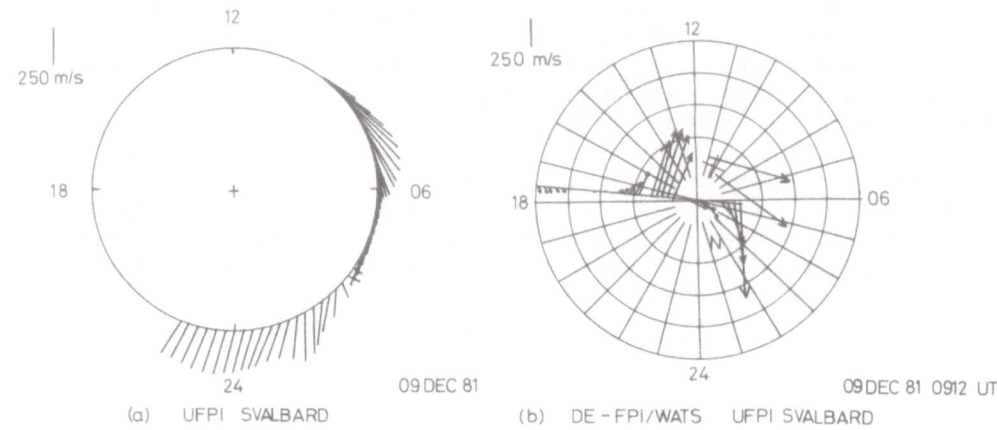

(a)   UFPI SVALBARD   (b)   DE - FPI/WATS   UFPI SVALBARD

Figure 1 (a).   Vectors of thermospheric wind at 230 km measured at LYR
on December 9, 1981 plotted on a geographic polar dial.   The circle is
at the geographic latitude of the station.
          (b).   Polar plot in geographic coordinates of the thermospheric
wind field at 230 km at 0912 on 9 December 1981 obtained by a
combination of measurements from Dynamics Explorer and LYR.   The ground
based station is marked with a cross indicating its approximate spatial
coverage.   The geomagnetic pole is marked with the symbol 'N'.

     Figure 1(b) shows the same day at an "instant" in the same day when
a DE2 pass provided the latitudinal variation of velocity along the
dawn-dusk meridian (Killeen et al., 1984).   The nature of the
instantaneous dayside clockwise vortex can be seen more clearly in this
figure.
     Figure 2(a) shows the same data as in figure 1 plotted in
geomagnetic coordinates on a circle of 75 degrees invariant latitude.
The cusp is now at a fixed position on this diagram.   In the
transformation, all vectors have been rotated by 45 degrees from figure
1.   It is now evident that the thermosphere circulates through the cusp
region which is situated a few degrees poleward of the locus of the
station near magnetic noon.
     Under the conditions of this day, the cross polar wind does not
blow parallel to the geomagnetic noon-midnight plane but somewhat
westward of it.   We shall see later how this is dependent on the sign of
IMF By.   Figure 2(b) shows the grand average of all days observed from
Longyearbyen on Spitsbergen.   The average cusp position is just poleward
of the station, and the average conditions are for the wind to have a
westward component when passing into the polar cap through the cusp
region.

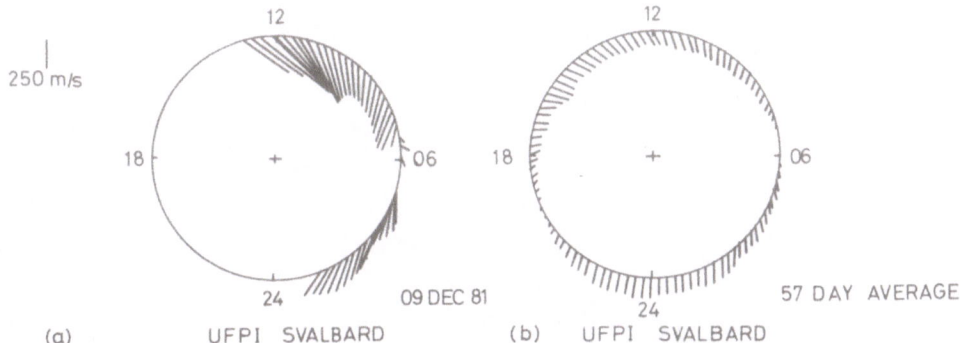

Figure 2(a). Vectors of thermospheric wind at 230 km for December 9, 1981 using the same data as for figure 1(a) plotted on a geomagnetic polar dial. The circle indicates the geomagnetic latitude (75) of the station.
        (b). Polar plot on a geomagnetic dial showing the grand average thermospheric wind at 230 km including 53 days of winter solstice data from LYR in the years 1979-1982 (McCormac, 1984).

    The data presented in most papers and in the figures above take no account of the vertical component since it is generally unknown. However, observations have been made from ground and space of the occurrence and nature of the vertical wind at high latitudes. There appear to be both long period (diurnal) and short period (impulsive) types of behaviour. Figure 3(a) shows the nature of impulsive events as recorded at Spitsbergen which are characterised by a short (less than 2 minutes) upward motion followed by a longer (5-30 minute) downward part. These seem to occur particularly on the nightside. In contrast the long period variations shown in figure 3(b) (Rees et al., 1984c) indicate downward flow by day and upward flow by night. Spencer et al. (1982) reported vertical wind observations made using the Wind and Temperature Spectrometer (WATS) instrument aboard Dynamics Explorer. Their results show that on each pass there was at least one event where the vertical wind exceeded 100 m/s. The events were of small spatial extent when "scanned" in the dawn-dusk plane.
    The implications are that the wind field is corrugated to a substantial degree and a comprehensive study of the vertical wind component is urgently required.
    Since the particle impact and the Pedersen currents in the thermosphere cause heating, it would be expected that the Doppler temperatures of the 630 nm emission should show a peak near the cusp. Although meridional passes of the DE satellite indicate a gradual increase of temperature (Killeen, private communication), the zonal

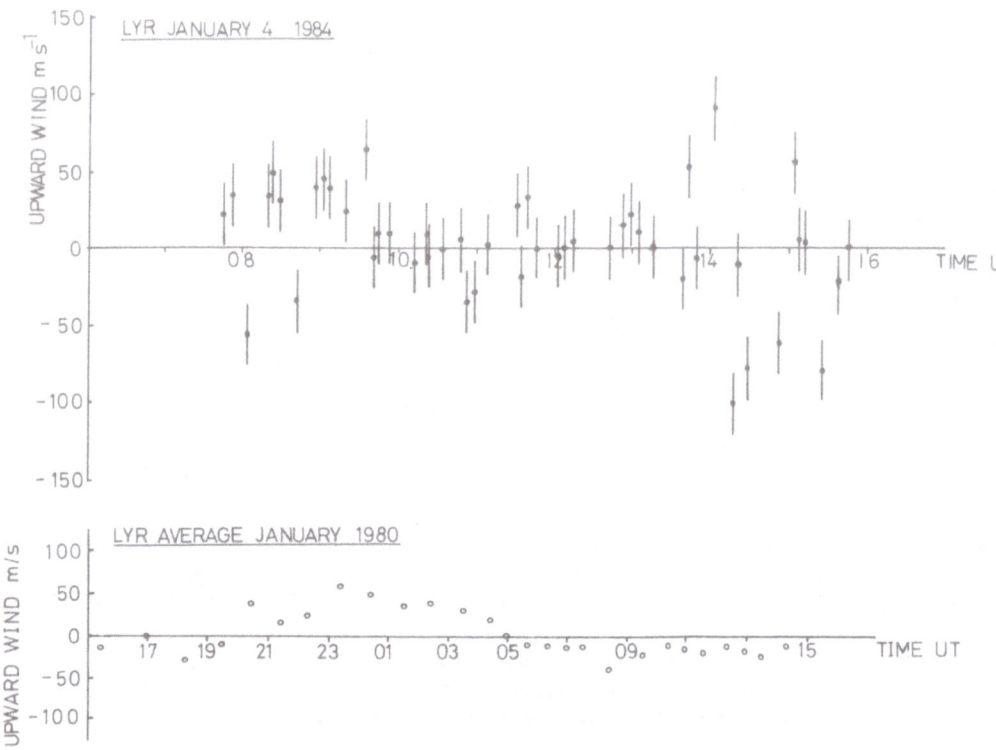

Figure 3(a). Measurements of the vertical component of the
thermospheric wind at 230 km at LYR on January 4, 1984.
        (b). Average vertical thermospheric winds at 230 km over LYR
for the month of January 1980.

gradient or temperature obtained from ground based observations appears
small or insignificant (Henriksen et al., 1982).
        Figure 4, taken from that paper, indicates spatially uniform
temperatures on the dayside with small temperature gradients. Elsewhere
in the polar cap, it is common to find apparently small scale structure
in the thermospheric temperature with a variation of 100K. Such
observations have been attributed to height variations of the centroid
of the source in a region of vertical temperature gradient rather than
horizontal gradients of temperature. It is probable that the centroid
of emission rises in the cusp because of a hot electron source of
excitation for oxygen atoms. Whenever a temperature change is observed
optically, care must be exercised in assessing its cause. In this case
the higher temperature observed in the cusp by Dynamics Explorer is
consistent with the expectation but has not been a noted feature in the
ground-based observations. Further studies in this area are needed.
        A change in geomagnetic activity involving an erosion of dayside
magnetic flux has the result that the cusp moves equatorward. The

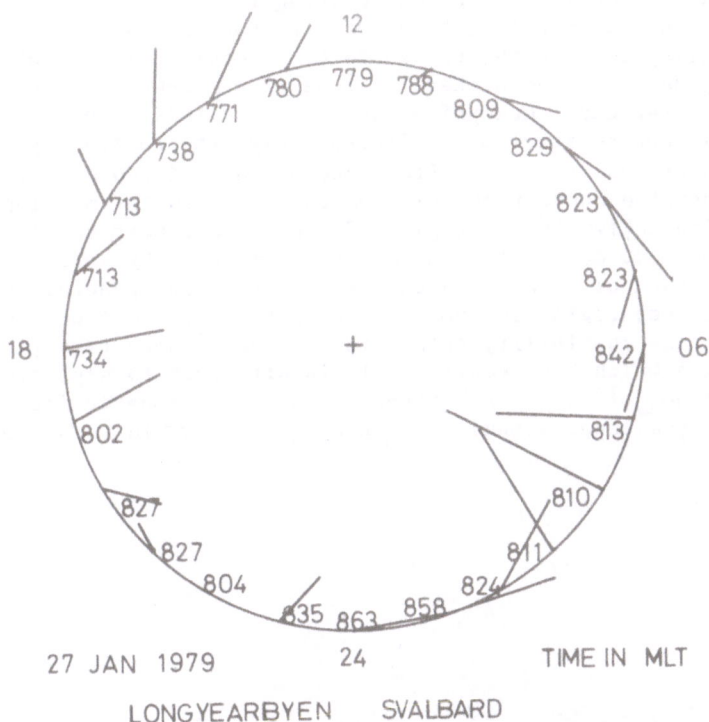

Figure 4. Polar plot in geomagnetic coordinates showing the mean thermospheric temperature at 230 km above LYR (averaging values obtained for all directions of view) and the vector temperature gradients for January 27, 1979.

flux has the result that the cusp moves equatorward. The Spitsbergen station then passes on its poleward side. The effect of increased reconnection is to increase the coupling between the magnetosphere and the solar wind. This often leads to an enhancement of auroral activity, better momentum coupling of neutrals and ions and the acceleration of neutral winds. Any feature of thermospheric circulation dependent on the cusp position will then be sampled in a different way by a ground-based station depending on the combined effects of different motion relative to the cusp and the increased auroral disturbance.
Observations show that there are two basic types of "day" near the winter solstice from the point of view of wind observations at Spitsbergen (Smith et al., 1984). The "A" day, which is most common, is for the condition when the cusp latitude is poleward of the station. Then the wind is spatially uniform at a few hundred m/s and approximately north westward in the geomagnetic frame at the time of passage. When the cusp passes the overhead position on the equatorward side the "B" day is observed. The wind is then weaker and more poleward. Also there is a systematic gradient of the wind showing that

it decelerates zonally whilst accelerating in the poleward direction.
The measurements do not lead to a definite conclusion as to the cause.
One possibility is that the existence of a local region of high pressure
at the cusp deviates the higher latitude flow poleward, launching it
across the polar cap. Meanwhile the lower latitude flow continues
westward without restriction. Alternatively, the reason may be the
interaction of two streams of flow, one of which is coupled to the
evening clockwise convection cell and the other to the morning
anticlockwise cell. At the region of convergence near the cusp, the
wind pattern must adjust to accommodate them, thereby setting up wind
gradients. Both of these arguments assume that the general features of
flow in the cusp region are not changes merely by the expansion of the
polar cap. The possibility exists that a change does take place with
the expansion which has occurred. It is difficult to separate the data
to take account of this. A cartoon of the thermospheric flow in the
vicinity of the cusp, assuming no change, is shown in figure 5.

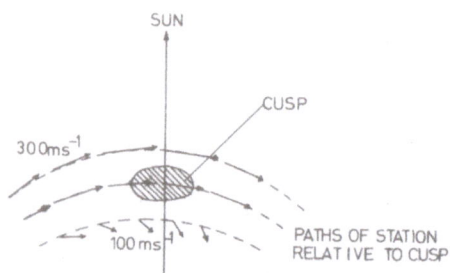

THERMOSPHERIC WINDS IN RELATION TO THE CUSP

Figure 5. Schematic diagram illustrating the apparent distribution of
thermospheric wind vectors in the vicinity of the cusp deduced from
observations at LYR.

Despite the exercise of great care in experimental technique to
avoid systematic errors, one is forced to conclude from the measurements
that the wind field is often substantially non uniform. Departures from
uniformity would be expected at a crossing through the eye of a vortex
in thermospheric flow. Since the introduction of an 8 direction scan
system in Spitsbergen, it has been possible to study such non
uniformities. Fourier analysis of the azimuthal distribution doppler
shifts yields both the uniform and the non-uniform components.
These are shown as a function of time in figure 6(a) for the
Longyearbyen station on 15/16 Jan. 1983. Prior to 0200 UT, the uniform
component dominates indicating a uniform wind field. Near 0300 UT it
can be seen that the uniform and non-uniform components have
approximately equal magnitudes. This may be interpreted as the passage
of the station near the eye of the morning circulation vortex when the
wind vector decreases and the wind field becomes substantially non-

uniform. At 0700 UT another non-uniform region occurs. In addition to
the Longyearbyen data, recordings were also made at Ny Alesund by the
UCL group. These have been analysed in a similar way and show in figure
6(b) that the uniform component is not significantly different in
amplitude and azimuth, as would be expected for two stations whose
sampling areas have more than 50% overlap.

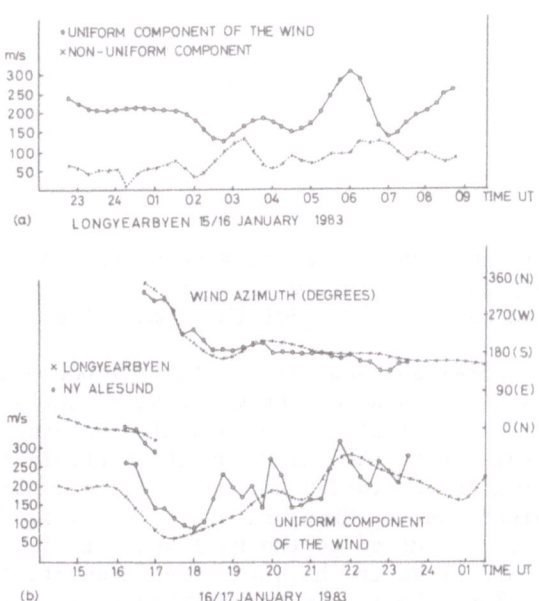

Figure 6(a). Results of the Fourier analysis of the azimuthal
distribution of Dopper shifts measured at LYR on 15/16 January 1983.
The lines show the magnitude of the fundamental frequency which
corresponds to the uniform component of the observed wind, and the
residual which is the magnitude of the combination of non-uniform
components.
        (b). Comparison of the uniform components for wind measurements
made at Ny Alesund and LYR which are neighbouring stations on
Spitsbergen separated by 110 km along the geomagnetic meridian.

3. POLAR CUSP ION DRIFT AND TEMPERATURE DATA

Since December 1980 it has been possible to make observations of the
732nm line of O^+ in all directions during the dayside passage of the
station from 0400 UT to 1700 UT at Spitsbergen (Smith et. al., 1982).
The interferogram can be processed to give the apparent Doppler shift
and line profile of the emission. Interpreting the Doppler shift data
as an averaged component of ion speed along the line of sight, it is
possible to process it into a form which gives a representation of the
mean ion convection pattern in the vicinity of the cusp. The resulting
sequence of vectors obtained may be plotted as a function of time, a

typical example of which is shown in figure 7.

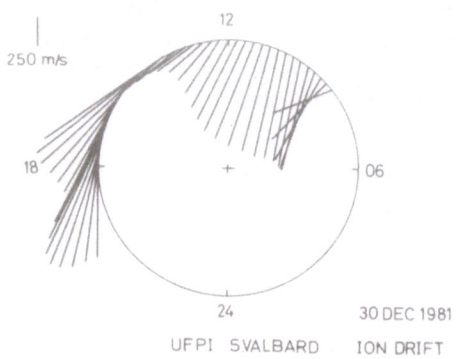

Figure 7. Polar plot in geomagnetic coordinates of the ion drift
vectors at approximately 240 km observed by optical Doppler
interferometry at LYR on December 30, 1981 (McCormac, 1984).

This has coarse temporal and spatial resolution, but contains
enough detail to relate to the wind measurements since the wind has a
tendency to respond most effectively to large scale features of the ion
convection. It is presumed that the height of the centroid of emission
of 732nm O^+ is in the region of 220-250 km, which is well above the
collision dominated height range. The ions are expected to be
travelling at a velocity very close to the Hall velocity for all heights
within the F-region at and above the height of measurement. Hence the
observations sample the ion velocity in approximately the same height
region as that of the neutrals measured using the 630 nm emission, and
may be used to estimate averaged ionospheric electric fields.
The grand average ion velocity plot is shown in figure 8 and should
be compared with figure 2(b), the grand average neutral plot.

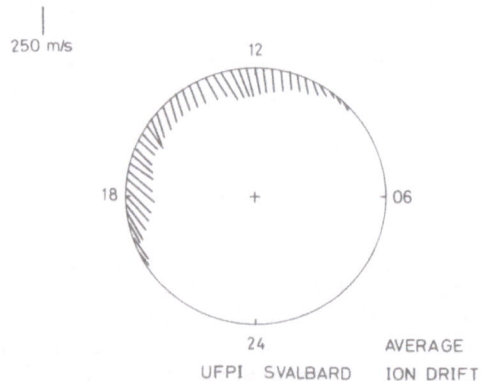

Figure 8. Polar plot of ion drift as in figure 7 but showing the grand
average over 20 winter solstice days between 1980 and 1982 (McCormac,
1984).

Many similarities can be seen in the afternoon sector whereas the morning sector shows less correspondence. Typically, it is found that neutral velocities approach 50% of the ion velocity on days when the flow is westward. When the flow has changed towards northward, neutral velocities are only 15% of the ions, although both flows may be very closely parallel. The coupling of eastward ion convection seems weaker than westward, probably for the reason suggested by Fuller-Rowell et al. (1984) that the Coriolis effect causes air parcels to deviate away from eastward convection channels, whereas it assists the entrainment of westward channels.

Considering the results from an individual day, as in figure 7, there is no evidence of a throat. This may well be due to the lack of spatial resolution. However, the general form of the "section" through the time evolving pattern has the same characteristics as the general pattern of satellite data and the empirical models representing snapshots of passes over an instantaneous condition (Heppner, 1977; Volland, 1978; Heelis, 1984). On the day concerned, the ion drift has an eastward trend in the vicinity of noon consistent with the existence of a dominant morning cell. This was an active day with Kp = 6 and IMF By negative. Hence an enlarged auroral oval with large or dominant morning convection cell would have been expected (Heelis, 1984).

The ion data has been organized to plot the geomagnetic zonal component of ion drift as a function of the IMF By component taken in hourly average values for the hour preceding the drift measurements.

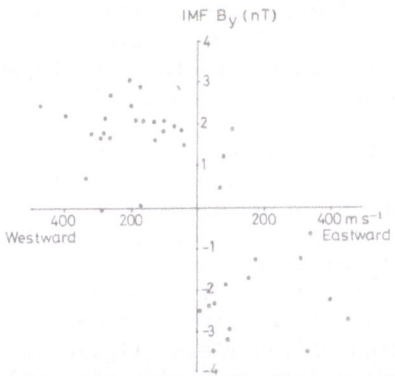

Figure 9. Hourly averaged zonal ion drift observed form LYR (the component parallel to the lines of constant geomagnetic invariant latitude) as a function of the similarly averaged IMF by component.

The resulting scatter plot (McCormac and Smith, 1984) in figure 9 shows how strong the relationship is between the ion drift direction in the region of the cusp and IMF By. This result is in agreement with the results of other workers (eg. Heelis, 1984) who have used satellite data. In this respect, the result is more of a validation of the technique than new information. Of more interest, however, is the corresponding diagram for neutral wind (McCormac and Smith, 1984) shown in figure 10.

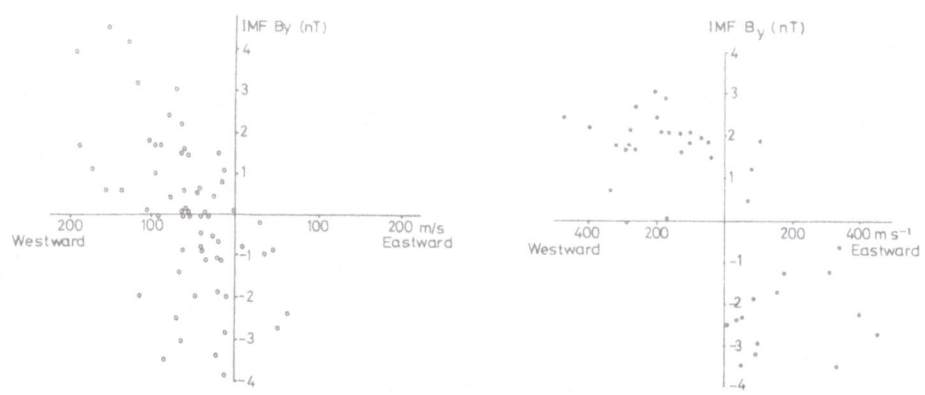

Figure 10. Hourly averaged zonal neutral wind speed observed from LYR
(the component parallel to lines of constant geomagnetic invariant
latitude) as a function of the hourly averaged IMF By component.

Since there is a By dependence of ion drift at the cusp then a
similar IMF By response should be seen in the neutral motion, if the
coupling were sufficiently strong. The result shows that strong
coupling does occur on a statistical basis, but that the effect is not
symmetrical with a bias towards positive By. The cause of this may also
be the effect described by Fuller-Rowell et al. (1984) which was
mentioned above. In effect clockwise neutral vorticity is more readily
set up than anticlockwise vorticity in the northern hemisphere.
      Combining long range UHF observations from EISCAT and our optical
Doppler techniques on ions, we have been studying the field parallel and
field perpendicular components of ion motion in the vertical plane
containing both EISCAT and the Longyearbyen station. The suprising
result of this work is the discovery of a region of apparent rapid
field-aligned downward flow of ions at a geomagnetic latitude of 71
degrees in the vicinity of the terminator in the afternoon sector. This
may not be directly related to the cusp itself, and may only be a
feature of the winter season if it is definitely terminator-related.
      Figure 11 shows the variation of the field aligned flow on 7
December 1981 and the corresponding data on 14 December 1983 when the
experiment was repeated. Figure 12 shows how an effect of a similar
nature but smaller in magnitude appears in the ionospheric model of
Quegan et al. (1982) as modified by Allen et al. (1984). A
latitudinally aligned, region of downward flow exists extending towards
the afternoon sector from magnetic noon. Our observing point traced a
path taking an arc passing through this region.
      Recent detailed studies of interferograms containing the O^+ 732 nm
emission have been carried out to test our presumption that they can all
be explained on the assumption that they consist of the 732 nm doublet
with an intensity ratio of 4:1 with some contamination from the OH
feature at 731.6nm which is the lambda split P2 line of the 8-3 band.

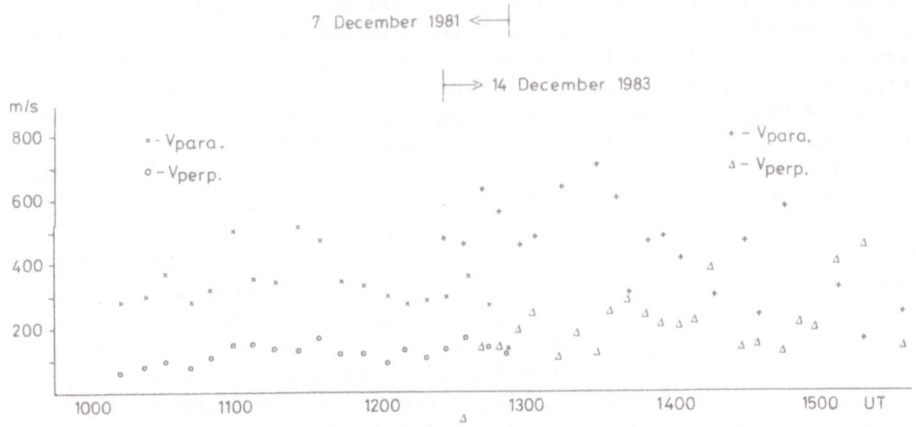

Figure 11. Field parallel and meridional field perpendicular components
of O+ ion drift measured in bistatic experiments using the Tromso beam
of the EISCAT radar and the optical interferometer at LYR. Note the
large downward field-aligned component which decreases slowly toward
zero some time after 1500 UT (1800 LMT).

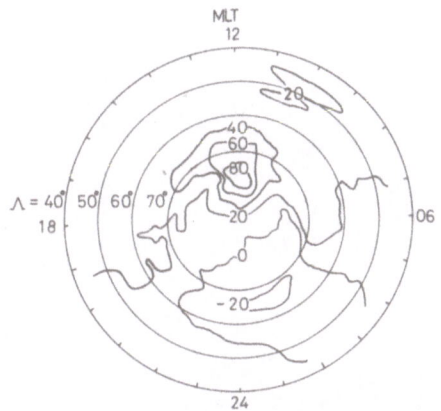

Figure 12. Contours of constant field-aligned ion flow for northern
hemisphere winter solstice conditions. Calculated using the Sheffield
ionospheric model (Allen et al., 1984).

Using this theory, and making the assumptions that the width and
wavelength of the OH feature is constant, a range of composite profiles
has been fitted, allowing the intensities of the O^+ and OH features to
vary independently. Also the background intensity and the width of the
732 lines in the doublet were incorporated as variables in the fitting
process. The result of this study indicates two important points about
the measurements which can be made by this method. Firstly, it became
necessary to admit that some of our line profiles were too wide to be
due to Doppler broadening considering realistic ion temperatures that

one might expect. However, the OH lines on the same interferograms had
the expected widths confirming that the interferometer had not suffered
a loss of adjustment. This has been attributed to the temporal and
spatial variation of Doppler shift which is present within the viewing
direction during the observation. Hence ion line profiles would be
systematically broadened leading to an upward displacement of the
retrieved temperature which is very difficult to estimate. Secondly,
leading from the first point, the possibility exists that at some times
grossly misleading averaging occurs in time and space in the
measurements of ion drift. It is conceivable that both blue and red
shifts could be combined in the same profile. Clearly, a degree of
caution is now essential in the conclusions drawn from optical ion
measurements. Fortunately, obviously excessive broadening is not a
frequent occurrence and is not believed to vitiate the measurements
presented below. If the broadening effect is not systematically
affecting the cusp region, then the ion temperature may be shown to be
greater in the cusp than outside as shown in figure 13.

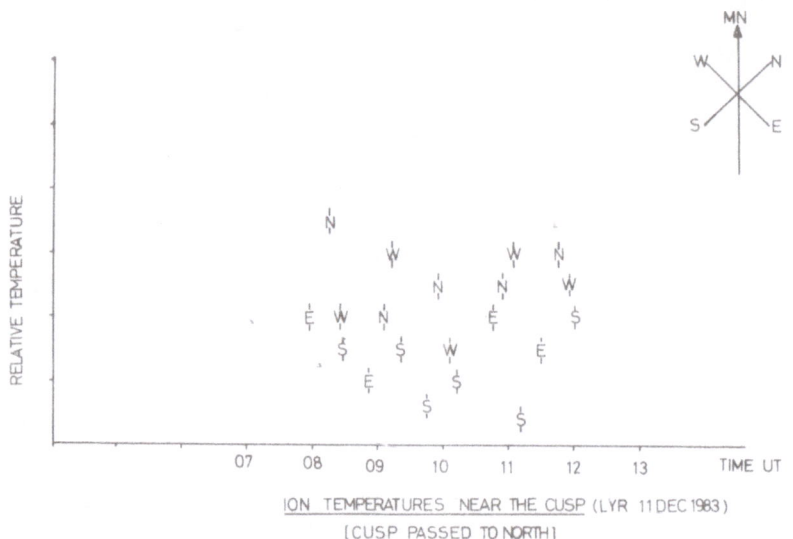

Figure 13. Relative ion temperature for the four geographic directions
of view of the LYR interferometer on December 11, 1983. North and west
directions are geomagnetically poleward of LYR. The data shows a
systematically higher ion temperature in the direction of the
geomagnetic pole. On this day the cusp passed on the poleward side of
LYR.

4.CONCLUSIONS

The optical technique is a powerful tool for the ground-based mapping of
a wind field in time evolving form, particularly where an array of
stations is involved. The data would be more useful if better spatial

and temporal resolution could be obtained, for example using a Doppler Imaging Spectrometer (Rees and Greenaway, 1983).

4.1   The polar cap dayside flow is frequently characterised (for a single latitude and a single UT-LT relationship) by a strong clockwise vortex which is consistent with an ion drag mechanism in which parcels of air are entrained in the dusk cell of high latitude ion convection.

4.2   The thermospheric flow is well coupled to ion motion which results in its response to IMF changes in a way less marked but yet similar to the ion convection. Effectiveness of the coupling in the production of a matching cell of neutral circulation is reduced for IMF By negative compared with By positive.

4.3   The circulation has irregular bursts of the vertical component of the wind away from the cusp but tends to be quite stable in close proximity. The apparent horizontal flow has two regions of particular non uniformity which correspond to passage through or close to large scale structure in the flow pattern. Non uniformities other than those have a generally smaller amplitude. Some of the smaller scale non uniformities may be associated with vertical velocity behavior.

4.4   The thermospheric temperature shows a systematic positive gradient when approaching the cusp along a meridian reflecting the high altitude particle heating. However, the zonal temperature gradients at cusp latitudes are much less obvious.

4.5   The cusp is a special feature which has observable effects on the thermospheric flow pattern. The cause of these is not certain, but the effects are consistent with either a high pressure island because of the high altitude heating or the complex reaction of the neutral thermosphere due to its coupling to tne convergence of ion convection in the "throat".

4.6   Substantial downward field aligned ion flow apparently exists below the F peak in the terminator region in the winter solstice. It does not seem reasonable to assume that accurate field perpendicular ion velocities can be obtained from monostatic ground-based observations without a detailed knowledge of it.

4.7   Distortions of 732nm O^+ interferograms may occur due to the change of ion drift with position along the line of sight and also with time during the accumulation of the data. This leads to overestimates of ion temperatures and misleading ion velocity measurements.

ACKNOWLEDGEMENTS

This work has been supported by the British Science and Engineering Research Council under grant SG/D/04299 and also by NATO research grant 0513/82.

REFERENCES

Allen, B.T., G.J. Bailey and R.J. Moffett, Ion distributions in the high latitude topside ionosphere, Ann. Geophys. (submitted) 1984.

Armstrong, E.B., Planet. Spa. Sci. 17, 957,1969.

Cocks, T.D., D.F. Creighton, and F. Jacka, J. Atmos. Terr. Phys., 42,499-511,1980.

Fuller-Rowell, T.J., D. Rees, S. Quegan, G.J. Bailey and R.J. Moffett, Planet. Spa. Sci. 32, 464-480, 1984.

Gault, W.A., and G.G. Shepherd, Adv. Spa. Res., 2, 111-114, 1983.

Hays, P.B. and R.G. Roble, Appl. Optics, 10, 193-200. 1971.

Hays, P.B. and R.G. Roble, J. Geophy. Res., 76, 5316-5321,1971.

Hays, P.B., Appl. Optics, 21, 1136-1141, 1982.

Heelis, R.A., J. Geophys. Res., 89, 2873-2881, 1984.

Henriksen, K., P.E. Sandholt, A. Egeland, R.W. Smith and P.J. Sweeney, J. Atmos. Terr. Phys. 44, 71-79, 1982.

Heppner, J.P., J. Geophys. Res., 82, 1115, 1977.

Hernandez, G., Planet. Spa. Sci., 20, 1309-1321, 1972.

Hernandez, G. and R.G. Roble, J. Geophys. Res., 81, 2065-2074, 1976.

Hernandez, G. and R.G. Roble, Appl. Optics, 18, 3376-3385, 1979.

Jacka, F., A.R.D. Bower and P.A. Wilksch, J. Atmos. Terr. Phys. 41, 397-407, 1979.

Killeen, T.L., P.B. Hays, N.W. Spencer and L.E. Wharton, Geophys. Res. Lett. 9, 957-960, 1982.

Killeen, T.L., R.W. Smith, P.B. Hays, N.W. Spencer, L.E. Wharton and F.G. McCormac, Geophys. Res. Lett. 11, 311-314, 1984.

Link, R., J.C. McConnell and G.G. Shepherd, An analysis of the spatial distribution of dayside cleft optical emissions, submitted to Planet. Spa. Sci., 1983.

McCormac, F.G., An optical investigation of ion and neutral motions in the polar thermosphere, PhD Thesis, Ulster Polytechnic, 1984.

McCormac, F.G., and R.W. Smith, The influence of the Interplanetary

Magnetic Field Y component on the ion and neutral motions in the polar thermosphere, Geophys. Res. Lett. (in press) 1984.

Nagy, A.F., R.J. Cicerone, P.B. Hays, K.D. McWatters, J.W. Meriwether, A.E. Belon and C.L. Rino, Radio Sci., 9, 315-322, 1974.

Quegan, S., G.J. Bailey, R.J. Moffett, R.A. Heelis, T.J. Fuller-Rowell, D. Rees and R.W. Spiro, J. Atmos. Terr. Phys. 44, 619, 1982.

Rees, D., P.A. Rounce, I. McWhirter, A.F.D. Scott, A.H. Greenaway and W. Towlson, J. Phys. E. Sci. Instrum., 15, 191, 1982.

Rees, D. and A.H. Greenaway, Appl. Optics,22, 1078, 1983.

Rees, D., P. Charleton, M. Carlson and P. Rounce, High latitude thermospheric circulation during the Energy Budget Campaign, J. Atmos. Terr. Phys. (in press) 1984a.

Rees, D., A.H. Greenaway, R. Gordon, L. McWhirter, P.J. Charleton and Ake Steen, Planet. Spa. Sci. 32, 273-285, 1984b.

Rees, D., R.W. Smith, P.J. Charleton, F.G. McCormac, N. Lloyd and Ake Steen, The generation of vertical thermospheric winds and gravity waves at auroral latitutes. I. Observations of vertical winds, Planet Spa. Sci. (submitted) 1984c.

Sipler, D.P., M.A. Biondi and R.G. Roble, Planet. Spa. Sci. 29, 1367-1372, 1984.

Smith R.W., Phys. E. Sci. Instrum., 6, 59-62, 1973.

Smith, R.W., "Exploration of the Polar Upper Atmosphere", pp. 189-198. Deehr and Holtet, eds., Reidel, 1980.

Smith, R.W. and P.J. Sweeney: Nature 284, 437-438, 1980.

Smith, R.W., G.G. Sivjee, R.D. Stewart, F.G. McCormac and C.S. Deehr: J. Geophys. Res., 87, 4455-4460, 1982.

Smith, R.W., K. Henriksen, C.S. Deehr, D. Rees, F.G. McCormac and G.G. Sivjee, Thermospheric winds in the cusp: dependence on the latitude of the cusp. Planet. Spa. Sci. (accepted) 1984.

Spencer, N.W., L.E. Wharton, J.C. Maurer and G.R. Carignan. Geophys. Res. Lett., 9, 953-956, 1982.

Tepley, C.A., R.G. Burnside, J.W. Meriwether, Jr., P.B. Hays and L.L. Cogger, Planet. Spa. Sci., 32, 493-501, 1984.

Volland, H., J. Geophys. Res., 83, 2695, 1978.

Zwick, H.H., and G.G. Shepherd, Planet. Spa. Sci., 21, 605, 1973.

NEUTRAL PARCEL TRANSPORT IN THE HIGH LATITUDE F-REGION

T. L. Killeen and *R. G. Roble
Space Physics Research Laboratory, The University
of Michigan, Ann Arbor, Michigan 48109-2143, USA.
*National Center for Atmospheric Research,
Boulder, Colorado 80307, USA.

ABSTRACT: The NCAR Thermospheric General Circulation model (TGCM) is
used together with a 'post-processor' diagnostic package to analyze
neutral parcel transport in the high latitude F-region. The diagnostic
package includes a feature that enables individual parcels of neutral
gas to be tracked by interpolation through the TGCM model grid in time
and space. Various trajectories have been calculated to illustrate the
predicted loci of neutral gas parcels that encounter large ion-drag
forces due to the ionospheric mapping of the magnetospheric convection
electric field. The forcing histories for such parcels have also been
calculated using the diagnostic package to decompose terms in the hydro-
dynamic (momentum) and thermodynamic (energy) equations along the parcel
trajectories. The trajectories, calculated for a "steady state",
diurnally reproducible run of the TGCM, show that the cusp region plays
an exceedingly important regulatory or nozzle role for the overall flow
pattern and has great influence on the energization of the F-region
polar cap. In particular, it is found that the trajectory of a
"typical" F-region neutral gas parcel passes through the cusp region
at least once and sometimes twice in a 24 hour period due to the
convergent geometry of the ion convection pattern.

1. INTRODUCTION

Thermospheric General Circulation models (TGCMs) have proved to be
powerful tools for the interpretation of experimental data of the high
latitude neutral wind field. Two such models are in an advanced state
of development, the NCAR-TGCM (Dickinson et al., 1981; Roble et al.,
1982) and the UCL-TGCM (Fuller-Rowell and Rees, 1980). These models,
which are run on large modern computers such as a CRAY-1 machine, have
recently been used to compare calculations with new, global, synoptic
measurements made available from the instrumentation on board the
Dynamics Explorer-2 spacecraft (Hoffman et al., 1981; Hays et al.,
1981; and Spencer et al., 1981). The comparisons have indicated
reasonable agreement between the predicted global F-region neutral
winds and the satellite observations (Roble et al., 1983, 1984;

J. A. Holtet and A. Egeland (eds.), The Polar Cusp, 261–278.

Rees et al., 1983; Hays et al., 1984) and have thereby raised the level
of confidence in the validity of the numerical procedures and parameterizations used in the TGCMs. The numerical models have also been
used to compare with wind and temperature measurements from groundbased
observatories (e.g. Rees et al., 1980; Hernandez and Roble, 1984), to
investigate the neutral thermospheric response to differing geophysical
inputs (e.g. Fuller-Rowell and Rees, 1981) and to study specific
characteristics of the predicted wind, temperature and compositional
structure of the thermosphere (e.g. Roble et al., 1983; Fuller-Rowell
and Rees et al., 1984).

    It is considered that the maturity of the TGCM is such that useful
physical insight can be gained from a detailed analysis of the processes
operating within the model. However, to extract such insight from a
large, sophisticated, non-linear numerical model is a non-trivial task
since most of the available computational resources are required for the
simultaneous numerical solution of the coupled thermodynamic and hydro-
dynamic equations. To alleviate the problem of 'interpretation' of TGCM
output, Killeen and Roble (1984) have recently developed a diagnostic
package (DP) for the NCAR-TGCM that provides detailed diagnostic
information on the physical processes operating within the model itself.
The philosophy behind the diagnostic package is to analyze the TGCM
'phenomenological' wind, temperature and composition predictions in
much the same way as experimental results, such as those from satellite
instruments, are themselves analyzed. The DP is exercised following the
basic TGCM calculations using, as input, the TGCM 'history' file containing the predicted winds and temperatures at selected Universal Times
(UTs) as well as other relevant input data to the model. The DP per-
forms a decomposition of the various terms incorporated in the TGCM
model equations for each UT recorded in the history file and for each
grid point along a given, selected, constant pressure surface. The
individual energy (heating) and momentum (forcing) terms are then
displayed to provide for a quantitative examination of their relative
importance in the overall thermospheric energy and momentum budgets.

    In the first application of the DP, Killeen and Roble (1984)
studied the balance of forces in the momentum equation of the TGCM that
are primarily responsible for maintaining the predicted high latitude
thermospheric wind system at the constant pressure levels $z = 1$ (300 km)
and $z = -4$ (120 km). Their results showed that, at F-region altitudes,
the largest forces in the "steady-state" are due to ion drag induced by
the sunward-drifting ions on the dawn and dusk sides of the auroral
oval/polar cap boundary. The F-region momentum balance was found to be
largely between the ion drag and pressure gradient forces, while, at
E-region altitude, the Coriolis force became important. Other forces
such as viscous and momentum advection were also, however, at times
significant. The momentum forcing discussed in the Killeen and Roble
study showed a strong UT dependence due to the diurnal oscillation of
the magnetospheric convection pattern with respect to the solar
terminator.

In this paper, we report on an extension to the DP that enables
the tracking, in time and space, of individual parcels of neutral gas
as they move through the TGCM model grid. Parcel 'trajectories' are
calculated from the TGCM horizontal wind predictions using a weighted
interpolation scheme to step forward in UT. Each calculated parcel
trajectory is subsequently used to provide a 'coordinate system' for
the decomposition of the momentum and energy terms discussed above.
Thus, the calculations provide both the locus of a given parcel of
neutral gas and its 'life history', describing the momentum forcing and
heating terms experienced by the parcel along its trajectory. The
parcels are generally tracked for one day (24 hours of model time) and
their trajectories updated at every hour of UT. The purpose of the
calculations is to study the effect of the ion convection geometry at
high latitudes on the transport of neutral parcels and, in particular,
to elucidate the role of the cusp region in controlling and regulating
the neutral flow.

Our analysis of neutral parcel trajectories is similar, in some
regards, to previous work that considered the loci of convecting
ionospheric flux tubes in the high latitude F-region (Spiro et al,
1978; Sojka et al. 1981). In the case of the ion transport studies, it
was found that many of the hitherto unexplained features of the high
latitude ionosphere (namely, the tongue of ionization, the main trough,
the polar hole, etc.) could be explained naturally through consideration
of the ion chemistry and photochemistry together with the dynamics
associated with the flux tube trajectories. It is to be expected that
similar insights concerning, for example, the morphology of long lived,
active trace constituents (such as NO and $N(^2D)$), will accrue from a
study of neutral parcel transport at high latitude. We note, in
passing, that the case of neutral transport is fundamentally different
from that of the ions in that there is a large divergent component to
the neutral flow and, consequently, the trajectories calculated are not
repeatable (i.e. do not close) and cannot be obtained using Lagrangian
techniques.

In section 2 we discuss in greater detail the TGCM model used in
our analysis of neutral parcel transport. In section 3 we present
parcel trajectories for a number of different cases to illustrate
examples of F-region parcel transport in the high latitude region for
differing geophysical circumstances. In section 4 we analyze the
forcing history for a representative parcel and finally, in section 5,
we discuss and summarize our findings with emphasis on the implications
for an improved understanding of the dayside cusp region.

2.   THE NCAR-THERMOSPHERIC GENERAL CIRCULATION MODEL

The NCAR Thermospheric General Circulation model (TGCM) is a three-
dimensional, time dependent model that has been described in detail by
Dickinson et al. (1981, 1984) and Roble et al. (1982). The model has
been used by Roble et al. (1982, 1983) to examine the influence of

magnetospheric convection on the high latitude circulation and temperature structure. It was also used by Hays et al., (1984) for comparison
with the average wind and temperature patterns observed over the
southern hemisphere polar cap by Dynamics Explorer. A more detailed
study of the model predictions for the southern (summer) hemisphere
polar region was performed by Roble et al, 1984 using a model run with
input parameters corresponding to the latter part of October, 1981 and
early part of November, 1981. The results of this work demonstrated
that the predicted temperature, composition and winds all show a strong
UT dependence that is related to the displacement between geomagnetic
and geographic poles.

    The model runs analyzed here used various empirical and semi-
empirical model prescriptions and parameterizations that have been
previously discussed in detail (Dickinson et al., 1981; 1984 and Roble
et al., 1982). The solar heating distribution and O_2 photodissociation
rates are calculated using the Hinteregger (1981) solar EUV flux values
and the Torr et al. (1980) solar UV flux values. The Chiu (1975)
empirical model of electron density is supplemented by the addition of
auroral particles according to the prescription of Roble et al., 1984.
The electron densities are used to calculate the ion-drag tensor. The
magnetospheric convection model of Sojka et al. (1979, 1980), including
displaced geomagnetic poles (north geomagnetic pole 73.8°N latitude,
291.0°E longitude, south geomagnetic pole 74.5°S latitude, 127°E
longitude), is used for the specification of the ion drift velocities.
Both the Joule heating and the ion-drag momentum source due to the
relative drift between the ions and neutrals are updated at each time
step in the model (typically 180 seconds) to determine the time
dependent Joule heating rate and momentum forcing. The heating due to
soft particle precipitation in the region of the dayside cusp is
parameterized by an energy flux of gaussian shape (5° latitude, 2 hours
local time width) with a maximum of 5 ergs/cm^2-sec centered near the
'throat' of the ion convection pattern. The auroral oval is a 5° wide
gaussian (maximum energy, 4 ergs cm^{-2} s^{-1}, characteristic particle
energy 24 eV) offset 3° from the auroral oval/polar cap boundary in the
sunward drifting ion region. For the simulations discussed here, the
TGCM was run until a 'steady-state' or diurnally reproducible neutral
wind field was obtained.

    Figure 1 shows the predicted neutral wind field over the northern
hemisphere polar cap for one such TGCM run using geophysical input
parameters appropriate to December 4th, 1981, and a cross cap potential
of 60 kV. The winds are shown for the z = 1 constant pressure surface
at four UTs. The coordinate system chosen for the presentation of
neutral winds and parcel trajectories is a latitude/local time polar
system, fixed with respect to the sun and not attached to the earth's
rotating surface. Thus, for a given season, the solar terminator is
invariant while the outline of the continents rotates with UT. The
component due to co-rotation has been removed from the plotted vectors
and these, therefore, show the wind directions that would be observed
from a ground station. The examples illustrate the tendency for the

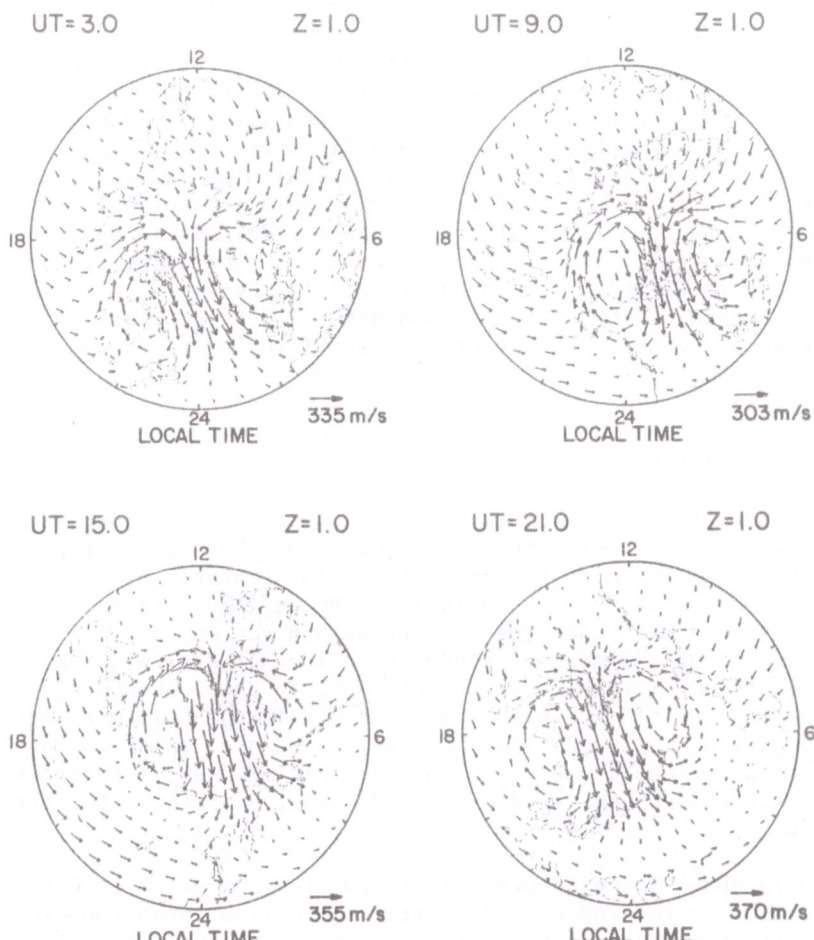

TGCM NEUTRAL WIND PREDICTIONS

Figure 1. Neutral wind vectors predicted by the NCAR-TGCM plotted in solar inertial polar coordinates (latitude and local solar time). The outer circle is at 40°N latitude. The wind field is calculated for a steady state run of the TGCM using geophysical parameters appropriate to 4th December 1981 and is depicted for four UTs, a) 3 UT, b) 9 UT, c) 15 UT, d) 21 UT. The maximum wind vector has the value given at bottom right for each polar dial.

predicted neutral wind field to follow the two cell vortex pattern of
ion convection; a tendency borne out by recent satellite measurements
(Killeen et al., 1982; Hays et al., 1984). There is a general anti-
solar flow over the geomagnetic polar cap bounded, on either side, by a
return (sunward) flow region. The sunward neutral wind velocities
predicted for the dusk cell of the flow pattern are of greater magnitude
than those predicted for the dawn cell. Maximum wind velocities for
this model simulation are of the order of 350 m/sec. There is a clear
UT dependence seen in the neutral wind pattern related to the diurnal
revolution of the geomagnetic pole and its associated ion convection
pattern about the geographic pole (compare Figs. la - ld). Since the
angular distance travelled in 1 hour by a typical high latitude F-region
neutral parcel is of the order of ten degrees only (see below), it is
obvious that the parcel trajectories will be controlled, in part, by
the UT-dependence of the ion drag forces caused by the displaced poles.
Thus, the trajectories will depend on the relative velocities between
the neutral parcels and the various boundaries and flow regimes of the
ion convection pattern as the latter move in UT to 'stir' the neutral
fluid.

3. PARCEL TRAJECTORIES

     A time step of one hour UT was used to update each trajectory for
the parcel transport calculations discussed here. This time step was
considered sufficiently short to enable the major features of the horizontal neutral circulation at high latitude to be illustrated. The
assumption was also made that vertical transport can be neglected.
Since the purpose of the present study is to investigate relatively
large-scale horizontal features of parcel loci in the convection-
dominated region and since, in general, the predicted horizontal winds
are of much greater magnitude than the vertical winds, this was considered to be a reasonable approach for our initial study. Future work
will extend the present analysis by considering transport in the vertical dimension.

     Figure 2a shows a trajectory calculated from the December 4th
TGCM run previously referenced. The trajectory illustrates the diurnal
path, in a reference frame fixed with respect to the sun, of an F-region
parcel starting from the geographic location 67.5°N latitude and -110°E
longitude at 0:00 hrs. UT. The starting location for this particular
trajectory calculation was chosen more-or-less arbitrarily to site the
parcel well within the high-latitude region. The numbers on the
trajectory refer to hours of UT. The dotted curve represents the UT
dependent motion of the centroid of the cusp region of the ion convection pattern used in the TGCM simulation. Thus, the symbol C_1 represents the location of the cusp centroid at a UT of 1:00 hrs. The solar
terminator is indicated in this and other figures by the curved hatched
line. This trajectory is typical of many calculated for the polar
F-region in that the parcel locus is a fairly tightly curved anticlockwise loop about a center that is offset toward the morning-side of

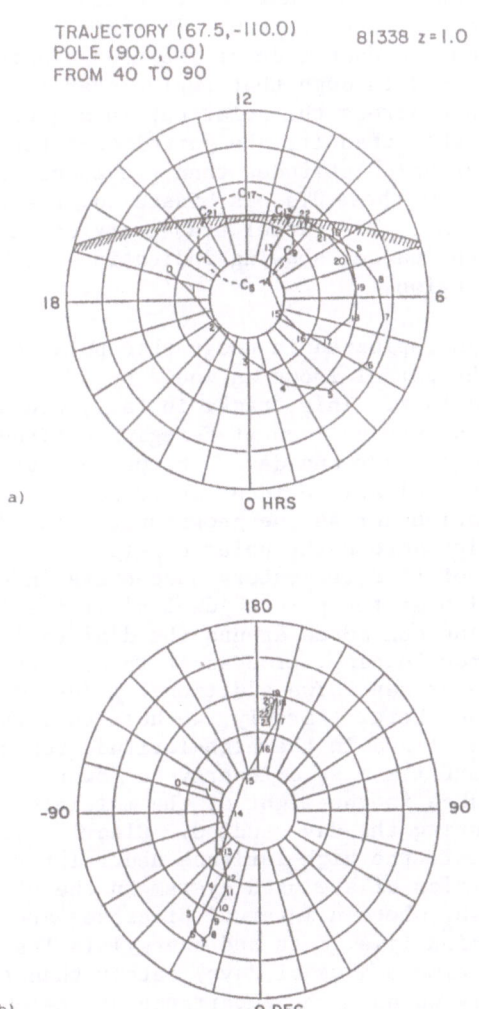

Figure 2.   Representative parcel trajectory. a) solar inertial polar
coordinates, b) geographic polar coordinates. The numbers on the
trajectory refer to hours of UT. The curved hatched line represents
the solar terminator. The dotted circle in a) represents the locus of
the centroid of the dayside cusp region, see text.

the polar dial. The parcel sets out at 0 UT from a location near the
cusp region, it moves rapidly (wind velocity ~350 m/sec) over the polar
cap towards the post-midnight sector. At about 5 UT, the parcel leaves
the region of strong electric fields and comes under the influence of
co-rotating ions. Between 5 and 10 UT the parcel basically co-rotates
until, near 10 UT, it encounters the sunward convecting ion stream in
the dawn cell of the ion convection pattern. The parcel is now drawn
towards the cusp region (on the dawn side of the noon-midnight meridian,
between C_9 and C_{13}), processed through that region near 11- 12 UT and
subsequently flung once more across the polar cap in a general antisunward direction. After this transit, the parcel co-rotates once more
(17- 21 UT) at higher geographic latitudes than previously but away from
the convection region that, at these UTs, is biased towards the dusk side
of the noon-midnight meridian. The parcel trajectory calculation is
terminated at 23 UT with the parcel showing indications of being drawn,
yet again, into the cusp region.

The same trajectory is replotted in geographic polar coordinates in
Figure 2b to illustrate the parcel locus as would be seen by a observer
attached to the earth's surface. This track, in fact, would represent
the locus of a constant pressure balloon at F-region altitude if one
could be deployed and tracked over one day. The periods of co-rotation
(5- 10 UT and 17- 22 UT) are clearly evident in these coordinates by
the lack of significant motion across the geographic polar dial. Also
evident are the two transits across the polar cap (1- 5 UT and 11- 16
UT). The geographic plot of this trajectory suggests a 'harmonic'
nature to the neutral motion as the parcel 'dwells' at the lowest geomagnetic latitudes until the sun moves around the dial to 'pick up' the
parcel with sunward ion drag forcing, process it through the cusp region
and fling it across the polar cap. Two additional points of particular
interest should be mentioned here. Firstly, we note that this particular parcel appears to be confined in the high-latitude region and
therefore does not transport its mass or energy to lower latitudes.
This point is of significance in the light of the multiple transits made
through the cusp region during the day studied. Clearly, if this parcel
trajectory is indeed typical, the high-latitude neutral F-region takes
on some of the characteristics of a closed system in the steady state,
diurnally reproducible case, wherein internal processes are relatively
fast (~hours) and information (i.e. mass and energy) is lost through
'leakage' processes (with time scales of days) rather than through
direct transport. Secondly we note the importance of the cusp region
due to the convergent geometry of the ion convection pattern. The sunward convecting ions in the wings of the convection pattern act to sweep
the parcel of air towards the cusp. Again, if this trajectory proves to
be typical, the cusp region clearly plays an important regulatory
(nozzle) role in the large scale neutral transport in the high latitude
F-region. Such a role could have important implications for the production and transport of chemically active and long lived trace constituents such as NO and $N(^2D)$, the latter being of particular interest due
to the reported observations of plumes of 5200Å emission downwind of the
region of particle precipitation (Gerard and Roble, 1982).

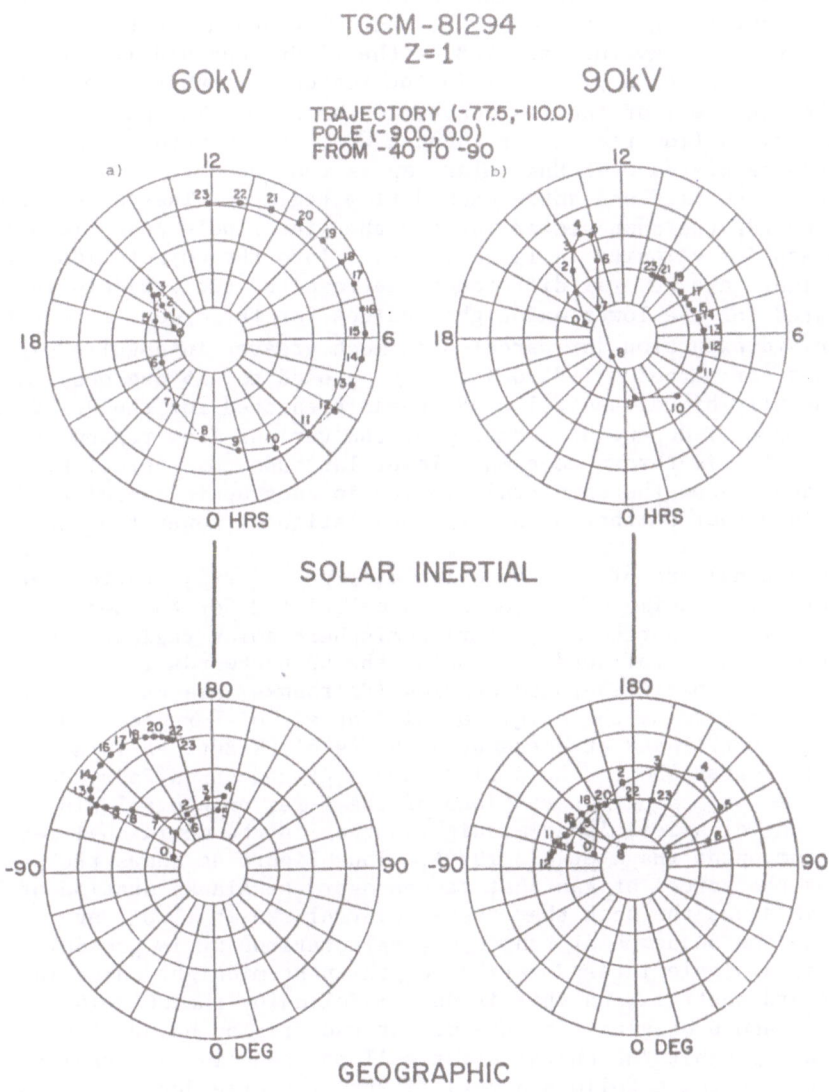

Figure 3. Parcel trajectories for a) 60 kV cross cap potential and b) 90 kV cross cap potential. The trajectories are shown in solar inertial polar coordinates (top) and geographic polar coordinates (bottom). The TGCM model run was for October 21, 1981. The numbers on each trajectory refer to hours of UT.

In Figure 3 are shown further trajectory examples calculated for
F-region neutral parcels in the southern (summer) hemisphere polar cap
for two TGCM runs for October 21st 1981. The two trajectories are
started from the same geographic location at 0 UT and are plotted in
both solar-inertial (top) and geographic (bottom) frames of reference.
The calculations differ only in the values of the cross cap potential
used as input to the TGCM (60 kV for parcel A on the left hand side of
the figure and 90 kV for parcel B on the right hand side). The starting
location for parcel A placed it in the region of sunward flow associated
with the dusk cell of the ion convection pattern. The parcel resides
in the sunward flow region for about 4 hrs. and then turns and is
rapidly accelerated over the polar cap as the boundary is crossed
between the sunward and anti-sunward flow regimes. The parcel then
turns towards the dawn sector (as for the winter pole case discussed
above) and its velocity is largely turned into the co-rotation direction. There are several differences between the two trajectories
calculated for the lower and higher values for the cross-cap potential.
The sunward excursion for parcel B is much greater in angular distance
than that for parcel A (~25 degrees as opposed to ~15 degrees) due to
the generally higher neutral wind speeds predicted for the 90 kV case.
However, the UT-dependent geometry of the various flow regions is such
that parcel A is thrown to a much lower latitude than parcel B. The two
lower panels show the same trajectories in geographic coordinates. It
can be seen that neither parcel reaches latitudes lower than 40°S.

As a final example of the calculated parcel trajectories, we show
in Figure 4, a family of trajectories calculated for the December 4th
model run in the northern (winter) hemisphere polar region. These
parcels have been followed by running the DP backwards in time from the
location of a specific ground station (Fairbanks, Alaska). The purpose
here is to determine the origin of six parcels of F-region neutral gas
that pass over Alaska at consecutive hours of UT Between 10 and 15 hrs.
These UTs correspond to the period when a ground-based Fabry Perot
interferometer making observations of the F-region neutral wind would
observe the classical 'post-midnight surge' in the meridional (equator-
wards) component (Hays et al, 1979). Thus Figure 4a shows the path
taken by the parcel of gas that passes over the Alaska station at 10 UT,
Figure 4b shows the path that passes over Alaska at 11 UT, etc. By
following the sequence of trajectory calculations it is possible to
arrive at a simple interpretation for the post-midnight surge in the
equatorward neutral wind that is well modelled by the NCAR-TGCM.
Parcels a and b originate in the midlatitude region below 40°N latitude
and do not exhibit the characteristic 'loop' seen in the previous
examples. Parcel c follows a very different course during its approach
to Alaska. Its trajectory is tightly looped showing two transits
through the cusp region during the day preceding the 12 UT passage over
Alaska. The remaining cases, d, e and f, show, interestingly, a
progressive 'unravelling' of the outer trajectory loop corresponding to
the parcel locus for the day preceding the Alaska passage. The inner
loop, corresponding to the 12 hours or so immediately preceding the
Alaska passage, on the other hand, becomes progressively tighter. In

PARCEL TRAJECTORIES PASSING OVER ALASKA

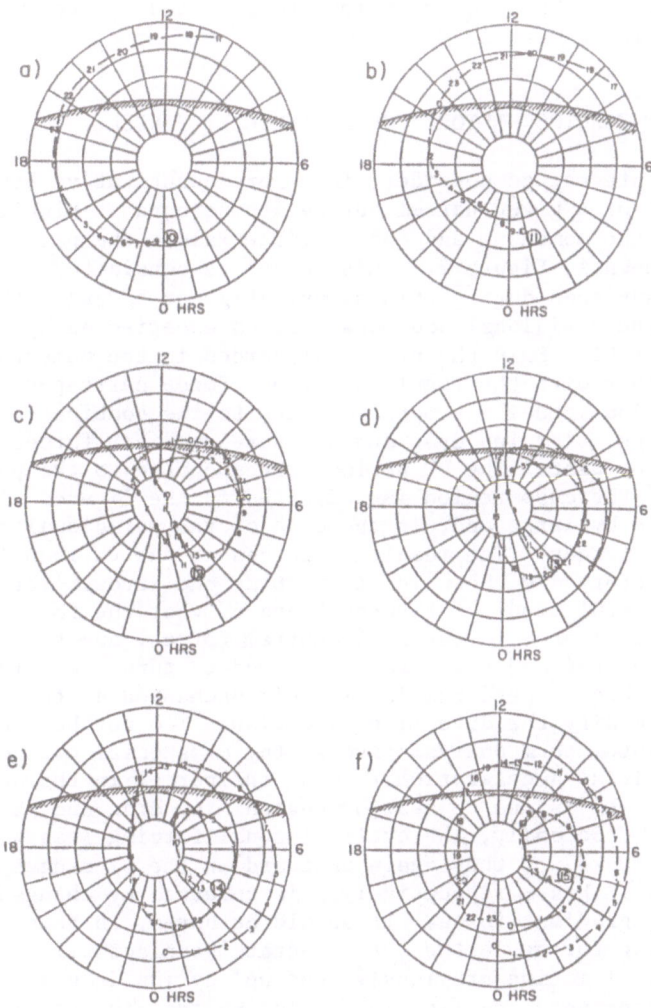

Figure 4. Family of six trajectories calculated backwards in time from passage over Alaska at a) 10 UT, b) 11 UT, c) 12 UT, d) 13 UT, e) 14 UT, and f) 15 UT. Coordinates are solar inertial, the numbers on each trajectory refer to hours of UT. The curved hatched line is the solar terminator.

all three cases, the parcel passing over Alaska has in its recent
history (36 hours) also passed twice through the dayside cusp region.
The postmidnight surge in the meridional wind measured by Hays et al
maximizes at around 13 UT and may be interpreted as being caused by the
acceleration experienced by the neutral gas 6 - 7 hours 'upstream' along
the parcel trajectory in the vicinity of the dayside cusp (see Fig. 4d)
coupled with the general tendency for the air to turn towards the co-
rotation direction. These effects lead to a maximum in the equatorward
wind speed at Alaska near 13 UT.

4.   ANALYSIS OF FORCING HISTORY FOR A 'TYPICAL' PARCEL

It is of interest to consider, in a more quantitative fashion, the
forcing history of a given parcel during its sojourn at high latitude.
Here we detail the momentum and energy forces along the parcel trajectory already shown in Figure 2. This parcel is considered to be representative of many seen during this study. Figures 5a and b show,
respectively, the meridional and zonal forces experienced by this parcel
as a function of UT. Only the principal forces in the momentum balance
are shown together with the resultant or net force corresponding to the
local acceleration actually experienced due to the combination of all
individual forces. The ion drag and pressure gradient forces, as
expected, are dominant. The UT period corresponding to the parcel
passage through the cusp region is indicated in the figure. This region
is characterized by large and, in the case of the meridional component,
rapidly varying forces. The resultant forces at the various locations
along the trajectory are shown in Figure 6 as the arrows. It can be
seen, from a careful study of Figures 5 and 6, that the forces experienced by the parcel are, in essence, central forces, due to a combination of both ion drag and pressure, that tend to turn the parcel towards
the left. The parcel speed remains largely unchanged as the forces act
principally to modify the direction of motion. Two further features of
the parcel momentum term analysis are worthy of special notice.
Firstly, the initial acceleration forcing the parcel in the anti-sunward
direction near 1 UT is due to near-orthogonal ion drag and pressure
gradient forces. Secondly, the cyclical zonal forcing (eastward
followed by westward and then again eastward in the UT ranges 1 - 6,
6 - 14, and 15 - 18 hrs., respectively), is governed by changes in sign
of the pressure gradient force. It should be noted, in this regard,
that the ion drag forces in the 'steady-state' can actually tend to
impede the neutral motion previously 'set up' by ion drag due to the
inherent time constants and lags in the ion-neutral interaction (Killeen
and Roble, 1984, Killeen et al., 1984). Thus care should be taken not
to interpret the depicted 'instantaneous' parcel forces as being those
responsible for the large scale neutral flow pattern since the latter
is, more correctly, a consequence of the time history of forcing during
the period of 'spin-up' in the TGCM.

The energy (heating) terms experienced by the parcel are indicated
in Figure 7. The terms are expressed in units of degrees per second

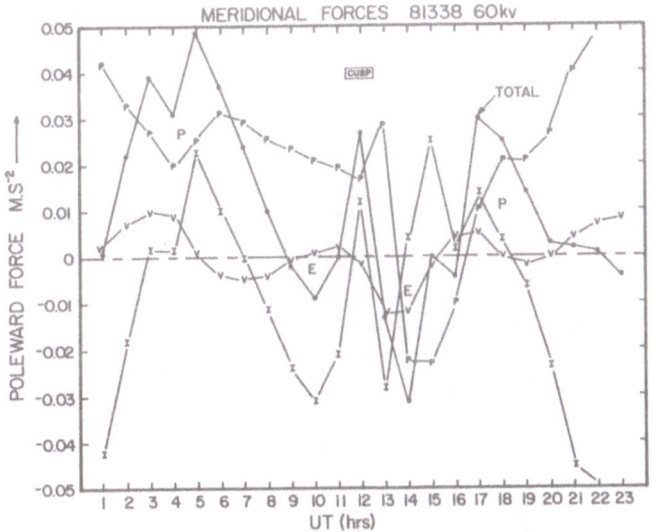

Figure 5a. Meridional force term analysis for the parcel whose trajectory is shown in Figure 2. Curves labelled P, I and V represent, respectively, the pressure gradient, ion drag and viscous forces. The foces are plotted as a function of UT and are per unit mass of atmosphere. The curve labelled 'total' represents the resultant force. The UT period corresponding to passage of the parcel through the cusp region is indicated.

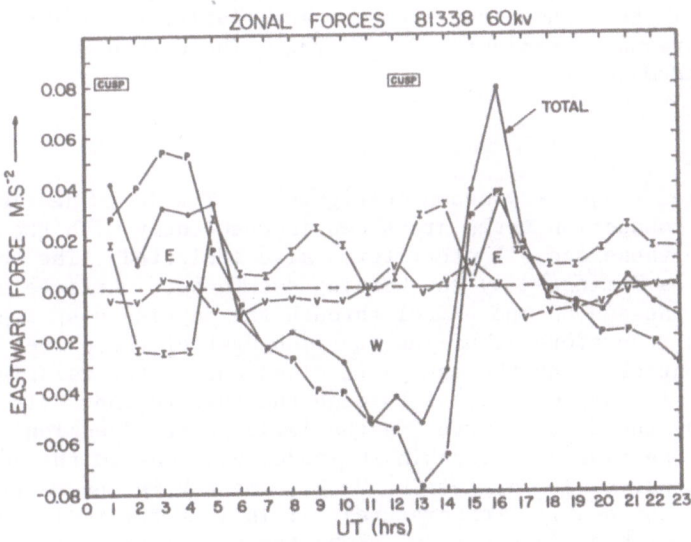

Figure 5b. Zonal force term analysis.

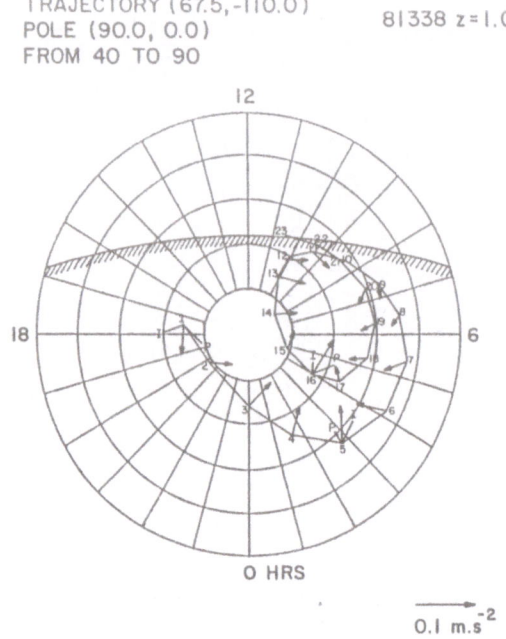

TRAJECTORY (67.5,-110.0)
POLE (90.0, 0.0)   81338 z=1.0
FROM 40 TO 90

Figure 6. Parcel trajectory in solar inertial polar coordinates as
for Figure 2. The arrows represent the resultant force per unit mass
on the parcel, scale given at bottom. The lines terminating in the
letters p and i where shown represent, respectively, the force components due to pressure gradient and ion drag. The curved hatched line
is the solar terminator.

along the parcel trajectory shown in Figure 2. The Joule heating and
downward heat conduction terms are shown independently with the total
heating due to these and all other terms also indicated. The total
heating curve is dominated by the heating due to soft particle precipitation during passage of the parcel through the dayside cusp region.
It can be seen, therefore, that the energy budget of the parcel is
principally controlled by the energy received during the relatively
short period (~1 hour) of transit through the cusp region. Also shown
in Figure 7, by the dotted curve, is the 'grid point advection' that
represents the heating occurring in the model grid due to the advection
of heat by the parcel (term given by $V \cdot \nabla T$, where V is the neutral wind
velocity and T the neutral temperature). This latter term illustrates
the deposition of heat 'downstream' along the parcel trajectory due to
thermal advection. It can be seen that a significant proportion of the
energy received by the parcel in the cusp region is deposited 2- 3 hours
downstream along its trajectory due to this process.

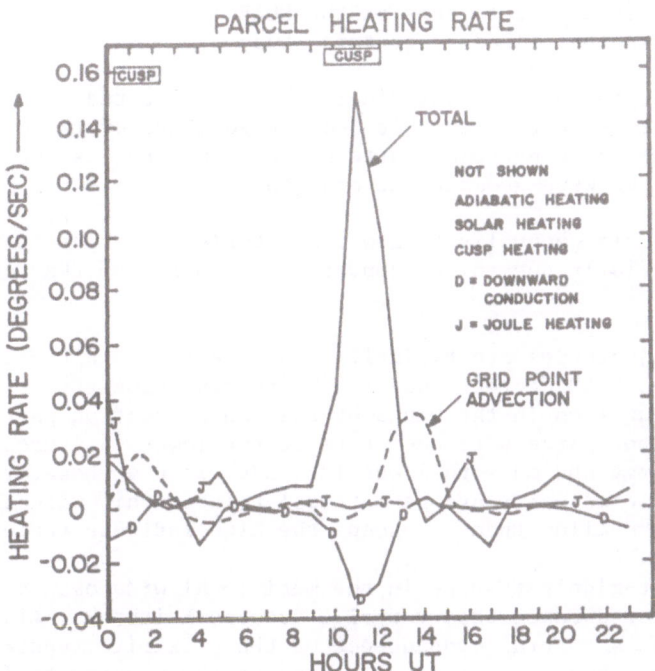

Figure 7. Energy term analysis for the parcel of Figure 2. Curves
labelled J and D represent, respectively, the Joule heating and downward heat conduction experienced by the parcel as a function of UT.
The dotted curve represents the thermal advection term (see text).
The total heating rate is shown and is dominated by heating due to
soft particle precipitation in the dayside cusp region, see text.

    The energy term analysis is taken to demonstrate the importance of
both the high altitude heating due to soft particle precipitation in the
cusp region and the neutral dynamics that can re-distribute the heating.
In this representative case, the cusp region preferentially heats the
neutral atmosphere 'downstream' or in the magnetic polar cap. Since, as
we have found, most parcels followed pass through the cusp region at
least once per day, this polar cap heating mechanism may be of particular importance and may help to explain the recent exospheric temperature
measurements from Dynamics Explorer that show an enhanced temperature in
the polar cap relative to the auroral oval (Hays et al, 1984).

5.   SUMMARY

    We have shown examples of F-region neutral parcel trajectories
calculated using steady state runs of the NCAR-TGCM and its diagnostic
package. The principal findings of this study are:

    1.   The F-region neutral parcels in the high latitude region follow
trajectories that are not repeatable and are governed by the UT-
dependence of the ion convection pattern that, in turn, is dependent on
the displacement between geomagnetic and geographic poles.

    2.   The parcels generally follow trajectories that pass through the
dayside cusp regularly due to the convergent geometry of the ion convection pattern.

    3.   The trajectories are typically confined to the high-latitude
region (latitudes > 40 degrees) due to the forces associated with the
sunward convecting ions in the wings of the ion convection pattern and
the tendency of the parcels to co-rotate at the lower latitudes. The
transport of energy and mass to lower latitudes is via those, relatively
few, parcels whose velocity and geometrical relationship with the ion
convection pattern allow them to escape the high-latitude region.

    4.   The post-midnight surge in the meridional wind observed from
high-latitude ground based observatories such as Fairbanks, Alaska, can
be interpreted as a natural consequence of the proximity, upstream, of
the acceleration region of the dayside cusp to the ground station at the
appropriate UT. Parcels of gas in the 'surge' are accelerated in the
anti-sunward direction through the cusp region and are then turned
towards the morning sector by a combination of ion drag and pressure
gradient forces.

    5.   The energy budget for a representative F-region neutral gas
parcel is dominated by direct heating in the region of the dayside cusp.
Cusp energy is transported 'downstream' along the parcel trajectory and
is advected into the geomagnetic polar cap. The cusp region, therefore,
acts as a 'processor' or 'factory' for cycling large volumes of F-region
gas and altering the thermodynamical state of the polar thermosphere.

ACKNOWLEDGEMENTS

    Useful discussion with Drs. E. C. Ridley, B. A. Emery, P. B. Hays,
J. W. Meriwether, Jr., and T. J. Fuller-Rowell are acknowledged. The
work was supported by NASA grant number NAS5-25691 to the University of
Michigan. The National Center for Atmosphere Research is sponsored by
the National Science Foundation.

REFERENCES

Chiu, Y. T., 1975, J. Atmos. Terr. Phys., 37, 1563.
Dickinson, R. E., Ridley, E. C., and Roble, R. G., 1981, J. Geophys.
    Res., 86, 1499.
Dickinson, R. E., Ridley, E. C., and Roble, R. G., 1984, J. Atmos. Sci.,
    41, 205.
Fuller-Rowell, T. J., and Rees, D., 1980, J. Atmos. Sci., 37, 2545.
Fuller-Rowell, T. J., and Rees, D., 1981, J. Atmos. Terr. Phys., 42,
    553.
Fuller-Rowell, T. J., and Rees, D., 1984, Planet. Space Sci., 32, 69.
Gérard, J.-C., and Roble, R. G., 1982, Planet. Space Sci., 30, 1091.
Hays, P. B., Killeen, T. L, and Kennedy, B. C., 1981, Space Sci. Inst.,
    5, 395.
Hays, P. B., Killeen, T. L., Spencer, N. W., Wharton, L. E., Roble,
    R. G., Emery, B. A., Fuller-Rowell, T. J., Rees, D., Frank, L. A.,
    and Craven, J. D., 1984, J. Geophys. Res., in press.
Hernandez, G., and Roble, R. G., 1984, J. Geophys. Res., 89, 327.
Hoffman, R. A., Hogan G. D., and Maehl, R. C., 1981, Space Sci.
    Instrum., 5, 349.
Hinteregger, H. E., 1981, Adv. Space Res., 1, 39 (COSPAR).
Killeen, T. L., Hays, P. B., Spencer, N. W., and Wharton, L. E., 1982,
    Geophys. Res. Lett., 9, 957.
Killeen, T. L., Hays, P. B., Spencer, N. W., and Wharton, L. E., 1983,
    Adv. Space Res., 2, 133, Pergamon Press, Oxford.
Killeen, T. L., and Roble, R. G., 1984, J. Geophys. Res., in press.
Killeen, T. L., Hays, P. B., Carignan, G. R., Heelis, R. A., Hanson,
    W. B., Spencer, N. W., and Brace, L. H., 1984, J. Geophys. Res.,
    in press.
Rees, D., Fuller-Rowell, T. J., and Smith. R. W., 1980, Planet. Space
    Sci., 28, 919.
Rees, D., Fuller-Rowell, T. J., Gordon, R., Killeen, T. L., Hays, P. B.,
    Wharton, L. E., and Spencer, N. W., 1983, Planet. Space Sci., 31,
    1299.
Roble, R. G., Dickinson, R. E., and Ridley, E. C., 1982, J. Geophys.
    Res., 87, 1599.
Roble, R. G., Dickinson, R. E., Ridley, E. C., Emery, B. A., Hays, P. B.,
    Killeen, T. L., and Spencer, N. W., 1983, Planet. Space Sci., 31, 1479.
Roble, R. G., Emery, B. A., Dickinson, R. E., Ridley, E. C., Killeen,
    T. L., Hays, P. B., Carignan, G. R., and Spencer, N. W., 1984a,
    J. Geophys., Res., in press.
Sojka, J. J., Raitt, W. J., and Schunk, R. W., 1979, J. Geophys. Res.,
    84, 5943.
Sojka, J. J., Raitt, W. J., and Schunk, R. W., 1980, J. Geophys. Res.,
    85, 1762.
Sojka, J. J., Raitt, W. J., and Schunk, R. W., 1981, J. Geophys. Res.,
    86, 609.
Spencer, N. W., Wharton, L. E., Niemann, H. B., Hedin A. E., Carignan,
    G. R., and Maurer, J. C., 1981, Space Sci., Inst., 5, 417.
Spiro, R. W., Heelis, R. A., and Hanson, W. B., 1978, J. Geophys. Res.,
    83, 4255.

Torr, M. R., Torr, D. G., and Hinteregger, H. E., 1980, J. Geophys.
    Res., 85, 6063.

OBSERVATIONS OF PLASMA STRUCTURE AND TRANSPORT AT HIGH LATITUDES

Edward J. Weber and Jurgen Buchau
Ionospheric Physics Division
Air Force Geophysics Laboratory
Hanscom AFB, MA 01731 USA

ABSTRACT. Radio and optical diagnostics from the AFGL Airborne Ionospheric Observatory are used to study the structure and motion of regions of enhanced F-region density at high latitudes. Plasma flow can be tracked from the poleward edge of the dayside cusp, across the polar cap and into the nightside auroral zone. Simultaneous satellite amplitude and phase scintillation measurements define the degree of structuring or intensity of sub-kilometer ionospheric irregularities within these regions. The combined measurements are used to track large scale plasma flow, and to infer plasma source regions.

1. INTRODUCTION

The plasma environment in the high latitude winter ionosphere is dominated by local production from precipitating particles and by transport from non-local source regions. This paper will discuss observations of both types of ionospheric plasma structures and dynamics, with emphasis on remote diagnostic techniques used to monitor this environment. The measurements are important for understanding basic high latitude processes, and for determining conditions which lead to the generation of ionospheric irregularities. Irregularities over the range of scale sizes from 100 m to 10 km cause amplitude and phase fluctuations (scintillation) of satellite radio links, and thus affect performance of communications systems. Kelley et al. (1982) have outlined the principles which govern transport and decay of irregularities at high latitudes. This type of framework or model requires realistic input parameters to track the subsequent evolution of plasma (and irregularities) which is caused primarily by convection.

Extensive optical, ionosonde and satellite scintillation measurements from the AFGL Airborne Ionospheric Observatory (AIO) have been conducted at auroral oval and polar cap latitudes. These have been combined with other coordinated remote and in-situ observations to provide a detailed measure of plasma structure and motion, particle precipitation regions, and characteristics of ionospheric irregularities.

279

J. A. Holtet and A. Egeland (eds.), The Polar Cusp, 279–292.
© 1985 by D. Reidel Publishing Company.

Previous measurements (Weber and Buchau 1981) have shown the existence of sub-visual, F-layer, polar cap auroras. These are aligned with the noon-midnight meridian, are produced by fluxes of low energy (100's eV) electrons, and drift predominately from dawn to dusk. The data base of low light level photometric measurements within the polar cap is only now becoming extensive enough to consider statistical analysis. However, there is a one to one correspondence of these auroras with soft, structured electron precipitation events, and the statistics of these events, derived from a large number of satellite passes (Hardy 1984) shows a strong correlation of occurrence (absence) of these fluxes with periods of Bz northward (southward). Polar cap F-layer auroras appear to be the dominant ionospheric feature during quiet magnetic conditions and, by inference, during periods of Bz northward. Although the auroras are commonly observed by all sky photometric optical techniques, the degree to which these auroras represent a significant increase in electron density has not been clearly established. Ionosonde meaurements during the peak of the solar cycle (1979-80) show a strong indication that these auroras are accompanied by increased electron density. Simultaneous scintillation measurements, which are proportional to $<(\Delta N)^2>$, the mean square electron density deviation, confirm the enhanced densities within the arcs. More recent measurements do not show definite indications of electron density enhancement using ionospheric sounders. Nor do the scintillation measurements show large amplitude and phase fluctuations. To address this apparent problem three solar cycle dependent possibilities are being investigated: variations in the precipitated electron energy flux in the range 20-600 eV; variations in the transport of ionization out of the production region; effects due to variations in the neutral composition. Future rocket programs are being planned to investigate some of these factors. During magnetically disturbed periods (Kp > 4), and during periods of Bz southward, large patches of F-region plasma are observed to convect in the anti-sunward direction. They are observed to originate within or equatorward of the dayside cusp region and have been tracked over essentially the entire polar cap (Buchau et al. 1983; Weber et al. 1984). The main objective of this paper is to demonstrate the effectiveness of remote optical and radio measurement as plasma diagnostics, and to show that transport of plasma over large distances from the production or source region is of major importance to the structure of the high latitude F-layer at all latitudes and local times.

2. OBSERVATIONS

Airborne and ground based measurements using the AIO have been conducted over a large fraction of the high latitude region.

In this paper, meausrements from three areas will be presented to illustrate the major results concerning structure and motion. These areas are: the high latitude edge of the dayside aurora or cusp near local noon, the central polar cap, and the nightside auroral oval.

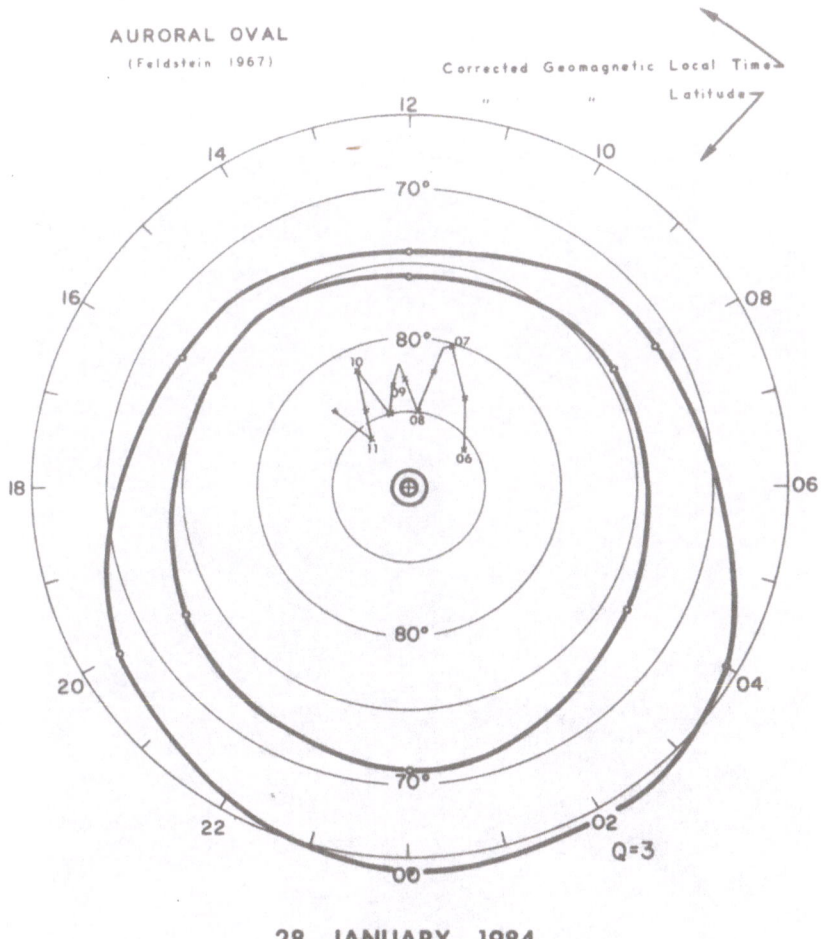

Figure 1. Corrected Geomagnetic Latitude/Local Time plot showing aircraft flight track on 28 January 1984.

Although the individual measurements are limited in spatial extent,
they investigate important areas of the overall high latitude convection
pattern.   In addition, these areas can be investigated during darkness,
thus allowing sensitive optical measurements.

2.1 Dayside Measurements

A flight was conducted on 28 January 1984 to investigate the structure of drifting F-layer patches at they separate from the poleward
edge of the dayside cusp region.   Figure 1 shows the flight track in

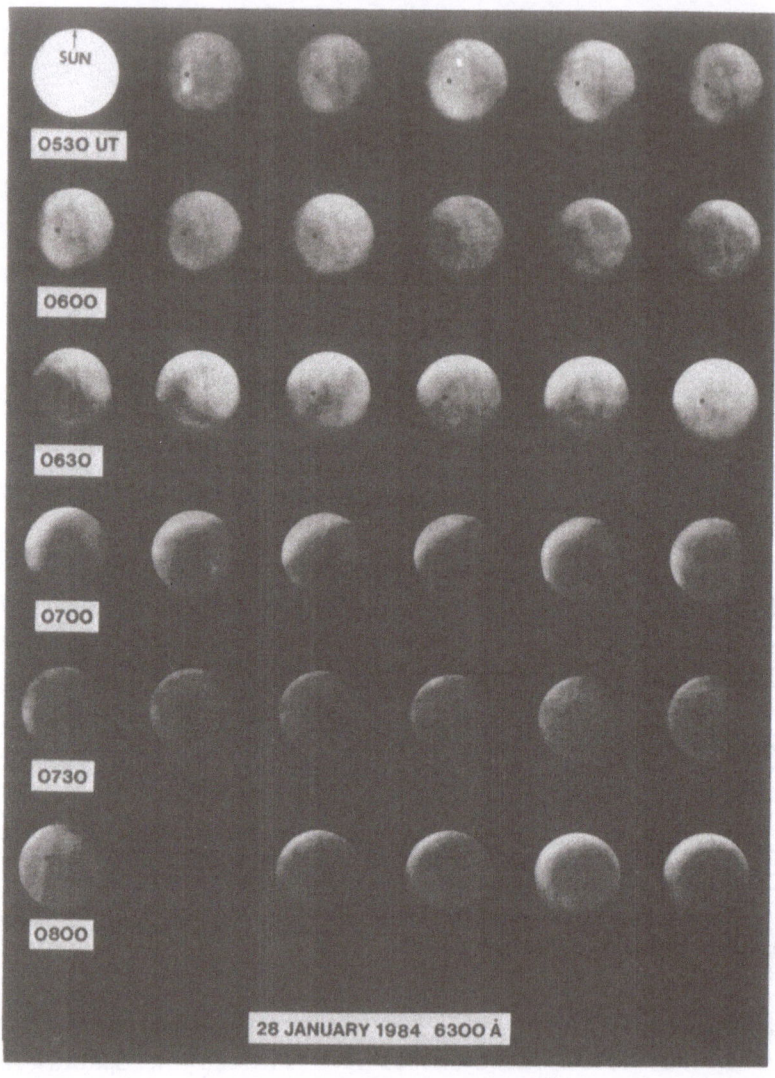

Figure 2. All Sky (155° field of view) images at 6300Å [OI] recorded at 5 minute intervals. The images show auroral emissions associated with cusp precipitation as well as drifting airglow patches. The dot in each image is the location of a satellite-aircraft 250 MHz link used for scintillation measurements.

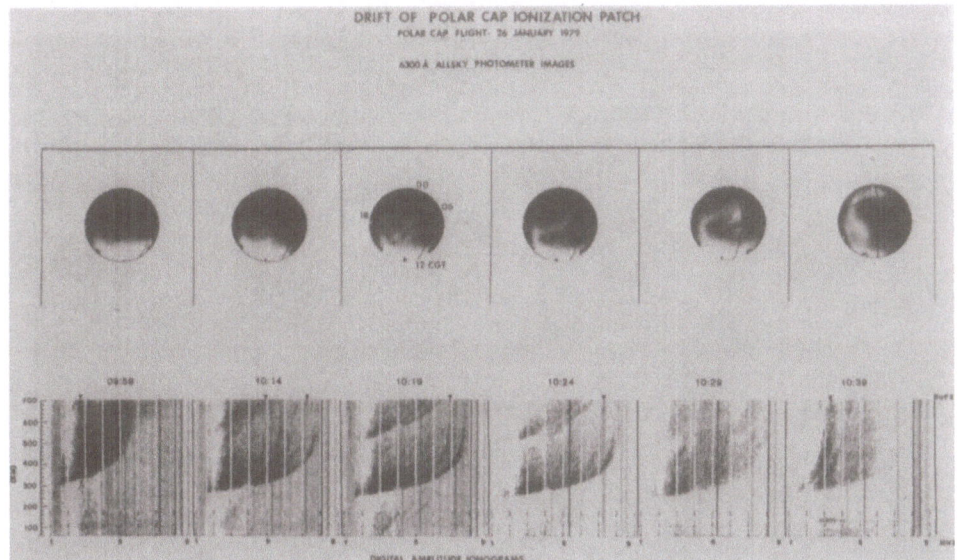

Figure 3. Sequence of 6300Å ASIP images and digital ionograms showing anti-sunward drift of an F-region ionization patch on 26 January 1979 near the dayside auroral zone.

Corrected Geomagnetic (CG) Latitude/Local time coordinates. This flight track illustrates the unique ability of the AIO to investigate the dark noon-sector ionosphere for approximately 5 hours. This flight was conducted in the geographic longitude region near Spitsbergen. Figure 2 shows 6300Å images at five minute intervals for a portion of the flight. The images are made using an All Sky Imaging Photometer (ASIP) with a 155° field of view lens, and map a region of 1200 km diameter at F-region altitudes. The aircraft flew from Thule AB, Greenland (86° CGL) toward the cusp region from 0530-0700 UT. The image at 0625 UT is the first to show the poleward edge of the cusp. At 0630UT, a patch of enhanced 6300Å airglow began to drift away from the cusp in the anti-sunward direction. The aircraft flight direction was opposite to the drift direction of the patch, which results in a rapid apparent patch velocity in the anti-sunward direction. By 0650 UT, the patch had drifted beyond the ASIP horizon. The dot in each image indicates the position of a satellite-aircraft signal raypath. Amplitude and phase scintillation measurements were made using a beacon signal at 250 MHz. During the passage of this patch through the signal raypath (0635-0650 UT) only weak fluctuations were observed. This absence of strong fluctuations was also observed on a similar flight on 26 January 1979. Figure 3 shows simultaneous All-Sky Images and digital ionograms recorded during the passage of a patch away from the noon-sector. In this case, a clear increase in the F-region elec-

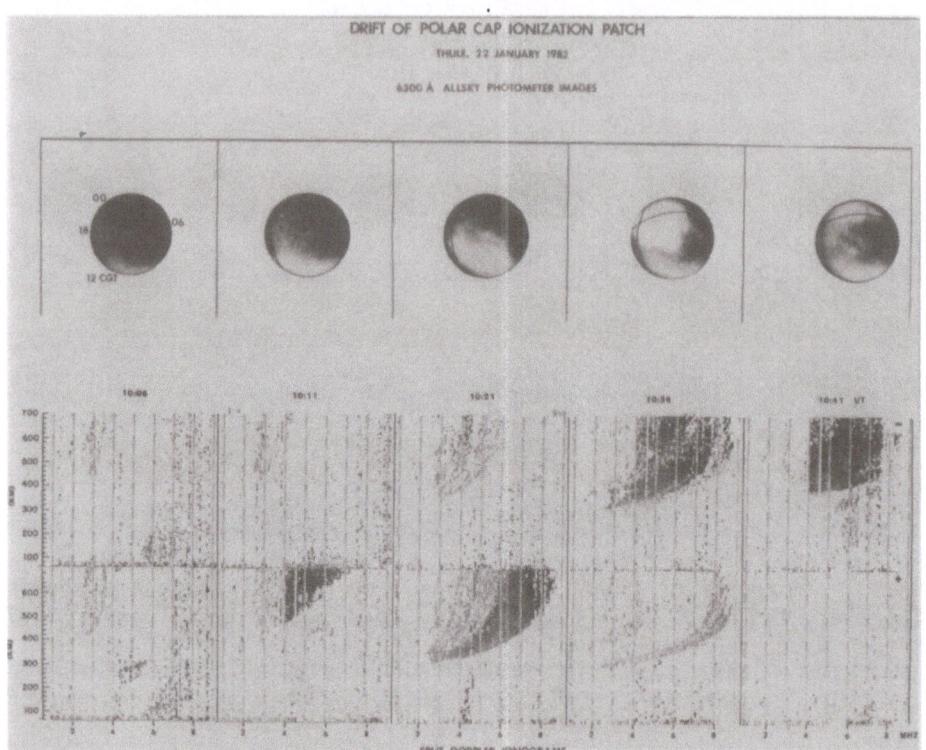

Figure 4.   Sequence of 6300Å ASIP images and Doppler ionograms showing the anti-sunward drift of an F-region patch from Thule AB.

tron density from outside to inside the patch was observed.   The F-region maximum plasma density increased from 9.7×10^4 to 6.4×10^5 el cm^{-3} (foF2 changed from 2.8 to 7.2 MHz).   Although scintillation from this patch occurred as a discrete event, fluctuation levels were low with the Scintillation Index (SI) of approximately 6 dB.

2.2 Central Polar Cap Measurements

Measurements have also been conducted with the AIO during several ground campaigns at Thule AB, Greenland (86° CGL).   An example of patch drift over Thule is shown in Figure 4.   A sequence of ASIP images and digital ionograms (separated into positive and negative Doppler) shows the anti-sunward drift of a region of increased plasma density (and increased airglow).   Amplitude and phase scintillation measurements were also conducted during this period.   Figure 5 shows the amplitude fluctuations and illustrates the highly structured character of this patch.   These scintillation measurements essentially map out the irregu-

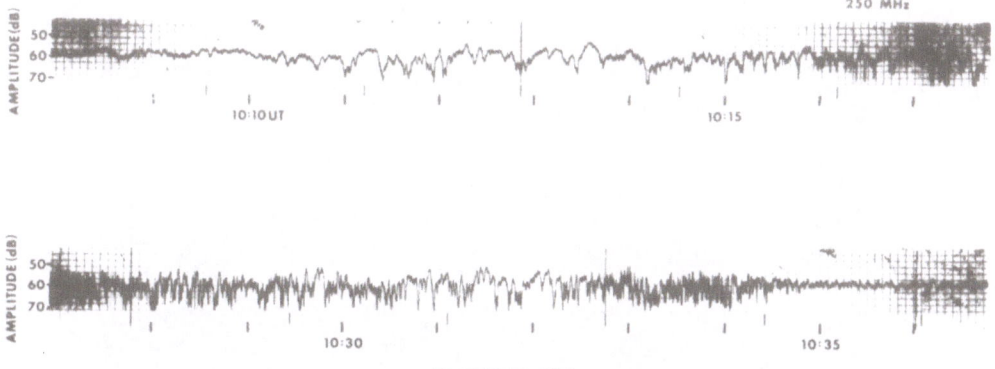

22 JANUARY 1982

Figure 5.   Amplitude fluctuations at 250 MHz satellite-ground propagation link showing highly structured plasma (100m - few km irregularities) within patch.

larity intensity along a noon-midnight cross section through the patch.
Since the patch drift velocity was constant at ~ 700 m/s, the higher
frequency fading rate in the trailing edge of this patch is suggestive
of more intense irregularities in this region. The more quantitative
phase scintillation measurements confirm this structure. This type of
structuring is consistent with an E x B or gradient drift type of
instability due to the motion of the patch (a region of steep horizontal Ne gradients) with respect to the neutral atmosphere.

    More recent measurements of several patch structures drifting over
Thule are shown in Figure 6. These images clearly show the passage of
several patches during the period of observation. The dot ($\bullet$) indicates the direction to a 250 MHz polar beacon satellite; the plus
(+) indicates the direction to a Global Positioning System (GPS) navi-
gation satellite operating at 1.2 and 1.5 GHz. These observations
(Klobuchar and Bishop, private communication, 1984) represent the first
measurements of Total Electron Content (TEC) and amplitude and phase
scintillation measurements using the GPS satellites in the polar cap.
They provide time-continuous measurements at high elevation angles (up
to 75°) for comparison with the coordinated ionospheric measurements.
Figure 7 shows a sequence of TEC variations using GPS as the patches
drifted through the raypath. Changes as large as 25 TEC units (1 TEC
unit = 1 x 10^{16} el/m^2) are associated with the patches. Simultaneous
scintillation measurements show large intensity irregularities associated with the patches, again indicating a highly structured medium.

2.3 Nightside Auroral Zone

    The final region to be discussed is the nightside auroral zone.
An aircraft flight was conducted on 29 January 1979 in conjunction with
Chatanika Radar measurements of plasma density, temperature and drift.

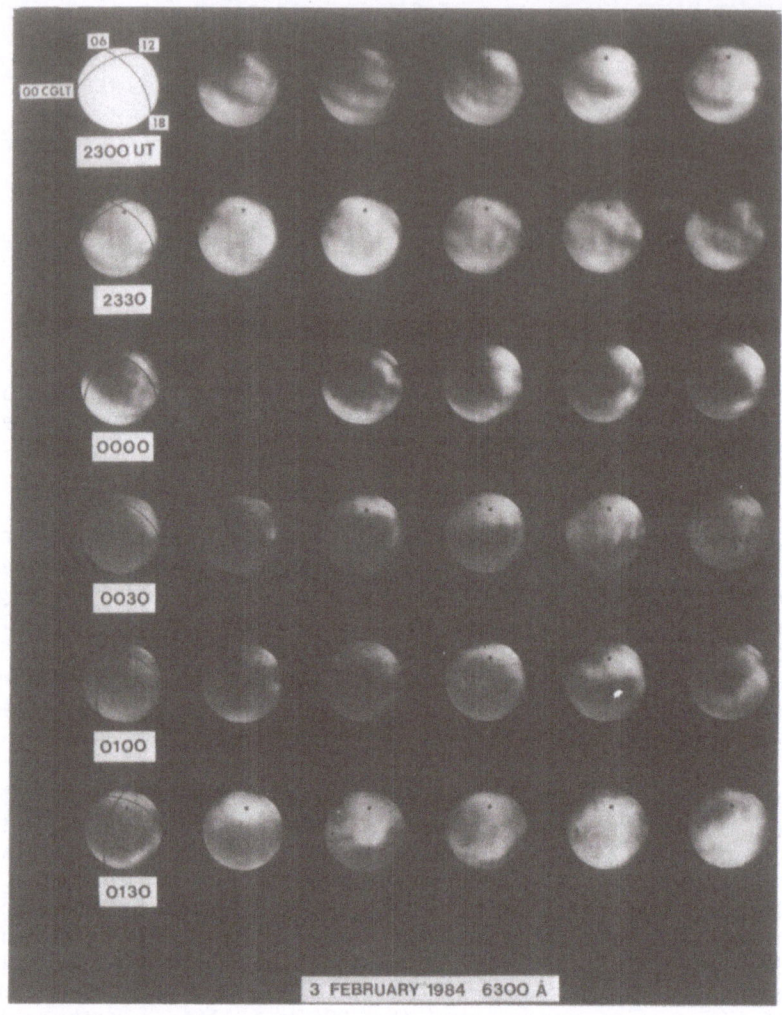

Figure 6. Sequence of 6300Å ASIP images on 3-4 February 1984 recorded at Thule AB. The dot (•) indicates the direction to a 250 MHz polar beacon satellite; the plus (+) indicates the direction to a GPS satellite.

Figure 8 shows the aircraft track along the radar magnetic meridian. Also shown is the sub-satellite track of the DMSP F-2 satellite at 15 second intervals (traced down along field lines from satellite altitude at 840 km to 250 km). A sequence of radar electron density contours made at 14 minute intervals along the magnetic meridian is shown in

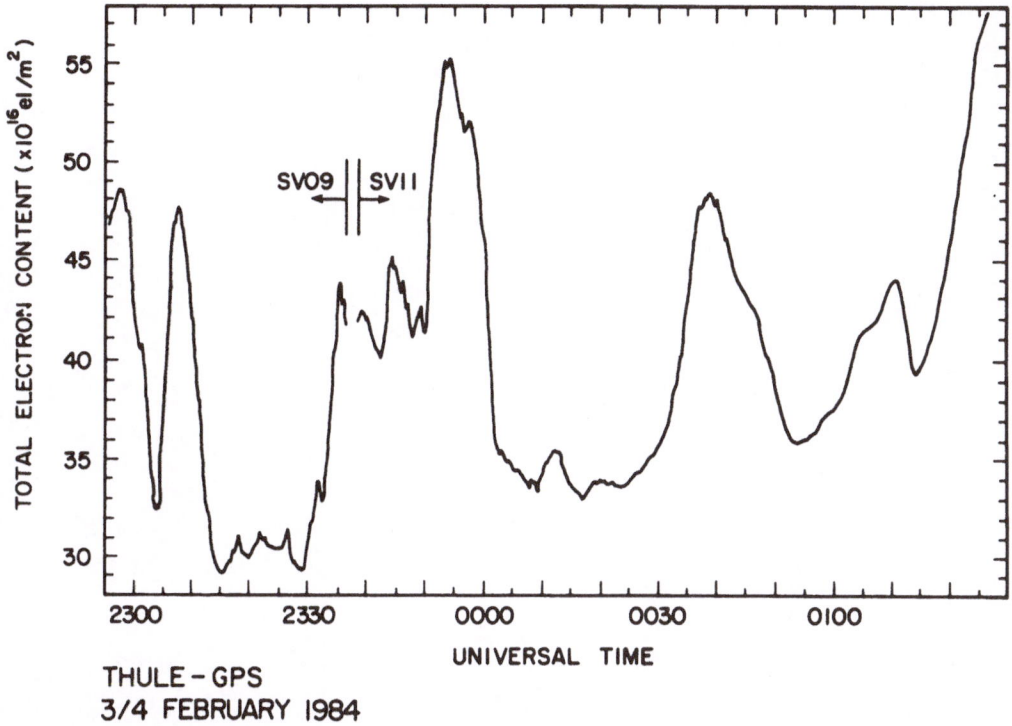

THULE – GPS
3/4 FEBRUARY 1984

Figure 7. Total Electron Content variations measured during the passage
of several F-region patches. The measurements were recorded at Thule
AB using 1.2 and 1.5 GHz signals from the GPS satellites.

Figure 9. Density contours of 1.6 and 2.0 x 10^5 el/cm^3 are highlighted.
The second map at 0617-0630 UT clearly illustrates major large scale
features of the nightside, high latitude ionosphere. The post-sunset
solar produced F-region extends to ~ 80 km north of the radar zenith.
A narrow trough separates this region from a region of enhanced F-region
density (the boundary "blob") located near the equatorward edge of the
diffuse aurora. Finally the auroral E-layer or diffuse aurora extends
over most of the northern half of the radar coverage area.

A pass of the DMSP F-2 satellite measured the latitudinal profile
of precipitating electrons within ~ 80 km of the radar meridian. These
electron fluxes are shown in Figure 10 for comparison with simultaneously measured ionospheric features. The electrons in the range 1-20
KeV show excellent agreement with the location of the auroral E-layer.
Quantitative estimates, using an ionization production code, also give
good agreement between precipitating electron fluxes, resulting E-layer
ionization profiles and N_2^+ optical emission. An important result is
the lack of agreement between regions of low energy (60-660 eV) electrons
and F-region ionization. Although the F-region structure exhibits a

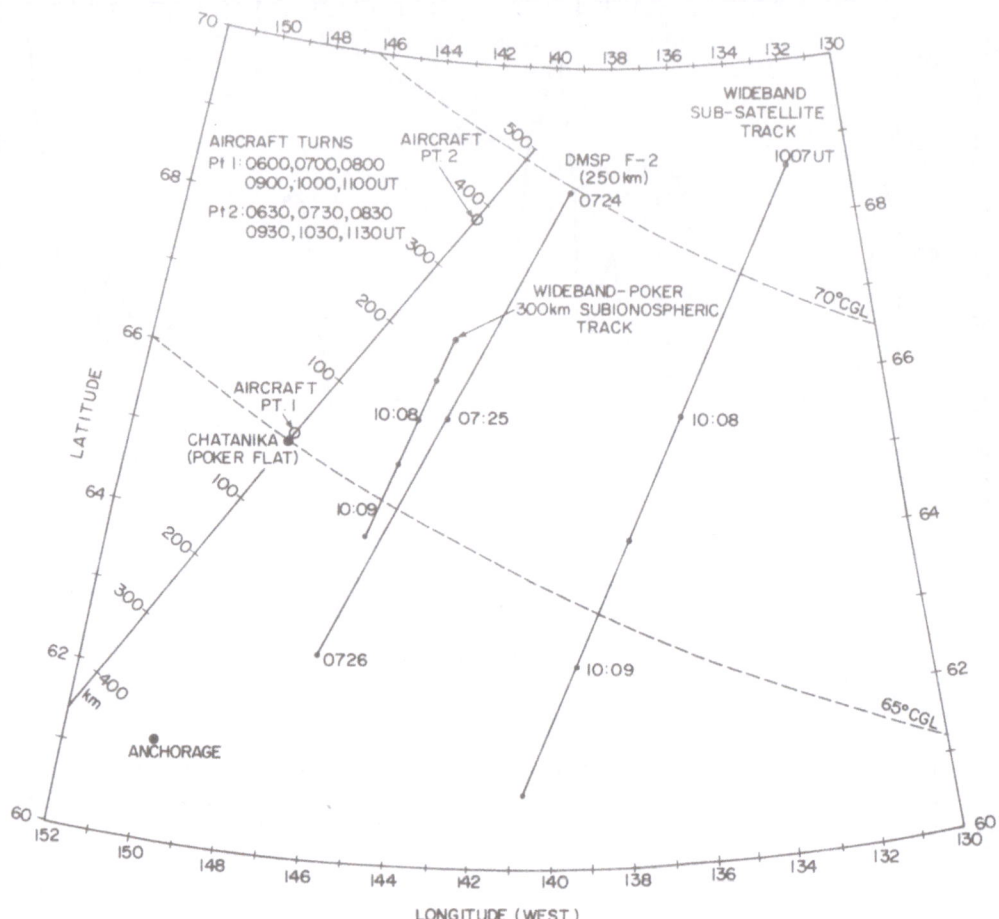

Figure 8. Map of Chatanika radar scan plane, aircraft flight track, and DMSP F-2 satellite track on 29 January 1979.

localized region of significantly enhanced density (boundary blob), no corresponding increase in the electron flux is present. This fact strongly suggests that the F-region plasma is not locally produced, but is convected into the radar field of view from a source region upstream in the E x B flow system.

   Amplitude scintillation measurements were conducted during the flight using the 250 MHz polar beacon satellite. Figure 11 shows the raypath location plotted into the radar maps for a period of intense scintillation. The aircraft motion moved the raypath from south to north (0617-0631 UT), and then from north to south (0634-0650) through the F-region boundary blob. Onset of significant scintillation occurred when the raypath encountered the 1.6×10^5 el/cm^3 contour on the equa-

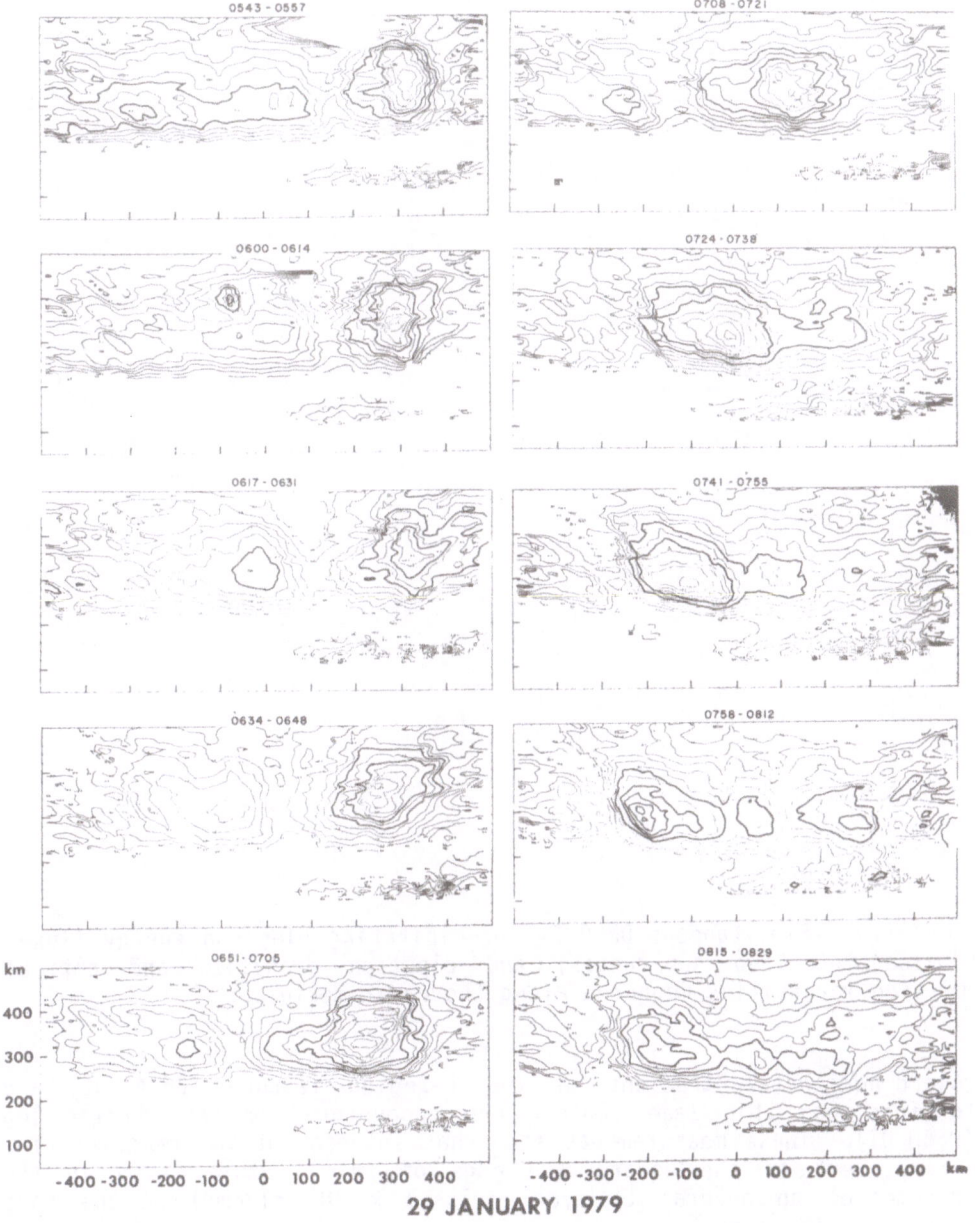

Figure 9. Sequence of Chatanika radar meridian scans showing electron density contours over ± 500 km ground distance and over the altitude range 80-500 km on 29 January 1979.

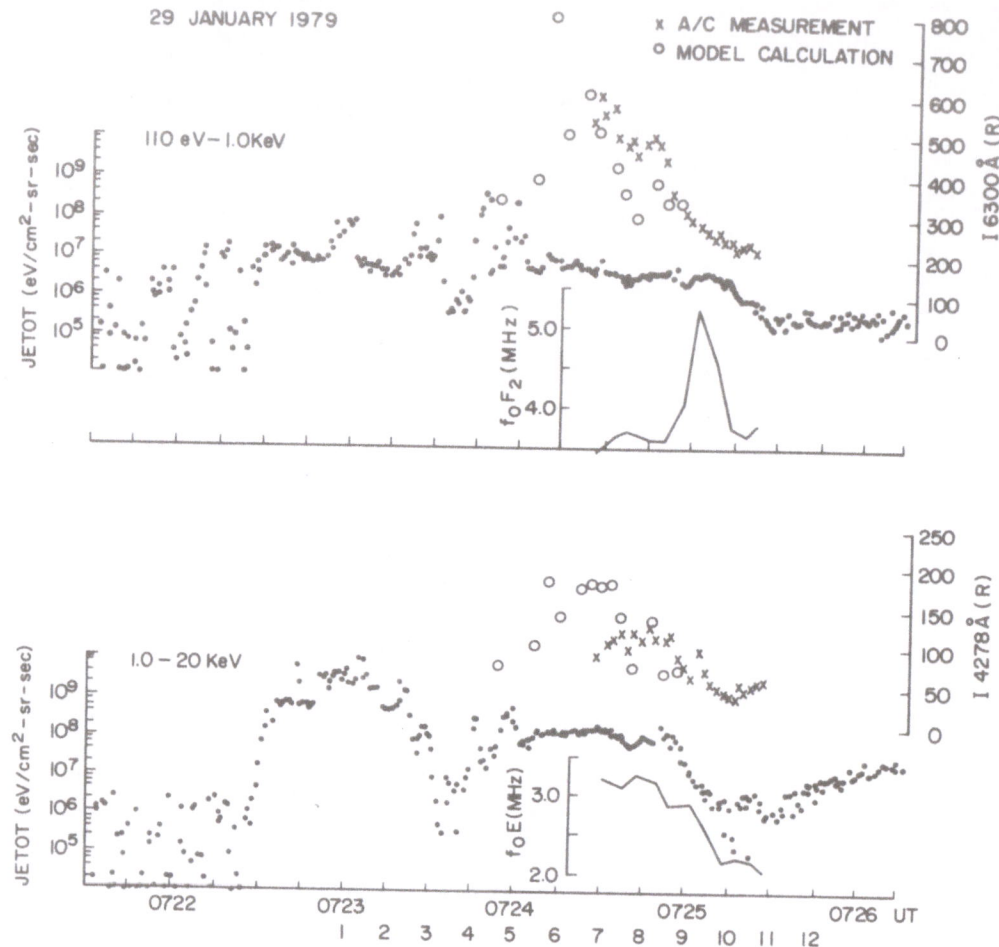

Figure 10. Simultaneous DMSP F-2 precipitating electron energy fluxes (divided into two sub-ranges) and electron densities and optical emission intensities measured along the radar meridian.

toward horizontal gradient of the F-region feature (0624 UT) and termined when this same contour was encountered on the reverse leg (0650 UT). These measurements show that intense sub-km irregularities were present on the equatorward edge of this feature, even in the presence of an auroral E layer (foE ~ 1 x 10^5 el/cm^3) on the same magnetic field lines.

3. DISCUSSION

This paper has attempted to describe plasma structure and motion

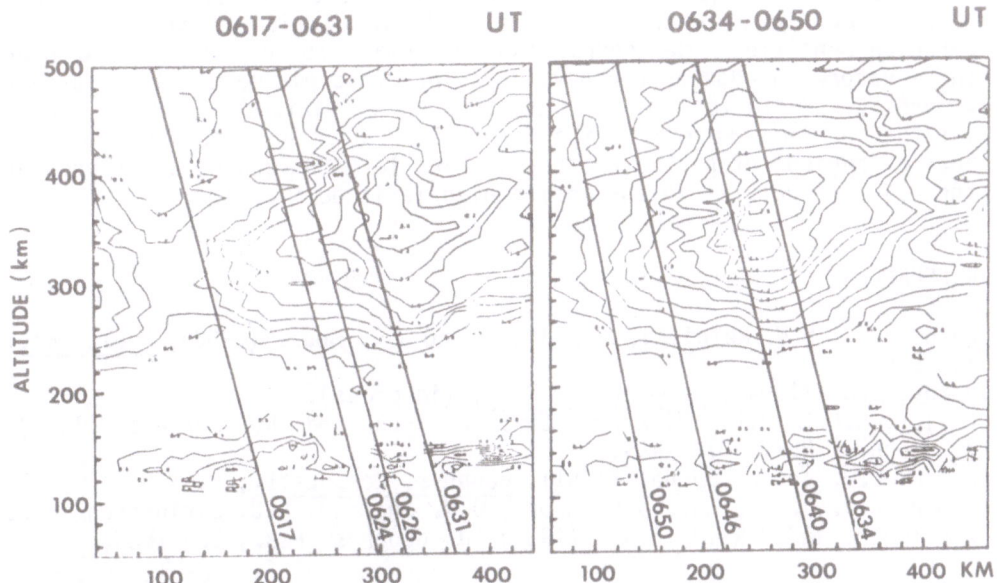

Figure 11.   Raypath to 250 MHz polar beacon satellite mapped into radar electron density contours for selected periods. Amplitude scintillation was observed when the raypath encountered the region of enhanced F-region density.

at three important locations in the high latitude ionosphere.   In all cases, regions of F-region plasma which were not locally produced (no local ionization source such as precipitating particles) were observed with radio and optical diagnostics on the AIO.   Plasma irregularity intensities were measured by observing satellite signal fluctuations (scintillation) produced by this F-region plasma.   The regions closest to the cusp were characterized by the lowest irregularity intensity levels, even though they were regions of enhanced density.   Within the polar cap and nightside auroral zone, irregularity intensities were larger.

A possible explanation of these observations is that solar pro-duced plasma equatorward of the dayside cusp region is convected through the cusp into the polar cap.   Some modulation mechanism is required to cause the (large scale) patchy nature of this plasma.   Once in the polar cap convection pattern, the plasma moves anti-sunward with respect to the neutral atmosphere and begins to structure.   Irregularity intens-ities grow as the process continues and results in more severe scintil-lation in the central polar cap than near the poleward edge of the cusp region.   Transit times between these two locations are approximately 30-50 minutes.   Observations of equatorward (followed by sunward) drift of plasma in the nightside auroral zone suggest that this may be a return flow of plasma which exits the nightside of the polar cap.

   The source region of this plasma is not definitely established. However, due to a recently discovered UT modulation of the polar cap F-region densities, the source may be solar produced ionization near the equatorward edge of the dayside cusp. This plasma displays correct UT density variation because of the large variation in geographic latitude of this source region during a 24-hour period. Whatever the source, regions of plasma are transported large distances and represent the major structures in the entire high latitude F-region.

REFERENCES

Buchau, J., B.W. Reinisch, E.J. Weber and J.G. Moore (1983), Radio Sci., 18, 995.
Hardy, D.A. (1984), J. Geophys. Res., (in press).
Kelley, M.C., J.F. Vickrey, C.W. Carlson and R. Torbert (1982), J. Geophys. Res., 87, 4469.
Weber, E.J. and J. Buchau (1981), Geophys. Res. Lett., 8, 125.
Weber, E.J., J. Buchau, J.G. Moore, J.R. Sharber, R.C. Livingston, J.D. Winningham and B.W. Reinisch (1984), J. Geophys. Res., 83, 1683.

INTERPLANETARY MAGNETIC FIELD EFFECTS ON
HIGH LATITUDE IONOSPHERIC CONVECTION

R. A. Heelis
Center for Space Sciences,
Physics Program
University of Texas at Dallas
Richardson, TX 75080, U.S.A.

ABSTRACT. A description of the dayside ionospheric convection
geometry is obtained from the study of satellite and ground-based
radar data. The most dramatic differences in the convection signature
are seen when comparing cases for northward-and southward-directed
interplanetary magnetic fields. When the IMF has a southward component it is found that for a normal garden hose orientation a dayside
merging region exists on the morning side of local noon for both signs
of B_y. The location of the merging region depends on the orientation
of the IMF in the X-Y plane. The convection pattern is then characterized by a circular convection cell partially surrounded by a
crescent-shaped convection cell. These cells occupy the dawn- and
dusksides respectively for $B_y < 0$ and reverse their positions for $B_y > 0$.
When the IMF has a northward component, two convection cells exist
entirely within the polar cap when $B_y \approx 0$. When $B_y < 0$ a single
anticlockwise circulating cell dominates the polar cap convection, and
a single clockwise cell dominates when $B_y > 0$. These cells are displaced a little to the dusk- or dawnsides respectively of the noon-
midnight meridian. Evidence exists for additional convection driven
by viscous interaction in all cases. The relationships between the
observed ionosphere convection and plasma flow in the magnetosphere
are discussed.

1.   INTRODUCTION

The equations of current continuity and Ohm's Law provide a system of
mathematical and physical equations by which the electric field and
the electric current in the ionosphere can be related. Much of the
synthesis of electric field and plasma velocity data in the F-region
is therefore made with the benefit of similar data sets derived from
field-aligned current and horizontal current measurements. The development of a self-consistent picture of the distribution and behavior
of these measurements has proceeded almost in parallel for the past
decade. In this paper the picture as it applies to the electric field
and plasma drift velocity and its dependence on the interplanetary mag-

293

J. A. Holtet and A. Egeland (eds.), The Polar Cusp, 293–303.
© 1985 by D. Reidel Publishing Company.

netic field will be described. An explanation of many of the features
seen in the observations appeals to the notion of the earth's magnetic
field lines at high latitudes being directly connected to the interplanetary magnetic field (IMF). Throughout this discussion the term
"polar cap" will be used to denote this region in the ionosphere.

The data sets from which the information on ionospheric convection patterns is obtained comes from measurements of the ambient
electric field and from the ambient ion drift velocity. In the F-
region the plasma moves at the $\underline{E}$ x $\underline{B}$ drift speed so that there is an
equivalence between the two measurements. The convection pattern will
therefore be described in terms of the plasma flow speed and direction, although this parameter may not be the measured one on all occasions. Most descriptions of the high latitude convection pattern
assume that the electric field is curl free and may thus be expressed
in terms of an electrostatic potential. The contours of equipotential
then describe the plasma flow paths. Both electric field and plasma
drift data can be easily integrated along the path that the data is
taken to derive the electrostatic potential, and it is from this technique that a signature of the convection pattern is derived. The most
dramatic differences in the high latitude convection pattern are seen
when comparing the observed plasma drifts when the IMF has a northward
component with those seen when it has a southward component. Within
these two broad categories there are significant variations that
depend upon the orientation of the IMF in the ecliptic plane (i.e. B_x
and B_y) as well as on magnetic activity.

2. SOUTHWARD INTERPLANETARY MAGNETIC FIELD

When the IMF has a southward component it has been well established
that the large-scale convection pattern is characterized by two cells
producing antisunward convection at highest latitudes and sunward convection at lower latitudes in the auroral zone (Cauffman and Gurnett
1972). Such a convection pattern leads directly to a boundary between
the sunward and antisunward convecting regions and, in fact, the
entire convection pattern can be well defined by describing the distribution of the electrostatic potential as a function of local time
around this boundary, and the distribution of potential as a function
of latitude both inside and outside the boundary. At latitudes
equatorward of the reversal boundary the ion velocity decreases
rapidly until the convection electric field is essentially shielded
from the ionosphere below about 55° invariant latitude. The penetration of the convection field to lower latitudes depends strongly on
magnetic activity as does the position of the convection reversal
boundary, but these variations will not be discussed here.

The distribution of the electrostatic potential around the convection reversal boundary and its distribution as a function of latitude poleward of the boundary are strong functions of the IMF
orientation and magnitude and thus significantly affect the dayside
high latitude convection pattern. It is toward a description of these
potential distributions that many studies of high latitude convection

have been addressed either qualitatively or quantitatively and to
which this paper is directed.

Perhaps the first dependence of the convection velocity on the
interplanetary field was observed near dawn and dusk. In the northern
hemisphere the magnitude of the electric field is found to be larger
near the dawn side or dusk side convection reversal, depending on
whether B_y is positive or negative, respectively, (Heppner 1972). At
that time it was supposed that the convection velocity was parallel to
the earth-sun line, both poleward and equatorward of the convection
reversal. Thus this asymmetry in the convection velocity would
require a similar asymmetry in the distribution of potential along the
reversal boundary. In the search for this asymmetry it was found that
the ion flow velocity is quite frequently parallel to the reversal
boundary at significant distances from dawn and dusk. Thus the idea
that the convection reversal boundary may be approximately an electric
equipotential with only a narrow local time region over which the flow
rotated from sunward to antisunward was advanced (Heelis et al. 1976).
The convection pattern in which all the sunward flow rotates to antisunward flow in a narrow "throat" region near local noon is highly
idealized and was constructed without the consideration of IMF orientation. Nevertheless it led naturally to the consideration of not
only an asymmetric potential distribution within the convection cells,
but to the possibility that the two convection cells may have quite
different shapes.

In terms of the electrostatic potential distribution the "zero"
potential contour separates flow in the two cells, and on the dayside
we can learn about possible asymmetries in the convection geometry by
observing its position as a function of IMF parameters. Both satellite and radar data suggest that flow across local noon from the
afternoon side is generally observed at lower latitudes. Such observations suggest that the center of the flow rotation region exists in
the pre-noon local time sector. Yasuhara et al. (1983) have shown
that a rotation of a symmetric convection pattern should be expected
when the ionospheric conductivity in the auroral zone exceeds that in
the polar cap. This rotation has the sense that would account for the
location of the zero potential line in the pre-noon hours, but combined data sets from satellites and ground-based radars have shown
that the apparent rotation may be due to convection cells having quite
different shapes (Heelis et al. 1983). From an examination of the ion
drift signatures from Atmosphere Explorer-C it was possible to quite
easily separate those passes that passed equatorward of the cusp and
those that passed poleward of the cusp. Those passing equatorward of
the cusp confirmed that flow across local noon toward dawn is almost
always observed. Poleward of the cusp, however, the direction of the
flow is found to be strongly dependent on the orientation of the IMF.
When $B_y<0$ and the IMF has a garden hose orientation the flow is
directed toward dusk. When $B_y>0$ and a garden hose orientation is pre-
served, then the flow is directed toward dawn. It was also found that
the ratio of the dawn-dusk component of the flow velocity to the noon-
midnight component was dependent on the ratio $|B_y|/|B_x|$, it becoming
larger as the ratio becomes larger. Figure 1 shows the dayside con-

vection signatures obtained from a synthesis of data from the
Atmosphere Explorer and Dynamics Explorer satellites (Heelis 1984).
It reflects the consistency of the data with two large-scale convection cells, one of which is approximately circular in shape and the
other that is crescent-shaped. When the IMF B_y is positive the dusk
cell takes on the circular shape, producing flow across local noon
from dusk to dawn at auroral latitudes. The dawn cell has a relatively small crescent-shape that lies alongside a section of the dusk
cell. This geometry was established in some detail by Heelis et al.
(1983). When the IMF B_y is negative the dawn cell has a circular
shape, but auroral zone flow across noon from dusk to dawn is maintained by a crescent-shaped dusk cell that encompasses a substantial
fraction of the dawn cell. In addition to the effects of auroral zone
conductivity, it is worth noting that the garden hose orientation of
the IMF also produces a tendency to shift the flow rotation region to
the dawn side of noon if B_x has some effect on the location of the
region where the IMF and the geomagnetic field are almost antiparallel. Such a model also predicts a dependence of the magnitude
of the shift on the ratio B_y/B_x, a trend that appears to be seen in
the data.

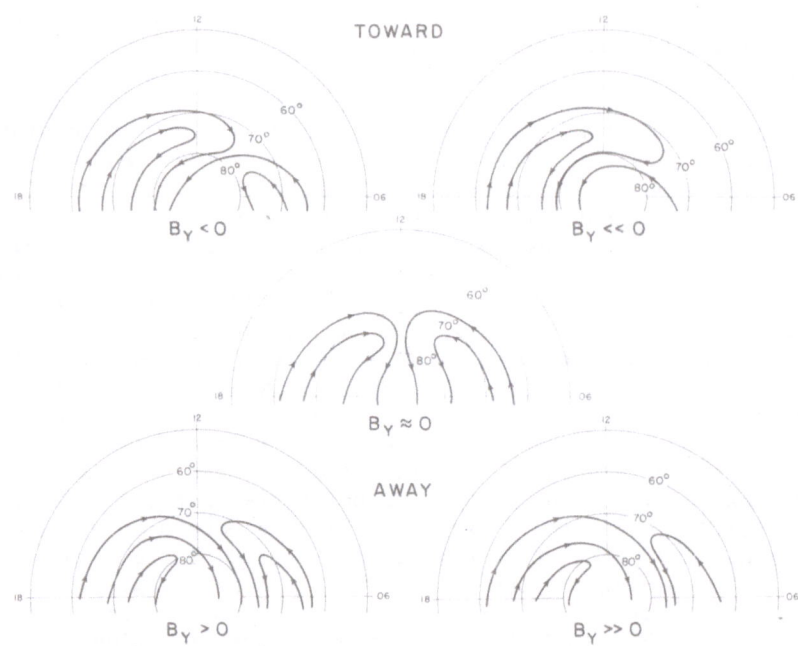

Figure 1.   Schematic diagrams of the dayside high latitude convection
pattern and its dependence on the IMF B_y component assuming B_z is negative and the IMF has a garden hose orientation. [Heelis, J. Geophys.
Res., <u>89</u>, 2873, 1984, copyright by the Am. Geophys. Union]

We have reached a fairly high degree of sophistication in our characterization of the dayside convection pattern when the IMF has a southward component; however, many problems remain. It seems clear that the concept of a confined region of local time over which the flow rotates from sunward to antisunward must be modified to include different cell geometries. The relative importance of magnetic activity (as it affects ionospheric conductivity) and IMF orientation on the convection geometry must be established as must the self-consistency between flow patterns such as those in Figure 1 with similar patterns of horizontal and field-aligned currents. The large degree of control on the convection pattern exercised by the IMF orientation leads naturally to the implication that the antisunward flow in the polar cap is derived from direct connection with the solar wind dynamo. However, studies of the relationship between the potential difference over the antisunward flow region and the IMF (Reiff et al. 1981) as well as the geometry of the convection cells near the reversal boundary (Foster 1984) do suggest a contribution to the antisunward flow from a viscous interaction mechanism on closed field lines. The effect of this electric field source on the geometry of the dayside convection pattern has not been addressed in any detail. In fact the location of a boundary between open and closed magnetic field lines in schematic illustrations such as Figure 1 will ultimately be required if we are to relate these convection patterns with motions in the outer magnetosphere and magnetosheath. Reiff and Burch (1984) have discussed the consequences of portions of the sunward flow region existing on open field lines and the possibility that the large circular convection is driven by reconnection at the dayside magnetopause and in the lobes of the magnetotail. Advances in this area may be expected from the study of the significance of multiple convection reversals and of simultaneous electric field and energetic particle data.

3.   NORTHWARD INTERPLANETARY MAGNETIC FIELD

When the IMF has a northward component the plasma flow speeds throughout the high latitude ionosphere are generally smaller than those observed when it is southward. In addition, the latitude region around the magnetic pole over which the electric field is applied is much smaller. These two events alone make the characterization of a convection pattern more difficult.
    The response of the ionospheric convection pattern to changes in the interplanetary magnetic field is not immediate. Indeed it has been shown that changes from a two-cell convection pattern may only be expected if the IMF remains northward for periods in excess of two hours (Wygant et al. 1983). It is the configuration of the dayside convection pattern in this situation to which we will refer here. On the nightside the signature of plasma drift or electric field is extremely erratic (Heppner 1972; Heelis and Hanson 1980) with multiple reversals preventing the identification of any large-scale features. On the dayside, however, large-scale convection features can quite often be recognized, perhaps due to the organized nature of the driver in this region and perhaps because the higher ionospheric conductivity

there compared to the nightside tends to short out some of the small-
scale structure. The most reproducible feature of the dayside convection in this case is the existence of at least some region of sunward
convection where antisunward convection would certainly be observed if
the IMF were southward. Most recent work has been directed toward the
identification of this sunward flow region with one or two convection
cells existing within the polar cap and consistent with models of the
solar wind-magnetosphere interaction that are used to explain the convection patterns for a southward IMF.

Figure 2 shows data from Dynamics Explorer 2 from three passes
taken during a day when the IMF was northward from 00:00 UT to 16:00
UT. The data show the ion drift velocity vector derived from the
retarding potential analyzer and the ion drift meter plotted on a
polar dial as a function of geographic latitude and local time. Also
shown are lines of constant invariant latitude at 80°, 70°, and 60° to
show the location of convection cells in magnetic coordinates. The
potential distribution along the satellite track is used to determine
the crossing points of equipotential flow lines and the observed flow
directions are used to develop the possible convection cells. The
data are shown to illustrate possible differences in the pattern
observed as B_y changes.

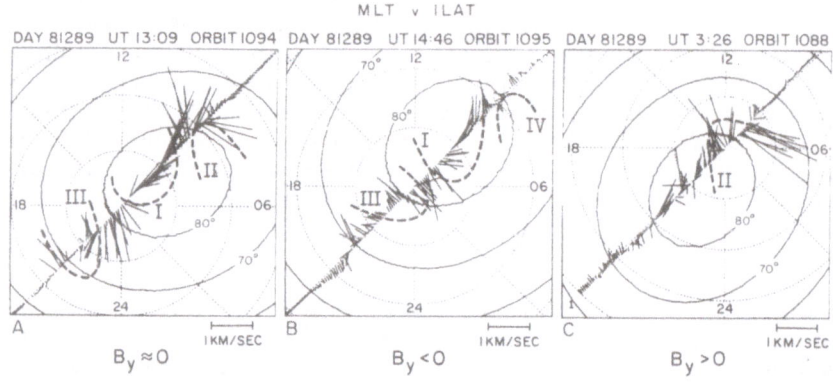

Figure 2. Ion convection velocities observed by DE-2 when the IMF B_z
component is positive. Cases are shown for different orientations of B_y.

A region of sunward flow, usually at the highest invariant latitudes reached by the satellite, is always seen on the dayside under
stable $B_z>0$ conditions. We note also that a considerable degree of
structure is seen within the large-scale convective flow even in a
stable IMF condition. Nevertheless, there are some notable differences in the flow geometry for different signs of B_y. In Figure 2a) B_y
is slightly negative but has just previously made a transition through
zero. In this pass there is a clear signature of two convection cells
on the dayside in which the plasma circulates to produce sunward flow
at the highest latitudes and antisunward flow at lower latitudes on

the dawn and dusk flanks. These cells appear to exist entirely within
the polar cap and are labelled I and II in the figure. Note that the
perimeter of the cells appears closest to the noon-midnight meridian
on the dayside. It should also be noted that at lower latitudes on
the nightside there exists the signature of a third convection cell
that circulates in a manner producing sunward flow at the lowest latitudes and antisunward flow at adjacent higher latitudes. A contour in
this circulation is labelled III.

When $B_y < 0$ [Figure 2b)] a single convection cell dominates the
dayside plasma drift signature. The circulation in this cell is
anticlockwise and again appears to occupy the entire polar cap. There
is also some evidence to suggest that this cell may be displaced
somewhat to the dusk side of the noon-midnight meridian, and that the
sunward flow may be almost parallel to the 10:00 hr MLT meridian. A
flow contour in this cell is labelled I. Below 80° invariant latitude
on the evening side there exist regions of convection with sunward and
antisunward components that are not part of the dayside convection
cell. This convection resides in the auroral zone and can be recon-
ciled with flow contours labelled III and IV that exist on the evening
and morning sides and that overlap near 23:00 hrs magnetic local time.
Figure 2c) shows the convection signature observed when $B_y > 0$. In this
case the most dominant feature is a single convection cell in which
the plasma circulates clockwise with sunward convection at the highest
latitudes. This cell labelled II exists entirely within the polar cap
and appears to be displaced to the dawnside of the noon-midnight meridian.

Other regions of sunward convection also exist within the polar
cap during this pass but the variability of the signature does not
allow us to determine whether they constitute a separate convection
cell or simply make up a single clockwise circulating cell.
Integration of this data to produce a potential distribution along the
satellite track suggests that they make up a single cell, but large
base-line errors prevent a conclusive result. At lower latitudes in
the evening there again exists some evidence for circulation, labelled
III, that has an antisunward component at higher latitudes and a sunward component at lower latitudes.

From studies of the high latitude convection pattern when the IMF
has a northward component, there is little doubt that the position and
precise direction of the flow with a sunward component is dependent on
the IMF orientation. Although statistical studies have not been conducted, it appears that the IMF sector structure determines the sense
of rotation of the cells and the existence (or dominance) of one cell
rather than two. Figure 3 shows schematically the convective flow
paths that are consistent with the data shown in Figure 2. These patterns may not have the statistical significance of Figure 1 but do
show characteristics that are seen in other convection and current
data. When $B_y \approx 0$ two cells exist possibly of about the same size and
distributed symmetrically on either side of the noon-midnight meridian. The dawn cell circulates in a clockwise manner, and both cells
exist within the polar cap. When $B_y < 0$ the dusk cell becomes much
larger than the dawn cell. Perhaps the dawn cell even disappears,

leaving a single cell that occupies the polar cap and protrudes across
the noon-midnight meridian onto the dawnside. The opposite configuration applies when $B_y > 0$. The sense of circulation in the convection
cells is consistent with the idea of merging of antiparallel fields on
open field lines in the cusp. The implications of this merging in
terms of field line geometry will be dealt with later.

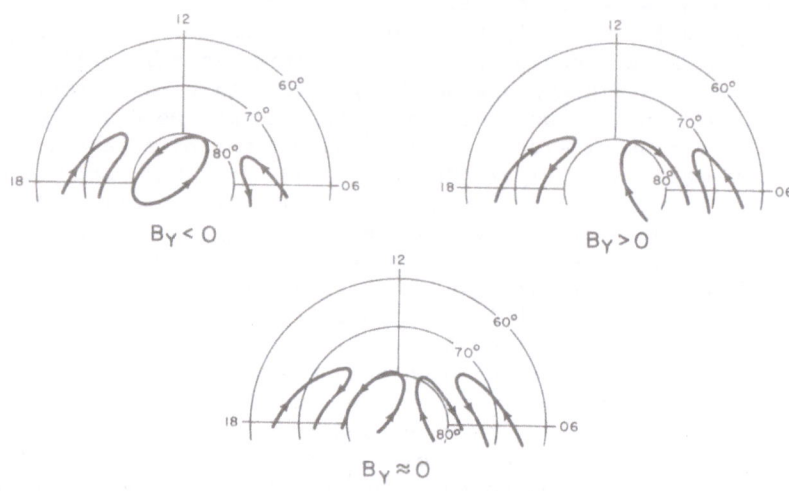

Figure 3. Schematic diagrams of the dayside high latitude convection
pattern and its dependence on the IMF B_y component assuming B_z is
positive.

In addition, the existence of a viscous interaction mechanism is also
supported by our data as an explanation of cells III and IV. These
cells have the same circulation sense and position as those found when
the IMF is southward and seem to persist at all times. We note that
the confinement of this organized flow to the dayside is consistent
with turbulent or stagnant flow on the nightside as is observed.

4.   SUMMARY

The direction of the convective flow near local noon for all orientations of the interplanetary magnetic field is certainly consistent
with the concept of merging of antiparallel magnetic fields (Crooker,
1979). In addition, convection driven by a viscous interaction mechanism in the magnetospheric boundary layers is required to explain the
global convection signature (Crooker, 1977). However, while the
magnetic field line topology and generation of the electric fields
required for convection is fairly straightforward in the case of
southward IMF, this is not true for a northward IMF.
     For a southward IMF the convection pattern can be well understood
by applying the "frozen in flux" concept. Of course this theorem must

break down where the magnetic field lines change from being closed to
being open and vice versa, but at all other points the concept becomes
a useful tool. Using this idea the antisunward convection electric
field is derived from the solar wind dynamo and thus exists on open
field lines in the polar cap. This same electromotive force in the
solar wind produces a potential difference across the magnetopause,
producing sunward convection on closed magnetic field lines in the
plasma sheet. The closed magnetic field lines become open on the
dayside magnetopause and closed at a neutral point in the magnetotail,
thus making a closed two-cell convection pattern. The relationship
between the flow on IMF lines and open field lines of the earth is
shown rather idealistically for this case in Figure 4A.

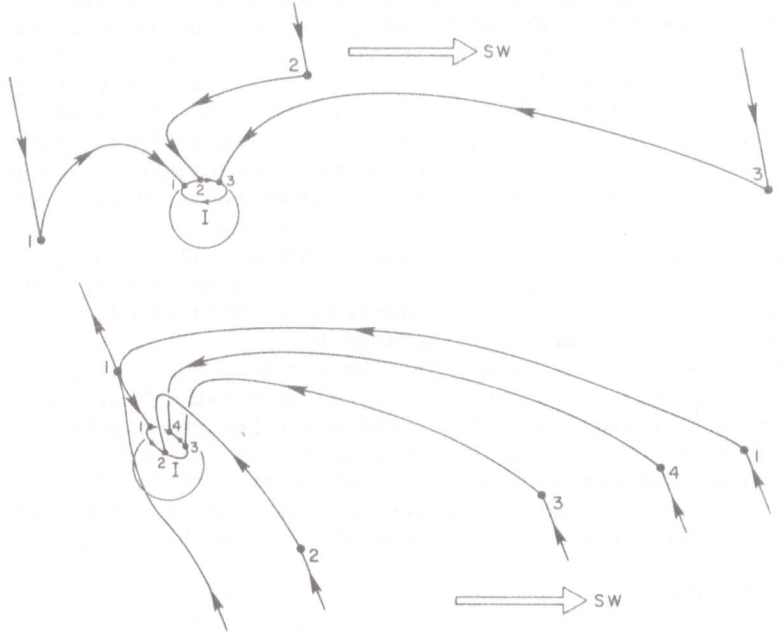

Figure 4. Schematic diagram of magnetic field topology and convection
directions associated with open field lines when the IMF has a southward (A) and a northward (B) component. In (B) only the field line
labelled 1 is in the plane of the figure. Others project toward the
reader.

This schematic is not drawn to scale in order that the ionospheric
flow associated with the field tubes can be more easily seen. The
heavy dots represent the point at which the field lines cross the magnetopause and the numbers represent a time sequence of frozen in flux
motion. The motion of the solar wind (S W) and the ionospheric plasma
(I) is shown. At point 3 the open field lines in the northern and
southern hemispheres reconnect and the subsequent sunward convection
in the auroral zones and plasma sheet occurs on closed field lines.

When the IMF has a northward component both sunward and antisunward convection in the most poleward cells occurs within the polar
cap. Such a convection pattern requires the recirculation of open
field lines in the magnetotail but cannot, of course, recirculate the
IMF (Reiff, 1982). The relationship between ionospheric flow and flow
in the magnetosphere and solar wind is shown schematically for the
case when $B_y<0$ in Figure 4B. This diagram is necessarily three dimensional because it involves both sunward and antisunward flow in the
ionosphere that corresponds only to antisunward flow in the magnetosheath and solar wind. Conceptually we could divide the flow into
two 2-D sketches in which the antisunward flow in the northern hemisphere is associated with antisunward flow down the flanks of the
magnetotail, and the sunward flow is associated with flow down the
tail toward the noon-midnight meridian. Again the numbers represent a
time sequence in the flow. This magnetic field topology implies that
open field lines emanating from the northern polar cap make their way
to the southern hemisphere and around the dawn or dusk flank, depending on whether B_y is positive or negative, respectively. This magnetic field topology also produces closed loop convection patterns on
open field lines within the polar cap. More recently the study of
visible auroral emissions with the polar cap has led to consideration
of magnetic field topologies for which some or all of the sunward
convection occurs on closed field lines (Reiff and Burch, 1984).
Observations of the convection velocities themselves are unable to
distinguish these alternative topologies.

At this point it is reasonable to say that the concepts in
explaining ionospheric convection for a southward IMF are well
accepted even though some differences about how the electric field is
generated might exist. The same statement may not be true for a
northward IMF, and further study of self-consistent patterns of convection and currents as well as theoretical modelling considerations
will be required to arrive at a generally acceptable picture.

ACKNOWLEDGMENTS

This work was supported by NASA grants NAG 5-305 and NAG 5-306 and by
Air Force Geophysics Laboratory grant F19628-83-K-0022 to The
University of Texas at Dallas. I thank P. H. Reiff for many useful
discussions.

REFERENCES

Cauffman, D. P., and Gurnett, D. A. 1972, Space Sci. Rev., **13**, 369
Crooker, N. U. 1977, J. Geophys. Res., **82**, 3629
Crooker, N. U. 1979, J. Geophys. Res., **84**, 951
Foster, J. C. 1984, J. Geophys. Res., **89**, 855
Heelis, R. A., Hanson, W. B., and Burch, J. L. 1976, J. Geophys. Res.,
      81, 3803
Heelis, R. A., and Hanson, W. B. 1980, J. Geophys. Res., **85**, 1995
Heelis, R. A., Foster, J. C., de la Beaujardiere, O., and Holt, J.
      1983, J. Geophys. Res., **88**, 10,111

Heelis, R. A. 1984, J. Geophys. Res., **89**, 2873
Heppner, J. P. 1972, J. Geophys. Res., **77**, 4877
Reiff, P. H., and Burch, J. L. 1984, ' IMF B$_y$-Dependent Plasma Flow and Birkeland Currents in the Dayside Magnetosphere,' to be published, J. Geophys. Res.
Reiff, P. H., Spiro, R. W., and Hill, T. W. 1981, J. Geophys. Res., **86**, 7639
Reiff, P. H. 1982, J. Geophys. Res., 8, 5976
Wygant, J. R., Torbert, R. B., and Mozer, F. S. 1983, J. Geophys. Res., **88**, 5727
Yasuhara, F., Greenwald, R., and Akasofu, S.-I. 1983, J. Geophys. Res., **88**, 5773

STRUCTURE IN THE DC AND AC ELECTRIC FIELDS ASSOCIATED WITH THE DAYSIDE CUSP REGION

Nelson C. Maynard
Laboratory for Extraterrestrial Physics
Goddard Space Flight Center
Greenbelt, MD 20771

ABSTRACT. The cusp region as seen in the AC and DC electric fields is one of intense variation. The intensity peaks within the soft particle precipitation. The only AC signal that appears to be unique to the cusp is broadband ULF-ELF magnetic noise. Other types of emissions are also found at other local times at high latitudes. The pattern of these signals, especially that of ULF-ELF broadband electrostatic noise (BEN), distinguishes the cusp region from other regions. BEN signatures are indicators of magnetosheath-like soft particle precipitation but not necessarily of open field lines. In addition, large spike-like features in the DC electric field are seen near local magnetic noon which appear to be related to the large convective electric fields that have been observed at the magnetopause. These features are not necessarily tied to convection reversals, but may appear within broader regions of zonal convective flow.

1. INTRODUCTION

The polar cusp region as seen in the electric field is one of intense structure. The term cusp region is taken in the broad sense to include that region where intense magnetosheath-like soft particle fluxes are observed, and is not restrictive to regions of direct access only. Ionospheric electric field magnitudes exceeding 100 mV/m are common with changes of 50 to 100 mV/m within 1 km often observable (see Maynard et al. 1982a). The general region is characterized by intense wave structures (both electrostatic and electromagnetic in nature) which have been observed at all altitudes from the magnetosheath to the ground. It is not intended to provide a comprehensive review of these phenomena. A detailed picture of the character of the AC electric and magnetic fields can be obtained from Holtet et al. (1983), Keskinen and Ossakow (1983), Curtis et al. (1982), Kintner and Temerin (1979), and Gurnett and Frank (1978), and the references found in each. Instead, a short topological road map will be presented of the AC phenomena and then the paper will concentrate on interplanetary magnetic field (IMF) effects on the low frequency end of the spectrum and on the structure in the DC electric field in the ionosphere.

J. A. Holtet and A. Egeland (eds.), The Polar Cusp, 305–322.
© 1985 by D. Reidel Publishing Company.

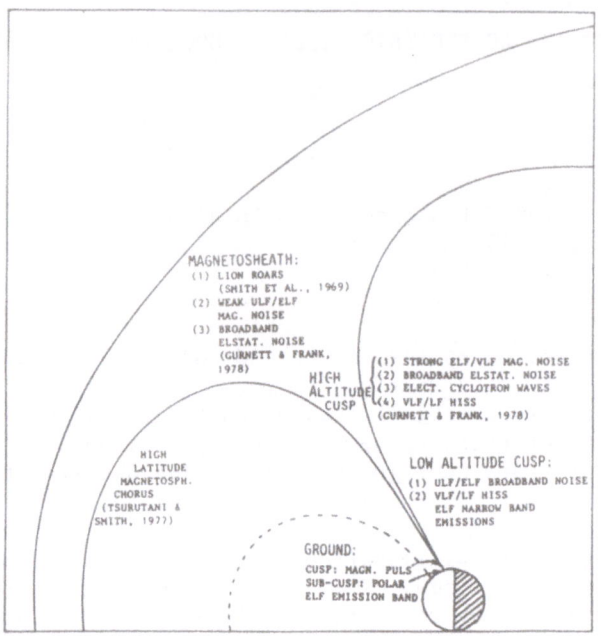

Figure 1: Schematic representation of the AC signals observed in the cusp region at different altitudes (from Holtet et al. 1983).

2. AC ELECTRIC FIELDS OCCURRING IN THE CUSP REGION

Figure 1 (from Holtet et al. 1983) summarizes the types of AC electric fields observed at different altitudes in the cusp region. Most of the emissions, while observable with particular characteristics in the cusp, are not exclusive to the cusp region.

Starting in the magnetosheath and extending throughout all altitudes, broadband electrostatic and electromagnetic noise is present in the ULF-ELF frequency ranges. From Hawkeye data in the high altitude cusp, Gurnett and Frank (1978) reported ULF-ELF magnetic field noise extending from below 1 Hz up to approximately the local electron gyrofrequency of a few hundred Hz. They interpreted these as whistler mode waves, and noted them as a good indicator of the polar cusp magnetosphere boundary. Similar waves but of weaker magnitude are also characteristic of the magnetosheath (Smith et al. 1967). This was the only emission that Gurnett and Frank uniquely associated with the cusp (and magnetosheath).

Broadband electrostatic noise is usually too strong to permit the electric field detection of the ULF-ELF magnetic noise. This type of noise is seen in the magnetosphere and ionosphere at all local times on auroral magnetic field lines and may be more intense on the nightside.

Curtis et al. (1982) suggested from DE-2 observations in the ionosphere
that this noise could be attributed to zero frequency spatial variations
from nonlocal field-aligned heating processes which reflect the
nonuniformity of the precipitating plasma penetrating into the
ionosphere.

Lion's roar (Smith et al. 1969) consisting of narrow band bursts of
whistler mode emission is characteristic of the magnetosheath only and
is almost never detected in the cusp (Gurnett and Frank, 1978). In
contrast, electrostatic electron cyclotron waves are common in the cusp
but not in the magnetosheath. These waves are not unique to the cusp
and often extend into the magnetosphere and polar cap. They are narrow
band emissions slightly above the local electron gyrofrequency.

In the VLF frequency range auroral hiss is observed in the
ionosphere and magnetosphere. These are whistler mode emissions which
are generally generated at lower altitudes, and, when seen in the
magnetosphere, are propagating upwards. Consequently, their occurrence
increases with decreasing altitude (Gurnett and Frank, 1978). Hiss is

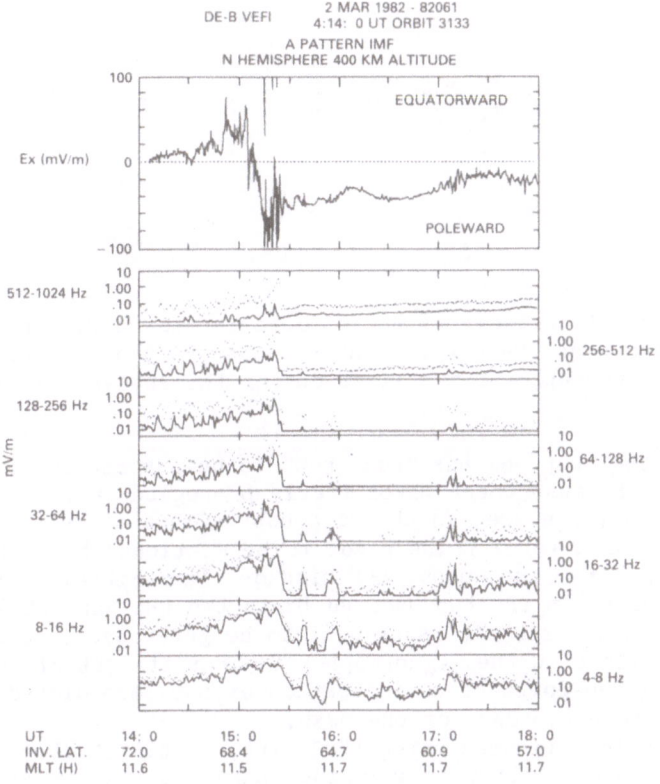

Figure 2: AC and DC electric field data from DE-2 for a pass through
the cusp region.

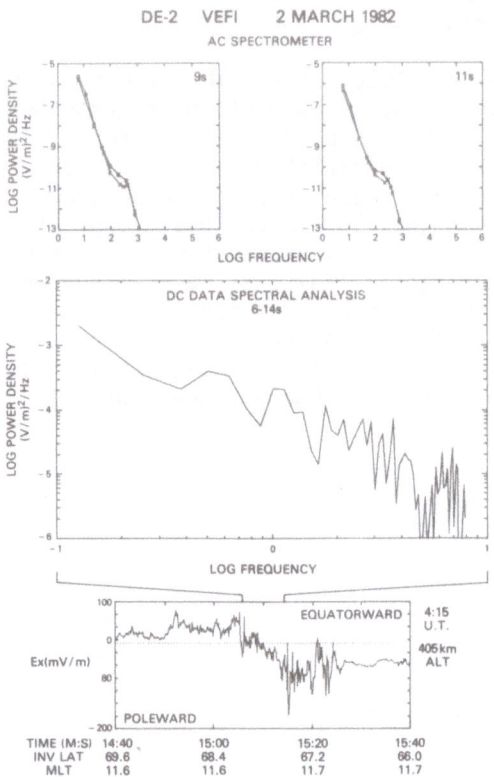

Figure 3:   Spectral analysis of the DC data in Figure 2.   The data and
interval are shown at the bottom.   At the top are two spectra from the
comb filter spectrometer at two times during the analysis interval.

not unique to the cusp and has been shown to be associated with particle
precipitation.   Because these waves freely propagate in the
magnetosphere they are seen in the adjacent regions as well.

    One additional type of noise needs to be mentioned because of its
expected absence from the cusp.   ELF hiss or plasmaspheric hiss (Thorne
et al. 1973) exists in the two hundred Hz to two kHz band at lower
latitudes than the cusp.   It is thought to be generated on closed
magnetic field lines in the magnetosphere.   Thus its cut-off with
increasing latitude could be an indication of the last closed field line
or the low latitude boundary of the cusp.

    At ionospheric altitudes using DE-2 data, Curtis et al. (1982)
identified broadband electrostatic noise (BEN) and the cut-off of ELF
hiss with the cusp.   An emission in the 16-64 kHz channel which was
between the lower hybrid and electron gyrofrequencies was identified as
whistler mode waves.   These waves were confined to the same general

region as the precipitating low energy fluxes associated with the cusp.
Since no magnetic receiver was available on DE-2, ULF-ELF magnetic noise
was not readily distinguishable. However, based on the spectral
characteristics, Curtis et al. identified the signals seen as broadband
electrostatic noise (BEN).

Figure 2 shows an example of the ULF-ELF signals seen by DE-2 and
variations in the DC electric field in the 11-12 hr MLT (magnetic local
time) region during a highly disturbed (K_p = 8) period (see Maynard et
al. 1981 for a description of the instrument). The satellite was moving
toward the equator. The soft particle precipitation ended near 4:15:30
coincident with the steep drop in the BEN. Note that at the lowest
frequencies the BEN extends to slightly lower latitudes. The sharpness
of the boundary in the 4-8 Hz channel is somewhat dependent on the
amount of structure in the DC electric field. In order to look at the
total spectrum of BEN, the DC signals near the peak of activity were
Fourier analyzed. Figure 3 shows the power spectra from the Fourier
analysis and from the spectrometer at two times during the analysis
interval. Note that the spectral index of the power spectrum fall-off
below 10 Hz is about -1.6 while above 10 Hz it is -3.7. A spectral
index of -3 at frequencies above the source frequency and -5/3 below the
source frequency from cascading is predicted by two dimensional
hydromagnetic theory. Gurnett and Frank (1977) found a -3 dependence
above a 10 Hz peak in the magnetosphere. Maynard et al. (1982b) found
slopes between nearly 0 and 5/3 below 10 Hz in intense magnetospheric
electric field turbulence. These last observations quoted are not from
the cusp region. As noted earlier, BEN emissions are found at all local
times in the auroral latitudes.

3. IMF EFFECTS ON AC ELECTRIC FIELDS IN THE CUSP REGION

The most unique quality of BEN in the cusp region is the sharp lower
latitude boundary (Egeland, private communication, 1982). As seen in
Figure 2 (and in some of the subsequent examples in this chapter), the
signals may change by over two orders of magnitude (four orders of
magnitude in spectral power density) in less than 20 km. Holtet et al.
(1983) suggested that this precipitous fall-off coupled with the cut-off
of the ELF hiss could be a good indicator of the cusp. The ELF hiss in
fact propagates to the ground and would become another indicator for
ground based cusp observations (Holtet et al. 1983).

In order to check the range of validity of this supposition, all
passes that have been processed on DE-2 for February, March, August, and
September 1982 were sorted according to the interplanetary magnetic
field (IMF) as measured by ISEE-3. A one hour time lag was used. The
sorting on B_y corresponded to that used by Heppner (1977) for his A and
B patterns. B_y negative (-140 < ϕ < 0°) for northern hemisphere passes
and B_y positive (40 < ϕ < 180°) for southern hemisphere passes comprised
the conditions for A pattern passes, while the opposite hemispherical
relationships were used to identify B pattern passes. Passes whose ϕ
value did not fall in the above ranges, which involved a sudden switch,

or which fell in a gap in the IMF data were excluded leaving a subset of
88 passes in the 8-14 hr MLT range.

Heppner (1984) updated the original OGO-6 patterns using DE-2 data
to better represent the dayside conditions. In the earlier study,
meridional dayside date was not available from OGO-6 and patterns in the
dayside cleft were completed in the simplest symmetrical form. The new
patterns labeled A-2 and B-2 are shown in Figures 4 and 5 respectively.
A comparison of these patterns shows that the dayside high latitude
convection configuration is extremely sensitive to the azimuthal
orientation of the IMF (Heppner 1984). Further discussion of the
effects of the IMF on the high latitude convection pattern can be found
in the chapter by Heelis.

The spatial pattern of the 16-32 Hz channel is taken to be
representative of the "cusp signature" of BEN. These patterns have been
separated according to B_y corresponding to A and B patterns and are
displayed in Figures 6 and 7. The passes are ordered horizontally with
magnetic local time and vertically with their approximate B_z value.
Examination of these figures shows that the sharp onset "cusp signature"
in general only occurs for B_z negative. An exception to this in the

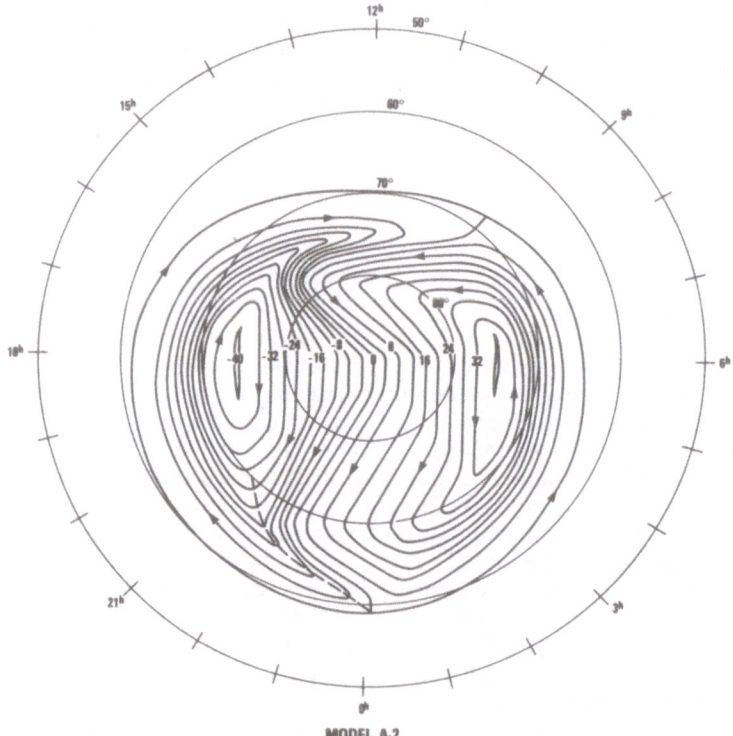

MODEL A-2

Figure 4: Heppner's (1984) A2 convection pattern for B_y negative
(positive) in the northern (southern) hemisphere originally presented at
the 1983 Irvington conference on Magnetospheric Currents.

9-10 hr column in Figure 6 involved a switch to B_z negative had a 45 minute delay been used instead of the 1 hour delay for the IMF effects to reach the magnetosphere from ISEE-3. More gradual onsets are seen for B_z positive and for all conditions at local times outside the range plotted. There is no clear B_y effect, although this may be a result of the small statistics. There is a slight tendency for the region of sharp onsets to be shifted to earlier local times for the A pattern and to later local times for the B pattern. An IMF $|B|$ effect may also be involved.

The relationship of the boundary of ELF hiss to the above patterns was also checked with no clear results. In some cases the ELF boundary was slightly equatorward, and sometimes it was slightly poleward. The proximity of the boundaries was closest when the BEN onset was sharpest. If we take the BEN onset to correspond to the onset of cusp-like soft particle precipitations (which it has in all cases checked), then a more poleward location of the ELF hiss boundary would indicate that some of the cleft precipitation must be on closed field lines (a conclusion earlier reached by McDiarmid et al. 1976). Vasyliunas (1979) has shown

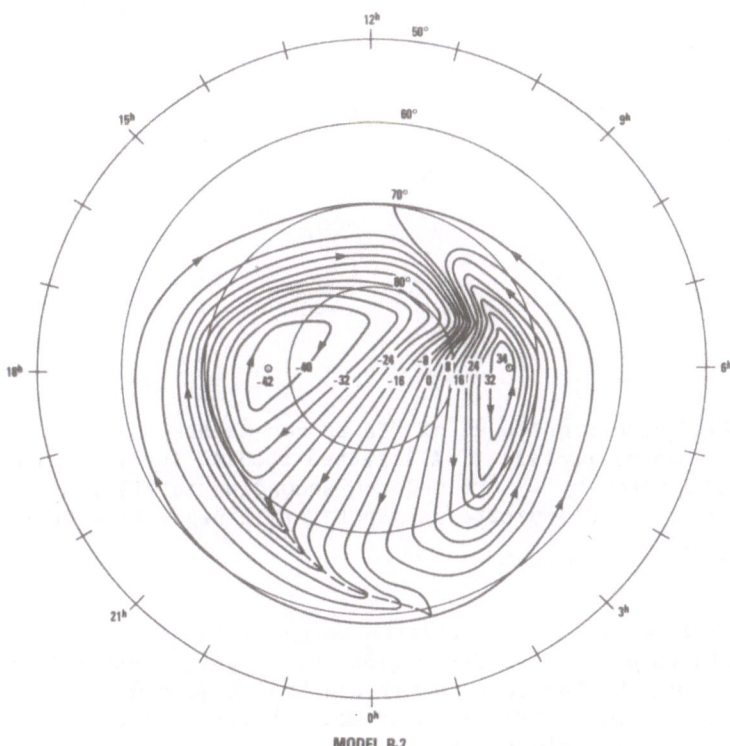

MODEL B-2

Figure 5: Heppner's (1984) B2 convection pattern for B_y positive (negative) in the northern (southern) hemisphere originally presented at the 1983 Irvington conference on Magnetospheric Currents.

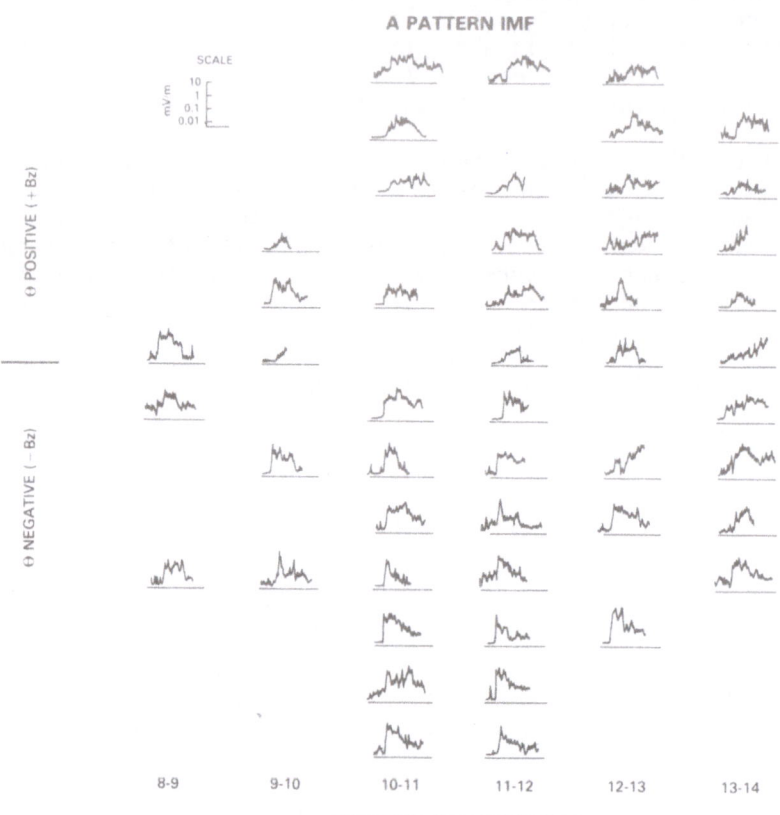

Figure 6: The "cusp signature" of BEN as represented by the 16–32 Hz
spectrometer channel profiles. All passes studied for which the A
pattern was indicated by B_y are ordered according to MLT and Θ.
Increasing values of Θ indicate larger B_z while negative Θ corresponds
to negative B_z.

a possible mapping of the low latitude boundary layer into the cusp
region, within which is imbedded regions of direct access referred to as
the "interior cusp." Particle energy spectra in each would be similar,
but only those in the "interior cusp" would be the nearest to
magnetosheath-like distributions.
     The above discussion shows that the sharp onset of BEN, when it
exists, is a good indication of the cusp region, the broad region of
soft particle precipitation associated with both the region of direct
entry and the mapping of the low latitude boundary layer. The broad
dayside local time range of this response is similar to that shown by

Figure 7:   Same as Figure 6 except that passes are for when the B pattern is indicated by the value of B_y.

Vasyliunas (1979) for these regions combined. The lack of a one to one agreement with the ELF hiss cut-off along with the more gradual profiles seen for B_z positive limits the usefulness of the confluence of these boundaries as a cusp identifier.

4.   STRUCTURE IN THE ELECTRIC FIELDS ASSOCIATED WITH THE CUSP

As can be seen from Figure 2, the electric field in the region of the cusp is extremely variable and turbulent. Large magnitude changes over short distances are common. This impulsive nature of the cusp electric field was reported by Maynard and Johnstone (1974) who suggested that small filament type regions of direct access to the magnetosheath existed within the broader region defined as the cusp. The large electric fields were associated with field lines that threaded out through the cusp into the magnetosheath region and were being dragged at the magnetosheath velocity.

   Since a large portion of the region where magnetosheath-like fluxes
are seen in the ionosphere may be on closed field lines (see above
discussion) which may be mapped to the low latitude boundary layer
(e.g., Vasyliunas 1979), the identification of the mapping topology of
the cusp is still an open question. Recent electric field observations
at the magnetopause may provide a clue as to what to expect.
   Paschmann et al. (1979) and Sonnerup et al. (1981) have offered
evidence for the existence of reconnecting magnetic fields at the
magnetopause by the observation of plasma jetting at magnetopause
rotational discontinuities. Aggson et al. (1983) showed that if the
rotation was less than 180° then there would be a significant component
of the velocity increase perpendicular to the magnetic field, and hence
a change in the convection electric field would be expected. Such
electric fields, sometimes as large as 20 mV/m, have been detected at
the magnetopause. An example is shown in Figure 8 (from Aggson et al.
1983). The indicated electric field is primarily in the sunward
direction. At rotational discontinuities de Hoffman and Teller (1950)
showed that there should be a coordinate frame moving with a velocity $\underline{V}_0$
in which the electric field would be zero on both sides of and within
the discontinuity. A $\underline{V}_0$ was found for the event in Figure 8 and the
resulting $\underline{V}_0 \times \underline{B}$ electric field is also shown. The agreement of the $\underline{V}_0$
$\times \underline{B}$ fit to the data gives credence to the validity of the data and

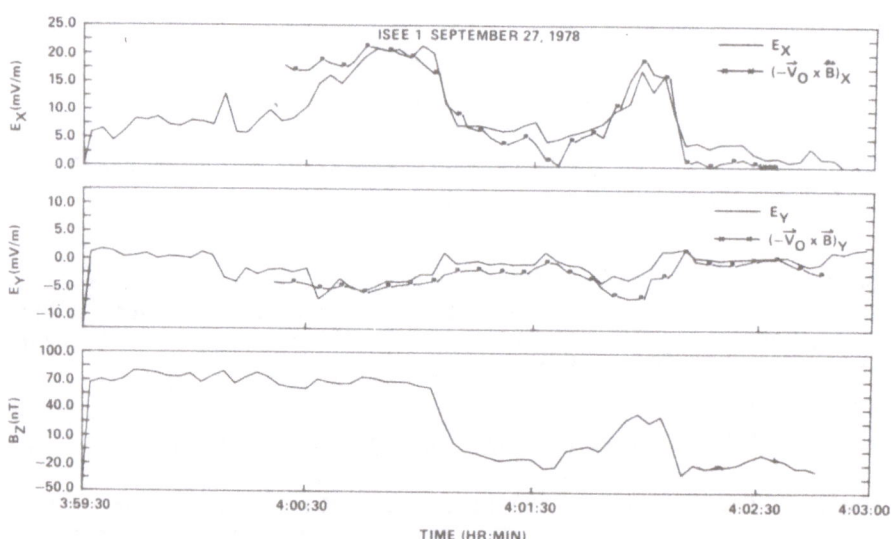

Figure 8: Electric fields at the magnetopause. The measured electric
field is compared to the electric field derived from the cross product
of the de Hoffman-Teller velocity (V_0) and the magnetic field (from
Aggson et al. 1983).

interpretation. The direction of the electric field will not always be sunward, but will be primarily perpendicular to the local plane of the magnetopause.

A large sunward electric field near noon at the magnetopause would map in some fashion to the cusp region of the ionosphere. While the details of that mapping are not understood, a significant poleward

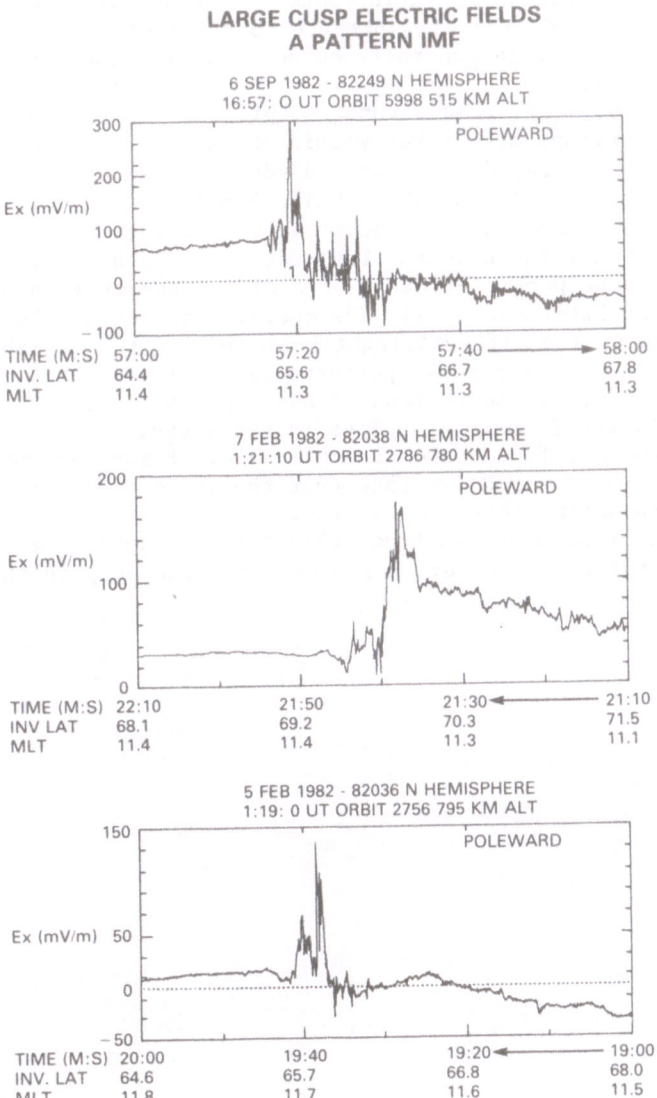

Figure 9: Examples of large poleward electric fields observed near the low latitude boundary of the cusp.

electric field component would be expected. Large localized poleward
electric fields are seen in the DE-2 data usually near the equatorward
boundary of the cusp region. Three examples are given in Figure 9. All
are in the 11-12 hr MLT region during disturbed conditions (K_p indices
of 7+, 5+, and 6 from top to bottom) so that the cusp was at low
invariant latitudes. These poleward directed fields are limited in
spatial extent and exceed 100 mV/m (reaching 300 mV/m in the first
case). The large electric field is close to but not necessarily at the
equatorward boundary of activity in the electric field. The onset of
electric field activity is generally coincident with the onset of soft
particle precipitation.

On September 6, 1982, passes through both the northern and southern
hemisphere cusp regions were taken within about 30 minutes of each
other. These are displayed in Figure 10 versus invariant latitude.
Both passes are between 11 and 12 hr MLT. Similar large poleward
electric fields were seen at the equatorward edge in both hemispheres.
The slight difference in invariant latitude (less than 1°) is easily
attributable to a temporal variation or a difference between actuality
and the invariant latitude model. The overall convection electric field
pattern differences from the different responses of the hemispheres to a
given B_y condition fit the basic patterns given in Figures 4 and 5.
Note that B_y is negative so that the A pattern applies in the northern
hemisphere while the B pattern applies in the south.

Large equatorward fields are also possible. Figure 11 shows such a
case between 10 and 11 hr MLT. Note that the large field is at the
onset of the BEN noise discussed earlier.

In order to study under what conditions these occur, the electric
field patterns for all 19 passes which observed the cusp in the 11-12 hr

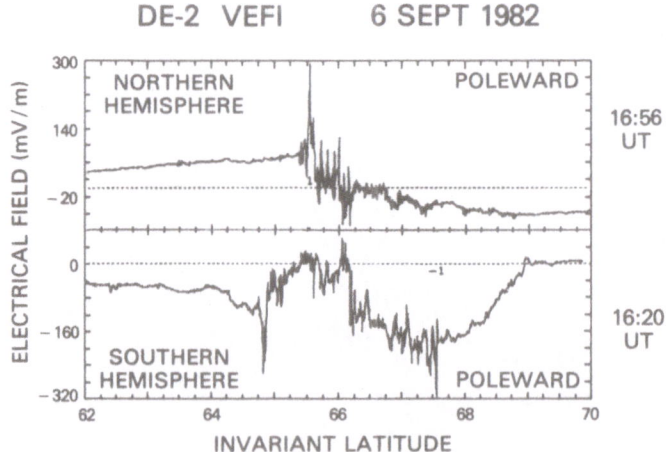

Figure 10: Conjugate passes showing large poleward electric fields near
the low latitude boundary of the cusp in both hemispheres.

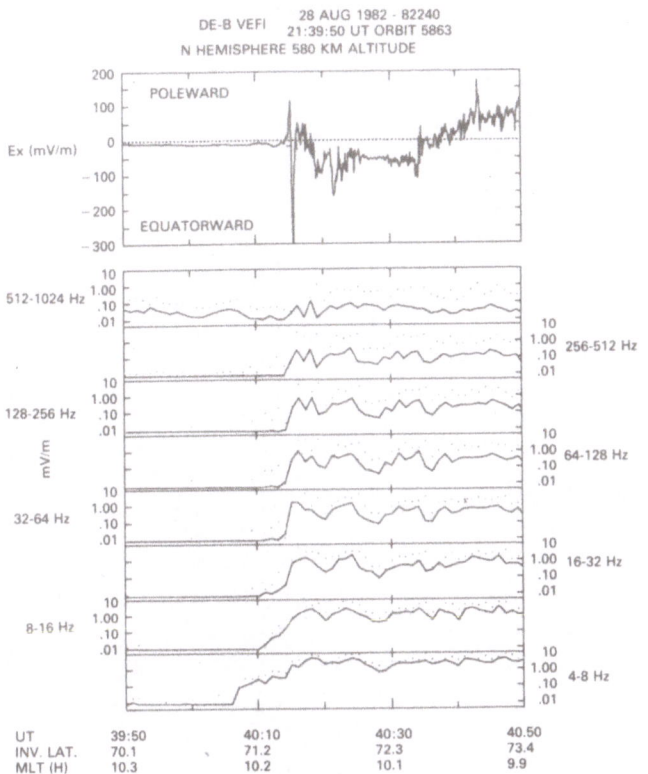

Figure 11: An example of a large equatorward electric field observed near the low latitude boundary of the cusp.

MLT range are displayed in Figure 12. The passes are grouped according to A or B pattern B_y conditions and according to B_z. The stronger negative B_y values are at the bottom of the columns while the stronger positive B_z values are at the top of the columns. Note that the large spike-like electric fields only occur on the equatorward side of the cusp for B_z negative. In one case (B pattern) a large field is seen on the poleward edge for B_z positive. However, the general condition for these fields to exist, as is the condition for merging in the low latitude regions of the dayside magnetopause, is for B_z to be negative.

The identification of the large spike-like ionospheric electric fields at the low latitude edge of the cusp region with magnetopause rotational discontinuities is by association only. This feature is most prevalent near noon, but large electric field features have been seen infrequently ±2 hr on either side of the 11-12 hr region. The actual mapping of this feature by two or more satellites could provide a direct indication of the cusp topology.

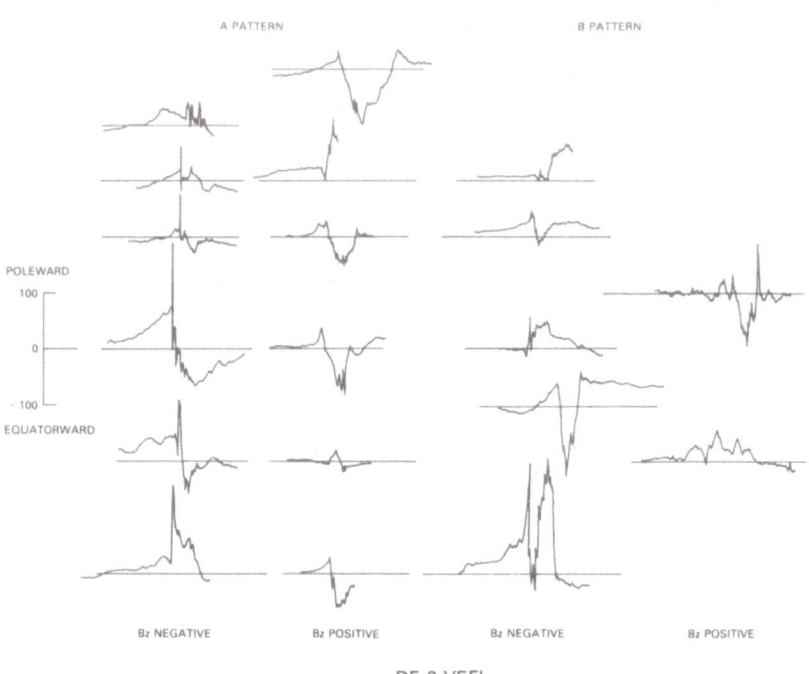

Figure 12:   Patterns of electric fields from all passes studied in the
11-12 MLT period.   The passes are ordered according to B_z and whether
they correspond to A or B type convection patterns.

5. THE RELATIONSHIP OF THE STRUCTURE IN THE ELECTRIC FIELDS TO DAYSIDE CONVECTION PATTERNS

The electric field away from the 11-12 hr region is also very structured
in the presence of cusp-like precipitation.   Figure 13a displays data
taken near 13 hr MLT from February 15, 1982, which shows the sharp onset
of BEN and structure in the electric field.   The IMF for this pass
indicated an A type convection pattern (Figure 4) with the polar cap
convection tipped toward the evening side.   This is reflected by the
reversal in the electric field as the cleft is crossed.   A corresponding
pass under B pattern IMF conditions from February 13, 1982, is shown in
Figure 13b.   Note that in this case the electric field remained poleward
well beyond the cusp-like precipitation and BEN onset near 05:11:30.
The largest electric field is in the center of the precipitation region.
Inspection of Figure 5 shows that the structure occurs in the middle of
the enlarged evening cell well away from any electric field reversal.

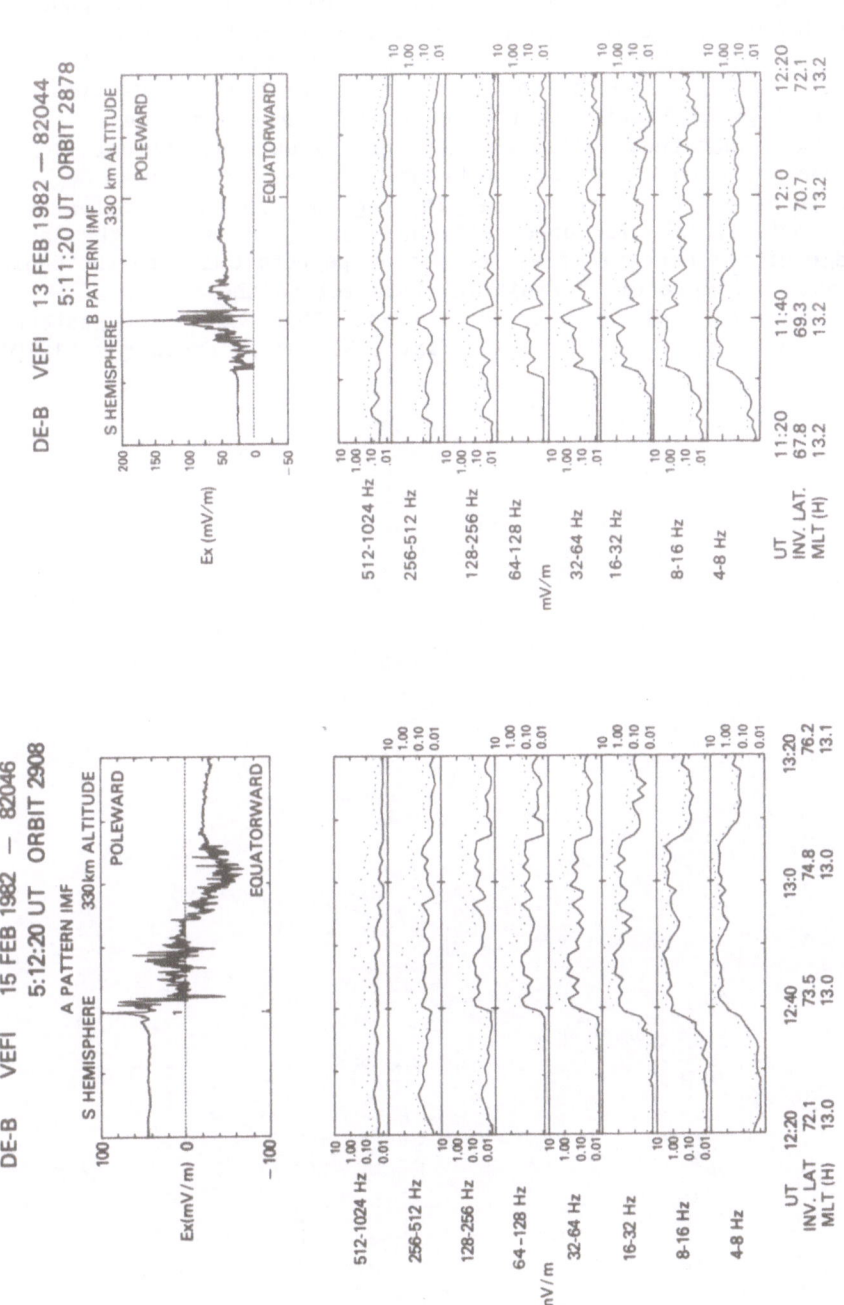

Figure 13: An example of an A type pass near 13 hr MLT showing the convection reversal associated with the intense variations in the electric field, and an example of a B type pass near 13 hr MLT where a large poleward electric field is seen imbedded within a broad region of zonal convection (poleward electric field).

Two other examples from B pattern IMF conditions are shown in Figure 14. In each, the meridional electric field remains poleward throughout and poleward of the region of soft particle fluxes. The bottom panel is near 13 hr MLT while the top panel is on the other side of noon at 10 hr MLT. Both passes are the result of an enlarged evening side cell. Heelis et al. (1976) found regions of shear reversal on the dayside which had cleft precipitation at the reversal but could not be identified with what they called the merging region. Rotational reversals where the vector direction changed at least 90° were considered to be connected to the merging region. In the examples in Figures 13b and 14, rotation appears to be minimal (note that the lack of knowledge of the electric field component perpendicular to the orbit plane prevents a completely definitive statement on the amount of rotation) in a region of cleft precipitation. This convection pattern is expected from Heppner's B pattern (Figure 5) and in Crooker's (1979)

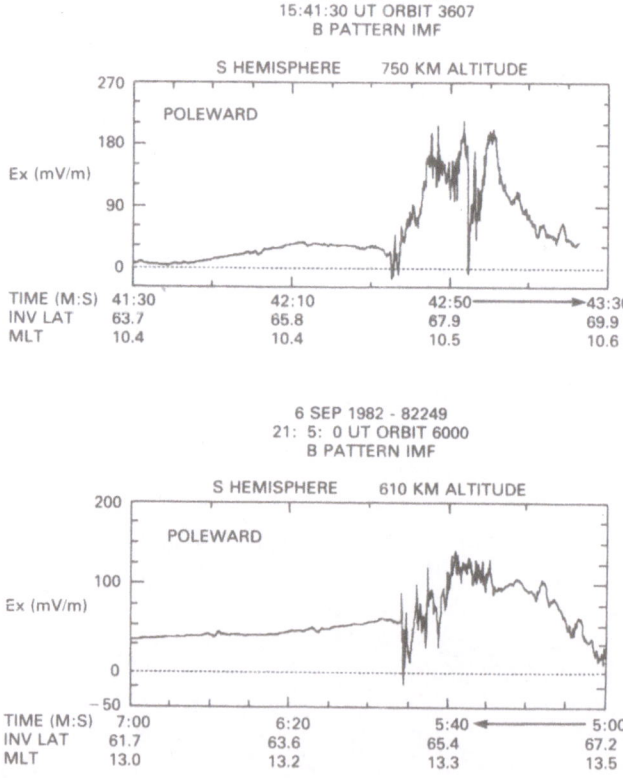

Figure 14: Examples of B type passes near 13 hr and 10 hr MLT in which the electric field activity is imbedded within a broader region of zonal convection (poleward electric field).

model for conditions where B_y is stronger than $-B_z$. Under these conditions closed loops appear within the polar cap which are enclosed in the evening side cell.

6. SUMMARY

There are few types of electric field signals that are unique to the dayside cusp region. Most are found elsewhere at high latitudes as well. The cusp region, however, is a region of intense variations in the electric field. What does appear to be unique is the pattern or structure of some of these signals. Two of these, the sharp onset of BEN and large magnitude narrow structures in the DC electric field near the low latitude boundary of the cusp region appear to be associated with southward B_z conditions only.

The sharp onset of BEN coupled with the termination of ELF plasmaspheric hiss seems to be a good indicator of the precipitation of magnetosheath-like particles and the cusp region. However, the lack of simultaneity of these boundaries emphasizes the overlap of these particle fluxes into regions of closed field lines. In this workshop the region of direct access of magnetosheath plasma was portrayed to be a small region near magnetic noon imbedded within a broader region of cusp-like particle fluxes which probably precipitated from the boundary layers after undergoing some acceleration. The BEN signature is indicative of or associated with this broader region of intense soft particle precipitation and occurs when the z component of the IMF is negative.

The signatures that may be tied directly to the magnetopause at the cusp are the large spike-like structures in the DC electric field which are seen during active periods when the z component of the IMF is negative. The association of these structures in the ionosphere with the large convective electric fields that have been observed at rotational discontinuities at the magnetopause is natural. Simultaneous or near simultaneous measurements of these features in the future would provide a direct mapping of the topology of open field lines.

The above features topologically appear within broad regions of basically zonal flow as well as at convection reversals dependent on the polarity of the y component of the IMF.

ACKNOWLEDGMENTS. I wish to thank Drs. T.L. Aggson, S.A. Curtis, and J.P. Heppner for helpful comments in the preparation of this paper. Dr. R.A. Hoffman provided comments on the precipitating particles from data from the LAPI instrument on DE-2 (Dr. J.D. Winningham, principal investigator).

REFERENCES

Aggson, T.L., Gambardella, P.J., and Maynard, N.C. 1983, J. Geophys. Res., 88, 10,000.

Crooker, N.U. 1979, J. Geophys. Res., 84, 951.

Curtis, S.A., Hoegy, W.R., Brace, L.H., Maynard, N.C., Sugiura, M., and Winningham, J.D. 1982, Geophys. Res. Lett., 9, 997.

De Hoffman, F., and Teller, E. 1950, Phys. Rev., 80, 692.

Gurnett, D.A., and Frank, L.A. 1977, J. Geophys. Res., 82, 1031.

Gurnett, D.A., and Frank, L.A. 1978, J. Geophys. Res., 83, 1447.

Heelis, R.A., Hanson, W.B., and Burch, J.L. 1976, J. Geophys. Res., 81, 3803.

Heppner, J.P. 1977, J. Geophys. Res., 82, 1115.

Heppner, J.P. 1984, Geophysica Norvegica, in press.

Holtet, J.A., Egeland, A., Doehl, J., Maynard, N.C., and Winningham, J.D. 1983, Radio Science, 18, 955.

Keskinen, M.J., and Ossakow, S.L. 1983, Radio Science, 18, 1077.

Kintner, P., and Temerin, M. 1979, P. 209, in Magnetospheric Boundary Layers, ESA SP-148, European Space Agency, Paris.

Maynard, N.C., and Johnstone, A.D. 1974, J. Geophys. Res., 79, 3111.

Maynard, N.C., Heppner, J.P., and Egeland, A. 1982a, Geophys. Res. Lett., 9, 981.

Maynard, N.C., Bielecki, E.A., and Burdick, H.A. 1981, Space Sci. Instru., 5, 523.

Maynard, N.C. Heppner, J.P., and Aggson, T.L. 1982b, J. Geophys. Res., 87, 1445.

McDiarmid, I.B., Burrows, J.R., and Budzinski, E.E. 1976, J. Geophys. Res., 81, 221.

Paschmann, G., Sonnerup, B.U.O., Papamastorakis, I., Skopke, N., Haerendel, G., Bame, S.J., Asbridge, J.R., Gosling, J.T., Russell, C.T., and Elphic, R.C. 1979, Nature, 282, 243.

Smith, E.J., Holzer, R.E., McLeod, M.G. and Russell, C.T. 1967, J. Geophys. Res., 72, 4803.

Smith, E.J., Holzer, R.E., and Russell, C.T. 1969, J. Geophys. Res., 74, 3027.

Sonnerup, B.U.O., Paschmann, G., Papamastorakis, I., Skopke, N., Haerendel, G., Bame, S.J., Asbridge, J.R., Gosling, J.T., and Russell, C.T. 1981, J. Geophys. Res., 86, 10,049.

Thorne, R.M., Smith, E.J., Burton, R.K., and Holzer, R.E. 1973, J. Geophys. Res., 78, 1581.

Vasyliunas, V. 1979, p. 387 in Magnetospheric Boundary Layers, ESA SP-148, European Space Agency, Paris.

LOW FREQUENCY WAVES AT THE DAYSIDE AURORAL OVAL

J.A. Holtet, S. Aasheim, A. Egeland and P.E. Sandholt
Institute of Physics
University of Oslo
P.O.Box 1038 Blindern
N-0315 Oslo 3
Norway

ABSTRACT. This paper is focused on naturally generated waves in the ULF to VLF range observed by ground based receivers near the dayside auroral oval. Generally, this region is characterized by its high level of emission activity over a wide frequency range. The magnetic field is in an interval ~ 3 hours on either side of magnetic noon perturbed by strong micropulsations. Superimposed on these "background pulsations" are intensive pulsation burts associated with enhancements in 6300Å auroral emissions. At extremely low frequencies the dominating emission is a structured hiss band below 1 kHz. Comparisons with satellite recordings show that these waves are generated and propagate on closed field lines. The high latitude cut-off of the emission may give information about the position of the last closed field line. Short, intensive brusts of VLF hiss appear during the midday hours. In the post-noon sector the occurrence of this emission type increases. It is supposed that this is connected to the increased occurrence of localized, short-living discrete auroral structures in this part of the oval.

1. INTRODUCTION

Particle and field measurements in the region of space magnetically connected to the dayside sector of the auroral oval - the magnetospheric cusp - are characterized by irregular structures and intense fluctuations (cf. the papers by Maynard and Muldrew in this volume, and references found here). In this paper we will concentrate on ground based observations of waves at frequencies from ULF to VLF and tie these to the wave conglomerate found above the polar ionosphere. Introductorily we will, however, briefly summarize the main features of magnetospheric plasma waves in the low altitude cusp in order to make the comparisons between ground and space observations more comprehensible. For a more complete picture of the wave activity in the high latitude dayside magnetosphere the reader is referred to e.g. Holtet et al. (1983), Curtis et al. (1982), Maynard (this volume) and references herein.

J. A. Holtet and A. Egeland (eds.), The Polar Cusp, 323–335.
© 1985 by D. Reidel Publishing Company.

2. LOW FREQUENCY WAVES IN THE LOW ALTITUDE DAYSIDE MAGNETOSPHERE

Figure 1 shows a spectrogram of the AC electric field measured in an early daytime pass of the Alouette 2 satellite. Following the spacecraft from the start of the recording near the plasmapause to its end in the cusp we notice the following features. A structured band of VLF chorus (f > 4 kHz) is present at mid latitudes. Another emission band appears at lower frequencies (< 2 kHz) around 60° magn. lat. and persists till the satellite enters the cusp. It should, however, be noted that although the ELF band is continuously present nearly from the plasmapause to the cusp, the character of the emissions changes significantly with increasing latitude. At the lowest latitudes only discrete emissions, chorus, are found. Gradually, the density of the discrete elements increases, and at approximately auroral zone latitudes a hiss component appears in the same frequency range. Approaching higher latitudes the hiss becomes dominating and a contraction of the band on the high frequency side appears. Approximately at 75° magn. lat. the band vanishes.

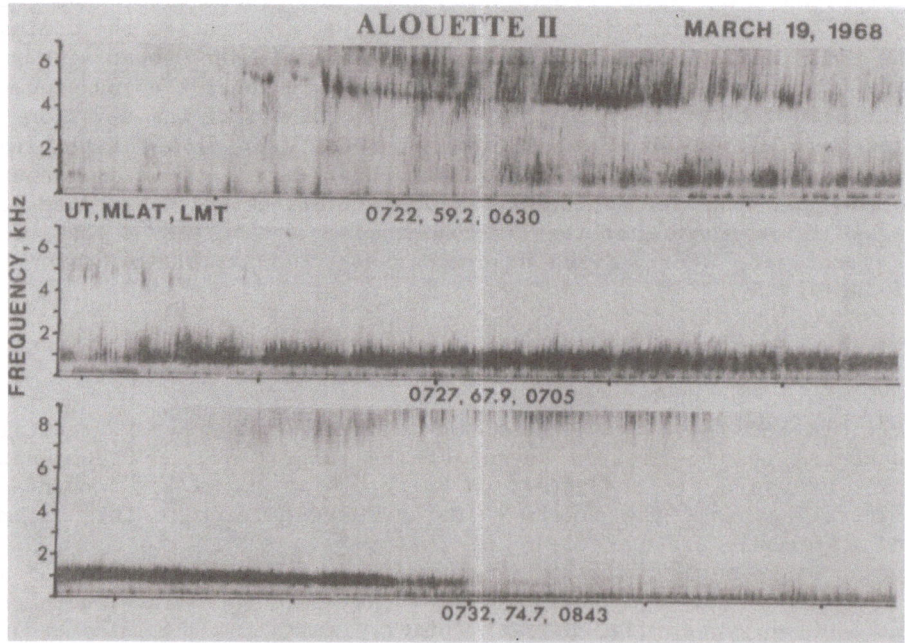

Figure 1. Spectrogram of ELF/VLF emissions recorded by the Alouette 2 satellite in a morning pass from the plasmasphere to the polar cusp. In the last 2 minutes of the pass the satellite remains at a latitude of ~ 75°, scanning the polar cap at the cusp. The tick marks on the time axis are minute marks (from Holtet et al. 1983).

   At high latitudes two other types of emissions are also present.
One is the structured VLF hiss above 7 kHz, the other a very dynamic
noise band below 1 kHz.
   Figure 2 shows AC and DC electric field data recorded by the
DE-2 in a crossing of the cusp near magnetic noon. Here the high
latitude ULF/VLF noise band can be studied in more detail. High
amplitude fluctuations are found in the 4 to 500 Hz channels between
$\sim$ 71.5° and 75° invariant latitude, i.e. a typical "cusp latitude".

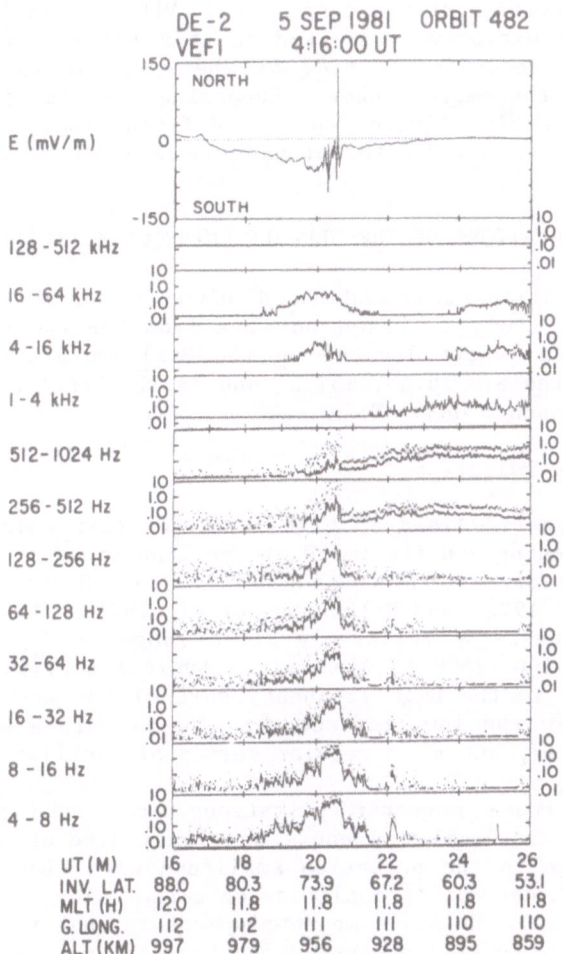

Figure 2. Electric fields observed in a southern hemisphere polar
cusp crossing by the DE-2 satellite. The DC data (top panel) are 0.5
s averages. The AC fields, measured by a comb filter spectrometer,
are rms averages (solid curve) and peak values (dotted curve), all in
millivolts per meter. (From Maynard et al., 1982.)

In this region intense and highly variable DC-fields are also observed (see also Maynard, this volume). The ELF emission band is observed
from the low latitude side of the turbulent region, while the appearance of VLF hiss coincides approximately to the E-field turbulence.

Three classes of waves are thus characteristic for the dayside
high latitudes at ionospheric altitudes. (1) Broadband electrostatic
noise below ~ 500 Hz, named BEN by Curtis et al. (1982). The occurrence of BEN coincides with the region of low energy particle precipitation. A high degree of similarity is observed in the variation in
field and particle structures (Holtet et al., 1983). It has been suggested (Curtis, loc.cit.) that the appearance of BEN can be used as
an indicator of the location of the cusp. (2) VLF hiss is found in
the cusp, but is not exclusively limited to this region. (3) ELF
hiss, observed on the equatorward side of the cusp, in the closed
field line region of the magnetosphere. Supposing that the propagation of these waves is also limited to closed field lines, the high
latitude cut off can be an indicator of the low latitude boundary of
the cusp.

3. GROUND BASED OBSERVATIONS OF THE ULF-VLF EMISSION ACTIVITY

The following presentation of ground based observations of low frequency waves near the cusp is based on data from the two stations
Ny-Ålesund and Hornsund, both located on the Svalbard archipelago.
The magnetic coordinates are 75.4°, 131.5° and 73.5°, 127.8°, respectively, and magnetic noon ~ 0830 UT.

3.1. Daytime Magnetic Pulsations

Recordings of the magnetic field at high latitudes have established
that the polar cusp region and its immediate neighbourhood is characterized by intense temporal fluctuations (cf. e.g. Samson et al.
1971, Rostoker et al. 1972, and Bolshakova et al. 1974). This micropulsation activity covers a relatively large range of frequencies
with periods from about 1000 to 1 sec, and typical amplitudes extending from << 1 nT in the high frequency part of the spectrum to
several tens of nT at the low frequencies. The pulsation patterns
are generally irregular, but more regular series of oscillations may
also be observed.

These high amplitude magnetic pulsations are found at cusp
latitudes in the time interval ~ 3 hours on either side of magnetic
noon. A smooth change in the pulsation amplitude with time, e.g. a
noon-time maximum with decreasing amplitude on either side, is usually not observed. The more normal time development is that the pulsations undergo several intensifications during the day hours (cf.Figs.
6, 7, and 8). Such enhancements in the amplitude of the magnetic
oscillations are often seen to be accompanied by simultaneous intensifications of the auroral luminosity, in particular in the 6300Å
line - the "cusp aurora".

An example of this is seen in Fig. 3. Peaks in the auroral
intensity, shown in the contour plot in the upper panel of the figure,

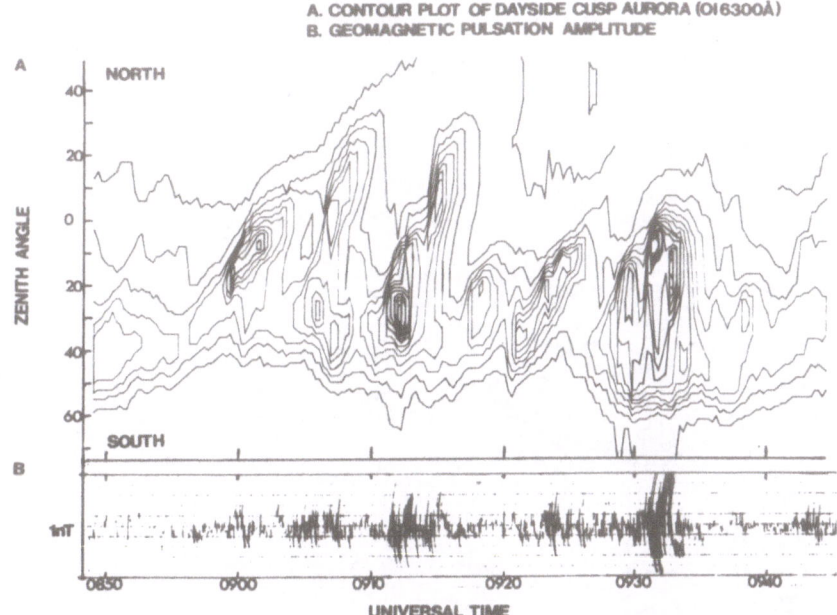

NY-ÅLESUND, SVALBARD NOV. 30 1979
A. CONTOUR PLOT OF DAYSIDE CUSP AURORA (OI 6300Å)
B. GEOMAGNETIC PULSATION AMPLITUDE

Figure 3. Upper panel: Contour plot of 6300Å auroral intensity measured by meridian scanning photometer as function of zenith angle and
time. Lower panel: Magnetic pulsation waveform measured in the frequency band 0.1-1 Hz. Notice the simultaneous intensifications in
auroral intensity and pulsation amplitude.

are seen to correspond to simultaneous pulsation bursts (lower panel).
In this figure the pulsation waveforms are limited to the frequency
band 0.1 - 1 Hz, but the enhancements are seen in the whole frequency
range down to some millihertz.
    Figure 4 shows a spectrogram of the magnetic field variations
during 6300Å auroral intensifications. The temporal and spectral development of the pulsation train shows a striking similarity with the
Pi bursts associated with substorm onsets and nighttime auroral arc
intensifications (e.g. Saito et al., 1976). It should, however, be
noted that the characteristic energies of the auroral particles connected to these two groups of events are significantly different, being a few keV at nighttime and only ~ 100 eV for cusp aurora.

3.2. ELF/VLF Emissions in the Midday Sector

An examination of ground based recordings of electromagnetic waves in
the range 0.2 - 10 kHz shows that the dominating daytime emission at
cusp latitude is band limited structured hiss below 1 kHz. This is

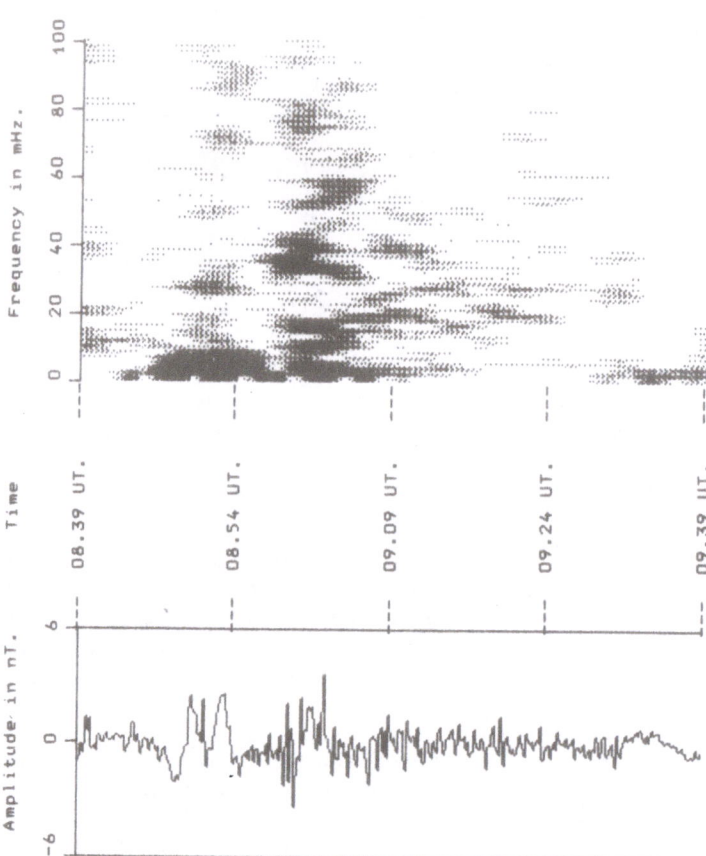

Figure 4. Spectrogram of magnetic pulsation bursts associated with
sudden intensification of the cusp aurora. The pulsation waveform is
plotted in the lower panel.

obviously the sub-ionospheric counterpart of the high latitude ELF
emission band observed by satellites.

    Statistically, the frequency of occurrence of this emission
peaks around magnetic noon. The occurrence curve is, however, slanted
to the pre-noon side, and maximum aplitudes are also often observed
slightly prior to noon. The emission activity is reduced during the

winter months. It is not clear whether this is a propagation or a source effect.

In a typical temporal development of an emission event the activity will start as a narrow band below 500 Hz (see Fig. 5). As the amplitude of the emissions increases the bandwidth will also be expanded. The low frequency border may also be lifted, but will usually stay below 500 Hz. In a weak event the whole band will in most cases be below 6-700 Hz, while in stronger events this limit may be shifted to ~ 1 kHz.

The weaker events are also usually rather unstructured with slowly varying amplitude, while the strong events can be very dynamic with a great deal of discrete structures. Dramatic amplitude changes can take place over a few minutes, and also on a shorter time scale (order of second) great variations in intensity are seen (Figs. 6 and 7). The amplitude variations are often associated with micropulsation activity, but a one to one correspondence between amplitude changes in the two data sets can usually not be recognized.

In recordings from South Pole Station (Rosenberg, data presented at this meeting) it has been observed that the intensity of the rapid amplitude variations is reduced around magnetic noon. It was suggested that this might be connected to the noon-time "gap" in the occurrence of discrete auroral forms (eg. Meng, 1981).

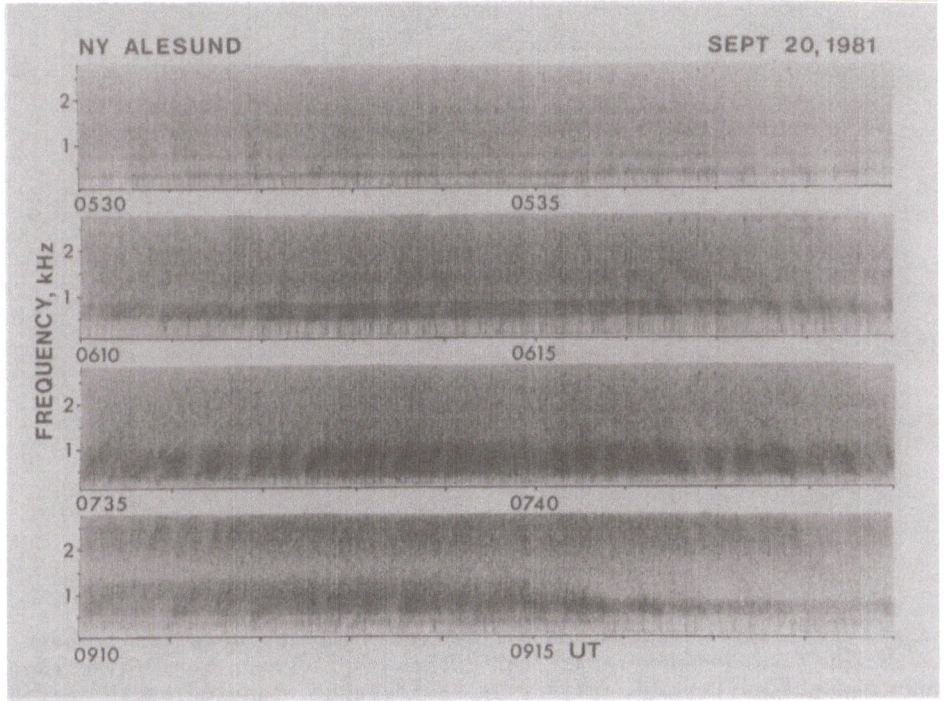

Figure 5. Spectrogram showing a typical development of a polar ELF hiss event.

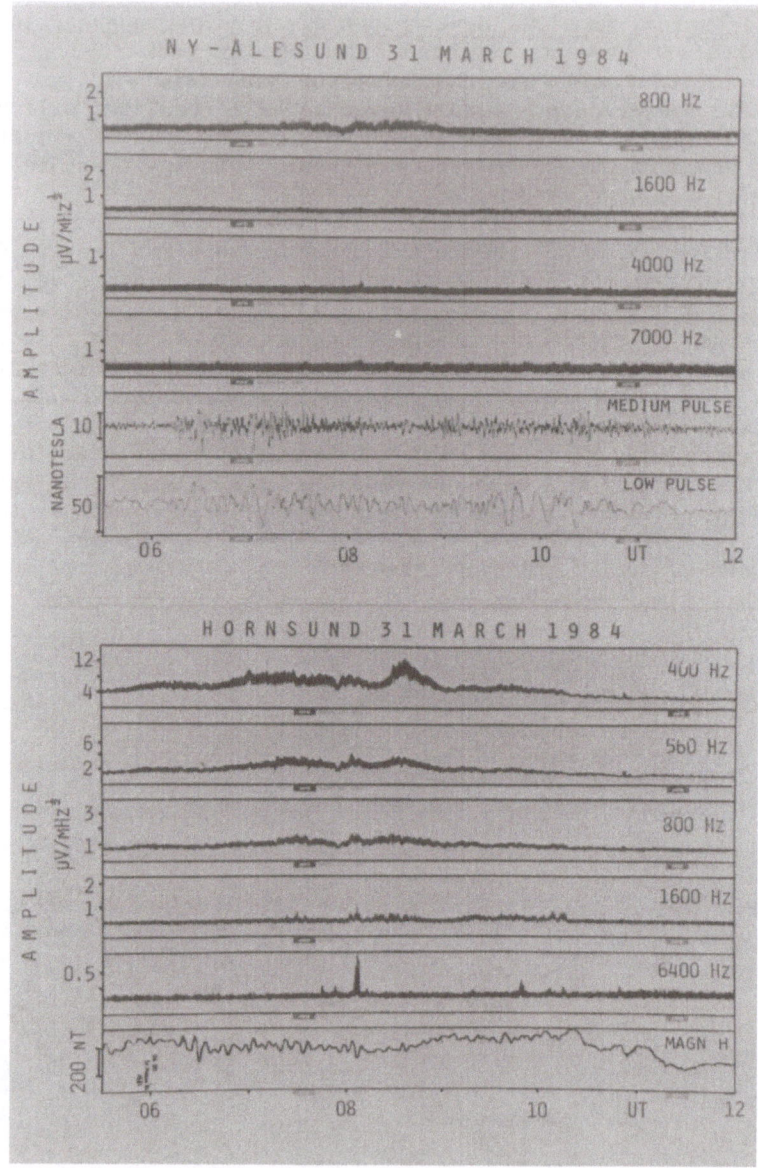

Figure 6. ELF/VLF emission amplitudes measured at selected frequencies at Ny-Ålesund (upper part) and Hornsund (lower). The figure also shows magnetic pulsation activity at Ny-Ålesund measured in the period ranges 100-600 s (low) and 10-100 s (medium), and the magnetic H-component at Hornsund.

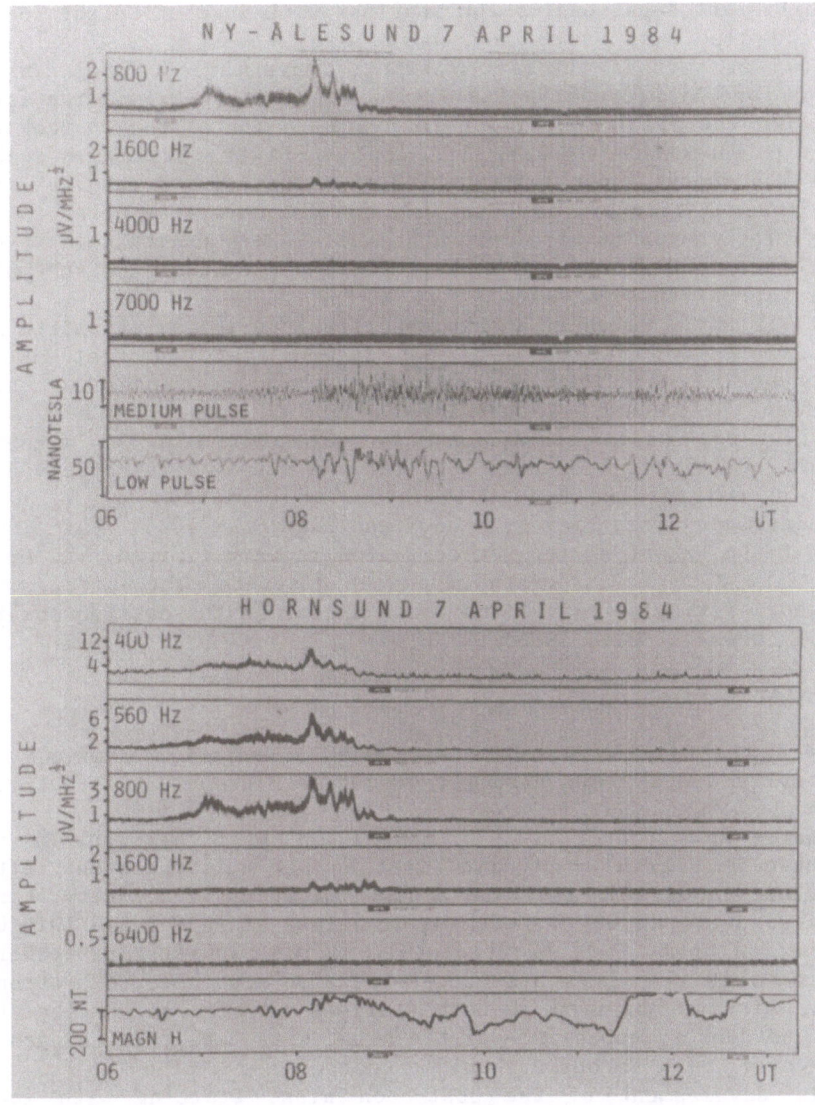

Figure 7. Simultaneous measurements at Ny-Ålesund and Hornsund of ELF/ VLF emission amplitudes and variations in the magnetic field. (See also text to Fig. 6.)

Comparison of recordings from the two receiving stations Hornsund (73.5°) and Ny-Ålesund (75.4°) will usually not show dramatic
differences in the ELF emission activity. Amplitude variations indicating different locations of the stations with respect to the source
region and also motions of the source are, however, observed. The
latter can be seen in the Fig. 7 event, where the 800 Hz intensity
ratio Hornsund/Ny-Ålesund varies from approximately 1.5 in the first
peak and in the "plateau period" to ~ 1.3 in the strongest peak and
then again increases to 1.4, 1.7, and 2 in the following peaks.

ELF/VLF chorus above 1 kHz, which is a very common morning emission at lower latitudes, is observed, but not at all as often as the
ELF hiss. The chorus is also more habitual at Hornsund than in Ny-Ålesund (see the 1600 Hz channel in Fig. 6), indicative of the sub-polar
source region of this emission.

The VLF hiss, which is a persistent feature in the satellite observations of cusp emissions is not at all that prominent in the
sub-ionospheric recordings in the noon hours. The noon-sector hiss
observed on the ground will usually be intense, impulsive bursts of
short duration, often accompanied by variations in the magnetic
field. The activity can be very localized. A strong event observed at
one station may not be seen at the other at a distance of ~ 230 km
(~2° of latitude) (see Fig. 6, ~ 0805 and Fig. 8, ~ 1005). This indicates that the extent of the source region is very limited. The emission is likely to be associated with the discrete, intensive, small
scale, short living auroral forms which appear in the dayside auroral
oval (e.g. Deehr et al., 1981).

3.3. Post-Noon Emissions

Going into the afternoon sector around 14-15 magnetic time the ELF
hiss activity fades away. Also the cusp-type micropulsation activity
will vanish.

Comming up stronger is a broadband VLF hiss emission. In the
event shown in Fig. 8 we observe first an ELF emission event. After
the conclusions of this emission a hiss "microburst" activity starts
at Ny-Ålesund around 09 UT, with major bursts at ~ 1005 and 1010 UT.
In the period up to ~ 12 UT absolutely no hiss is recorded at Hornsund. From this time hiss structures are also seen here, and strong,
impulsive hiss is observed at both stations in the last part of the
figure. The dynamic appearance of the hiss, with amplitude changes of
15-20 dB over a few seconds, should be noted.

This development of the post-noon hiss, starting with local
structures at the northernmost station, which later on move southward
and spread out over a wider latitude range at the same time as the
emission becomes more vigerous, is typical.

The hiss bursts are often associated with weak micropulsation
activity. The pulsation trains do, however, not resemble the Pi
bursts accompanying night-time auroral activity. The hiss is presumably connected to the discrete auroral forms which appear with increased
frequency in the post-noon sector of the oval. A connection with the
post-noon "hot-spot" observed in the particle precipitation (Evans,
this volume) may be suggested.

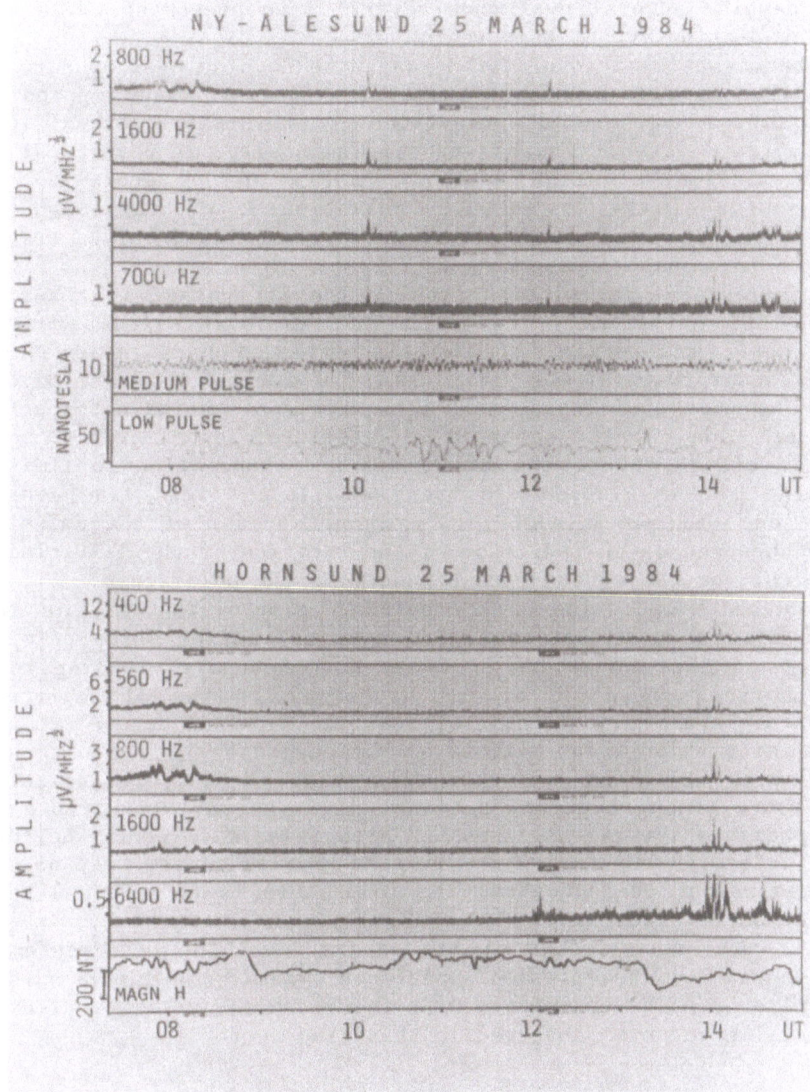

Figure 8. Simultaneous measurements at Ny-Ålesund and Hornsund of ELF/ VLF emission amplitudes and variations in the magnetic field. (See also text to Fig. 6.)

4 SUMMARY AND CONCLUSIONS

Observations of low frequency waves by ground based receivers in and near the dayside auroral oval show several wave phenomena which can be noted as distinctive of this region.

In the magnetic micropulsation range the dayside oval appears as a region of enhanced, broadband pulsation activity. Superimposed on this cusp-sector "background" pulsations are burstslike intensifications associated with increased auroral luminosity, in particular in the 6300Å emission, and rapid poleward motions of the auroral forms.

The source of the "background" pulsation is probably magnetospheric. It is not clear whether the auroral-associated bursts are of ionospheric or magnetospheric origin.

The characteristic emission type in the ELF range is a band of structured ELF emissions below ~ 1 kHz. These emissions are the sub-ionospheric counterpart of the ELF emission band observed by satellites equatorward of the polar cusp. The source of the emissions has to be found in the closed field line domain of the magnetosphere. The high latitude cut off observed by satellites is believed to be a propagation effect, where the magnetospheric propagation is limited on one side by the gradient at the last closed field line, which serves as one wall in a duct. The poleward border of the emission could thus be used as an indicator of the location of the last closed fieldline, or, of the equatorward boundary of the polar cusp.

The ducted propagation on the poleward side of the emission region will also give directions of wave normal angles which permits penetration of the ionosphere near the duct boundary, leading to a sub-ionospheric propagation of the waves. Monitoring of amplitude variations along a meridian chain of stations may then be used to get a (rough) information about motions of the cusp.

VLF hiss, which is a recurrent feature in the cusp emission spectrum above the ionosphere, does not have the same occurrence in sub-ionospheric observations. When VLF hiss appears it is in the form of short, intensive bursts, with a limited spatial range. Most of the VLF hiss emissions present above the ionosphere is therefore likely to be electrostatic waves.

In the post-noon sector of the auroral oval the occurrence of VLF hiss increases. The emission appears as dynamic, intensive bursts, and are believed to be associated with the increased activity of discrete auroral structures observed in this time sector.

ACKNOWLEDGEMENTS

The Institute of Geophysics of the Polish Academy of Sciences is acknowledged for opening their research station at Hornsund for our recordings and also for making the magnetic data available to us. We will also express our thanks to the staff at the Hornsund station for their assistance. The support by the Norwegian Polar Institute and the staff at the NPI Research Station at Ny-Ålesund has been of inestimable value for the observation program. The pulsation magnetometer at Ny-Ålesund has been provided by Dr. V.A. Troitskaya, Institute of Physics of the Earth, Academy of Sciences of the USSR, Moscow.

REFERENCES

Bolshakova, O.V., I.N. Menshutina, and M.I. Podovkin, Bull. Antarctica 13, 5, 1974.

Curtis, S.A., W.R. Hoegy, L.H. Brace, N.C. Maynard, M. Sugiura, and J.D. Winningham, Geophys. Res. Lett., 9, 997, 1982.

Deehr, C.S., G.J. Romick, and G.G. Sivjee, p. 259 in "Exploration of the Polar Upper Atmosphere" (Eds. C.S. Deehr and J.A. Holtet), D. Reidel Publ. Comp., Dordrecht, Holland, 1981.

Evans, D.S. This volume.

Holtet, J.A., A. Egeland, J. Doehl, N.C. Maynard, and J.D. Winningham, Radio Science, 18, 955, 1983.

Maynard, N.C., This volume.

Maynard, N.C., J.P. Heppner, and A. Egeland, Geophys. Res. Lett., 9, 981, 1982.

Muldrew, P.B. This volume.

Meng, C.I., J. Geophys. Res. 86, 2149, 1981.

Rostoker, G., J.C. Samson, and Y. Higushi, J. Geophys. Res., 77, 4700, 1972.

Samson, J.C., J.A. Jacobs and G. Rostoker, J. Geophys. Res., 76, 3675, 1971.

Saito, T., K. Yumoto, and Y. Koyama, Planet. Space Sci., 24, 1025. 1976.

INCOHERENT-SCATTER RADAR OBSERVATIONS OF THE CUSP

J. D. Kelly
SRI International
Radio Physics Laboratory
Menlo Park, California   94025
USA

ABSTRACT. With the establishment of the incoherent-scatter radar at
Sondre Stromfjord, Greenland, ground-based observations of the dayside
auroral oval are now being made. The initial experiments conducted were
designed to identify the ionospheric manifestation of the polar cusp in
terms of plasma convection and structure in plasma density and temperature. We have identified repeatable features of these parameters.
Plasma convection within the field of view of the radar (70 to 80° Λ) is
generally two-celled, its flow is parallel to the auroral oval, and
shear reversals and rotational reversals are frequently observed on the
dayside. The F region shows structures in both density and temperature
that are associated with specific features of convection patterns.
Along the shear reversal, we have observed enhancements in electron density and electron temperature. Other F-region density structures have
been observed without a corresponding increase in electron temperature.
These structures occur most frequently when there is a strong antisunward component in ion velocity and structures are generally aligned
in the direction of the flow.

1.   INTRODUCTION

The incoherent-scatter radar in Sondre Stromfjord, Greenland, has been
in operation since February 1983. Sondre Stromfjord is on the west
coast of Greenland at an invariant latitude of about 74°. This radar
system had been located at Chatanika, Alaska for eleven years where it
was used extensively for auroral zone research. At its new location,
the Sondrestrom radar can be used to measure ionospheric parameters in
the dayside polar cusp region.

One of the first tasks was designing experiments to examine the
noontime sector and identify an ionospheric "cusp" signature. The incoherent-scatter radar technique provides measurements of electron number
density, plasma temperatures, and ion E X B drift velocity all as a
function of time, altitude, and distance (latitude/longitude) from the
radar.

J. A. Holtet and A. Egeland (eds.), The Polar Cusp, 337–348.
© *1985 by D. Reidel Publishing Company.*

Convection patterns have been deduced by examining ion-drift velocities measured from spacecraft [Heelis et al., 1976]. These measurements show variations in the ion velocity near magnetic local noon, i.e., 90° rotation of the vector and 180° reversals.

Heikkila and Winningham [1971] have reported observations of soft-particle precipitation using ISIS-1 satellite data. They suggest that the cusp ionization source will effect the ionosphere above 200 km so that new layers will be formed. Knudsen [1977] also reports on calculations based on ionosonde data that the F region will be enhanced as plasma convects through the cusp ionization source region, and that the result will be a tongue of enhanced ionization extending poleward into the polar cap. Estimates of plasma temperature enhancements have been made by Potemra et al. [1978], Shepherd [1979], and others. Elevated electron temperatures are expected to result from ionization caused by soft-particle flux.

With this background we examined our noon-sector radar data. Our intent was to (1) identify the magnetospheric convection pattern associated with the cusp, (2) identify the F-region electron-density struc-

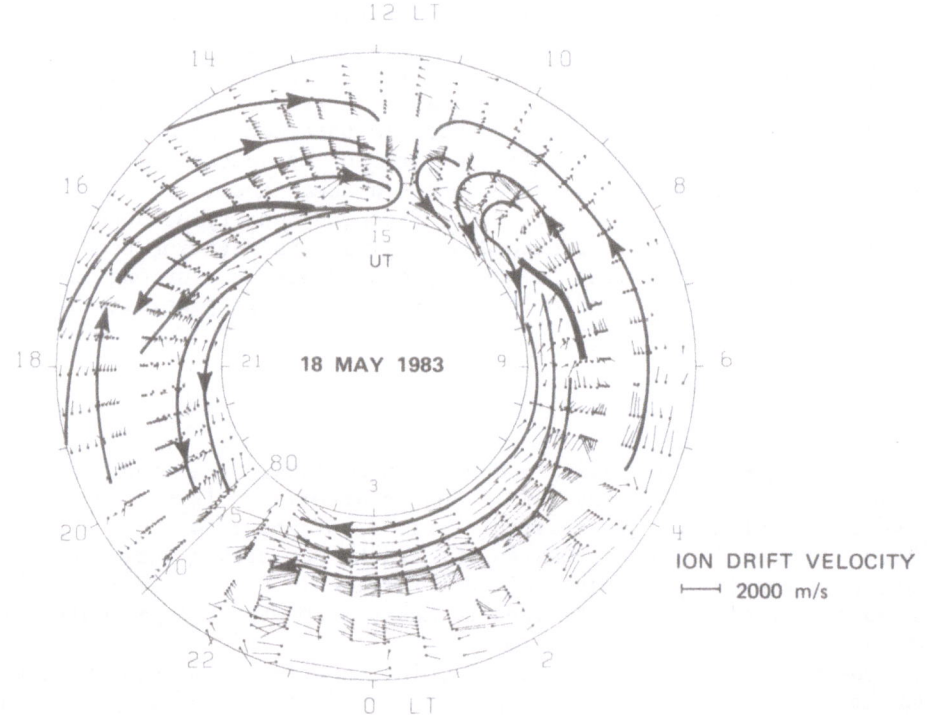

FIGURE 1   CLOCKDIAL PLOT OF ION VELOCITY VECTORS FOR 18 MAY 1983 COVERING 70 TO 80° Λ. The dayside convection pattern is characterized by two cells with rotational reversal near 1100 LT, and shear reversal is indicated in the dawn and dusk cells.

ture in the vicinity of the cusp, and (3) to examine the plasma-tempera-
ture variations in the polar cusp region.

2.   DAYSIDE CONVECTION PATTERNS

The convection patterns measured during both routine monthly experiments
performed on the Incoherent-Scatter Radar World Day and experiments
designed specifically for observations of the noon sector phenomena will
be presented. Figure 1 shows the convection pattern as measured by the
radar over a 24-hour period. This figure shows ion-drift velocities as
a function of time and invariant latitude. The measurements of $V_\perp$ are
determined from pairs of antenna positions symmetric about the magnetic
meridian for the lower antenna elevation angles. For antenna positions
more overhead, a triad of positions was used--the third position being
parallel to $\bar{B}$. As a result, measurements are obtained in the region
between 70° and 80° invariant latitude. The features are (1) two-cell
convection pattern that is symmetric around magnetic noon (1100 local
time, LT), (2) flow is parallel to the auroral oval, (3) a shear
reversal is seen between the east-west flow, and (4) a rotational

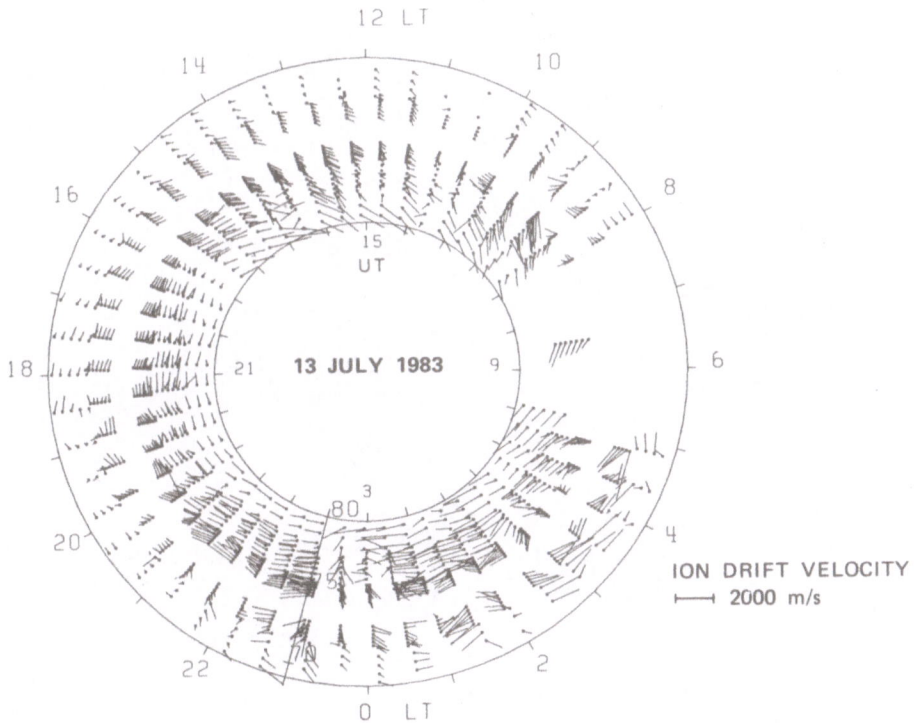

FIGURE 2   ION VELOCITY VECTORS FOR 13 JULY 1983. In this case, no shear reversal is
seen within the field of view of the radar — all flow in the dusk cell is sunward.

reversal occurs near 1100 LT. This figure is not a "snap-shot" of the
convection pattern--the measurement takes 24-hours and consequently
temporal variations will affect our picture.

Figure 2 is another example of convection pattern. In this case,
no shear reversal is seen in the dusk cell. Instead, the flow is entirely sunward indicating that the polar cap boundary is poleward of our
field of view. A rotational reversal exists over a very broad region
(0900 to 1300 LT).

Figure 3 shows an example of convection in which the flow in both
the dawn and dusk cells is predominantly antisunward. Also, the flow
is strictly east-west throughout the noon period (i.e., no rotation is
observed). This is not the expected convection pattern for the noon
sector, i.e., there is no throat at all or it is extremely narrow and
cannot be resolved within our 30-min temporal resolution.

3.   F-REGION DENSITY STRUCTURE

We have also examined the F-region electron density around noon and have
identified two repeatable features. These features are enhancements in

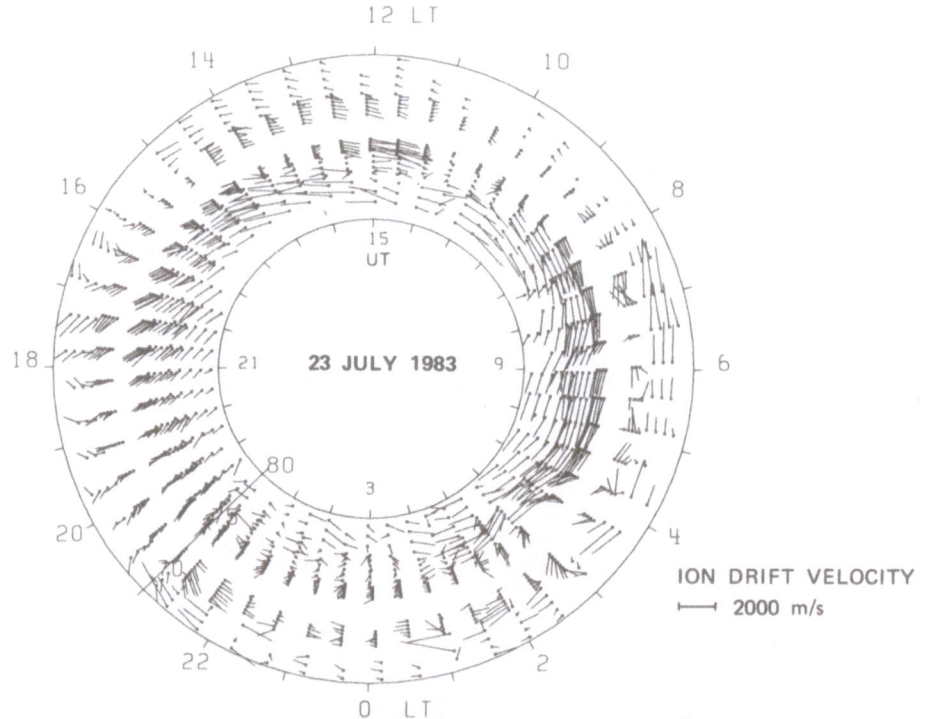

FIGURE 3   ION VELOCITY VECTORS FOR 23 JULY 1983. In this convection pattern, the
flow in both cells is antisunward and no rotational reversal is observed.

the electron number density and each is associated with a specific convection feature.

Figure 1 showed the convection pattern from 18 May 1983 with the
shear reversal indicated. Figure 4 shows the corresponding electron
densities at 250-km altitude. The poleward edge of the density enhancement is collocated with the shear reversal. Figure 5 shows the electron
temperature (T_e) at 350-km altitude. The figure indicates T_e is elevated
within the enhanced region. Robinson et al. [1984] have investigated
similar signatures in the radar data. Figure 6 shows data from a scan
in the magnetic meridian on 9 September 1983. This figure shows the
orientation of the density structure relative to the reversal of the
electric field. The peak F-region density occurs just equatorward of
the reversal. In this region, T_e was also elevated.

In addition to the radar data, data were obtained from the TRIAD
and NOAA 7 satellites. These data indicated that a strong upward field-
aligned current existed along the region of electric-field reversal.
The energy spectrum of the precipitating particles is consistent with
the altitude of the density enhancement.

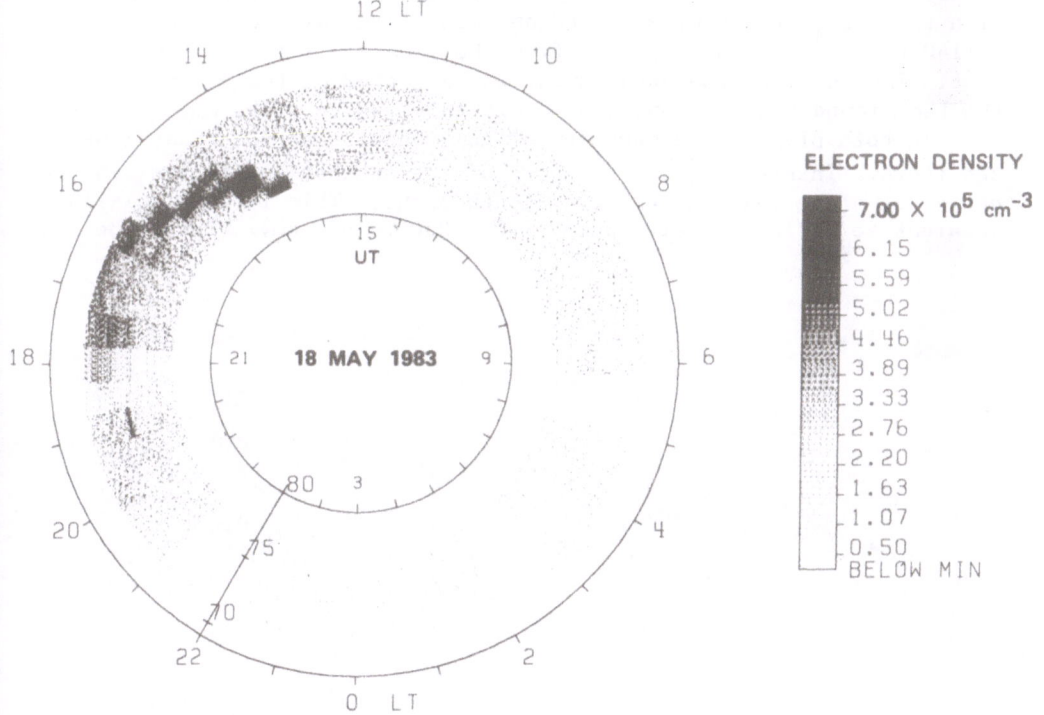

FIGURE 4   CLOCKDIAL PLOT OF ELECTRON NUMBER DENSITY AT 350-km ALTITUDE
ON 18 MAY 1983. An enhancement in density is seen in the dusk cell collocated
with the şhear reversal indicated in Figure 1.

   The other type of repeated F-region density enhancement is shown in
Figure 7.   This figure shows the electron number density (N_e) measured
at 350-km altitude on 13 and 14 September 1983.   The period of interest
here is postnoon (1200-1400 LT) when drift velocities are large with a
significant antisunward component as shown in Figure 8.   An enhancement
of approximately a factor of 3 in the electron density can be seen
during this period.   The local time extent of the enhancement decreases
with increasing latitude and is about one hour in east-west extent over
Sondrestrom.   The electron and ion temperatures (T_i) are approximately
equal within the enhancement, but T_e is somewhat elevated over T_i outside of the enhancement.   This anticorrelation between T_e and N_e is
expected in the absence of a heat source because the electron energy
loss rate increases with N_e.

   Cross-sections of the ionization ridge were obtained during a different experiment designed to provide high spatial resolution.   This
experiment consisted of alternate east-west and north-south scans.
Figures 9 and 10 show consecutive contour maps of electron density as a
function of altitude and distance from Sondrestrom.   The period covered
is from 1014 to 1242 LT on 27 February 1983.   The feature of interest is
the ionization enhancement at 300-km altitude.   The enhancement was first
detected to the east of the radar at about 1030 LT.   As the radar rotated
with the earth, the ionization enhancement eventually was overhead at
$\sim$ 1140 LT and later disappeared from the radar's field of view to the
west.   The level of the enhancement was a factor of two or more over
the background and was confined to altitudes above $\sim$ 250 km.

   In both planes, the enhancement only appears at the F-layer peak
and above.   Interestingly, the width of the enhancement is less in the
east-west plane than in the north-south plane.   This is particularly
apparent at the higher altitudes (i.e., 350 to 400 km).   Thus, the

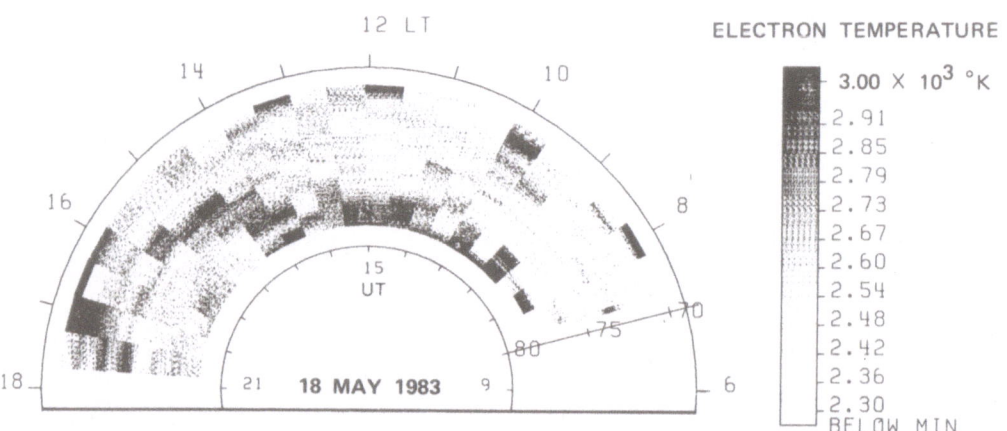

FIGURE 5   CLOCKDIAL PLOT OF ELECTRON TEMPERATURE AT 350-km ALTITUDE FOR
            18 MAY 1983.   The temperatures are elevated along the poleward edge of the
            electron–density enhancement seen in Figure 4.

geometry of the enhancement is anisotropic and "ridgelike" extending further in the north-south than in the east-west direction.

The convection pattern in the postnoon period was similar to the September data just described in that it had a strong northwest flow. The electron temperature within the density enhancement was approximately equal to the ion temperature. This is possibly a source of the "cold" polar cap ionization reported by Weber et al. [1984].

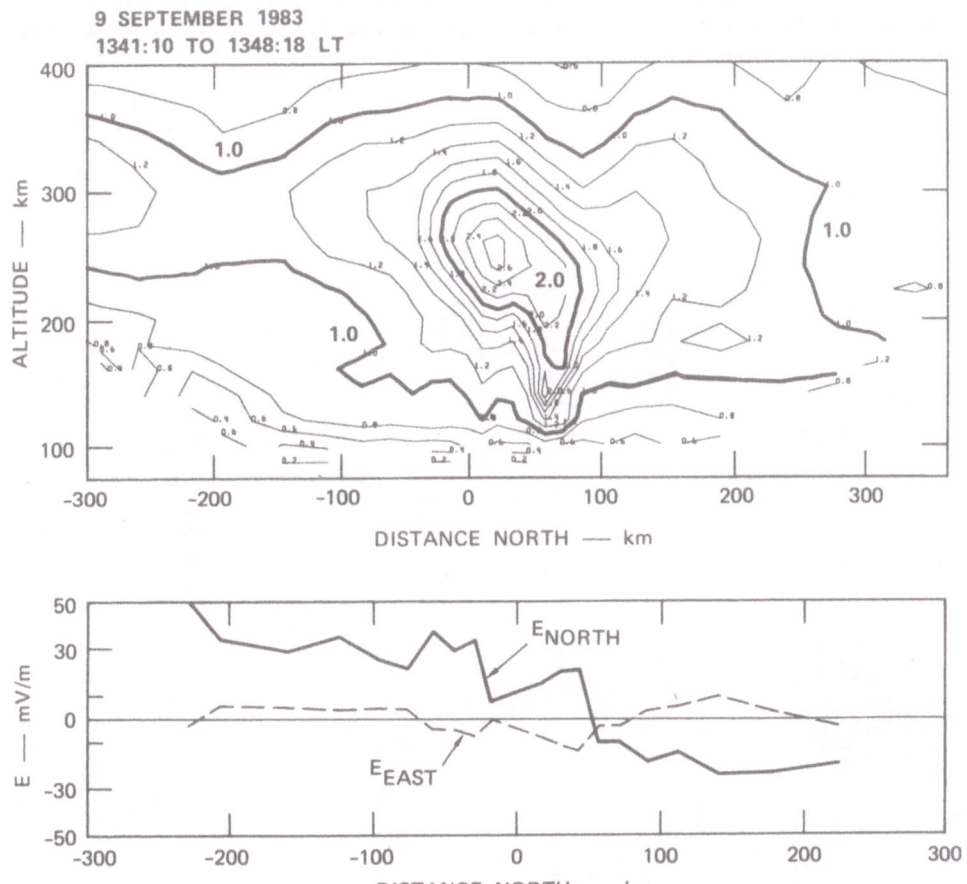

FIGURE 6   CONTOURS OF CONSTANT ELECTRON-DENSITY AND ELECTRIC-FIELD COMPONENTS AS A FUNCTION OF DISTANCE FROM SONDRESTROM IN THE MAGNETIC MERIDIAN FOR 9 SEPTEMBER 1983. The reversal in the north-south component of the electric field occurs at the poleward edge of the electron-density enhancement.

4.   PLASMA TEMPERATURE STRUCTURE

As expected, the dayside temperature structure is strongly related to
the convection pattern.   In the previous section describing density
structures we pointed out that T_e is elevated in the regions of soft-
particle flux.   Another example of temperature structure reported by
Kofman and Wickwar [1984], is seen in Figure 11.   The region of enhanced
T_e is collocated with the enhanced F-region electron densities, as shown
in Figure 12.   As in the data discussed earlier, these enhancements are
associated with the shear reversal in the convection.   The ion temperature in this region is not enhanced, as seen in Figure 13.   Enhancements
in T_i do occur as a result of Joule heat input from large electric
fields (i.e., drift velocities).   In this example, T_i is enhanced in the
dawn sector and along the poleward edge of the field of view where the
ion drift velocities are large ($\sim$ 1.5 km/s).

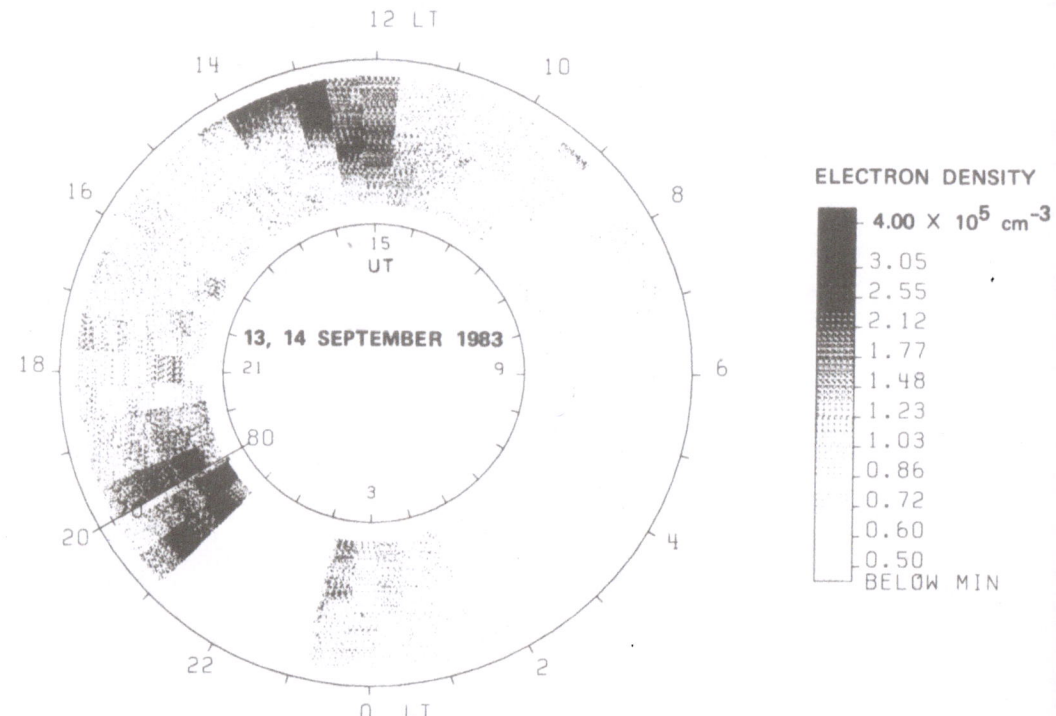

FIGURE 7   CLOCKDIAL PLOT OF ELECTRON NUMBER DENSITY AT 350-km ALTITUDE
ON 13 AND 14 SEPTEMBER 1983, SHOWING A RIDGE OF ENHANCED DENSITY
BETWEEN 1100 AND 1400 LT

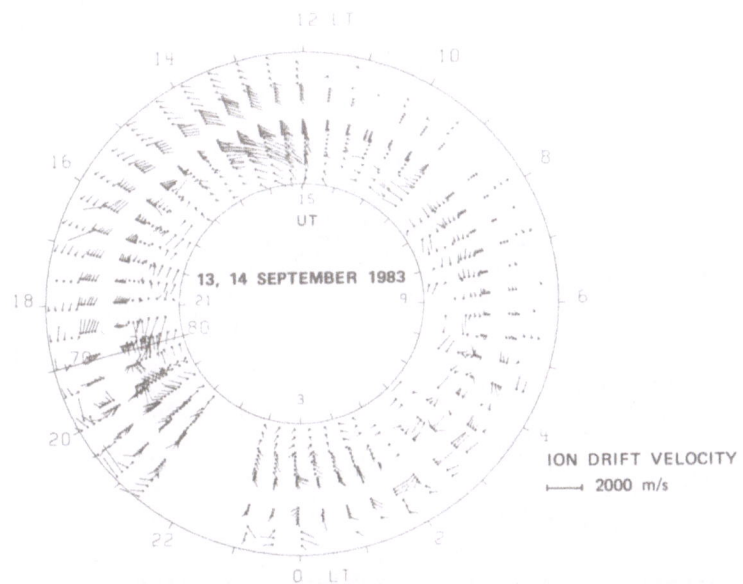

FIGURE 8   CLOCKDIAL PLOT OF ION VELOCITY VECTORS FOR 13 AND 14 SEPTEMBER
1983, SHOWING A STRONG NORTHWEST FLOW CORRESPONDING TO THE
ENHANCEMENT IN ELECTRON DENSITY SHOWN IN FIGURE 7

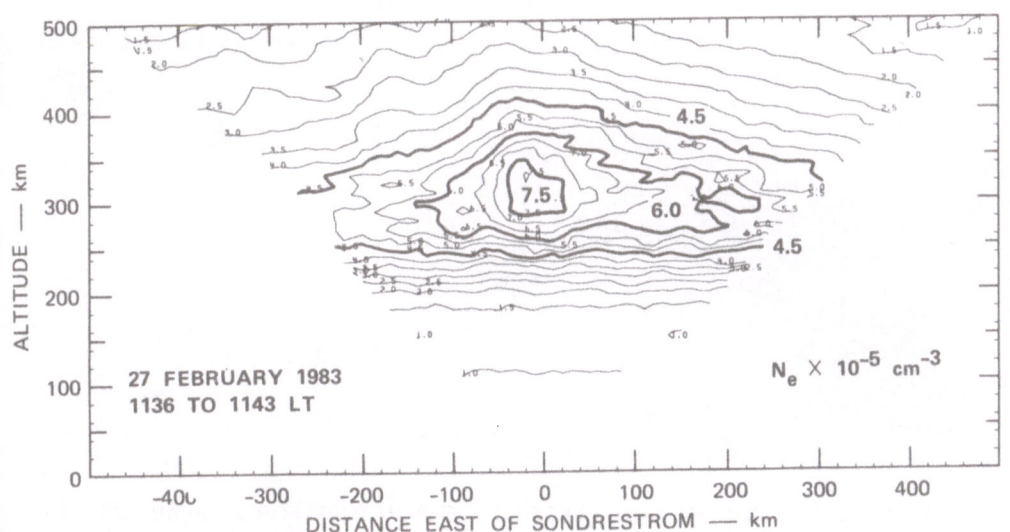

FIGURE 9   CONTOURS OF CONSTANT ELECTRON NUMBER DENSITY FOR 27 FEBRUARY
1983 IN THE EAST-WEST PLANE.   The enhancement extends approximately one
hour in local time.

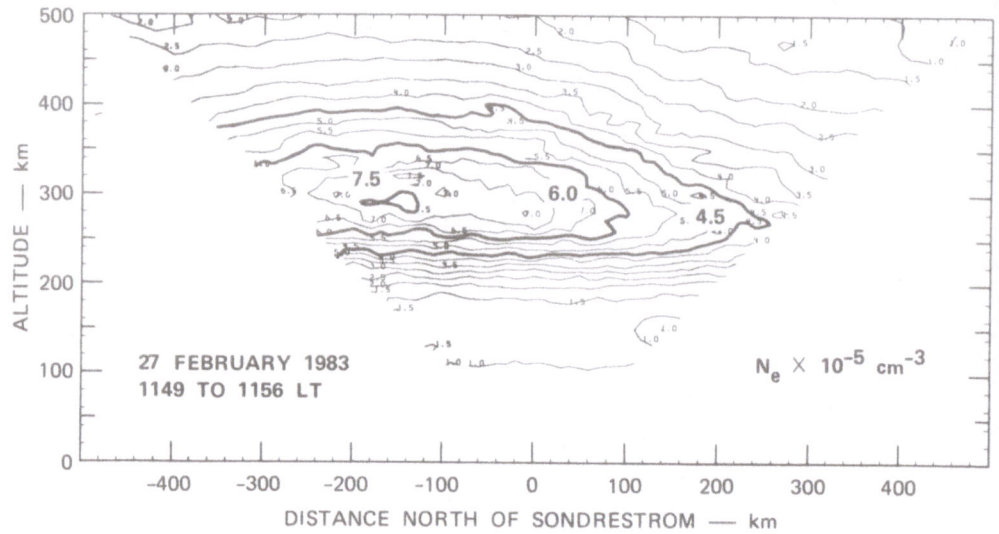

FIGURE 10   CONTOURS OF CONSTANT ELECTRON NUMBER DENSITY FOR 27 FEBRUARY
1983 IN THE NORTH-SOUTH PLANE.   The enhancement extends further in the
north-south direction than the east-west direction.

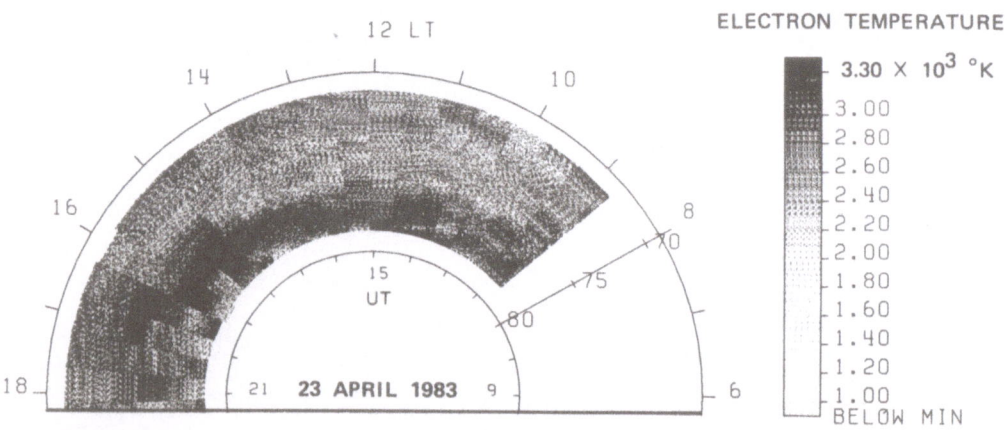

FIGURE 11   CLOCKDIAL PLOT OF ELECTRON TEMPERATURE FOR 23 APRIL 1983 AT
350-km ALTITUDE.   An enhancement in the electron temperature can be seen
extending from noon to 1700 LT.   This is collocated with a shear reversal in the
ion velocity.

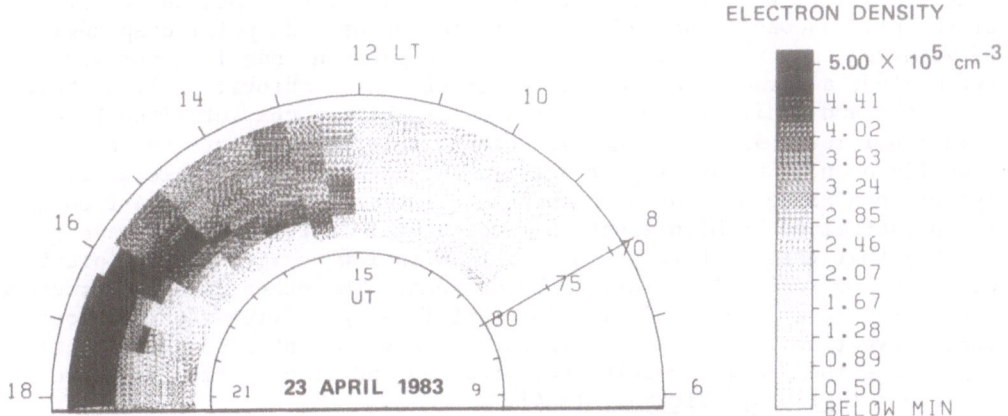

FIGURE 12   CLOCKDIAL PLOT OF ELECTRON NUMBER DENSITY ON 23 APRIL 1983 AT
350-km ALTITUDE.   This enhancement is associated with the elevated electron
temperature shown in Figure 11.

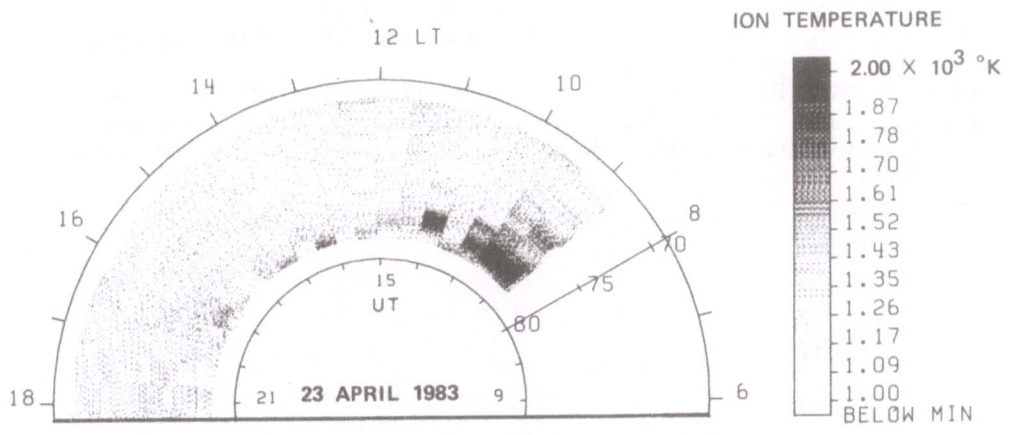

FIGURE 13   CLOCKDIAL PLOT OF ION TEMPERATURE FOR 23 April 1983 AT 350-km
ALTITUDE.   The prenoon enhancement is associated with large electric fields and
the resulting Joule heating.

5.   SUMMARY

From the first year of operation of the Sondrestrom incoherent-scatter
radar facility, we have identified what appear to be repeated features
of the plasma convection and structure in the dayside polar cusp region.
Between 70° and 80° Λ, we see a convection pattern that is generally
two-celled; a plasma flow that is parallel to the auroral oval; a shear
reversal; and a division between the cells that is characterized by a
rotational reversal occurring over one to three hours centered on
magnetic noon. The density structures seen on the dayside are of two
types: one is coincident with the shear reversal in the plasma flow and
is characterized by high electron temperature. Satellite data have indicated that this feature is associated with low energy electron precipitation and region 1 field-aligned current. The other density structure
seen often is a ridge of ionization that is aligned with a strong northwest convection. The plasma temperatures are not enhanced in this case,
suggesting the plasma is solar EUV produced and transported from lower
latitudes [Kelly and Vickrey, 1984].

6.   REFERENCES

Heelis, R.A., W.B. Hanson, and J.L. Burch, J. Geophys. Res. 81, 3803,
    1976.
Heikkila, W.J., and J.D. Winningham, J. Geophys. Res. 76, 883, 1971.
Kelly, J.D., and J.F. Vickrey, Geophys. Res. Lett. (in press), 1984.
Kofman, W., and V.B. Wickwar, Geophys. Res. Lett. (in press), 1984.
Knudsen, W.C., J. Geophys. Res. 79, 1046, 1974.
Potemra, T.A., J.P. Doering, W.K. Peterson, C.O. Boxtrom, R.A. Hoffman,
    L.H. Brace, J. Geophys. Res. 83, 3877, 1978.
Robinson, R.M., D.S. Evans, T.A. Potemra, and J.D. Kelly, Geophys.
    Res. Lett. (in press), 1984.
Shepherd, G., Rev. Geophys. Space Phys. 17, 2017, 1979.
Weber, E.J., J. Buchau, J.G. Moore, J.R. Sharber, R.C. Livingston,
    J.D. Winningham, and B.W. Reinisch, J. Geophys. Res. 89, 1683,
    1984.

HIGH-RESOLUTION OBSERVATIONS OF ELECTRIC FIELDS AND F-REGION PLASMA PARAMETERS IN THE CLEFT IONOSPHERE

J. C. Foster, J. M. Holt
M.I.T. Haystack Observatory
Westford, Massachusetts 01886
U.S.A.

J. D. Kelly, V. B. Wickwar
SRI International
Menlo Park, California 94025
U.S.A.

ABSTRACT. Details of the high-latitude ionosphere in the vicinity of the noon sector cleft have been examined using the Sondrestrom radar on April 18, 1983. An experiment combining azimuth and elevation scans was used to reveal the patterns of convection and F region temperature and density over a region of 1000 km diameter centered on the radar at $75°\Lambda$. High-resolution line-of-sight observations have been used to produce detailed maps of the convection pattern and plasma transport each 20 minutes as the radar rotated through the noon sector during the 8 hour experiment. Patterns of electrostatic potential across the radar field of view reveal strong poleward and eastward convection in the morning sector several hours pre-noon, a region of low speed convection associated with the velocity convergence region, and high speed poleward and westward plasma transport from the afternoon sector into the region of cleft precipitation and polar cap entry near noon. The divergence of plasma flow at polar latitudes away from the cleft is seen in both the convection velocity and F region density data (poleward flow through the ionospheric cleft populates the polar cap with enhanced plasma density). A topside density trough is associated with the poleward edge of the strong afternoon convection toward noon where flow velocities approach 2000 m/s and a potential drop of 50 kV is seen across a 500 km region. The region of high-speed plasma entry into the polar cap was characterized by predominently northwestward directed flow on this day on which the IMF was directed away from the sun and Bz was near 0. Plasma entered polar latitudes across a rotational convection reversal which spanned at least 3 hours of local time in the noon sector. Low energy particle precipitation and an enhancement in the ionospheric electron temperature and density were observed at this convection reversal. High-resolution redar data revealed a region of multiple soft arcs aligned with the reversal and polar cap boundary in the post-noon sector.

J. A. Holtet and A. Egeland (eds.), The Polar Cusp, 349–364.
© *1985 by D. Reidel Publishing Company.*

1. Introduction

There is increasing evidence for the persistence and day-to-day repeatability of the magnetospheric processes which govern the configuration of the dayside auroral oval. Magnetic merging between the solar and the geomagnetic fields occurs at the dayside magnetopause during times of southward directed interplanetary magnetic field, the viscous interaction of the IMF with the magnetosphere drives portions of the plasmasheet antisunward along the flanks of the magnetotail, and parallel electric fields accompany the field aligned currents which penetrate to low altitudes as a result of these processes. The signatures of these mechanisms are observed at lower altitudes near magnetic noon at invariant latitudes, Λ, near 78° in that region termed the ionospheric cleft. There precipitating particles, field aligned currents, electric fields, and waves extend downward along magnetic field lines producing the optical emissions, ionospheric perturbations, and dynamics which characterize the noontime cleft.

Near noon at high latitudes a general softening of the characteristic energy of precipitating electrons to less than 100 eV is accompanied by a "gap" in the 557.7 nm emissions and a marked increase in the 630 nm / 557.7 nm ratio. Micropulsations and ELF hiss are associated with field lines near this region as is a reduction in riometer absorption. These noontime signatures define a central region several hours of local time wide which is roughly coincident with the region in which the flow of the ionospheric plasma convecting sunward from the night sector at auroral latitudes rotates poleward and enters the polar cap. The interrelationship of these signatures and their implications for our understanding of the magnetospheric processes which cause them are the concern of many recent investigations of the high latitude dayside cleft.

At latitudes below 73°Λ the plasma convection pattern is well documented by incoherent scatter radar observations from Chatanika (Foster et al. 1981) and Millstone Hill (Evans et al. 1980) and that at higher latitude has been inferred from the combination of observations by balloon, barium release, rocket, and satellite passes. Atmospheric Explorer passes near noon have been studied by Heelis et al. (1976) and interpreted in terms of a confined region of relatively high speed plasma flow into the polar cap at latitudes near 80°Λ which Reiff et al. (1978) have described as a throat in the dayside convection pattern. More recently, Heelis (1984) has characterized the dayside convection pattern in terms of its IMF dependences and has concluded that a throatlike region of poleward flow may only exist during times of reconfiguration of the pattern. For a steady orientation of the IMF, plasma convection entering the polar cap crosses noon from dusk toward dawn (westward in northern hemisphere) with a more gradual poleward component.

If the bulk of the sunward flowing plasma is confined to enter the polar cap in a longitudinally narrow region near noon the potential drop

across such a throat would amount to a large portion of the cross polar
cap potential, which is of the order of 30 - 100 kV, and a large
eastward electric field component and high speed poleward plasma
convection would characterize its vicinity. The Joule heating
associated with this high field region would constitute a significant
high latitude heat source. Plasma velocities in excess of 1000 m/s are
often seen in the vicinity of the cleft (Heelis et al. 1976; Foster and
Doupnik 1984) and, on the average a region of enhanced Joule heating (5
ergs cm-2 sec-1) is seen extending 2 to 3 hours in local time on either
side of the noon meridian at latitudes above 75°Λ (Foster et al. 1983).
That study found an electric field of 50 mV/m total magnitude
characteristic of the high latitude dayside heating region. This report
describes the findings of a high spatial resolution investigation of the
dayside patterns of ionospheric convection electric fields, plasma
transport, and precipitation-produced ionization which has been
conducted with the Sondre Stromfjord, Greenland incoherent scatter radar
under moderately active conditions on April 18, 1983.

2. Experimental Technique

During 1982 and 1983 the Chatanika, Alaska incoherent scatter
radar, which had been used in studies of the auroral ionosphere since
1971, was relocated to Sondre Stromfjord on the west coast of Greenland
(65.8° N, 50.2° W). At its present location at 75.2°Λ the radar is
well situated for studies of the dayside auroral oval and the signatures
of the ionospheric cleft. In addition, its location in Greenland places
the facility at the northern end of a chain of incoherent scatter radars
roughly aligned along a magnetic meridian connecting Sondrestrom with
Millstone Hill, Arecibo, and Jicamarca. Multi-radar experiments to
monitor the latitudinal propagation of energy deposited in the high-
latitude ionosphere and thermosphere are conducted on a regular basis
from these sites. The experimental results reported herein were made
early in the operational lifetime of the Greenland facility as a test of
techniques now being applied in the multi-station observations.

A 360° scan in azimuth from magnetic south through east to south
was employed at a constant 45° elevation angle. Measurements were made
throughout the E- and F-region with a usable signal for spectral
analysis being obtained at altitudes below 500 km. Each scan about the
radar's position was completed in 15 minutes and surveyed an area of 8°
latitudinal extent centered on 75°Λ. The east-west coverage afforded
amounted to approximately two hours of local time. Use of a 320 μs
transmitted pulse resulted in 50 km spatial resolution along the line of
sight from the radar and the return signal was integrated over 5 sec,
giving 2° azimuth resolution. An elevation scan from south to north
along the magnetic meridian was performed each hour to provide details
of the latitude-altitude structure of the features surveyed in the
azimuth scans. Similar azimuth scanning experiments are run at
Millstone Hill on a regular basis and analysis techniques developed for

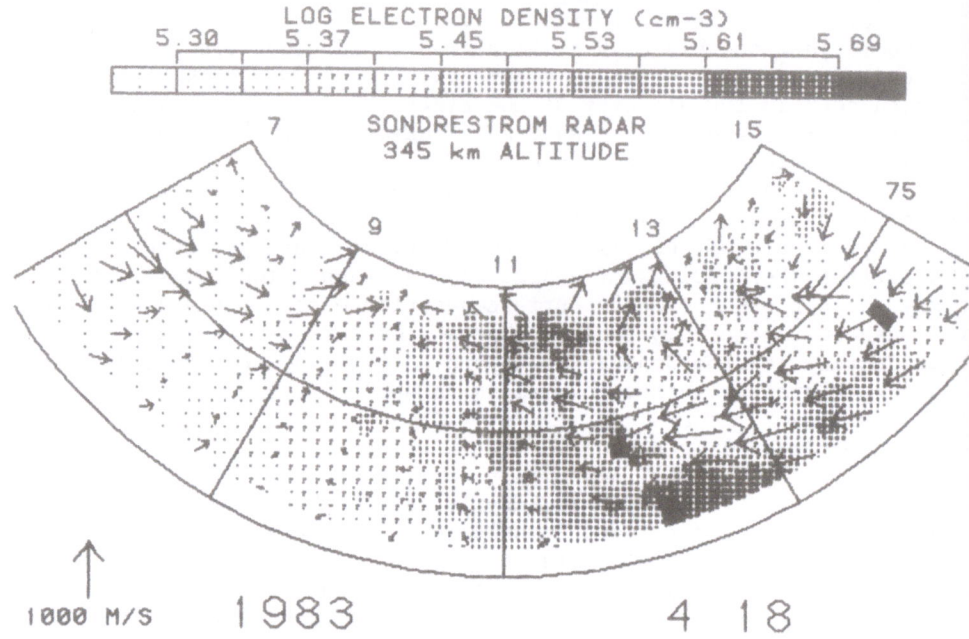

Figure 1. Patterns of F-region density and plasma convection derived by averaging data acquired in 22 separate azimuth scans about the location of the Sondrestrom radar at 75Å, presented in magnetic latitude local time coordinates. The predominantly northwestward directed convection toward noon from the afternoon sector carries a tongue of enhanced plasma density through the cleft ionosphere and into the polar cap.

use at that radar facility have been applied to the data acquired in this experiment.

     In order to discuss the large-scale characteristics of the patterns of dayside plasma transport and electric fields an averaged picture of the experimental observations has been prepared. Figure 1 presents a composite picture of the cleft ionosphere on April 18, 1983 as viewed from Sondrestrom. Data from 22 individual azimuth scans have been averaged into bins encompassing one-half degree of magnetic latitude and 15 minutes of local time. The F-region density data at 345 km altitude have been gray-scale coded on a logarithmic scale and are presented in the figure. Individual density measurements at altitudes between 200 and 550 km were used in preparing the figure by normalizing each each sample to that at 345 km using a density-altitude profile derived as an average over the data from an entire azimuth scan. Line of sight velocities were similarly averaged into local time-latitude bins to determine the average pattern of plasma convection throughout the experiment. Although an individual Doppler measurement samples only that component of the plasma velocity along the line of sight to the radar, by combining all the measurements falling within a given bin with information on the orientation of the beam in magnetic coordinates for each, a vector plasma velocity can be determined within that sampling bin. This technique assumes that the general features of the convection pattern do not change over the 1 to 3 hours that the radar samples a given local time-latitude region. The results of this calculation are presented as the vector convection arrows superimposed on the average F-region density data in Figure 1. The general features of the patterns revealed with these averaging techniques are elucidated when the individual high-resolution azimuth scans are examined, as is presented in Section 3.

During the course of the experiment the convection reversal between sunward flow at lower latitudes and antisunward flow at high latitudes fell within the radar's field of view until 2 hours before magnetic noon and again beginning 1 hour after noon. During a three hour interval near noon convection proceded generally toward the northwest, flowing from low to high latitudes without a perceptible reversal at latitudes below 79°Λ. As will be seen in Figure 3, the northwestward flow increased in magnitude rather uniformly with latitude across this region, reaching a maximum of 2000 m/s immediately equatorward of the post-noon sunward-antisunward reversal.

The average F-region densities were relatively lower both before and after local noon where a factor of 2-3 enhancement was associated with a tongue of ionization extending poleward from lower latitudes. This density feature was seen to be co-located with the region of northwestward directed plasma convection from lower latitudes toward the polar cap. The patterns of convection and F-region density revealed in this averaged figure are qualitatively similar to those observed in radar experiments which probed the cleft ionosphere from Chatanika during disturbed conditions when the cleft was observable from that auroral-latitude station (see Foster and Doupnik 1984, figure 2). An F-region depletion, or topside plasma trough, lay immediately equatorward of the post-noon convection reversal. Evidence of this trough is seen in Figure 1 near 74°Λ at 14:00 LT, although the position of this narrow feature is blurred by smearing in this average picture. Such a feature was reported by Foster et al. (1980) immediately poleward of a region of enhanced soft particle precipitation during an excursion of the cleft to latitudes near 70°Λ prior to a substorm expansion phase. Comparisons with satellite overflights (eg Figure 6) confirm that a region of enhanced electron precipitation, which produced an increase in the local ionization to altitudes as low as 150 km, was co-located with the post-noon convection reversal. In this experiment a similar region of precipitation was not observed at the pre-noon convection reversal. In the sections below selected single scan "snapshots" encompassing the entire radar field of view are presented to illustrate the various high-resolution features observed during the course of the experiment.

3. High-Resolution Observations

   Individual azimuth and elevation scans provided high-resolution
images of the cleft ionosphere at various local times through the course
of the event. Figure 2 presents plasma density and convection electric
field patterns within the radar's field of view near local noon (solar
local time is UT - 3 hours). Each scan frame covers 8° of magnetic
latitude and nearly 2 hours of local time. Data are presented in
azimuth and ground distance coordinates with geodetic north at the top.
Constant altitude circles are at intervals of 150 km, although the data
are projected onto a surface at a constant 325 km altitude. Magnetic
latitude and longitude grid lines are superimposed on the data. The
electron densities are gray-shaded after having been normalized to their
value at 350 km as was done for Figure 1.

   The potential distribution over the field of view of the radar was
computed using the method developed for Millstone Hill azimuth scans
(Holt et al. 1984). Each azimuth scan was treated individually,
assuming that the potential pattern did not change during the scan and
that the pattern could be represented by a function of range and
azimuth. The potential is determined by least squares fitting the
assumed function to the measured line-of-sight plasma velocities, given
that

$$\bar{E} = -\bar{\nabla}\Phi$$

$$\bar{V} = (\bar{E} \times \bar{B})/B^2.$$

where $\bar{E}$ is the electrostatic field, Φ is the electrostatic potential, $\bar{V}$
is the ion velocity, and $\bar{B}$ is the geomagnetic field. The component of
the electric field perpendicular to the radar line-of-sight is directly
determined by the measured Doppler velocity. The azimuthal derivative
of the electric field component along the line-of-sight is also
determined because

$$\bar{\nabla} \times \bar{E} = 0$$

If the field is electrostatic, these derivatives provide information on
the radial component of the electric field, which is not directly
measured. However, the azimuthal variation of the radial electric field
component only determines the value of that component to within a
constant of integration. For Millstone Hill data, this constant may be
determined by requiring that the potential be zero at Millstone's
latitude. Clearly this assumption is not valid for Sondrestrom which
usually is situated within the region of high-speed convection.
Instead, we have assumed that the line integral of the line-of-sight
component of the electric field along circles centered on the radar is
zero. In other words, of all the potential patterns consistent with the
velocity measurements, we choose the pattern for which the average
radial electric field is zero. This assumption is not valid if there is

substantial convection of plasma around the radar, as might happen if
the radar were in the middle of a closed convection cell associated with
localized field-aligned currents in the near vicinity of the radar.
However, it is inherently impossible for a monostatic radar to measure
such circulation (the Doppler velocity would be zero), and our
assumption is exactly satisfied for a constant electric field within the
field of view of the radar. Potential contours are shown at 2 kV
intervals and have been derived using only the data obtained below 450
km altitude.

The data presented in Figure 2 depict the conditions observed as
the radar rotated into the tongue of enhanced F-region ionization
associated with the northwestward directed convection from the afternoon
sector cell (cf Figure 1 at 10 - 12 LT). During the scan at 14:10 UT a
32 kV potential difference was observed across a span of 4° of latitude
giving an average field strength of 50 mV/m. The flow rotated northward

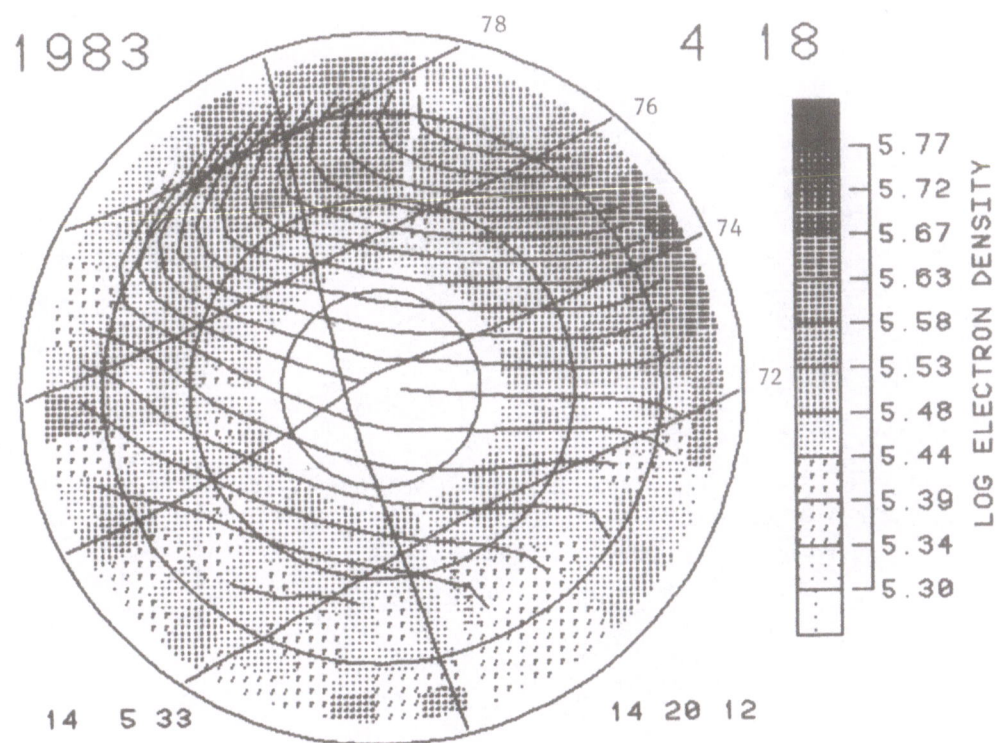

Figure 2. Patterns of F-region electron density at 345 km altitude
and convection observed from Sondrestrom 1 hour before local noon.
Contours of the electrostatic potential derived from the line-of-
sight velocity observations are shown at 2 kV intervals. The 50 mV/m
average electric field carries plasma northwestward to a rotational
reversal at 77°Λ.

above 77°Λ where the radar observed a density enhancement related to soft particle precipitation, as will be discussed in the Section 4.

Presented in Figure 3 are data at solar noon which show additional features of the region of high-speed plasma entry to polar latitudes. By this time the complete rotation of the convection from northwestward to northeastward fell within the experimental field of view and some 60 kV of potential were associated with that portion of the cleft entry region. It is noteworthy that the flow patern even in this region of 100 mV/m average electric field is not constricted, but plasma entry into the polar cap occurs across the entire 2 hour span of local times being surveyed. The electric field strength increases regularly up to the latitude at which the flow rotates poleward and eastward. An enhancement in the density at this point is again related to the region of intensified low-energy precipitation, and a marked F-region depletion, or topside trough, is seen immediately equatorward of the precipitation feature and the flow reversal.

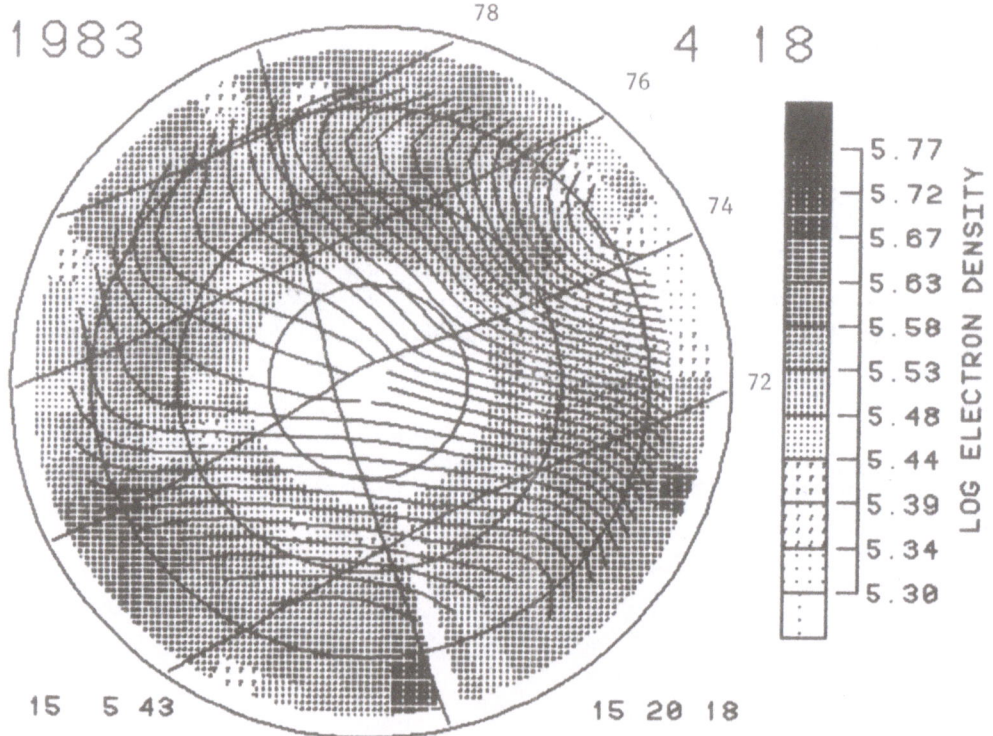

Figure 3. Patterns of F-region electron density at 345 km altitude and convection observed from Sondrestrom at local noon. The electric field strength exceeds 100 mV/m immediately equatorward of the convection reversal at 76°Λ.

Figure 4 presents high-resolution data depicting conditions along
the post-noon boundary between the auroral and polar cap ionosphere.
Plasma convection is predominantly westward and cuts off sharply at 74°Λ
at 16:15 UT (14:15 MLT). High-density F-region plasma is seen
equatorward of 72°Λ associated with the equatorward portion of the
noonward convection region. The high-latitude trough is very pronounced
along the poleward edge of the sunward convection region and an east-
west aligned precipitation feature is seen directly at the convection
boundary. There is evidence of plasma transport across this boundary,
but the bulk of the flow is parallel to the precipitation feature and
proceeds toward noon. The average electric field exceeds 100 mV/m over
a 400 km latitudinal extent immediately equatorward of the sunward
convection boundary.

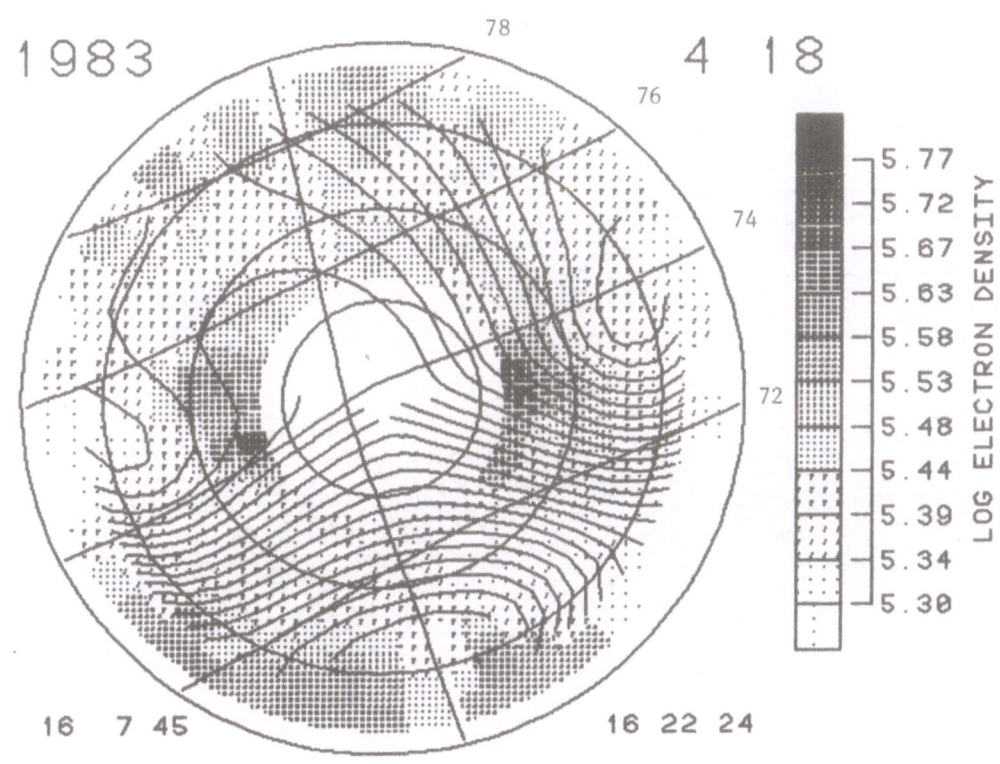

Figure 4. Patterns of F-region electron density at 345 km altitude and
convection observed from Sondrestrom 1 hour after local noon.
Precipitation-enhanced densities mark the poleward edge of the region of
high-speed plasma transport toward noon from the afternoon sector.

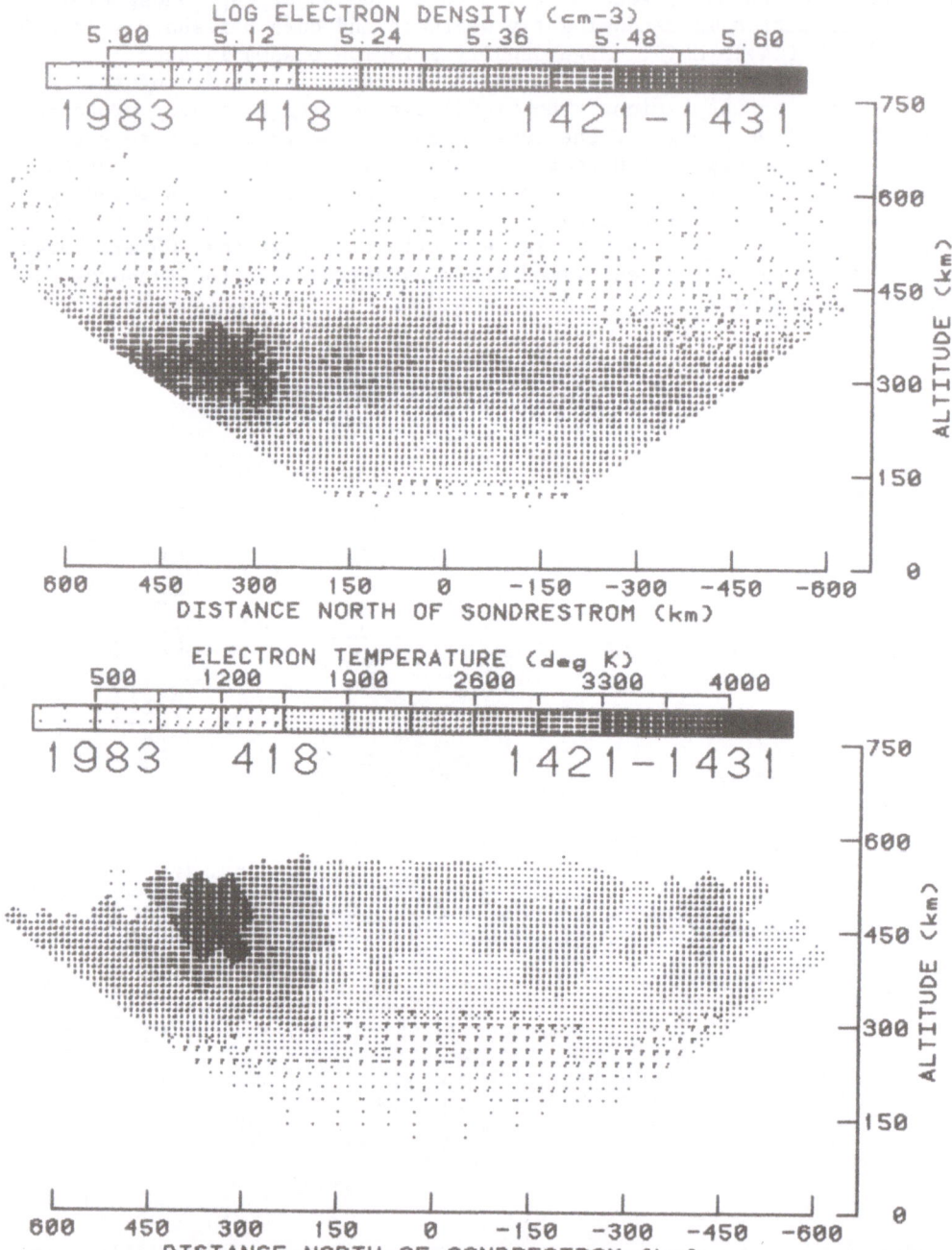

Figure 5. Latitudinal distribution of electron densities (top) and temperatures (bottom) corresponding to the data shown in Figure 2.

4. Relationship to Particle Precipitation

During the course of the experiment polar orbiting satellites equipped with particle detectors made several overflights within the radar field of view. At 14:11 UT the P78-1 satellite made a pass directly along the west coast of Greenland, bisecting the region surveyed by the radar azimuth scan shown in Figure 2. The satellite detectors observed a narrow region of precipitation with 200 eV average energy at 76.5°Λ and at a bearing of 329° azimuth from Sondrestrom. This region was sampled by the radar within one minute of the overflight, at which time the radar observed enhanced ionization and the poleward edge of a region of strong westward convection at that point. Ten minutes later the radar executed a south to north elevation scan parallel to the satellite path and observed the enhancement in

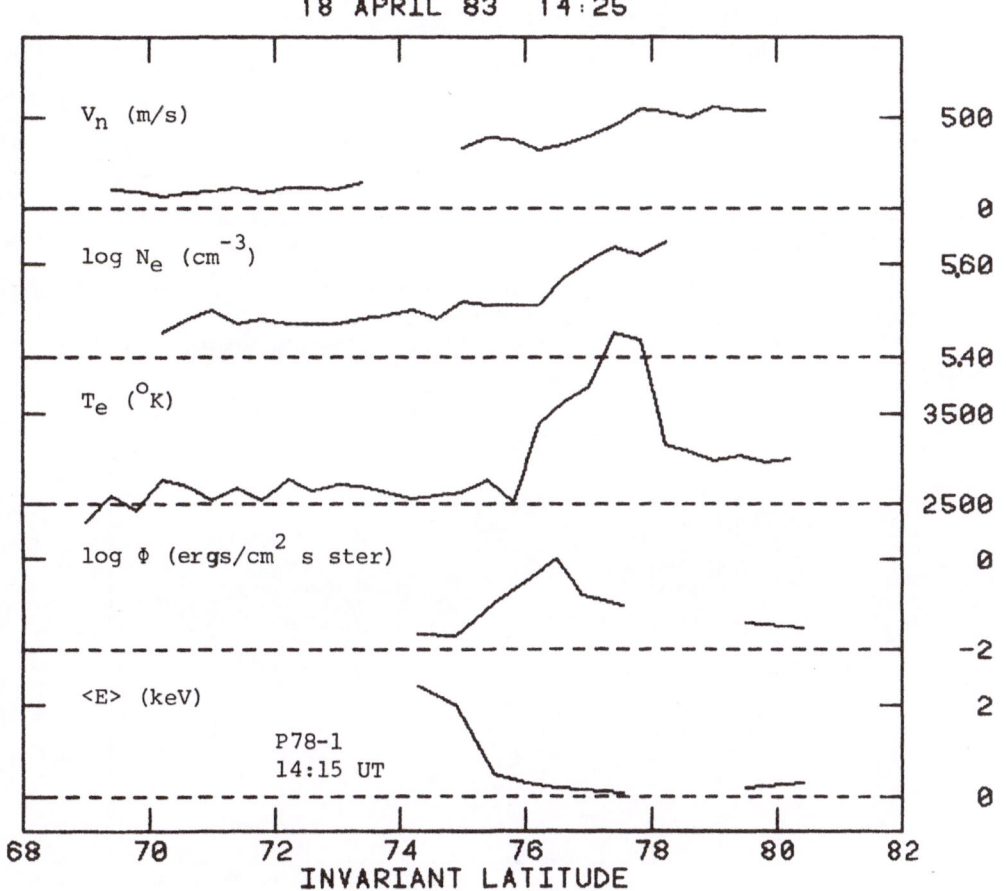

Figure 6. Latitudinal variation of electron precipitation and ionospheric parameters corresponding to the data in Figure 5.

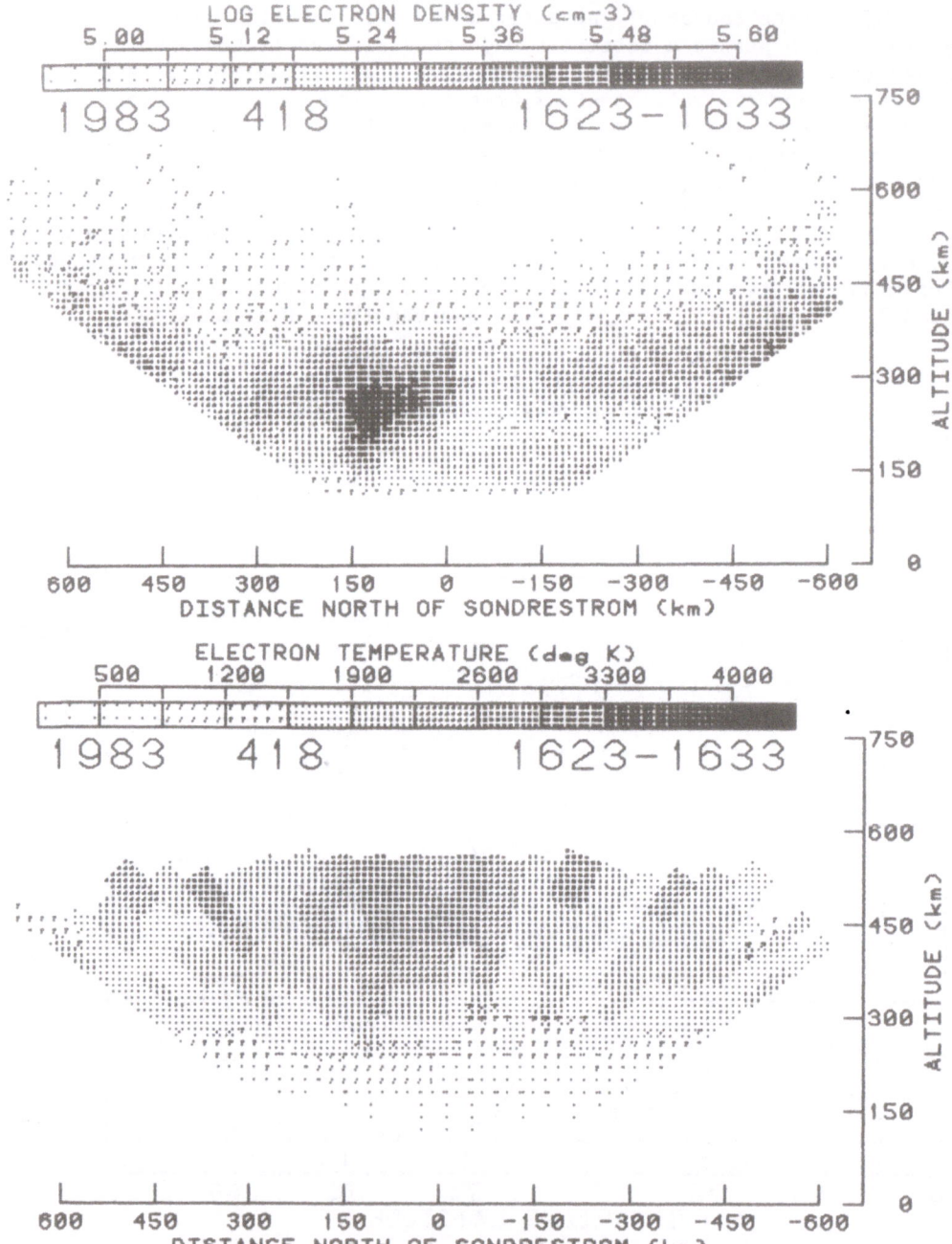

Figure 7.   Latitudinal distribution of electron densities (top) and temperatures (bottcm) corresponding to the data shown in Figure 4.

ionization and electron temperature associated with the precipitation in
detail. Figure 5 presents the latitude-altitude variation of plasma
density and electron temperature measured by the radar scan and Figure 6
presents an intercomparison of the satellite precipitation data and
radar-deduced parameters as functions of invariant latitude.
Immediately poleward of 75°Λ the satellite observed a drop in the average
electron energy from 2 keV to 200 eV while the energy flux increased by
two orders of magnitude. On the radar scan 10 minutes later, the
electron temperature rose abruptly by 1000 K at 76°Λ accompanied by a
50% enhancement of the plasma density between 200 km and 400 km
altitude. The northward component of the plasma flow increased from 250
m/s to 550 m/s at the precipitation feature and, with reference to the
azimuth scan data (not shown), the direction of the plasma motion
changed by 90° , from N-W to N-E, at that point. The enhanced soft

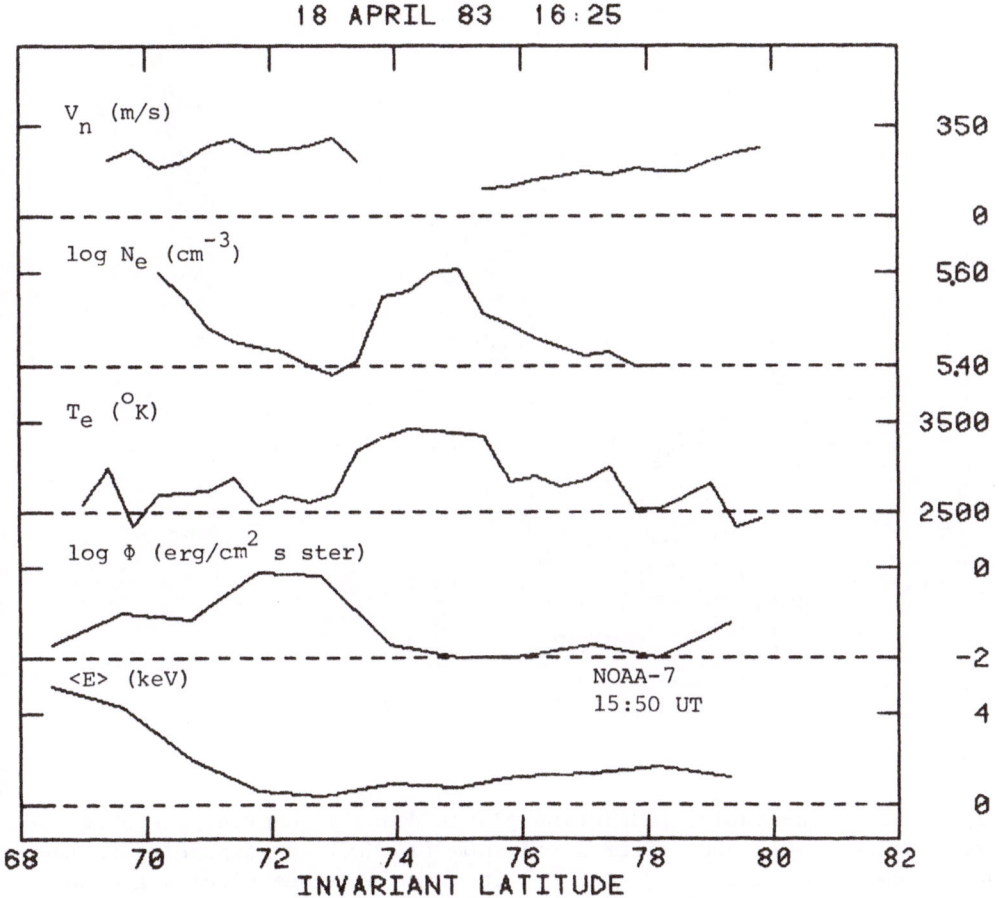

Figure 8. Latitudinal variation of electron precipitation and
ionospheric parameters corresponding to the data in Figure 7.

precipitation region was thus located directly on the poleward edge of
the region of intense N-W directed plasma flow.

    At the time of the post-noon azimuth scan presented in Figure 4 the
NOAA-7 satellite overflew the eastern portion of the scan region and
observed the a narrow precipitation region similar to that reported by
D. Evans (this proceedings). An east-west aligned precipitation feature
at the poleward edge of the westward convection region was inferred from
the enhanced densities at 200 km altitude seen during the radar scan.
That feature was more clearly seen during a south to north elevation
scan at 16:24 UT, which is presented in Figure 7. The E- and F-region
ionization enhancement is easily seen 1 north of the radar as is a
topside trough on its equatorward edge which was associated with a
narrow band of elevated electron temperature. Precipitation data from

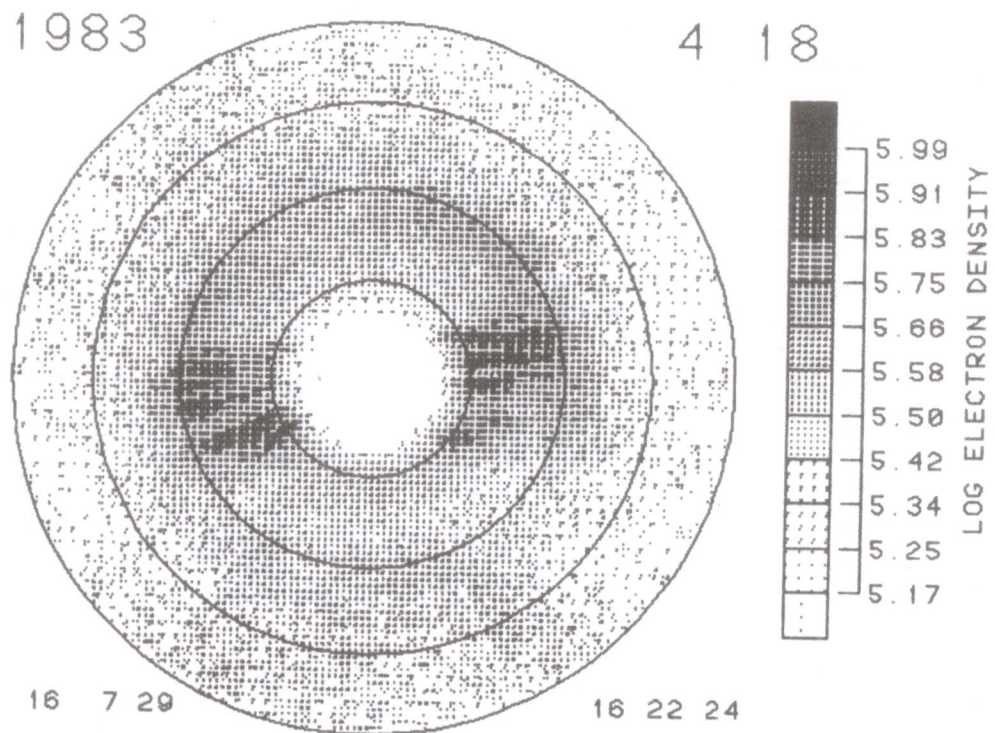

Figure 9. High-spatial resolution plasma density observations observed
from Sondrestrom 1 hour after local noon (14 MLT) corresponding to the
data shown in Figures 4, 7, and 8. Constant altitude circles are shown
at 150 km intervals. Precipitation-enhanced plasma density was observed
as narrow east-west aligned bands at the poleward edge of the post-noon
region of sunward convection. A topside density trough is apparent
immediately equatorward of the precipitation structures.

NOAA-7 are presented in Figure 8 along with the latitudinal variation of plasma parameters observed along the radar's meridian 30 minutes later. The latitudinal displacement between the sharp features observed by the radar (73.5°- 75.5°Λ) and the precipitation region (71.5°- 73.5°Λ) is readily explained in terms of the poleward motion of the feature as observed during the intervening period. The azimuth scan at the exact time of the overflight showed a density enhancement at 200 km altitude whose poleward edge occurred at 73.5°Λ, in excellent agreement with the satellite data. By 1608 UT (eg Figure 4) the enhancement had moved poleward to cover the span 72.7°Λ to 74.3°Λ and at 1635 UT (from the scan following the data presented in this figure) the equatorward edge of that region was at 75.8°Λ.

High-resolution density observations during the azimuth scan shown in Figure 4 are presented in Figure 9. These data are presented with 2° azimuth and 25 km altitude resolution. The precipitation-produced density enhancement on the poleward edge of the region of strong convection toward noon was striated in the form of parallel, narrow, east-west aligned features extending in altitude from 150 km to 350 km. As the 1° full width radar beam looked parallel to the precipitation features at 90° and 270° azimuth the multiple topside density enhancements associated with the arcs were clearly resolved. The ionization enhancements produced by this post-noon precipitation extended in altitude from 120 km to 350 km. Radar observations of a latitudinally narrow band of precipitation-enhanced ionization with a similar altitude structure near noon were reported by Foster et al. (1980). It is suggested that these arcs are associated with the band of elevated electron temperature seen in recent Sondrestrom surveys of the dayside ionosphere and that they carry field-aligned currents associated with the rotational reversal of the convection pattern which is observed at their location.

ACKNOWLEDGMENTS. This work was supported by the National Science Foundation through grant ATM1819812 to the Massachusetts Institute of Technology and cooperative agreement ATM8121671 to SRI International.

REFERENCES
Evans, J. V., J. M. Holt, W. L. Oliver, and R. H. Wand, Millstone Hill incoherent scatter observations of auroral convection over 60°< Λ <75°, 2, Initial results, J. Geophys. Res., 85, 41-54, 1980.
Foster, J. C., Ionospheric signatures of magnetospheric convection, J. Geophys. Res., 89, 855-865, 1984.
Foster, J. C., G. S. Stiles, and J. R. Doupnik, Radar observations of cleft dynamics, J. Geophys. Res., 85, 3453-3460, 1980.
Foster, J. C., J. R. Doupnik, and G. S. Stiles, Large-scale patterns of auroral ionospheric convection observed with the Chatanika radar, J. Geophys. Res., 86, 11357-11371, 1981.
Foster, J. C., J.-P. St. Maurice, and V. J. Abreu, Joule heating at high latitudes, J. Geophys. Res., 88, 4885-4896, 1983.
Foster, J. C., and J. R. Doupnik, Plasma convection in the vicinity of the dayside cleft, J. Geophys., Res., 89, in press, 1984.

Heelis, R. A., The effects of interplanetary magnetic field orientation on dayside high-latitude ionospheric convection, J. Geophys. Res., 89, 2873-2880, 1984.

Heelis, R. A., W. B. Hanson, and J. L. Burch, Ion convection reversals in the dayside cleft, J. Geophys. Res., 81, 3803-3809, 1976.

Holt, J. M., R, H, Wand and J. V. Evans, Millstone Hill measurements on 26 February 1979 during the solar eclipse and formation of a midday F-region trough, J. Atmos. Terr. Phys., 46, 251, 1984.

Reiff, P. H., J. L. Burch, and

R. A. Heelis, Dayside auroral arcs and convection, Geophys. Res. Lett., 5, 391, 1978.

OBSERVATIONS OF THE UNSTABLE E-REGION IN THE POLAR CUSP

P. Stauning, J. K. Olesen
Division of Geophysics, Meteorological Institute
Lyngbyvej 100, DK-2100 Copenhagen - Denmark

R. T. Tsunoda
Radiophysics Laboratory, SRI International
Menlo Park, California 94025 - USA

ABSTRACT. The incoherent scatter radar facility (ISR) recently installed at Søndre Strømfjord, Greenland (inv. lat. = 74°), has been used, along with ionosondes, riometers and other equipment, to investigate the polar Slant E Condition (SEC). SEC is a common term for the occurrence in ionograms of a diffuse, slanted backscatter trace (slant E_s) emerging from the E-region and an abnormally high damping of the return signal at frequencies close to the E-region critical frequency (lacuna). ISR observations of E-region densities, electron and ion temperatures and F-region plasma convection through 5 days in August 1983 have shown, that SEC occurs during events of substantially elevated E-region electron temperatures related to the occurrence of strong horizontal electric fields. It has also been found that a peculiar type of weak, slowly fluctuating cosmic noise absorption commonly observed on riometers during SEC events could be caused by simple, non-deviative absorption resulting from the strongly enhanced electron-neutral collision frequencies in the heated E-region. This effect may also explain the ionogram lacuna in terms of an abnormally high level of E-region deviative absorption of the usual collisional type.

1. INTRODUCTION

1.1. The Slant E Condition

'Slant E_s' is the name given by Heppner et al. (1952) to a nearly straight, slanted scatter trace extending from the high frequency end of a normal or sporadic E-layer in ionograms from vertical incidence sounding stations. Slant E_s is primarily observed from stations in the auroral regions or close to the magnetic equator.

Olesen and Rybner (1958) investigated the characteristics of slant E_s as observed at Godhavn, Greenland (inv. lat. = 77o) close to the latitude of the polar cusp region. They noted that in cases of slant E_s traces the ionograms were usually characterized by weak or very often missing reflections from the higher part of the E-layer and the lower part of the F1-layer (lacuna).

Olesen (1972, 1975) used the two features together to characterize the more general term 'Slant E Condition' (SEC) defining a specific, disturbed state of the

365

J. A. Holtet and A. Egeland (eds.), The Polar Cusp, 365–376.

E-region. Figure 1 shows an example of a distinct SEC case. The upper part is the original ionogram, while the lower part is a schematic reproduction that serves to show the sounding frequency and virtual height scales and to illustrate the basic SEC identification criteria, the slant E_s trace and the lacuna, used for the ionogram scaling.

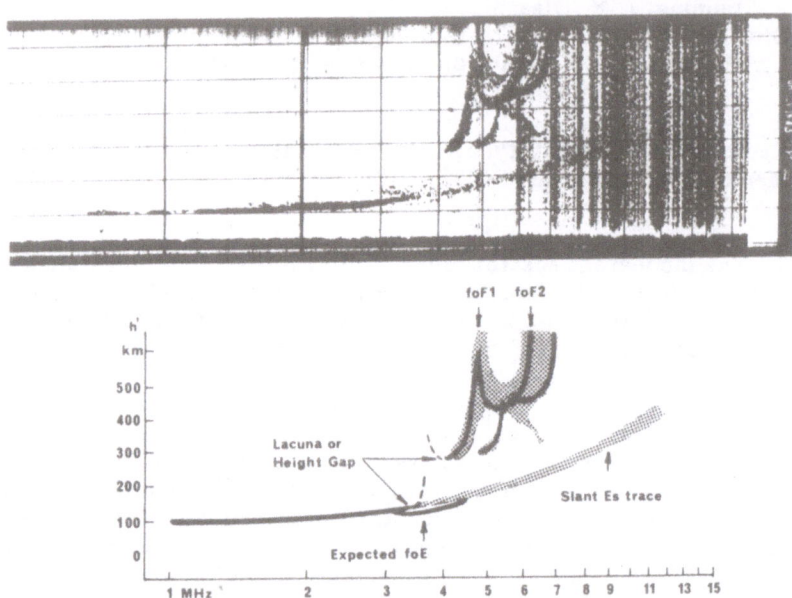

Figure 1.   Day time ionogram with the SEC phenomenon at an auroral zone station. Narssarssuaq June 15, 1969 at 1559 LMT

1.2. Backscatter from E-region plasma instabilities

The association of the polar slant E condition with E-region plasma instabilities, which was suggested by Olesen (1972), has been further substantiated by observations of strong horizontal electric fields occurring in the ionosphere during SEC events as reported e.g. by Primdahl et al. (1974), Iversen et al. (1975), Olesen et al. (1976), Bahnsen et al. (1978).

The ionosonde observations of slant E_s traces are readily explained in terms of backscatter from the long wavelength ($\lambda = 10-100$ m) part of the spectrum of field aligned electron density fluctuations in the E-region. The sounding signals responsible for the slant E_s trace propagate off-vertical and the increase in virtual height with increasing frequency simply reflects the increase in horizontal distance to the scattering region. The density fluctuations are presumably caused by unstable plasma waves driven by strong electric fields.

A review of the theory of such plasma instabilities, their relation to ionospheric fields and plasma parameters and relevant experimental studies in both equatorial and auroral regions has been given by Fejer and Kelley (1980).

1.3. Lacuna and cosmic noise absorption

It was shown by Vassal (1971) and Vassal et al. (1976), that F-lacuna events are usually accompanied by weak, fluctuating cosmic noise absorption as observed by riometers e.g. at 30 Mhz. This type of absorption is not the usual D-region non-deviative collisional absorption produced e.g. during energetic particle pre-cipitation events since ionosonde signals reflected from the lower part of the E-region remain largely unaffected by lacuna conditions.

The SEC definition by Olesen (1972), implies that the absorption mechanism responsible for the E-F lacuna effect observed during SEC events (not necessarily all F-lacuna cases) is located in the E-region. The use of typical electron density and collision frequency profiles to compute the collisional absorption, however, has previously led to the conclusion that E-region absorption intensities are too small to explain the lacuna conditions as well as the absorption intensities of 0.1-0.5 dB typically observed by 30 Mhz riometers during SEC events not accompanied by energetic particle precipitation.

In an alternative approach D'Angelo (1976, 1978, 1981) and Mehta and D'Angelo (1980) have suggested scattering of the radio signal by plasma wave structures to be a possible explanation of such absorption events. Farley et al. (1981) from calculations of scattering cross sections and Siren (1982) from high temporal resolution absorption observations argue, that such a scattering mechanism is, at best, not effective enough to explain tenths of dB cosmic noise absorption intensities at 30 Mhz.

1.4. SEC and electron heating events

Like the SEC phenomena presumably are the manifestations of rather intense E-region plasma instability events associated with strong electric fields, so are the events of highly enhanced E-region temperatures reported by Wickwar et al. (1981), Schlegel and St.-Maurice (1981). St.-Maurice et al. (1981) have suggested the excessive heating of electrons to several times the ion or neutral tempera-tures at heights around 110 km to be caused by unstable plasma waves driven by strong horizontal electric fields.

To correlate SEC and electron heating events seems an obvious idea, but is, however, no easy task at auroral latitudes since events of high electric fields are usually accompanied by intense energetic particle precipitation. The resulting D-region absorption will in many cases inhibit SEC scaling by causing partial or complete blanking of the ionograms. In riometer recordings the possible occur-rence of E-region absorption will be inseparable from the usually much stronger D-region auroral absorption contributions.

For correlation studies of SEC and electron heating events the summer cusp region provides much better observing conditions which include a steady, rather dense, horizontally stratified (mainly solar UV produced) E-region,frequent occur-rence of large, homogeneous, horizontal electric fields and the absence of ener-getic particle precipitation.

2. OBSERVATIONS

2.1. Observational schedule

A general description of the incoherent scatter radar at Sdr. Strømfjord and a summary of initial observations have been given by Kelly (1983).The data reduction programs PRISIS, EPEC and ACFIT used to analyse the radar data have been described by De La Beaujardiere et al. (1980).

The radar was operated for SEC studies during the mid-day hours from approx. 10 to 18 UT (08-16 MLT) on August 8, 12, 13, 15 and 18, 1983. The signal mode used for the experiment was composed of a sequence of 320 μsec, 60μsec and 3 · 60μsec pulses with a separation of 9 msec. The antenna pointing program started with an 11-position N-S scan followed by a S-N elevation scan and then alternately an extended dwell along the B-field or dwells in three different, near-vertical directions to end the sequence.

The ionosonde at Sdr. Strømfjord was operated on a routine basis to provide ionograms each quarter of the hour throughout the campaign period. Riometers, magnetometers and other ground based equipment were in continuous operation at Sdr. Strømfjord and at the net of geophysical observatories located in Greenland.

2.2. Plasma convection pattern on August 12

Figure 2 shows the F-region plasma convection velocities derived from the radar observations through the day hours of August 12. The convection velocities have been plotted in a polar diagram using invariant latitude and magnetic local time coordinates. The location of Sdr. Strømfjord is shown in the plot as a half-circle with MLT hour marks. The magnetic local time used in the diagram is approx. 2 hours behind UT time.

Having been made on basis of a time sequence of data the diagram is not an instantaneous map of the plasma convection. Still it gives the impression of a rather stationary two-cell convection pattern having an enlarged afternoon cell. The magnetic activity was quite high and rather steady throughout the day with K_p ranging from 4- to 5o through the hours of radar operation.

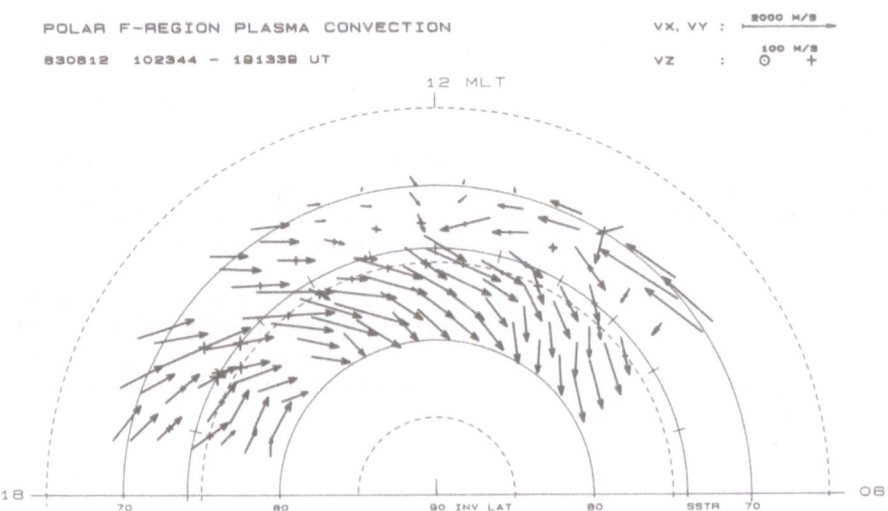

Figure 2. Polar plot of F-region plasma convection observed by the incoherent scatter radar at Sdr. Strømfjord at 1023-1913 UT on August 12, 1983.

A spectacular event started at Sdr. Strømfjord shortly after 1400 MLT (16 UT) at the afternoon side of the cusp region. The event lasted until approx. 1600 MLT (1800 UT) and was characterized by quite high and rather steady convection velocities approaching 2000 m/s. Flow direction was close to NW during the event.

2.3. Summary of data for August 12

In Figure 3 radar, magnetometer, riometer and ionosonde observations from Sdr. Strømfjord have been summarized for the period from 12 to 20 UT on August 12. The upper panel shows the electric field component perpendicular to the magnetic field in the ionosphere above Sdr. Strømfjord. The values have been derived from F-region plasma drift velocities and the bars indicate the duration of each observation. The connecting lines have been drawn for convenience only and should not be taken to represent intermediate electric field intensities. During the time from 12 to 15 UT (10-13 MLT) the electric field intensities were moderately high being around 50 mV/m. The event of particular interest started a little after 1600 UT. The field direction was roughly geomagnetic NE during the event while its magnitude reached a peak value of approx. 100 mV/m shortly after 1700 UT.

The magnetogram in the second panel shows the H and D components. Between 12 and 15 UT the magnetic field was moderately disturbed with excursions in H and D of 100-200 nT from the quiet level. At 1615 UT large positive bays in both H and D started. The magnetic disturbances reached peak values of 500 nT in H and 400 nT in the D component.

The dots in the middle panel show samples of electron density and temperature and ion temperature derived from the multipulse data for a selected gate at a height of 105 km. The electron density at the 105 km level, like at the other E-region gates, changed little during the time from 12 to 19 UT. During the special event lasting from approx. 16 to 18 UT the density at the 105 km level stayed close to $2.0 \cdot 10^5$ el/cm^3.

The ion and particularly the electron temperatures showed much greater variability. During the time from 12 to 15 UT the electron and the ion temperatures were both moderately enhanced above the quiet level of approx. 220°K. During the event that started a little after 1600 UT the electron temperature increased an order of magnitude above the normal level to reach peak values of more than 2500°K around 1700 UT. This is the most spectacular feature of the event and justifies the use of the term 'electron heating event' to describe the conditions.

At the selected height of 105 km the ion temperature is rather closely tied to the temperature of the neutrals representing a much larger mass and the changes were relatively small. At greater heights the Joule heating effect produced much greater excursions in the ion temperatures during the intervals of intense electric fields.

The next lower panel shows 30 Mhz riometer absorption intensities. From 12 to 16 UT the absorption was very small and fluctuating. At 1615 UT, simultaneously with the onset of the magnetic disturbance, a larger and more steady absorption event started. The absorption intensity stayed at a level of approx. 0.6 dB from 1620 to 1730 and then decreased slowly.

The panel below shows the values of f_{min} from ionogram scaling. There was some change in equipment sensitivity due to receiver adjustments made between 1530 and 1600 UT so f_{min} values measured before the change should not be compared to those measured after. The values of f_{min} were rather small during the heating event from 16 to 18 UT.

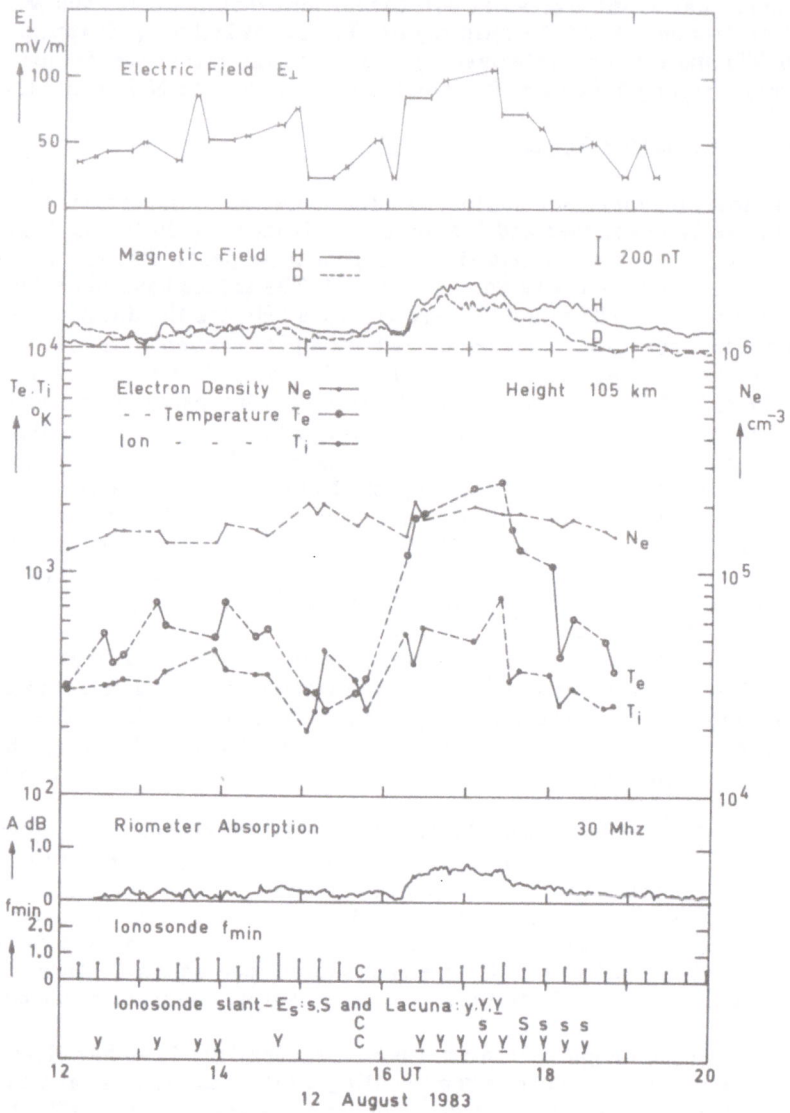

Figure 3. Summary of radar, magnetometer, riometer and ionosonde observations from Sdr. Strømfjord at 12-20 UT on August 12, 1983.

The bottom panel shows signatures originating from the SEC scaling. The letter S marks the occurrence of a slant E_s trace in the ionogram using a small letter to describe a weak, short trace and a capital letter for an intense slant E_s trace. The letter Y marks a lacuna event using again a small letter for a weak case, while a capital letter describes a distinct lacuna with an extended gap. Capital letter with underscore ($\underline{Y}$) means total lacuna i.e. no F-region echoes.

A few lacuna events occurred during the moderately disturbed period from 12 to 15 UT. During 2 hours following the onset at appr. 1615 UT of the intense heating event all ionograms showed lacuna conditions. In the beginning of the event the lacuna was total and slant E_s traces were not discernible while later the lacuna conditions degraded to partial lacuna and slant E_s traces appeared.

2.4. Temperature and density profiles

For a more detailed examination of changes in E-region plasma parameters related to the intense heating event starting appr. 1615 UT Figures 4 and 5 show height profiles of temperatures, density and absorption at 1547 UT and 1622 UT respectively. For the height range from 90 to 150 km the left-hand section of these figures show electron and ion temperatures while the middle sections show electron density profiles.

By comparing Figures 4 and 5 it can be seen that the electron densities showed no appreciable changes related to the heating event. The temperature profiles, however, did change considerably as the event developed. At 1547 UT, just before the onset of the heating event, the electron and ion temperature profiles were almost identical and also rather close to the expected neutral temperature profile (US Standard Atm.Suppl. 1966; CIRA 1972). At 1622 UT, when the event was fully developed, both ion and electron temperatures were substantially increased above the corresponding values at 1547 UT throughout the selected height range. The largest increase in the electron temperature occurred at 105 km at the E-region density peak where, at 1622 UT, the temperature was almost an order of magnitude higher than at 1547 UT. For the ions the largest temperature increase occurred at a height of 130 km.

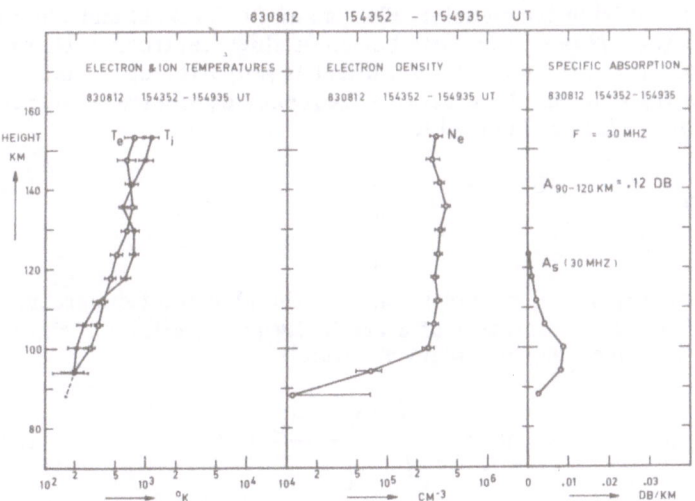

Figure 4. Height profiles of measured electron and ion temperature and electron densities and profile of computed specific 30 Mhz absorption at Sdr. Strømfjord at approx. 1547 UT on August 12, 1983.

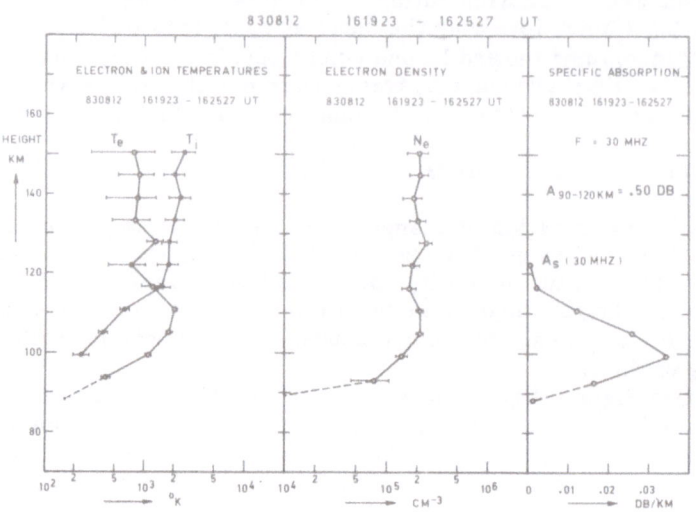

Figure 5. Same as Figure 4 at approx. 1622 UT.

3. DISCUSSION OF OBSERVATIONS

3.1. Riometer absorption during electron heating events.

The observed values of electron densities and temperatures have been used along
with a model for the neutral atmosphere to compute 30 Mhz absorption intensities. An important point for these calculations is the temperature dependence of
electron-neutral collision frequencies. Phelps and Pack (1959) and Phelps (1960)
established that for a Maxwellian distribution of slow electrons in Nitrogen, the
frequency of collisions for momentum transfer is proportional to electron energy.
Accordingly the formula for the effective frequency of electron-neutral collisions
given by Boström (1973) has been adopted:

$$\nu_{en} = 1.5 \cdot 10^{-17} \cdot N_n \cdot T_e \tag{1}$$

where N_n is the neutral number density and T_e the electron temperature.
     The non-deviative absorption of a vertically propagating EM signal is to a
fair approximation given by the Appleton formula :

$$A \ (dB) = 4.5 \cdot 10^{-5} \cdot \int \frac{N_e \cdot \nu_e}{\nu_e^2 + \omega^2} \cdot dh \tag{2}$$

where $N_e(h)$ and $\nu_e(h)$ are the height dependent values of electron density
and collision frequency respectively while ω is the angular wave frequency. For
the E-region and the regions below the collisions with ions may be neglected

(Boström 1973) so there ν_e in (2) equals ν_{en} given by equation (1). The integration of (2) should be performed over all ionospheric heights.

With the exception of very intense precipitation events the radar at Sdr. Strømfjord, however, is not capable of measuring electron densities and temperatures below heights of appr. 90 km . The contributions to the total absorption from the different height ranges should therefore be considered very carefully for comparisons of measured and calculated values to be meaningful. It is a major requirement that D-region absorption, produced by energetic particle precipitation, is very weak. During the time of the intense electron heating event i.e. 16-18 UT on August 12 there are several indications (see below) that the energetic particle precipitation at the station was indeed at a very low level.

Using the values of T_e derived from the radar observations and the atmospheric model for 60⁰ N and the month of July from US Standard Atm. Suppl. (1966) makes it possible to calculate values of ν_{en} according to (1). With these values and the observed values of N_e the integration of (2) could be performed over the height range where the radar measurements produce reliable data - in this case from approx. 90 km to 150 km.

Height profiles for the calculated specific absorption A_s at 30 Mhz (in dB/km) are shown in the right-hand sections of Figures 4 and 5 at times of approx. 1547 and 1622 UT respectively. From Figure 4 to Figure 5 the increase in total absorption related to the electron heating event is readily observed.

The measured 30 Mhz absorption intensities and the calculated values derived from integration over the E-region from 90 to 120 km for all available temperature and density profiles in the period from 15 to 19 UT are shown in Figure 6.

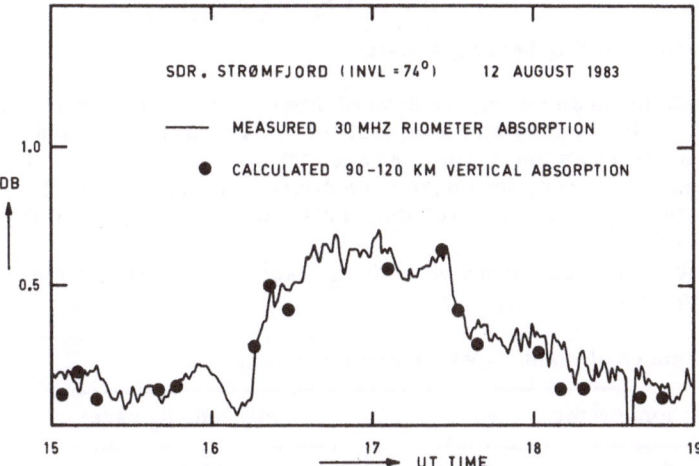

Figure 6. Comparison of measured 30 Mhz riometer absorption intensities and calculated absorption values derived from observed electron density and temperature profiles in the 90 to 120 km height range.

The calculated values fit remarkably well to the measured riometer absorption intensities. The largest deviations are seen after 18 UT, where a small solar flare event adds a D-region contribution of approx. 0.1 dB to absorption intensities.

For the absorption values shown in Figure 6 it should be noted, that the steady amount of D-region absorption (appr. 0.5 dB) produced by the normal solar

UV and X-ray emissions is neglected in both cases.

The assumed absence of energetic particle precipitation during the heating event is substantiated by the observations of the low and rather uniform values of f_{min} from 16 to 20 UT shown in Figure 3. In addition 30 Mhz riometer recordings (not shown) from Thule close to the magnetic pole show less than 0.1 dB of absorption from 12 to 18 UT indicating the absence of PCA activity. Riometer recordings (not shown) from Angmagssalik and Narssarssuaq in the auroral zone also show less than 0.1 dB of absorption at 30 Mhz from 12 to 18 UT so there is most likely no auroral absorption event in progress. Furthermore, two passes of the NOAA-7 satellite over Greenland occurred during this time. Table I shows the satellite positions and values of the measured energy flux of electrons in the 0.3-20 keV range at the orbital intersect of the inv. latitude of Sdr. Strømfjord (67.0°N, 309.1°E.)

TABLE I   NOAA-7   0.3-20 keV Electron Energy Flux. August 12, 1983

UT Time	Latitude	Longitude	Inv. Lat.	Energy Flux
1547	69.80°N	325.51°E	74.00°	.20 ergs/cm^2 sec.
1727	65.42°N	305.20°E	74.00°	.11 ergs/cm^2 sec.

The second pass was almost overhead Sdr. Strømfjord in the middle of the intense heating event at a time where the riometer absorption was 0.58 dB. The measured energy flux of 0.11 ergs/cm^2 sec is very weak indeed compared to the intensities of several ergs/cm^2 sec usually measured during moderate auroral events.

3.2. SEC observations during heating events

From the observations made during the days of August 12, 13, 15 and 18 a total of 144 samples of E-region density and temperature profiles were recorded by the radar system. In 66 cases an ionogram was taken within the radar observing period of 6 minutes. These 66 observations have been divided in groups according to electron temperature at a height of 105 km where the electron heating appears to be most pronounced.

Table II shows the occurrences of slant E_s traces related to the electron temperature at 105 km.

TABLE II Occurrences of slant E_s vs. electron temperature

T_e (105 km)	Ionograms	s	S	all slant E_s cases	
200-300°K	30	3	0	3	10%
300-400°K	17	3	3	6	35%
400-800°K	13	3	1	4	31%
800-1600°K	4	1	2	3	75%
>1600°K	2	0	1	1	50%

There is a clear tendency for the slant E_s traces to occur more frequently during heating events than during quiet conditions. Some cases, however, do occur at temperatures close to the undisturbed level (appr. 220°K) while on the other hand in some of the more intense heating events the slant E_s traces were not discernible.

The occurrences of lacunae divided according to 105 km electron temperature as shown in Table III are clearly related to the degree of electron heating.

TABLE III. Occurrence of lacunae vs. electron temperature

T_e (105 km)	Ionograms	y	Y	Y̲		all lacunae
200-300°K	30	0	0	0	0	0%
300-400°	17	4	0	0	4	24%
400-800°	13	9	0	0	9	69%
800-1600°	4	1	1	2	4	100%
>1600°	2	0	0	2	2	100%

For temperatures less than 300°K the lacuna effect is not discernible. Above 300°K the lacuna is seen in more than 50% of all ionograms. Above 800° all ionograms show lacuna usually with a strong effect as indicated by capital Y and Y̲ with underscore.

The lacuna phenomenon thus seems to be very directly related to the E-region electron temperature. Also the grading of lacuna intensities from weak events, showing damped signals at the E-F transition or a small gap only, through cases with a distinct gap to events of total lacuna, corresponds well to increasing levels of electron temperatures.

This behaviour suggests that enhanced deviative E-region collisional absorption occurring during electron heating events is responsible for the lacuna phenomenon associated with SEC events.

4. CONCLUSIONS

Use of the incoherent scatter radar at Sdr. Strømfjord has revealed E-region heating events associated with the occurrence of intense horizontal electric fields in the noon sector at cusp latitudes. The observed temperatures were similar to or larger than those found by Schlegel and St.-Maurice (1981) and Wickwar et al. (1981) in heating events observed in the morning and evening sectors of the auroral oval.

Slant E_s traces and lacunae which are common signatures of SEC events showed a markedly increased occurrence frequency during electron heating events. The intensity of the lacuna condition was well correlated with the electron temperature at the E-layer density peak at a height of appr. 105 km.

Weak, fluctuating cosmic noise absorption typical of SEC events was recorded during all observed electron heating events. In a particular case an intense heating event was observed while NOAA-7 satellite observations and other measurements indicated that the energetic particle precipitation was very weak. Collisional E-region absorption intensities computed from the observed electron density and temperature profiles and a standard neutral atmospheric model showed for this case a remarkable agreement with the observed cosmic noise absorption intensities.

The close correlation of ionogram lacuna intensities with E-region electron temperatures and the agreement between calculated and observed cosmic noise absorption intensities suggest that excessive absorption in the E-region caused by the enhanced electron-neutral collision frequencies during electron heating events may explain cases of lacuna and events of E-field related cosmic noise absorption observed in polar regions.

ACKNOWLEDGEMENTS. The use of NOAA-7 data generously supplied by D.S. Evans at Space Environment Laboratory, NOAA, Boulder and magnetometer data supplied by E. Friis-Christensen, Danish Meteorological Institute, is gratefully acknowledged. Also we wish to express our sincere appreciation of the support and assistance received from J. Kelly and members of the staff at Radiophysics Laboratory at SRI International and from the staff headed by F. Steenstrup at the incoherent community at Sdr. Strømfjord.

REFERENCES
Bahnsen, A., Ungstrup, E., Fälthammar, C.-G., Fahleson, U., Olesen, J.K., Primdahl , F., Spangslev, F. and Pedersen ,A. 1978, J. Geophys. Res., 83, 5191
Boström, R. 1973, p. 181 in Cosmical Geophysics , (Eds. A. Egeland, Ø. Holter and A. Omholt). Universitetsforlaget, Oslo, Norway.
D'Angelo, N. 1976, J. Geophys. Res., 81, 5581.
D'Angelo, N. 1978, Ann. Geophys., 34, 51.
D'Angelo, N. 1981, Ann. Geophys., 37, 417.
De La Beaujardiere, O. 1980, The Software System for the Chatanika Incoherent-Scatter Radar. SRI International, Menlo Park, California.
Farley, D.T., Ierkic, H.M. and Fejer, B.G. 1981, J. Geophys. Res., 86, 1569
Fejer, B.G. and Kelley, M.C. 1980, Rev. Geophys. Space Phys., 18, 401.
Heppner, J.F., Burne, E.C. and Belon, A.E. 1952, J. Geophys. Res., 57, 121.
Iversen, I.B., D'Angelo, N. and Olesen, J.K. 1975, J. Geophys. Res., 80, 3713.
Kelly, J. 1983, Geophys. Res. Lett., 10, 1112
Mehta, N.C. and D'Angelo, N. 1980, J. Geophys. Res., 85, 1779
Olesen, J.K. 1972, p. 27.1 in Radar Propagation in the Arctic, - AGARD-CP-97, (Ed. J. Frihagen), NATO, Paris.
Olesen, J.K. 1975, p. 227 in High Latitude Supplement to the URSI Handbook on Ionogram Interpretation and Reduction, UAG-50, (Ed. W.R. Piggott) WDC-A, NOAA, Boulder.
Olesen, J.K. and Rybner, J. 1958, p. 37 in Sporadic E Ionization, AGARDOGRAPH 34, (Ed. B. Landmark), NATO, Paris.
Olesen, J.K., Primdahl, F., Spangslev, F., Ungstrup, E., Bahnsen, A., Fahleson, U., Fälthammar, C.-G. and Pedersen, A. 1976, Geophys. Res. Lett. 3., 711
Phelps, A.V. 1960, J. Appl. Phys., 31, 31, 1723.
Phelps, A.V. and Pack, J.L. 1959, Phys. Rev. Letters, 3, 340.
Primdahl, F., Olesen, J.K. and Spangslev, F. 1974, J. Geophys. Res., 79, 4262.
Schlegel, K. and St.-Maurice, J.P. 1981, J. Geophys. Res., 86, 1447.
Siren, J.C. 1982, J. Geophys. Res., 87, 6271.
St.-Maurice, J.P., Schlegel, K. and Banks, P.M. 1981, J. Geophys. Res., 86, 1453.
Vassal, J. 1971, Etude de certaines Phenomenes D'Absorption ionospherique anormale, (These), Laboratorie de Geophysique Externe, St. Maur, France.
Vassal, J., Berthelier, J.J., Lavergnat, J. and Sylvain, M. 1976, J. Atm. Terr. Phys., 38, 1289.
Wickwar, V.B., Lathuillere, C., Kofman, W. and Lejeune, G. 1981, J. Geophys. Res., 86, 4721.

OBSERVATIONS OF THE CUSP WITH TOPSIDE SOUNDERS

D.B. Muldrew

Communications Research Centre
Department of Communications
P.O. Box 11490, Station H
Ottawa, Ontario K2H 8S2

ABSTRACT. Topside sounders observe the most intense spread F anywhere in
the ionosphere when they are within the cusp region. In terms of spread
F, the cusp usually has a sharp equatorward and poleward boundary.
Aspect sensitive scatter occurs from the field-aligned irregularities
within these boundaries as the satellite approaches, traverses, and
departs from the cusp region. This scatter is from small-scale irregularities with cross-field periodicity of about one half the radio frequency wavelength, i.e. tens of meters. Occasionally when the satellite
is within medium-scale irregularities (hundreds of meters) ducted echoes
are observed which indicate that some of these irregularities are
continuous along the magnetic field from the satellite to the reflection
height — typically a distance of hundreds of kilometers. Electron
density profiles obtained from ionograms recorded through the cusp region
show an increase in scale height which implies a large increase in plasma
temperature. Above about 600 km these profiles usually show a significant electron density increase extending over a few degrees in latitude.
Enhanced vertical temperature gradients at 1000 km height may imply a
heat source, other than precipitating particles, located above this
height. The soft electron precipitation in the cusp is responsible for
emission of electromagnetic radiation by wave-particle interaction.

1. CUSP SPREAD F (SMALL-SCALE IRREGULARITIES)

Severe spread F in the cusp region was first observed with the Alouette 1
sounder (Petrie 1962). Dyson and Winningham (1974) note that the
small-scale irregularities responsible for the spread F occur up to
altitudes of at least 3000 km and that the equatorward edge is quite
abrupt. Using the particle detectors aboard ISIS they found that the
equatorward boundaries of spread and of the ≤ 300 eV electron
precipitation coincide. The scatter is aspect sensitive with respect to
magnetic field-aligned irregularities. For backscatter occurring near
the satellite the irregularities have a Fourier wavelength component
perpendicular to the magnetic field equal to one half the radio
wavelength, i.e. tens of meters (small-scale irregularities). Scatter

J. A. Holtet and A. Egeland (eds.), The Polar Cusp, 377–386.
© *1985 by D. Reidel Publishing Company.*

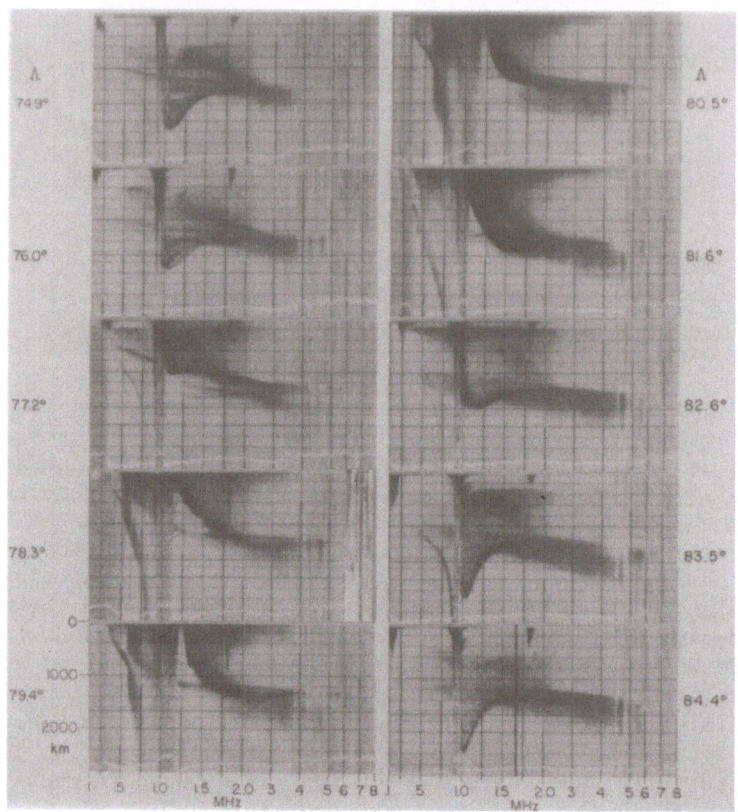

Figure 1.   ISIS 2 ionograms recorded as the satellite approaches, passes through, and departs from the cusp region.   After Whitteker (1976).

occurs from throughout the region and not just at the boundary.

Whitteker (1976), in a review of topside sounding of the cusp region, presents a series of ionograms recorded by ISIS 2 as it approaches, traverses, and recedes from the cusp region.   This series is reproduced in Figure 1.   In the first ionogram it can be seen that there are almost no small-scale irregularities for a horizontal distance of a few hundred kilometers away from the satellite.   This ionogram also shows that there is a sharp (presumably field-aligned) boundary followed by an extensive region of irregularities.   As the satellite moves poleward the next two ionograms show the boundary closer to the satellite.   Between about 78° and 82° invariant latitude, the satellite is within the region of small-scale irregularities.   Beyond about 82°, the satellite moves away from a sharp boundary corresponding to the poleward side of the cusp region.

2.   MEDIUM-SCALE IRREGULARITIES

Figure 2 is from Muldrew (1970) and shows an ISIS 1 ionogram recorded
when the satellite was inside the cusp irregularity region. In addition
to the intense aspect-sensitive scatter occurring at and near the
satellite height (2010 km) there is intense spread along the X wave
reflection trace up to about 5.0 MHz. This is probably due to a
combination of aspect-sensitive scatter from irregularities near
reflection and multimode propagation of trapped waves in magnetic
field-aligned irregularities hundreds of kilometers long (Calvert and
Schmid 1964). These irregularities are probably several wavelengths
(i.e. from about 100 m to about 1 km) in diameter and can be considered
to be medium-scale irregularities. The cleaner trace extending from 5.0
to 8.4 MHz results from echoes reflected near the peak of the F layer,
equatorward of the irregularity boundary. Here the penetration frequency
of the F layer is higher than inside the cusp irregularity region.

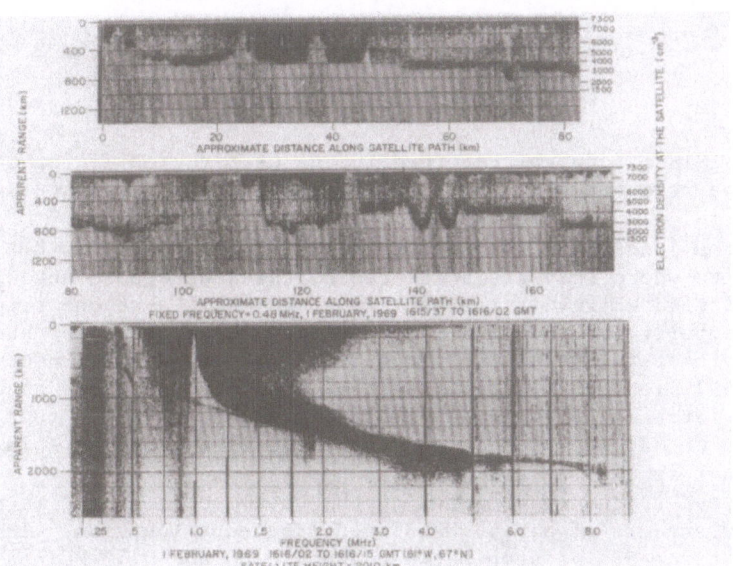

Figure 2. Fixed-frequency and swept-frequency ISIS 1 ionograms.
Variations in the apparent range of the Z wave trace indicate variations
in electron density along the satellite path. After Muldrew (1970).

The continuity of the field-aligned medium-scale irregularities for
at least hundreds of kilometers along the magnetic field is further
established by a remarkable feature which occurs in the nighttime auroral
zone (Muldrew 1983) and in the cusp region. Figure 3 shows a cusp
ionogram with a two-hop trace (Muldrew and Hagg 1969). This trace can be
seen from 3.3 to 4.2 MHz and from 600 to 1000 km. Waves are trapped in
irregularity ducts, propagate downward, are reflected from the denser

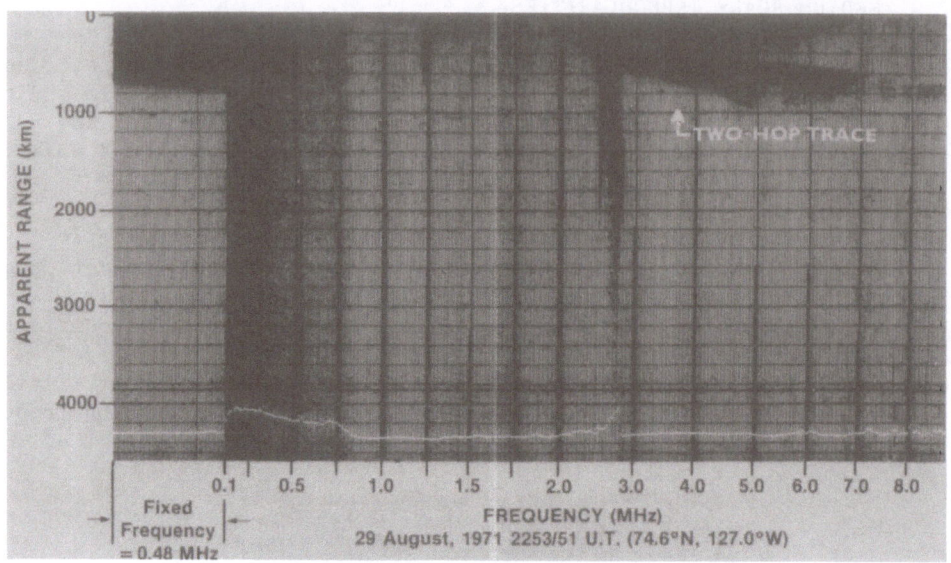

Figure 3.   ISIS 1 ionogram recorded in the cusp region.   A two-hop X wave trace can be seen between 3.3 and 4.2 MHz.   Satellite height is 663 km.

region of the ionosphere below the satellite, and return in the ducts to the satellite where they are scattered by the long dipole antenna; some of the scattered energy is trapped again and makes a second trip.   The ducts are tubular, or rod like, and probably less than a few hundred meters in diameter (Muldrew and Hagg 1969).   By using the theory of Farley (1959), it can be determined that the electric field associated with irregularities having cross-field dimensions of a few tens of meters or more can be mapped upward along the field line from the F layer peak to the magnetosphere.

The swept-frequency ionogram at the bottom of Figure 2 is preceded by about 30 seconds of a 0.48 MHz fixed-frequency ionogram.   The Z-wave reflection trace can be seen between about 0.40 and 0.50 MHz on the swept-frequency ionogram and this trace can also be observed on the fixed-frequency ionogram.   The variations in the apparent range of the Z trace are caused by variations in the electron density encountered by the satellite.   The magnitude of these variations can be estimated by making certain assumptions (see Muldrew 1970).   The estimated electron density at the satellite height, as determined by the apparent range of the Z trace, is given along the right-hand side of Figure 2.   Superimposed on a slowly-varying background density are variations which can be as great as 75% in 200 m (e.g. at 35 km distance along the satellite path) or greater than 90% in 400 m (e.g. at 126 km).   The 200 m resolution is determined by the distance the satellite travels between transmitted pulses (i.e. between scan lines).   The density excursions can be positive or negative; however the positive excursions are more numerous.

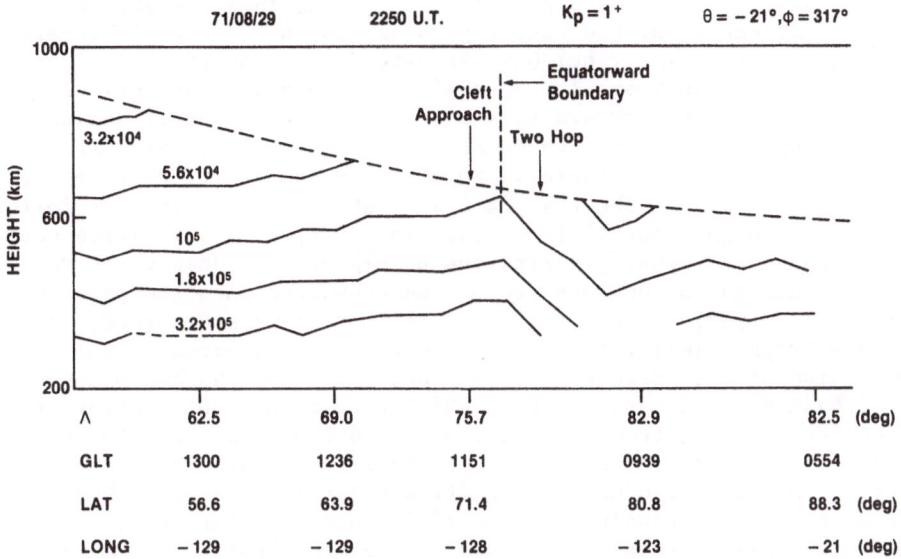

Figure 4. ISIS 1 electron-density (cm^{-3}) contours as a function of
height and invariant latitude. The geomagnetic local time, geographic
latitude and longitude are given along the abscissa. The dashed line
gives the satellite height. The date, time, Kp and (θ,ϕ) for the IMF
direction (King 1975) are given at the top. Figure 3 was recorded at the
point labelled "two hop".

3.   LARGE-SCALE IRREGULARITIES

The electron density at a given height through the cusp region varies on
a scale of hundreds of kilometers. Figures 4-6 are not typical, e.g. see
Nelms (1966) and Hruška ('1973). Electron density contours are plotted as
a function of height and latitude throughout the cusp region.
       In Figure 4, the equatorward boundary of intense irregularities is
shown by the dashed line. On the ionogram recorded at the point labelled
'cusp approach', there is a sharp spread F cut off corresponding to the
irregularity boundary. In Figure 4, the large-scale irregularity
structure (hundreds of kilometers) associated with the cusp seems to
start near this boundary and extend poleward. Between 300 and 600 km
height there is a considerable decrease in electron density extending
over about 7° of invariant latitude, especially at the lowest heights.
The decrease between about 300 and 450 km is roughly a factor of 3.
       Figure 5 was obtained using ISIS 2 data. This satellite has a
circular orbit near 1400 km. At this height, the equatorward spread F
irregularity boundary is sometimes difficult to determine. It occurs
somewhere between about 70° and 72° invariant latitude. Poleward of the
boundary and below 600 km it can be seen that again there is a decrease
in electron density. The density decreases by about a factor of two at
300 km. However, above about 800 km the density increases and near 1300

km the increase is about a factor of two. At the lower heights the
large-scale structure of the cusp region extends over about 4° invariant
latitude and at the higher heights, it extends over about 9°.

Figures 4, 5 and 6 are for Kp's of 1+, 2+ and 4o, respectively. In
Figure 6 the sharp equatorward spread F boundary was clearly defined and
from the ionograms was found to be at 73° invariant latitude. Again just
poleward of this boundary there is an increase in density above about 500
km height and extending about 4° in invariant latitude. This is followed
by a density trough about 4° in width. In addition to this large-scale
structure there are density variations at higher latitudes which were
probably generated in the cusp region and convected into the polar cap
(Kelley et al. 1982). Two other cases of Kp = 4o were examined; in one
of these a trough similar to that of Figure 6 was observed on the
poleward side of the increase. This trough could be the one observed by
Foster (these Proceedings) using the Millstone incoherent scatter radar.

The large-scale structure in Figures 4-6 cannot be said to be
typical. In some cases almost no large-scale structure is observed.
However, more often than not, a significant density increase above about
600 km is observed extending about 4-7° in invariant latitude.
Large-scale density variations of the F layer peak (near 300 km) are well
illustrated by Whitteker (1976).

It is difficult to say if the increase usually observed above 600 km
is related to the one observed by Kelly (these Proceedings) using the
Sondrestrom incoherent radar. He observed an increase over about 5° of
latitude which is in good agreement with the above. However, the
observed east-west extent with the radar is about 1 hour in local time
near noon whereas the satellite observations show increases at various
times throughout the daytime.

4.   AVERAGE CUSP CHARACTERISTICS

Titheridge (1976) was able to determine the mean plasma temperature $\frac{1}{2}$(Te
+ Ti), the vertical plasma temperature gradient, and the transition
height at which oxygen and hydrogen ions have equal densities, by
matching observed and theoretical electron density profiles. Good
matching could only be obtained if a temperature gradient was assumed; it
could not be obtained, for zero temperature gradient, from the effects of
helium ions or by assuming vertical ion fluxes.

Figure 7 shows Titheridge's results for winter daytime. The plasma
temperature in the cusp region shows an increase of about 25% at 400 km
and about 50% at 1000 km height. The vertical temperature gradient,
transition height, and plasma frequency at 1000 km, also show a
significant increase. Titheridge concludes, presumably by assuming Te =
Ti, that the total downward heat flux from a source above 1000 km is
increased by a factor of 3 in the cusp region.

The results for the summer daytime cusp are similar to Figure 7.
However, the peaks in the cusp region in the temperature, vertical
temperature gradient, transition height and plasma frequency have
somewhat smaller amplitude and are wider than winter daytime. The
increase in plasma temperature occurs at all local times throughout the

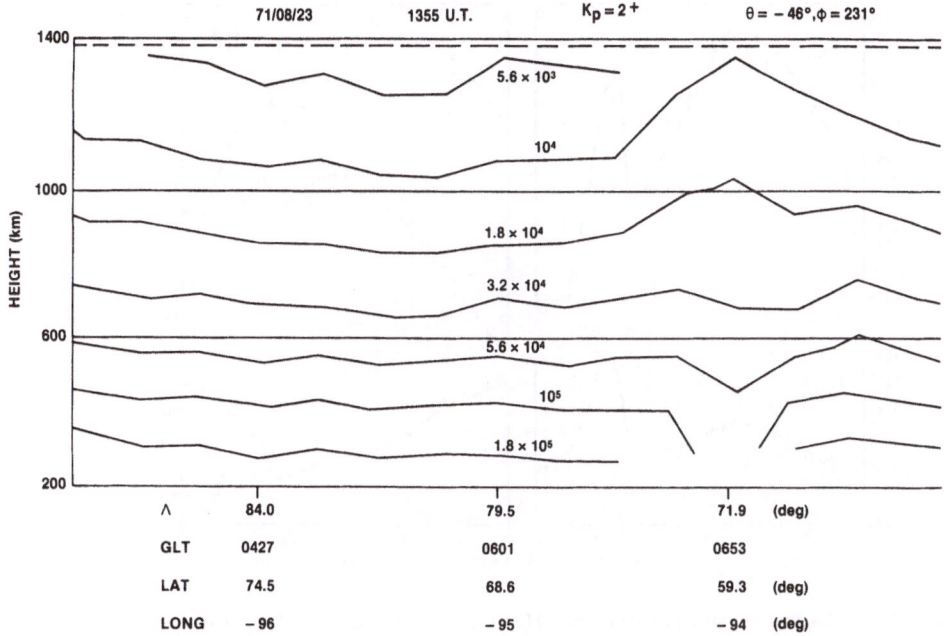

Figure 5.   ISIS 2 electron-density contours for Kp = 2+.   See Figure 4.

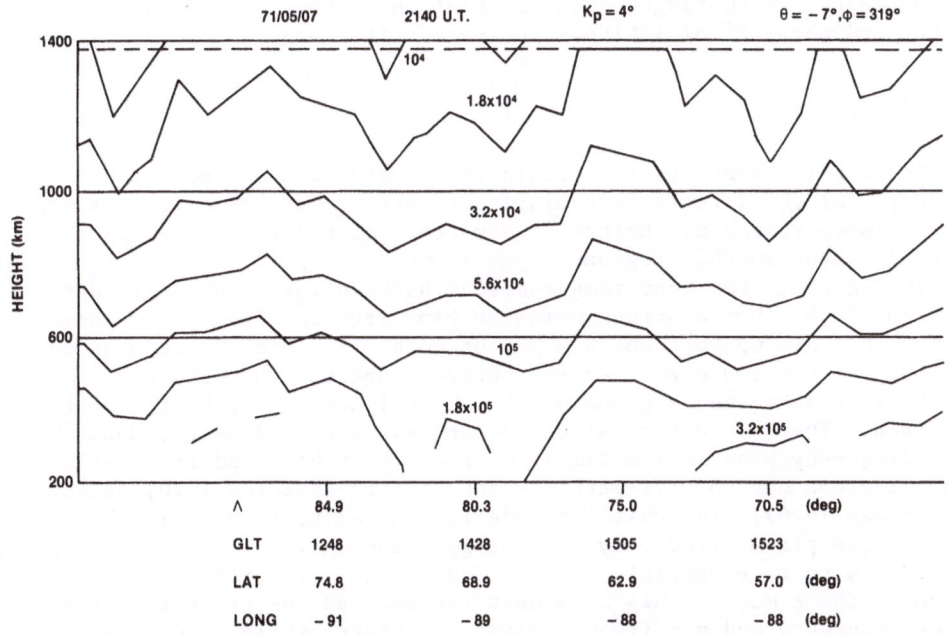

Figure 6.   ISIS 2 electron-density contours for Kp = 4o.   See Figure 4.

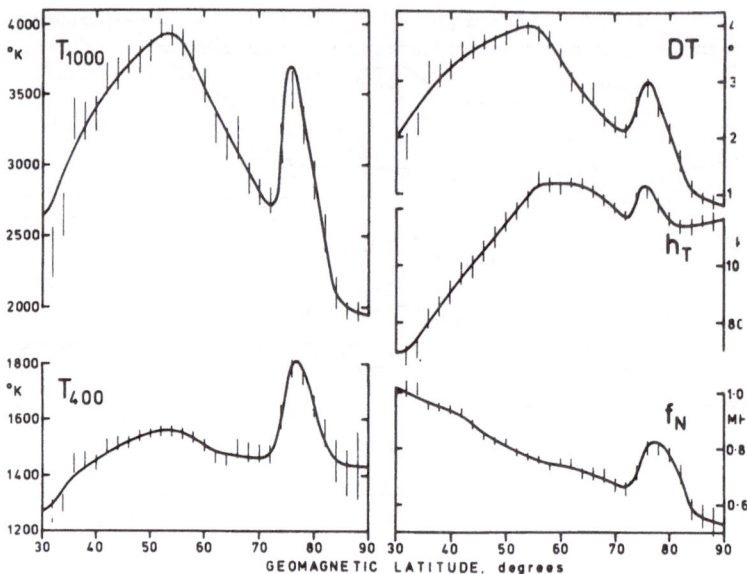

Figure 7. Plasma temperature at 1000 km and at 400 km, vertical temperature gradient, oxygen-ion/hydrogen-ion transition height, and plasma frequency at 1000 km as a function of latitude for winter day conditions and Kp ≤ 3. After Titheridge (1976).

daytime hours. Titheridge estimates that the cusp region moves equatorward about 2° of latitude for each unit change in Kp.

5.   NATURAL EMISSIONS

When the topside sounders are within the equatorward and poleward boundaries of the intense irregularities, strong natural emissions are nearly always received. Holtet (these Proceedings) also observed low frequency waves in this region.

In Figure 1, the ionograms recorded between invariant latitudes of 77.2° and 81.6° show a strong emission band from the lower frequency limit of 0.1 MHz up to about 0.3 MHz or more. This can be seen as a darkening of the ionogram over the entire range of apparent range and in the AGC voltage which is given by the white lines at the bottom of the ionograms. These ionograms also show an emission band near 1 MHz. The lower frequency band is sometimes referred to as hiss and is whistler mode emissions due to wave-particle interaction (Muldrew 1970; James 1973; Maggs 1976). The whistler mode is limited to frequencies below the lower of the plasma frequency and the gyrofrequency. The upper frequency band is due to wave-particle interaction causing emission in the upper branch of the Z mode. This band must lie between the greater of the plasma frequency and gyrofrequency and the upper-hybrid frequency.

Whistler-mode hiss can also be observed in Figure 2. These

emissions are somewhat unusual in that three bands can be observed in the
swept-frequency ionogram: 0.10 - 0.25 MHz, 0.37 - 0.45 MHz and 0.53 -
0.57 MHz. If these bands are harmonically related, the harmonics are
generated naturally since an examination of the amplitude scans and of
the AGC indicate that the signal in the lower frequency band is not
sufficiently intense to generate harmonics in the receiver. The
emissions at 0.48 MHz are relatively weak but can be seen in the
fixed-frequency ionogram. Variation with distance and/or time can be
seen as the satellite moves through the cusp region.

6.   DISCUSSION

Energetic particles precipitating into the ionosphere can generate an
increase in electron density by direct ionization. However, the heating
produced by these particles probably has a greater effect on the electron
density distribution (Whitteker 1977). The variability, in time and
space, of the particle precipitation will produce horizontal (or
cross-magnetic field) gradients in electron density. Examples of these
gradients are shown in Figure 2. The particle-produced gradients can
generate small-scale irregularities, responsible for spread F, through
the gradient-drift instability (Kelley et al. 1982) since strong
convection occurs throughout this region.
    The energy of the precipitating electrons in the cusp region is of
the order of 100 eV (Winningham 1972; these Proceedings). Precipitation
is thought to be responsible for the emission of whistler and Z-mode
radiation. Enhanced whistler emission bands can be seen near 37 and 143
km distance in the fixed-frequency ionogram of Figure 2. The darkening
of the background (e.g. see near 37 km) indicates a true increase in the
natural radiation level at 0.48 MHz and is not due to a decrease in the
AGC voltage of the receiver. Increased signal levels seem to correlate
with density variations. This may be due to ducting of the whistler
waves along the field aligned variations, it may be that the radiation
intensity is a function of plasma frequency, or it may be that the
increased electron precipitation is responsible for both the enhanced
radiation and the density deviations. For the last case it should be
noted that the formation of density variations lags the causative
precipitation by about 10 minutes (Whitteker 1977).
    The increase in plasma temperature gradient at 1000 km height
deduced by Titheridge (1976) is another intriguing characteristic of the
cusp region. If both electron and ion temperature gradients are
increased, this indicates that the major heat source for the cusp region
is above 1000 km. Whitteker (1977) has shown that the energy deposition
by the soft cusp electrons occurs below 1000 km. Lemaire (1979; these
Proceedings) proposes that plasma irregularities in the solar wind
(plasmoids) penetrate into the magnetosphere. Their excess kinetic
energy would produce a polarization field across the irregularity and
consequently field-aligned currents and Joule heating in the E region
would result. The thermal particles of the plasmoids (about 10 eV
electrons and 100 eV protons) would precipitate along field lines at cusp
latitudes and also deposit their energy below 1000 km. At greater

heights where the electron density is low, the electron velocity
responsible for the field-aligned currents could exceed the electron
thermal velocity resulting in current-driven instabilities (Kindel and
Kennel 1971). This would occur at a height of a few thousand kilometers
where densities of less than 3×10^7 m^{-3} are common (Timleck and Nelms,
1969) and plasma temperatures are about 10^4 K. A streaming velocity of
the electrons relative to the ions of about 6×10^5 m/s (i.e. a current
less than about 3×10^{-6} A/m^2) would be required. At still greater
heights this instability would not occur because the drop in density
would not be sufficient to compensate for the increased temperature and
reduced current density (due to magnetic field divergence). Landau
damping of the waves generated by the instability would heat the plasma
electrons and ions and this heat would be conducted downward.

In conclusion, various cusp phenomena such as ionization
irregularities, wave emissions and heat flow are apparently related to
the intense precipitation of soft electrons and ions in the cusp region.

REFERENCES

Calvert, W., and Schmid, C.W. 1964, J. Geophys. Res., 69, 1839.
Dyson, P.L., and Winningham, J.D. 1974, J. Geophys. Res., 79, 5219.
Farley, D.T., Jr. 1959, J. Geophys. Res., 64, 1225.
Hruška, A., McDiarmid, I.B., and Burrows, J.R. 1973, J. Geophys. Res.,
78, 2311.
James, H.G. 1973, J. Geophys. Res., 78, 4578.
Kelley, M.C., Vickrey, J.F., Carlson, C.W., and Tobert, R. 1982, J.
Geophys. Res., 87, 4469.
Kindel, J.M., and Kennel, C.F. 1971, J. Geophys. Res., 76, 3055.
King, J.H. 1975, Interplanetary Magnetic Field Data, Report UAG-46, World
Data Center A., NOAA, Boulder, Colorado.
Lemaire, J. 1979, Proc. Magnetospheric Boundary Layers Conf., Albach,
11-15 June 1979, ESA SP 148, 365.
Maggs, J.E. 1976, J. Geophys. Res., 81, 1707.
Muldrew, D.B. 1970, p. 786 in Space Research X, North-Holland, Amsterdam.
Muldrew, D.B. 1983, Radio Sci., 18, 1140.
Muldrew, D.B., and Hagg, E.L. 1969, Proc. IEEE, 57, 1128.
Nelms, G.L. 1966, p. 358 in Electron Density Profiles in Ionosphere and
Exosphere, (Ed. Jon Frihagen), North-Holland, Amsterdam.
Petrie, L.E. 1963, Can. J. Phys., 41, 194.
Timleck, P.L., and Nelms, G.L. 1969, Proc. IEEE, 57, 1164.
Titheridge, J.E. 1976, J. Geophys. Res., 81, 3221.
Whitteker, J.H. 1976, J. Geophys. Res., 81, 1279.
Whitteker, J.H. 1977, Planet. Space Sci., 25, 773.
Winningham, J.D. 1972, in Earth's Magnetospheric Processes (Ed. B.M.
McCormac), Reidel, Dordrecht.

DEFINITION OF THE CUSP

Walter J. Heikkila
Center for Space Sciences
The University of Texas at Dallas
P.O. Box 830688
Richardson, Texas 75083-0688   USA

ABSTRACT. The history of the dayside or polar cusp is reviewed
briefly, beginning with the work of Chapman and Ferraro, Axford and
Hines, and Dungey. Solar wind plasma comes down the cusp to low
altitudes; the remarkably broad area over which this penetration
occurs prompted the use of the term cleft, apparently at the foot of
the entry layer inside the magnetopause. The consensus of discussion
at the workshop was that cusp should refer mainly to the narrow
feature in the magnetic profile. Cleft would refer to the broader
region defined by the solar wind plasma (perhaps somewhat modified
by acceleration or retardation processes); distinctive features
observable in the ionosphere or on the ground are photo-emission,
enhanced plasma density and temperature, plasma convection,
electromagnetic waves, and turbulence.

1. HISTORY
     The dayside or polar cusp (or cleft) plays a key role in the
physics of the high latitude dayside ionosphere and magnetosphere,
as well as providing a window to the magnetopause.
     The concept of a cusp in the magnetic field topology on the
dayside magnetopause began with the work of Chapman and Ferraro
(1931), whose illustration is shown in Figure 1. They assumed no
magnetic field in the interplanetary medium (no IMF) on the right
side of their figure; consequently, the geomagnetic field lines
inside the magnetopause will have to be diverted away from two
points labelled Q, with field lines at low latitudes closing on the
dayside while field lines at higher latitudes close over the poles
in the magnetotail. Thus, a cusp is formed in the magnetic field
line profile.
     Since there is no magnetic pressure at the points Q, a solar
wind with finite pressure can force penetration across the magnetopause. Chapman and Ferraro did not discuss this aspect, instead
taking the solar plasma stream as a plasma with perfect conductivity as befits a magnetohydrodynamic model of a plasma.

387

J. A. Holtet and A. Egeland (eds.), The Polar Cusp, 387–395.
© *1985 by D. Reidel Publishing Company.*

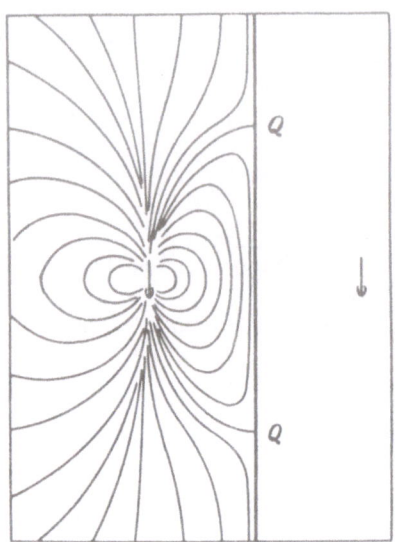

Figure 1. Chapman and Ferraro's (1931) figure of a perfectly
conducting plasma with no magnetic field advancing on a magnetic
dipole, producing two cusps QQ.

     Axford and Hines (1961) in proposing a viscous model of solar
wind-magnetosphere interaction also assumed no IMF, and qualitatively they had a similar cusp but in three dimensions. They assumed
that a viscous-like process (which they left undefined) somehow
imparted solar wind particles, momentum, and energy to drive magnetospheric processes such as large scale plasma convection and
energization of particles. They referred to the region at high
latitudes on the dayside as the zone of confusion,aptly put in view
of recent observations and discussions!
     Dungey (1961) in a paper which appeared in the same year
pointed out that an interplanetary magnetic field, quite rightly in
retrospect, would affect solar wind-magnetosphere interactions.He
surmized that a southward IMF would be conducive for such interactions because there would be X-type magnetic neutral lines near the
equatorial plane near the sub-solar point and in the magnetotail, as
shown in Figure 2 (after Cowley, 1980). Radiating out from this X-
line there is a multiply connected surface (the separatrix)
dividing the magnetic field lines into (1) interplanetary (no feet
on the ground),(2) closed, with two feet on the ground, and (3)
open, with only one foot on the ground while the other end reaches
into the solar wind, as indicated in Figure 2 by the solid lines.
The open field lines cross the magnetopause, with a small normal
component. The electric field is everywhere directed from dawn to
dusk, as appropriate for plasma convection into the dayside X-line
in the equatorial plane from both sides; the only possible exit is
onto open field lines over the poles. In this reconnection process
the plasma gains momentum in high speed jets, but also energy

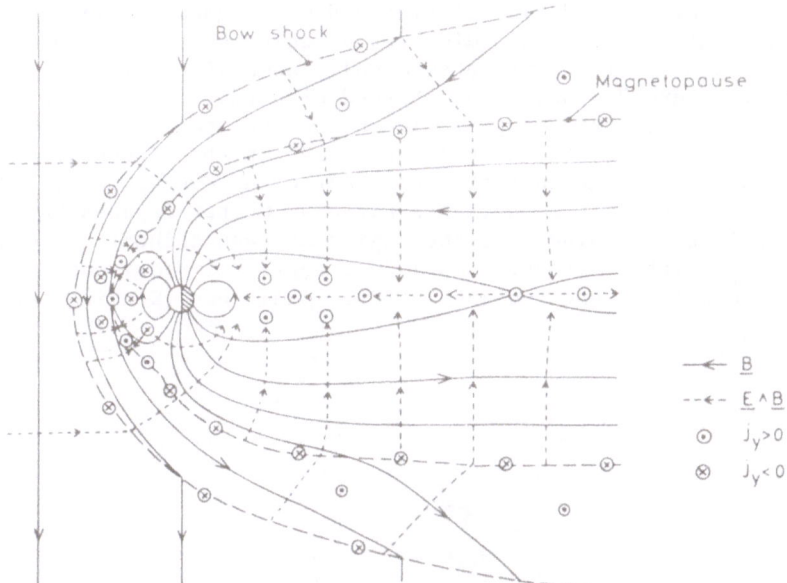

Figure 2. The reconnection model of the magnetosphere (after Cowley, 1980) for a southward interplanetary magnetic field. The magnetic field lines are shown solid. The electric field is everywhere from dawn to dusk, and the direction of $\underline{E} \times \underline{B}$ is shown by the short-dash lines. The circled dots and dashes indicate the direction of various currents, and also $\underline{E}.\underline{J} > 0$ (dots) and $\underline{E}.\underline{J} < 0$ (crosses), where the plasma gains or loses energy.

since $\underline{E}.\underline{J}$ is positive near the X-lines. Over the lobes of the magnetotail the direction of the magnetopause current is reversed, and thus the plasma loses energy, the idea being that this generator supplies the reconnection load, both at the dayside magnetopause as well as in the tail.

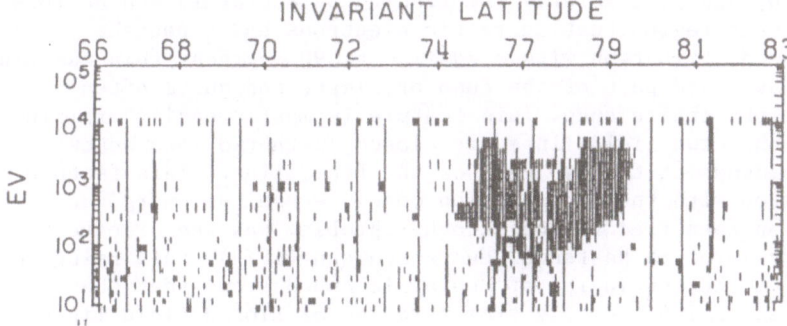

Figure 3. A spectrogram of magnetosheath ions penetrating to low altitudes over the invariant latitude range 75°-79.5°.

Heikkila et al. (1970), Heikkila and Winningham (1971), and Frank and Ackerson (1971) did discover a region that appeared to be shocked solar wind (i.e., magnetosheath) plasma with low altitude satellites; Figure 3 shows data obtained with ISIS-1. The ion flux appeared to be energized, especially at the low and high latitudes edges, somewhat as predicted by MHD theories of the reconnection process. This led me at this time to endorse the Dungey model, as shown by Figure 4. This is a rather naive view, but it does have some merit in emphazising that the region of magnetosheath plasma penetration is rather broad, some 1 to 5 degrees in latitude and 8 hours in local time centered just before local noon (Winningham, 1972).

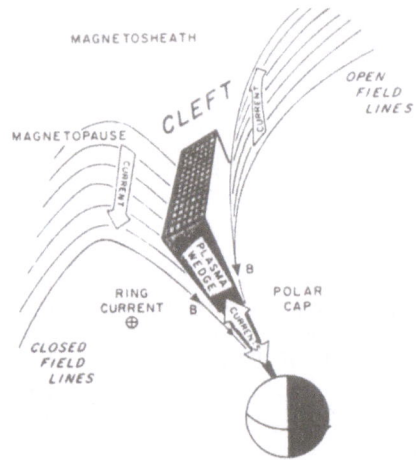

Figure 4. Introduction of the term cleft was prompted by the rather broad region of magnetosheath plasma penetration with apparently free entry. We now know that this entry comes by way of the entry layer, as indicated in Figure 5.

Shortly thereafter, the Dungey model suffered some apparent setbacks. One was in fact the very large region of solar wind penetration, and particularly the finding by McDiarmid and Burrows (1972) in that region that energetic electrons had a pancake pitch angle distribution with a maximum at 90 degrees; this was true in the equatorward part of the cusp or cleft, and quite often throughout the entire cusp. This feature is mostly easily explained by saying that the field lines are closed, with two loss cones formed by losses at the two ends of the field line. This feature did not agree with the reconnection model, where the energized plasma is on open field lines. Another problem was the discovery of an entry layer by Haerendel and Paschmann (1975), this being con-sistent with the conclusion of McDiarmid and Burrows that the cleft plasma comes at least partly from a region of closed field lines inside the magnetopause as shown in Figure 5. Finally there was the observation by Heikkila (1975) that the energization predicted by the reconnection model was not apparent in satellite observations such as HEOS-2.

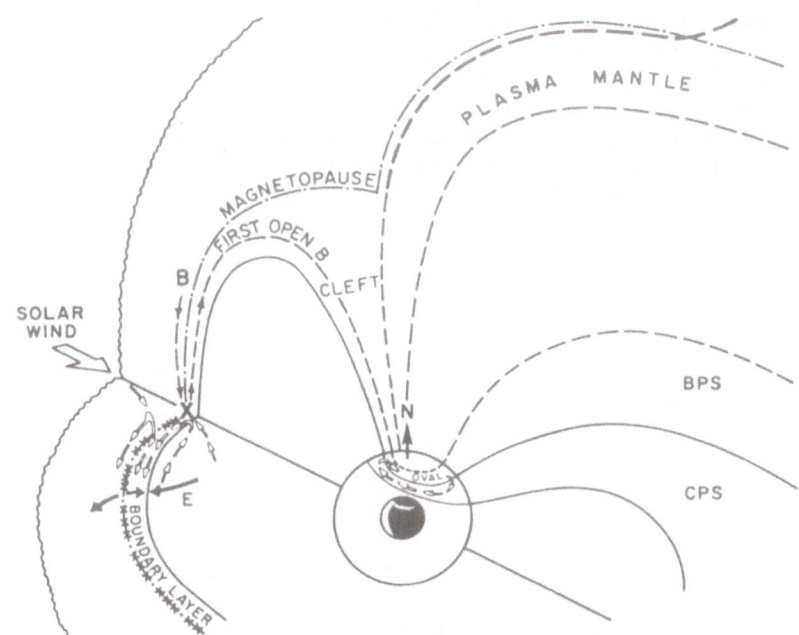

Figure 5.   A three dimensional view of magnetosheath plasma inside
the magnetopause.   It is assumed here the entry layer, which is
partly on closed magnetic field lines, maps into the cleft region at
low altitudes. Convection patterns are indicated by open arrows.

An attempt to resolve the difficuties was made by Reiff et al.
(1977) who invoked a combination of reconnection and viscous
interaction models to explain the observation. The dispersion seen
in the ion spectrograms, as in Figure 3, with the highest energies
at the equatorward edge of the cleft followed by lower energies on
going poleward, could be explained by the difference in field
aligned velocities of ions of different energies in combination
with a constant $\underline{E} \times \underline{B}$ convection velocity due to a constant dawn-
dusk electric field at the separatrix. Almost concurrenmtly, Heelis
et al (1976) using AE drift meter data reported poleward plasma convection through a throat near local noon, again agreeing with a
dawn-dusk electric field. In fact, Mozer et al. (1979) reported the
measurement of such an electric field at the magnetopause with the
ISEE satellite, although their interpretation of the data has been
questioned by Heikkila (1982a) on the basis of the model used for
the analysis, whether it should be steady state and two dimensional
or transient and three dimensional. Finally, Paschmann et al. (1979)
and Sonnerup et al. (1981) using ISEE data provided what they considered as overwhelming proof (to quote a phrase used by Cowley,
1982) of the jets of plasma at the dayside magnetopause predicted
by reconection theories.

However, even then, there was some question as to the reconnection intrepretation; Johnstone (1979), noting that the jets were
easily detectable on some orbits, enquired as to why they were not
always present when the IMF was favorable. Also, the other questions
raised above have not been adequately answered (Heikkila, 1984).

The ISEE satellites made another discovery, that of flux
transfer events reported by Russell and Elphic (1978). An FTE is a
localized and transient phenomenon, where a magnetosheath magnetic
flux tube and the plasma it contains becomes connected to the
geomagnetic field, as shown in Figure 6. This finding was

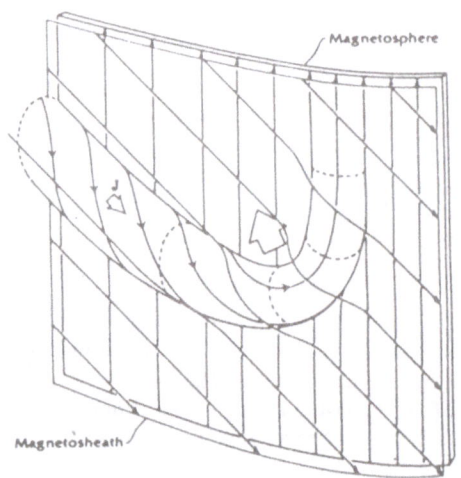

Figure 6. A sketch of a flux transfer event in the northern
hemisphere; a similar counterpart would occur in the south. A field
aligned current produces a twist of the field lines.

foreshadowed by the rocket data of Torbert and Carlson (1976),
which showed that the ions in the cleft showed evidence of time
dispersion, and suggested a transient source some 12 earth radii
away, perhaps at the dayside magnetopause. This led Lemaire (1977)
and Heikkila (1982b) to devote theoretical attention to impulsive
transfer across the magnetopause. It now seems clear that the FTE's
(or impulsive events) may play a key role at the magnetpause, and
once again calls into question the role of steady state reconnection versus FTE or impulsive processes.

It is clear that the cleft is a very important region for
magnetospheric and ionospheric physics by providing a window to the
magnetopause and the solar wind.

2. DEFINITIONS

To my knowledge, every pass of a low-altitude polar-orbiting satellite near noon shows evidence of magnetosheath plasma if the satellite is equipped with suitable particle detectors. This plasma is rather easy to identify due to the characteristic latitude profile. Proceeding poleward, the outer van Allen belt electrons and ions of 10 keV energies have a sharp cutoff (usually less than 1° in latitude), although there is a weak dribble beyond that. Suddenly, the flux of low energy electrons becomes much stronger; the electron energy is typically about 100 eV, but can vary from a peak at 20 eV to 500 eV. The ion energy is usually higher, by about an order of magnitude. These are just the kinds of energies of the plasma particles in the magnetosheath, being solar wind particles after crossing the bow shock. Furthermore, the flux intensities also agree. This early identification has been amply borne out by more recent work. Further poleward, the fluxes of all particles become much weaker, signifying entry of the satellite into the polar cap.

This intense flux of particles produces many features which can readily be observed. The most obvious is the photo-emission, being especially high in the 630 nm line of atomic oxygen due to the low electron energies near the noon meridian. Television pictures from the ground (Pamela Rothwell, private communication) shows that the emission is highly structured and varying with time. Away from noon, both toward dawn as well as dusk, the green line of oxygen increases in brightness, and arcs become evident along the auroral oval. There seems to be some field-aligned acceleration process away from noon, perhaps associated with the plasma convection reversal. That convection is poleward near noon, but away from that local time sector it is wrapped along the auroral oval. Plasma waves of many kinds are also produced, often with a highly localized nature. The plasma density and temperature show characteristic increases.

There was considerable discussion at the workshop on possible and useful definitions. The consensus was that the term cusp should denote a small region, and that consequently it is appropriate for the magnetic topology, as shown by the HEOS-2 data of Hedgecock and Thomas (1975), and in Figure 5.

On the other hand, there is no denying the fact that the region of plasma penetration can be quite broad, as indicated in Figure 4 and 5 and implied by the term cleft. Consequently, there was agreement at the meeting that the following definitions (mostly due to a suggestion by Gordon Shepherd) would be useful to the community:

"The cleft is the low altitude region around noon of about 100 eV electron precipitation associated with 6300A emission, but containing also structured features of higher energy. The cusp is a more localized region near noon within the cleft characterized by low energy precipitation only, having no discrete auroral arcs, but often displaying irregular behaviour, presumably associated with the magnetic cusp."

ACKNOWLEDGEMENT
    I am grateful to NATO for financial assistance at the Workshop, and to NASA under grant NSG 5085.

REFERENCES

Axford, W.I. and C.O. Hines, A unifying theory of high-latitude geophysical phenomenon and geomagnetic storms, Can. J. Phys. 38, 1433, 1961.

Chapman, S., and V.C.A. Ferraro, A new theory of magnetic storms. Part I. The initial phase, Terr. Mag. Atmos. Electr., 36, 171, (1931).

Cowley, S.W.H., Plasma populations in a simple open magnetosphere, Space Sci. Rev. 26, 217, 1980

Cowley, S.W.H., The causes of convection in the Earth's magnetosphere: A review of developments during the IMS, Rev. Geophys. Space Phys. 20, 531, 1982.

Dungey, J.W., Interplanetary field and the auroral zones, Phys. Rev. Lett. 6, 47, 1961.

Haerendal, G. and G. Paschmann, Entry of solar wind plasma into the magnetosphere, from Physics of the Hot Plasma in the Magnetosphere, eds. B. Hultqvist and L. Stenflo, Nobel Foundation Symposium, 1975.

Hedgecock, P.C. and B.T. Thomas, HEOS observations of the configuration of the magnetosphere, Geophysics Journal of the Royal Astronomical Society 41, 391, 1975.

Heelis, R.A., W.B. Hanson and J.L. Burch, Ion convection velocity reversals in the dayside cleft, J. Geophys. Res., 81, 3803, 1976.

Heikkila, W.J., The morphology of auroral particle precipitation, Space Research XII, 1344, 1972.

Heikkila, W.J., Is there an electrostatic field tangential to the dayside magnetopause and neutral line? Geophys. Res. Lett., 2, 954, 1975.

Heikkila, W.J., Inductive electric field at the magnetopause, Geophys. Res. Lett., 9, 877, 1982a.

Heikkila, W.J., Impulsive plasma transport though the magnetopause, Geophys. Res. Lett., 9, 159, 1982b.

Heikkila, W.J., The electromagnetic field for an open magnetosphere, in Magnetic Reconnection in Space and Laboratory Physics, E.W. Hones, Jr. Ed., Geophysical Monograph 30, 39 1984

Heikkila, W.J., J.B. Smith, J. Tarstrup, and J.D. Winningham, The soft particle spectrometer in the ISIS-I satellite, The Review of Scientific Instruments 41, 1393, 1970.

Heikkila, W.J. and J.D. Winningham, Penetration of magnetosheath plasma to low altitudes through the dayside magnetospheric cusps, J. Geophys. Res., 76, 883, 1971.

Johnstone, A., The solar wind connection, Nature 282, 229, 1979.

Lemaire, J., Impulsive penetration of filamentary plasma elements into the magnetospheres of the Earth and Jupiter, Planet Spa. Sci., 25, 887, 1977.

McDiarmid, I.B., J.R. Burrows, and M.D. Wilson, Solar particles and the dayside limit of closed field lines, J. Geophys. Res. 77, 1103, 1972.

Paschman. G., B.U.O. Sonnerup, I. Papamastorakis, N. Sckopke, G. Haerendel, S.J. Bame, J.R. Asbridge, J.T. Gosling, C.T. Russell, and R.C. Elphic, Plasma acceleration at the earth's magneto-Evidence for reconnection, Nature, 282, 243, 1979.

Reiff, P.H., T.W. Hill and J.L. Burch, Solar wind plasma injection at the dayside magnetospheric cusp, J. Geophys. Res., 82, 479, 1977.

Russell, C.T. and R.C. Elphic, Initial ISEE magnetometer results: Magnetopause observations, Space Sci. Rev., 22, 681, 1978.

Russell, C.T., Reconnection at the earth's magnetopause: magnetic field observations and flux transfer events, in Magnetic Reconnection in Space and Laboratory Physics, E.W.Hones, Jr., ed., Geophysical Monoghraph 30, 124, 1984,

Sonnerup, B.U.O., G. Paschmann, I. Papamastorakis, N. Sckhopke, G. Haerendel, S.J. Bame, J.R. Asbridge, J.T. Gosling, and C.T. Russell, Evidence for magnetic field reconnection at the earth's magnetopause. J. Geophys. Res. 86, 10049, 1981.

Winningham, J.D., Characteristics of magnetosheath plasma observed at low altitudes in the dayside magnetospheric cusps, Earth's Magnetospheric Processes, 68-80, Ed. B.M. McCormac, D. Reidel Publ. Co., Dordrecht-Holland, 1972.

FUTURE CUSP EXPERIMENTS AND THEIR COORDINATION

H.C. CARLSON, JR.
Air Force Geophysics Laboratory
Ionospheric Physics Division
Hanscom Air Force Base, Massachusetts 01731

T.L. Killeen
Space Research Building
Ann Arbor, Michigan 48109

R. Heelis
University of Texas at Dallas
Richardson, TX 75080

A. Egeland
University of Oslo
P.O. Box 1038-Blindern
Oslo 3, NORWAY

J.D. Kelly
Radio Physics Laboratory
Menlo Park, CA 94025

D.J. McEwen
University of Saskatchewan
Saskatoon Canada
S7N 0W0

T.J. Rosenberg
University of Maryland
Institute of Physical Science
College Park, Maryland 20742

ABSTRACT.
    Our future progress will hinge upon: comprehensive complementary
diagnostics at selected sites, coordination among selected chains of
stations as well as with satellites, and effective interactive exchange
between models and experiment. The locations, diagnostic capabilties,
motivation, future research plans, and rationale for coordination is
presented here for major sites, diagnostic complexes, and chains of
stations key to future cusp experiments. The role of individual satellites is interwoven into the discussion, and the need for future
multi-spacecraft programs underscored.

INTRODUCTION.

    Plasma stresses and particle populations unique to the geomagnet-
ically singular region of the cusp foster a host of specific studies of
processes peculiar to it. The cusp also is a window on processes coupling solar wind, and magneto-iono-thermo-spheric processes. Further,
the cusp processes both ionospheric plasma and parcels of thermosphere
flowing through it, so that its properties are of direct consequence to

J. A. Holtet and A. Egeland (eds.), The Polar Cusp, 397–410.

quasiglobal scale properties in and near the polar cap. All of these aspects, which have been addressed elsewhere in this proceedings, drive future cusp experiments.

More than ever before, future progress depends on thoughtful coordination of: (a) complementary diagnostics clustered at selected sites, (b) combined ground and spaceborne diagnostics for space-time discrimination, (c) chains of ground-based stations to trace geophysical system responses, and (d) multi-spacecraft programs.

Sites or diagnostic complexes key to future cusp studies are located around Sondrestromfjord, Svalbard, South Pole, and Cape Parry. Diagnostics, planned operations, motivation and planned studies at each of these are discussed below. Specific efforts coordinated with spacecraft are interleaved.

Propagation, penetration, or coupling of polar region effects to lower latitudes can in part be studied by chains of incoherent scatter radars, particularly in the east coast U.S. longitude, but also in the European sector.

Tracing plasma flow over the polar cap is accomplished by polar orbiting satellites (Heelis, this proceedings) and incoherent scatter radars (Kelly, Foster, this proceedings), and eventually by Fabry Perots and a variety of spaced receiver (cross-correlation) techniques.

Polar Thermospheric circulation is traced by Fabry Perot Interferometers (FPI), (Kileen, Smith, Rees this proceedings). As expected from estimates of plasma drag on the thermosphere, this circulation can be dominated by the polar plasma flow. The FPI locations and objectives are described below. Optical signatures seen on All Sky (Intensified) Imaging Photometers (ASIP's) map and trace virtual motion of plasma features. They show much polar plasma comes into the polar cap from lower latitudes, through the cusp (Weber, this volume), and can delineate regions of differing flow velocity (Carlson, et. al., 1984). A chain of these stations is being established over the polar cap (coordinated by Carlson and Weber). The significance of multi-spacecraft programs is highlighted separately below.

FUTURE OPTICAL STUDIES OF CUSP RELATED DYNAMICS

The past several years have seen an unprecedented influx of new experimental information on the dynamic state of the high latitude thermosphere. The most important sources of data have been satellite measurements from instruments on board Dynamics Explorer (e.g. Killeen et al. 1982, 1984; Spencer et al. 1982; Hays et al. 1984) and from the array of ground-based Fabry-Perot interferometers located in the North American and Scandinavian sectors (e.g. Smith and Sweeney, 1980; Meriwether et al 1983; Rees et al. 1980; Hernandez and Roble 1976). Recent measurements and associated theoretical studies have been reviewed by

Roble (1983) and Meriwether (1983) and have indicated that the cusp region has a critical role in the overall polar thermospheric neutral circulation due to the convergent geometry of the ion convection near the dayside auroral zone and the dominant importance of ion drag forcing the thermospheric momentum balance. New experimental results reported

Table 1
Future Studies of Polar Cusp Dynamics

Description	Comments	Techniques
Cusp region as a heat source	Expected from theoretical arguments. Particle and Joule heating, direct insertion at F-region altitudes.	FPI chain temperature TGCM predictions WAMDI measurements Radar comparisons
Cusp region as a momentum source	Strong ion/neutral coupling seen in both previous observations and theoretical predictions. Relative importance for thermospheric momentum budget requires study. Coupling can play a role in the formation and dynamics of enhanced plasma density regions.	FPI chain, ion and neutral winds. TGCM predictions. WAMDI measurements. Radar comparisons. AFGL ionospheric laboratory results.
Lower Thermospheric response	Largely unexplored as yet. Strong signatures expected from theoretical work with large scale vortices and induced compositional perturbations.	FPI 5577A measurements. DE data base, TGCM predictions. Radar observations.
Seasonal behavior	Extended coverage needed to test TGCM predictions.	FPI chain and radar
Universal Time effects	Due to diurnal modulation of momentum and energy forcing functions.	Collation of FPI chain data required. DE data base. Radar measurements.
IMF dependencies including Bz +ve effects	Neutral dynamics clearly filters ion convection forcing functions with ~'1 hour, 1000km' averaging. Measurements of neutrals therefore provide a conveniently 'smoothed' monitor for ion convection and hence magnetospheric forcing.	FPI chain, DE data base, TGCM predictions. Radar measurements.

in this volume by Smith indicate an IMF (By) dependence directly to the interplanetary magnetic field. Theoretical results also reported here (Killeen and Roble), showing the cyclical 'processing' of neutral gas through the cusp region, attest to the great importance of the magneto-spheric - ionosphere - neutral atmosphere coupling in the dayside auroral zone in determining the thermodynamic state of the polar thermosphere.

Table 2
High- and Mid-Latitude Fabry-Perot Interferometer Observatories

Observatory	Lat	Long	Investigator(s)	Comments
Svalbard	78.2N	15.6E	R.W. Smith	Direct cusp measurements of both neutral and ion velocities. 24 hour coverage during polar winter.
Thule	76.0N	70.0W	J.W. Meriwether T.L. Killeen	New instrument to be deployed November 1984. Measurements as for Svalbard in opposite longitude (UT) sector. Collaborative University of Michigan and AFGL ionospheric lab FPI program.
Sondrestrom	67.0N	51.0W	J.W. Meriwether	Colocated with radar facility Northern edge of auroral zone
Fairbanks	64.8N	147.8W	G. Hernandez M.H. Rees	Auroral zone station. 6300 and 5577A observations.
Kiruna	67.8N	21.2E	D. Rees	Auroral zone neutral dynamics. New Doppler imaging system (DIS) installed.
Calgary	51.0N	114.0W	J.W. Meriwether L.L. Cogger	Sub-auroral station.
Ann Arbor	42.3N	83.7W	P.B. Hays J.W. Meriwether	Mid-latitude neutral response
Fritz Peak	39.9N	105.0W	G. Hernandez	Mid-latitude neutral response
Laurel Ridge	40.0N	79.0W	M.A. Biondi	Mid-latitude neutral response

Though much recent progress has been made, many new questions have been raised and there is a clear need to quantify further the various coupling processes occurring in the region associated with the polar cusp. In particular, there is a requirement to relate local dayside observations of the neutral and ion dynamics to more global scale phenomena using both extended (high and mid-latitude) measurements and appropriate modelling techniques. Table 1 indicates a sample of the main areas of future research required to improve the present understanding of polar cusp dynamics and its role in the global circulation.

As indicated in Table 1, future studies of 'cusp' dynamics will utilize heavily the North American and Scandinavian 'chains' of Fabry-Perot interferometers. These instruments, listed in Table 2, provide nighttime vector measurements of the neutral wind field by direct determination of the Doppler shift of the 6300A thermospheric emission feature. Ion drift velocities can also be provided using the 7320 O+ emission feature in regions where the surface brightnesses are sufficiently high (Smith, this volume). Incoherent scatter radar data from the Sondrestrom facility, where available, will provide invaluable information on ionospheric parameters. The new WAMDI (field widened Michelson interferometer, Shepherd, this volume) instrument has the potential for improving significantly the ground-based coverage of fine structure in the neutral wind and temperature field. Finally, the Dynamics Explorer data base, which contains full vector neutral wind obseravations for various seasons and levels of geomagnetic activity, will continue to provide a rich source of detailed information on the ion/neutral coupling and resultant neutral wind phenomenology in the dayside auroral zone. The satellite observations made during the period August 1981-February 1983 serve to complement the ground based measurements by providing detailed latitudinal coverage. Unfortunately, however, there is no new mission planned for the near future that will continue the thermospheric neutral wind measurements from an orbital platform.

The studies outlined in Table 1 serve to illustrate both the multivariate nature of the basic problems and the multiple observational techniques that will be required to provide the key information for their solution. We draw attention, in particular, to the fact that progress will be most readily made through a synthesis of available measurements rather than through a series of 'single station' studies. There is a basic recognition within the community of Thermospheric Dynamicists that the principal requirement is for joint observations and interpretations of the composite data sets. This conclusion derives principally from the relatively long time constants for changes in the thermospheric neutral fluid dynamics and thermodynamics ($\sim$ 10's minutes to hours) and the consequent 'coherence' of the global scale circulation with large scale features, reminiscent of 'weather patterns', persisting for hours. In such a fluid, the cusp region, which undoubtedly plays a critical role in both the momentum and energy budgets, should not be considered in isolation from the rest of the neutral circulation pattern but should be studied in the context of the global scale wind structure.

An important element in the future work will be the use of large scale Thermospheric General Circulation models (TGCMS) in the interpretation of the experimental data. Models such as the NCAR and UCL TGCMs (Dickinson et al. 1981; Roble et al. 1982; Fuller-Rowell and Rees 1980) can provide important diagnostics that can conveniently relate local and fragmentary data sets to global scale forcing mechanisms (Killeen and Roble 1984). Conversely, the new data sets will provide increasingly stiff constraints for the model parameterizations and will, thereby, lead to increasing realism in the model inputs.

The outlook for continued progress in the study of high latitude thermospheric dynamics appears promising with new experimental and theoretical tools becoming available and the network of ground-based Fabry-Perots in place. The prospects for fruitful collaborative work are also good with the formation of a new IAGA working group (II-b Thermospheric Dynamics) designed to foster interchange of data and modelling results as well as the efforts supported by the National Science Foundation to encourage collaborative work within the US optical aeronomy community. The next few years should see a significant increase in the understanding of how the neutral atmosphere acts as a sink for magnetospheric energy and momentum. Perhaps more significantly, as the basic coupling mechanisms become better understood, measurements of neutral dynamics will become useful as a monitor of the basic magnetosphere/solar wind interaction in a time-averaged sense.

FUTURE POLAR CUSP EXPERIMENTS WITH THE SONDRESTROM INCOHERENT-SCATTER RADAR

The primary reason for establishing an incoherent-scatter radar in Sondrestromfjord, Greenland, was to provide information concerning the dynamical state of the ionospheric plasma in the vicinity of the polar cusp. The radar is capable of measuring the plasma densities, temperatures, and drift velocities. These data and data from other instruments, both ground-based and satellite, will be extremely useful in interpreting the complex phenomena associated with the polar-cusp region. The general subjects that have been identified for future research include:

• Determination of the ionospheric signature associated with the cusp as measured by incoherent-scatter radar, and the relationship of this signature to other measurements--such as soft particle flux, field-aligned current systems, optical emissions, and ionograms.
• Identification of the general convection pattern in the auroral oval and the polar cap, and determination of its relationship to solar-wind parameters.
• Examination of the noontime period in order to determine the nature of the "throat" region (spatially constricted plasma flow into the polar cap)--does it exist and under what conditions?
• Measurement of the convection-reversal region with high temporal resolution, and examination of its relationship to ionospheric-current systems.

The radar is operated for 24 hours each month on the Incoherent-Scatter World Day. This has provided an extensive initial data base and will continue. In addition, this experiment was expanded to 72 hours in January 1984, and again in June 1984. We will expand the experiment to 72 hours in January 1985 to study the global effects of substorms. These experiments involve all of the incoherent scatter radars--Sondrestrom, EISCAT, Millstone Hill, St. Santin, Arecibo, and Jicamarca.

In addition to the World Day experiments, we will continue to schedule frequent experiments in conjunction with the Millstone Hill radar. These experiments are usually 24 hours in duration, and are especially useful in studying high-latitude convection because of the overlapping coverage. By combining the data, the convection pattern can be measured from 80° invariant latitude to mid-latitude.

In addition, other experiments designed to measure convection in the dayside-cusp region have been performed and will continue. These experiments are scheduled at a rate of about 200 hours per year and involve scientists from SRI International, Stanford University, and the Danish Meteorological Institute.

Experiments are scheduled each month to coincide with satellite passes. These experiments involve HILAT, NOAA 7, NOAA 8, and DMSP satellites, and are designed to examine F-region plasma-density structure in relation to particle flux, field-aligned currents, and electric fields.

The Sondrestrom radar will support four rocket campaigns in Greenland in December 1984, and in· January and February 1985. The December campaign involves Cornell University and the University of California at Berkeley. The experiment consists of two rockets that will overfly the dayside convection-reversal boundary. Data from the radar and rockets will be combined to examine structure in particle precipitation and in plasma density, temperature, and flow. In January 1985, two rockets will be flown for the University of New Hampshire. The intent is to overfly nighttime aurora. One of the two February 1985 camaigns will be an Air Force Geophysics Laboratory experiment. This experiment will use two rockets to place five barium clouds over an F-region, sun-aligned auroral arc in order to examine convection patterns with high spatial resolution. The fourth rocket campaign will be conducted by Cornell University and the Danish Meteorological Institute. This experiment also involves chemical releases--barium for F-region observations and tri-methyl-aluminum (TMA) for E-region observations. By combining the radar and chemical release data, it is hoped to examine the magnetospheric-forcing function and the resulting thermospheric response.

It is expected that additional rocket campaigns will be planned for Sondrestrom that will take advantage of the capabilities of the Sondrestrom radar and the other instruments in the area.

THE DAYSIDE CUSP PROGRAM AT SVALBARD

Measurements of the Earth's magnetic field and auroral observations have been carried out at Svalbard, more or less continuously, since the turn of the century. Not until the International Geophysical Year (IGY) 1957-58, when cameras were extensively used in polar regions, was it concluded that daytime aurora occur frequently at Svalbard. Since then standard magnetic and all-sky recordings have been carried out at Svalbard. These observations have been maintained by the Auroral Observatory, University of Tromso.

Naturally occurring electromagnetic emissions and magnetic micro-pulsations have been recorded continuously at Svalbard for more than ten years. A multinational dayside auroral project started at Long-yearbyen during the winter 1978/79. Table 3 gives a list of the four observatories in operation today, their coordinates and the instrument setup. A large amount of Svalbard data, both optical and geomagnetic as well as electromagnetic wave measurements, are presented in this volume (Sandholt et al.; Holtet et al.; Sivjee, Henriksen; Smith).

A new auroral station was built at Adventdalen, Longyearbyen in 1983. Currently plans project upgarding to modern working and living quarters for scientists in Longyearbyen (a House of Science) within a few years.

The cusp research program at Svalbard will be continued and in several aspects extended during the next five years. Some of the proposed plans and ideas are briefly summarized in the following.
(1) In order to map the location, latitudinal width and dynamics of day-side cusp auroras vs. time and ionospheric/magnetospheric activity, multichannel meridian scanning photometers and/or all-sky imaging photo-meters, as well as auroral TV-cameras will be operated at all stations. (This is done in cooperation with AFGL/USA, and University of Tokyo, Japan.)
(2) The study of the midday auroral optical spectrum will continue and be extended into the infrared (i.e. from 0.3 to 3 μm) (Sivjee, this volume). The future program includes: the $O^+(^2 D) - N_2$ charge ex-change process, variations in the N, and O abundances in the polar atmosphere, non-LTE vibration distributions of thermospheric molecules, and wave-induced perturbations of N, OH and O_2-emissions.
(3) Helium abundance in the polar atmosphere using the He I emission at 10830Å, will be correlated with the 5876Å (He I). Spectrometric measurments will be extended to the OH-bands which will be used to monitor the temperature of the mesosphere. The slowly varying solar depression angle at Svalbard is excellent for twilight studies of resonance emissions, especially the alkali metals in the mesopause. The program will also be extended to include measurements of the solar spectrum from 290 to 900 nm for biological, climatical and medical purposes. New optical instruments such as auroral laser and/or infrared Michelson interferometer have been proposed.
(4) As noted in Table 3, the net of stations for recordings of local geomagnetic activity, magnetic micropulsations and emissions in the audiofrequency range will be extended. The ELF/VLF receiver at Ny-Ålesund will be upgraded and will be included in the Global net for monitoring ULF-VLF noise and wave activity. In addition, there are plans to install a digital ionosonde at Ny-Ålesund.
(5) Of particular importance to the description of the polar upper

atmosphere is the thermospheric circulation. Optical interferometric observations yielding ion and neutral winds will continue during selected periods. Particular studies to be undertaken include atmospheric heating, conductivity and the relationship to auroral and magnetospheric processes.

(6) The OH and O_2 airglow near the mesopause shows hgh intensities and large changes in temperature at 78°N latitude. These changes are related to transport processes associated with atmospheric disturbances. Three spectrometers will be used to determine the directional onset and decay of atmospheric disturbances such as gravity waves at the mesopause. It appears that these observations can be made in less than five minutes at the high levels of airglow intensity found in the winter polar region.

(7) The Polar Geophysical Institute, Appatity, USSR, operates a geophysical observatory at Barentsbery, roughly 40 km from the station at Longyearbyen. A very large USSR cusp program including three new sites at Svalbard have been mentioned, but details are not known.

(8) Coordination will be particularly significant with the new VHF incoherent scatter radar at EISCAT, and polar orbiting satellites such as HILAT and VIKING will be important. Coordination of ground and satellite data have already started. In addition, a dayside auroral-cusp sounding-rocket program from Svalbard has been proposed.

Those overseeing the comprehensive Svalbard studies of processes unique in the polar cusp region, welcome participants at Svalbard from other research groups.

FUTURE CUSP EXPERIMENTS IN THE SOUTH POLAR REGION

The southern hemisphere polar region offers several advantages for measurements of the dayside polar cusp and related phenomena. Identification of the cusp relies primarily on optical signatures of the precipitating particles. Optical studies from the ground are possible at South Pole for 4 1/2 months of the year, about three times the observing time available from sites at comparable magnetic latitudes in the northern hemisphere. Also, because the Geomagnetic 75° latitude cusp is at the geographic South Pole, polar-orbiting satellites pass over the geographic South Pole on each orbit, providing opportunities for regular coordinated observations. Another important advantage is the extremely low noise (electromagnetic) environment of much of Antarctica which contributes to high signal-to-noise ratios for measurements over the frequency range from ULF to HF.

Measurements of the cusp and related phenomena in the southern hemisphere are important because then can provide information on hemispheric asymmetries or similarities, as the case may be. For example, the orientation of the interplanetary magnetic field as it sweeps past Earth may lead to hemispheric asymmetries in the boundary between open and closed field lines. Seasonal effects may influence the detectability of ionospheric signatures of the cusp. "Conjugate" measurements thus offer the possibility of investigating these various influences.

The United States upper atmosphere physics program in Antarctica has been carried out primarily at Siple, South Pole, and McMurdo

(Rosenberg 1982; Eather 1983)). Of particular interest for high latitude observations in the dayside polar cusp are the measurements made at South Pole and McMurdo (Table 3)

Table 3
Instrumentation at Svalbard (Mag. Noon 08:30 UT): NyÅlesund (NY Å79.0 geog. lat, 75.4 geomag), Longyearbyen (LYR 78.2, 74.4), Hornsund (HSD 77.0, 73.5), Bjornoya (BJA 74.5, 71.1). Instrumentation at Antartic: South Pole (SP -90.0, -75.0), McMurdo (MM -77.5, -79.0)

INSTRUMENT	LOCATION					
	SVALBARD				ANTARTIC	
	NYA	LYR	HSD	BJA	SP	MM
Ionosonde	P				X	
Riometers	X		X	X	X	X
Broadband VLF	X	P	X		X	
Magnetometer	X		X	X	F,I	F
Micr. Puls.	X		P			
Keogram(6300Å, 4278Å)					X	
Photometer (6300Å, 4278 Å	S	S	P		Z	Z
Intensified ASC (6300Å)	X				X	
Conventional ASC	X	X	X		X	
Fabry-Perot Interferometer	X	X				
Ozone		P				
Auroral Spectrometer	P	X				
Auroroal TV Imager	P	X				

D - Fluxagate Magnetometer (f < 0.5 Hz)
I - Induction Magnetometer (f < 5 Hz)
Z - Zenith   S - Scanning
X - Existing   P - Planned

South Pole (invariant latitude ~ 75°), when near magnetic local noon, occupies a transition region between the polar cap and auroral oval for moderate levels of geomagnetic activity. McMurdo (invariant latitude ~ 79°), on the other hand, is deeper in the polar cap and thus better suited for observing the cusp during geomagnetically quiet conditions (e.g. see Eather 1983). Not to be ignored, however, is Siple and other stations at similar invariant latitude (~ 61°) which can observe the cusp during some geomagnetic storms, as has been reported Currently, Siple station is closed until November 1985 when it will reopen for a two-year period.

Possible future cusp experiments in the south polar region relate to the following programs: (1) Automatic Geophysical Observatories; (2) Expansion of upper atmosphere physics facilities at McMurdo; (3) Balloon measurements of electric fields and electron precipitation at South Pole; (4) Multiple narrow-beam riometer measurements at South Pole; and (5) HF radars. Each will be described briefly below. Some are relatively firm at this time (1,2). Others have not yet been formally approved. Although not directly related to the programs listed above, it is worth mentioning that a program of fluxgate magnetometer and riometer measurements from the South Pole conjugate location near Frobisher Bay, Canada has been approved and should become operational in Mid-1985.

(1) Automatic Geophysical Observatories (AGOs) - An AGO is being developed (Lockheed Palo Alto Research Laboratory) to allow a variety of geophysical experiments at remote locations of special scientific interest. It will have the capability to operate untended for a year at a time. It is presently planned to deploy the first AGO at the nominal conjugate point of Sondrestrom, Greenland for operations beginning in 1987. The full complement of geophysical instrumentation remains to be identified. Photometers, magnetometers, riometers, and VLF receivers are likely to be included. Contact the National Science Foundation DPP regarding experimental opportunities.

(2) McMurdo Expansion - In 1985 the instruments at McMurdo plus new ELF/VLF noise survey instrumentation will be moved to a specially protected low noise site (Arrival Heights) nearby. A dedicated U.S. upper atmosphere physics facility will be newly erected for this purpose. Additional optical and magnetic instrumentation is likely.

(3) Balloon Measurements at the South Pole - The geographic South Pole is roughly conjugate to Sondrestrom, Greenland. Comparison of measurements made simultaneously at Sondrestrom and South Pole would provide the opportunity to study the conjugacy of cusp phenomena. Knowledge of the degree of magnetic conjugacy exhibited by electric fields and particles at cusp latitudes is important for deriving accurate models of magnetospheric structure. Light winds at high latitudes (~ 10 mb) over the Pole in austral summer, and the extremely slow variation of the solar depression angle, should permit long duration balloon flights which remain essentially overhead. The Universities of Houston and Maryland are proposing to carry out a series of balloon measurements of electric fields and particles (i.e., x-rays) from South Pole during the 1985-86 austral summer. The principal aim is to obtain a nearly continuous data base for more than one solar rotation. Major emphasis would be placed on comparison with conjugate northern hemisphere radar and related data.

(4) Multiple Narrow-Beam Riometry at South Pole - Riometer measurements of cosmic noise absorption are sensitive to the precipitation of electrons above a few keV energy. Although the riometer does not respond to the soft cusp electron precipitation, it is able to examine the harder precipitation at the equatorward boundary of the cusp and structures that may temporarily be embedded within the cusp precipitation itself. The present broad-beam riometer measurements at South Pole are limited in their ability to resolve the size and motion of precipitation regions. University of Maryland proposes to augment the broad-beam measurements at South Pole with a multiple narrow-beam array.

(5) HF Radars - Means to establish joint U.S.-U.K. proposal to establish HF radars at Siple and Halley Bay with overlapping coverage of the polar ionosphere over South Pole and the general region conjugate to Sondrestrom and other northern hemisphere radar facilities are being explored.

CANADIAN CLEFT STUDIES IN PROGRESS OR PLANNED

A HILAT ground station was installed in January 1984 at Churchill, Manitoba and is expected to operate through the life of the satellite. A second telemetry station is scheduled to be set up at Inuvik, N.W.T. (see map) and operate from August 1984 for one year. These will record the experimental data for all polar passes over Northern Canada.

Two chains of multichannel meridian scanning photometers are scheduled to be installed beginning in late 1984, through Rankin Inlet and Contwoyto Lake, as shown on the map. They will record and transmit to a central station one meridian scan each minute of 4709, 4861, 5577, and 6300Å emissions over an invariant latitude span from approximately 78° to 48°. These stations also include three-component magnetometers and riometers. They form a part of a Canadian CANOPUS array which is expected to remain in operation for the remainder of the decade. Another component is a bistatic auroral radar system (BARS) which will begin operation in late 1984. This 50 MHz radar system will be very similar in operation to the European STARE system. The region over which electric field measurements will extend is centered on Churchill and is depicted by the cross-hatched region on the map extending through the auroral zone up to 73°Λ.

The CANOPUS network is primarily intended for studies of the night-side auroral oval but will be useful for correlative studies of dayside auroral phenomena.

An expeditionary rocket base exists at Cape Parry, N.W.T. (Λ = 75°) and a down range station is located at Sachs Harbour (Λ = 77°). This station and a more northerly site at Mould Bay (Λ = 81°) are expected to be used as ground stations in December 1986 at solar minimum for optical studies of dayside and polar cap aurora. These studies may be complemented by one or more rocket flights from Cape Parry poleward These constitute a second cleft campaign now in preliminary planning stages.

FUTURE SATELLITE EXPERIMENTS

One of the values of satellite data at ionospheric heights is the short time required to traverse the high latitude region. If it is assumed that the high latitude convection pattern is stable over a period of about twenty minutes then essentially a snap-shot of that pattern can be obtained from a satellite. At much higher altitudes satellites are able to provide data in a region that is largely inaccessible by other means.

In both cases, however, the inability to distinguish between space and time variations can be a significant barrier to the interpretation of data. Additionally only very limited regions of the ionosphere-magnetosphere-solar wind system can be probed at any one time, making the construction of a global picture from the data set a somewhat arbitrary procedure. The future for satellite diagnosis of the high latitude region therefore, lies in the use of multiple satellite data sets and the combination of simultaneously obtained satellite, rocket and ground based data. With such a combination of data both of the difficulties mentioned above can be removed.

The location of a radar facility at Sondrestromfjord now makes available the high latitude ground-based data on ion concentration temperature and dynamics that can be combined with data from high inclination satellites. Such a data combination allows the separation of space and time variations when the sampled volumes are almost the same and allows the construction of an accurate two dimensional convection pattern when the sampled volumes are slightly separated. The opportunity to perform this type of collaborative experiment should be provided by satellite borne instrumentation from the United States Defense Meteorological Satellite Program, the HILAT program and the European VIKING program. All these satellites will include instrumentation for measuring the ambient electric field or ion drift velocity, the energetic particle environment and the magnetic field perturbations produced by field aligned currents.

Space and time variations that include the response of the ionosphere-magnetosphere-solar wind system to changes in the primary internal and external drivers are most easily achieved with multiple satellites that sample the same region within a few minutes of each other. In the ionosphere for example, the detection of moving electrodynamic features associated with the aurora and the changes in the convection pattern during substorms could be handled with such data sets. The deployment of multiple satellite payloads also provides some flexibility in their location in a given plasma volume. This presents perhaps the opportunity to measure the real "curl" of a vector field. Opportunities in this area may be forthcoming from consideration of small satellites with limited instrumentation that could be deployed from the space shuttle for ionospheric studies or placed in much higher orbits for studies of the variabilities in the outer magnetosphere. It is likely that any mission of this type will foster international cooperation in its ultimate success.

Finally, while pressing forward with studies of space and time variations and the construction of two dimensional pictures in particular regions of space, we must continue to pursue the goal of understanding the connections between the different regions of space and their effects on each other. Studies in this area require the deployment of appropriately instrumented satellites to simultaneously probe the inner magnetosphere, the outer magnetosphere, the solar wind, the outer plasmasphere and the ionosphere. The International Solar Terrestrial Physics program is directed toward this goal and will hopefully be operational while other ionospheric satellites programs are active.

We have reached an understanding of the information provided by a satellite instrument complement to the point where additional simultaneous data is now required to make the next steps. In some cases data bases are already available and need careful examination to find the simultaneous data that exists. In other cases well planned campaigns involving current satellite and ground based facilities have and will be planned to address questions concerning the real two dimensional relationships between different parameters. The presentations at this workshop have readily illustrated the power of such an investigative

approach by utilizing data from PROGNOZ, GEOS, ISEE, DE, DMSP along
with ground based and interplanetary magnetic field data.

It is clear that some questions concerning the temporal evolution
of the ionosphere-magnetosphere-solar wind system can only be answered
by successfully conducting some of the new satellite missions currently
under consideration.

REFERENCES

Carlson, H.C., E.J. Weber, J. Buchau, J.G. Moore, Plasma characteristics
of polar cap F layer arcs, Geophys. Res. Lett., 1984.
Dickinson, R.E., E.C. Ridley and R.G. Roble, A three-dimensional
general circulation model of the thermosphere, J. Geophys. Res., 86,
1499-1512, 1981.
Eather, R.H., Dayside auroral dynamics, J. Geophys. Res., 89, 1695,
1984.
Fuller-Rowell, T.J., and D. Rees, A three-dimensional, time dependent
global model of the thermosphere, J. Atmos. Sci., 37, 2545-2567, 1980.
Hays, P.B., T.L. Killeen, N.W. Spencer, L.E. Wharton, R.G. Roble, B.A.
Emery, T.J. Fuller-Rowell, D. Rees, L.A. Frank and J.D. Craven, Observations of the dynamics of the polar thermosphere, J. Geophys. Res.
in press, 1984.
Hernandez, G. and R.G. Roble, Direct measurements of nighttime thermo-
spheric winds and temperatures 2. Geomagnetic storms, J. Geophys. Res.,
81, 2065-2074, 1976.
Killeen, T.L., P.B. Hays, N.W. Spencer and L.E. Wharton, Neutral winds
in the polar thermosphere as measured from Dynamics Explorer, Geophys.
Res. Lett., 9, 957-960, 1982.
Killeen, T.L. and R.G. Roble, An analysis of the high latitude thermo-
spheric wind pattern calculated by a Thermospheric General Model:
I. Momentum Forcing, J. Geophys. Res., in press, 1984.
Killeen, T.L., R.W. Smith, P.B. Hays, N.W. Spencer, L.E. Wharton and
F.G. McCormac, Neutral winds in the high latitude winter F-region:
Coordinated obseravations from ground based and space, Geophys. Res.
Lett., 11, 311-314, 1984.
Meriwether, J.W., Jr., Observation of thermospheric dynamics at high
latitudes from ground and space, Radio Sci., 18, 1035-1052, 1983.
Meriwether, J.W. Jr., C.A. Tepley, P.B. Hays and L.L. Cogger, Remote
ground-based observations of terrestrial airglow emissions and
thermospheric dynamics at Calgary, Alberta, Optical Engineering, 22,
128-131, 1983.
Rees. D., T.J. Fuller-Rowell and R.W. Smith, Measurements of high
latitude thermospheric winds by rocket and ground-based techniques
dynamical model, Planet. Space Sci., 28, 919-932, 1980.
Roble, R.G., Dynamics of the Earth's thermosphere, Rev. Geophys. and
Space Phys., 21, 217-233, 1983.
Roble, R.G., R.E. Dickinson and E.C. Ridley, Global circulation and
temperature structure of thermosphere with high latitude plasma
convection, J. Geophys. Res., 87, 1599-1614, 1982.
Rosenberg, T.J., Research at United States Antarctic stations during
the International Magnetosphere Study, in The IMS Source Book,
American Geophysical Union, 1982.

SUMMARY

Vytenis M. Vasyliunas
Max-Planck-Institut für Aeronomie
D-3411 Katlenburg-Lindau
Federal Republic of Germany

This paper is intended as a coherent summary of the topics and results covered at this Workshop, of the questions that have been answered (or else have not been answered) and of the controversies that still remain, to be presented in a rather broad-brush fashion ("stich-wortartig" as one says in German); detailed presentations are to be found in the individual papers of this volume. It is natural, in doing this, to compare the status of polar cusp studies at this Workshop with previous meetings devoted to the subject, in particular with the first magnetospheric cleft symposium held in Dallas in 1973 for which I also published a summary (Vasyliunas, 1974). The most immediately apparent result of such a comparison is that enormous amounts of additional data, and qualitatively new data as well, on the polar cusp have become available in the 11 years since the Dallas symposium. With the new data, inevitably, many new questions arise; of the old questions - those that were the main topic of discussion at Dallas - there are im-portant ones that still remain unanswered and some that have received a clear answer. It becomes clearer at each successive meeting that the polar cusp is not an isolated topic but one that has close connections to many other aspects of magnetospheric physics; several presentations did not deal directly with the cusp but were nevertheless obviously relevant to the subject of the Workshop.

On the name of the region in question, the earlier lively dis-agreement between those who call it "polar cusp" and those who prefer "magnetospheric cleft" seems now to have been more or less resolved, at least insofar as the leading proponents of "cleft" have switched to "cusp". The definition of the polar cusp was the subject of much dis-cussion at the Workshop. Many ways of defining it were put forth, not always with a clear distinction between a conceptual definition as such and a method of observational identification or determination when a given spacecraft or instrument is located within the cusp.

If we leave aside the subtle questions of definition and adopt an operational approach of looking at what people do instead of listening to what they say, the primary and most widely used method of identifying the cusp is by means of charged particle observations: one finds a region with a population of particles of magnetosheath-like energies and

411

J. A. Holtet and A. Egeland (eds.), The Polar Cusp, 411–418.
© 1985 by D. Reidel Publishing Company.

(equally important) magnetosheath-like intensities precipitating into
the atmosphere at high latitudes somewhere within or near the noon
local time sector. At the Workshop we heard reviews of the observations
of such particles at low altitudes (Peterson, Johnstone) and at high
altitudes (Sckopke, Lundin); more recently, particle observations at
altitudes intermediate between the ionosphere and the outer magnetosphere have become available and, as Winningham convincingly showed,
constitute an enormously detailed and rich data set. Nevertheless,
when we consider the relation of the cusp to the structure and topology
of the magnetosphere, there are old and important questions that remain
unanswered. The outstanding question is the relationship of open and
closed magnetic field lines to the plasma structure defined by the low-
altitude polar cusp and the associated boundary layers in the outer
magnetosphere. That was already a major topic of discussion at the
Dallas meeting; the two extreme positions presented there - that the
cusp lies entirely on closed field lines, or that it is formed entirely
by simple direct plasma entry along open field lines - have both been
abandoned since then, but it is still not clear where within the plasma
structure the boundary between open and closed field lines lies. It is
generally accepted that field lines that thread the so-called "entry
layer" (for which I prefer the name "interior cusp;" cf. figure and
discussion in Vasyliunas, 1979) also pass through the low-altitude
polar cusp, but it is uncertain whether field lines from the low-
latitude magnetospheric boundary layer go directly to the polar cusp,
passing equatorward of the "entry layer," or whether they go to the
"entry layer" first. It is also uncertain what fraction of the field
lines that thread the polar cusp comes from the plasma mantle. All
these questions remain unsettled after more than 11 years of research.

On the other hand, there were several conclusive results reported
at the Workshop. First, the cusp contains, in addition to its usual
plasma of magnetosheath origin, a particle population from an ionospheric source, accelerated to magnetosheath-like energies (which by
ionospheric standards are very high) and observed both at low altitudes (Johnstone) and within the plasma mantle (Lundin). Second, the
source of magnetosheath plasma for the cusp appears not to be steady
or uniform but rather localized and time-varying, as indicated by the
observed energy - time as well as energy - pitch angle dispersion of
ions within the cusp, consistent with a localized source at the
magnetopause (Peterson); this brings to mind theories of the so-called
impulsive entry of plasma into the magnetosphere, two controversial
versions of which were reviewed at the Workshop by their authors
(Lemaire, Heikkila) without anything substantially new being said on
either side of the controversy. Third, nearly simultaneous observations of the northern and southern polar cusps are now available
(Meng); they show that the two cusps are reasonably similar - the
conjugacy may not be perfect but, considering the limitations and
uncertainties of the data, it is remarkably good.

Other attempted correlations of high and low altitude cusp
phenomena were reported with less conclusive results. Candidi and Meng
compared the intensity of precipitating low energy electrons in the
cusp with the electron concentration in the solar wind, finding that the

two tended to increase and decrease together. Nielsen et al. attempted
to find, in ionospheric flow observations, signatures of simultaneously
observed flux transfer events at the magnetopause. Maynard associated
intense electric fields observed near the magnetopause with similar
intense electric fields in the ionosphere near the equatorward edge of
the polar cusp; if firmly established, this would be an important
result showing that the low-latitude boundary layer maps into the
equatorward edge of the cusp, but at present there is no independent
support for the association and the best one can do is argue the other
way round - if the mapping is assumed, then one has evidence that the
two electric fields, near the magnetopause and in the ionosphere, are
related.

On the local-time distribution of charged particles from the cusp
precipitating into the ionosphere, a reasonably clear picture has now
emerged. There is a broad region of precipitation extending essentially
over all daytime hours and merging with the auroral oval, and there is
a clear smaller feature within a few hours of local noon characterized
primarily by softer (i.e. lower energy) precipitation, as reported by
Johnstone and (on the basis of a vast statistical study) by
Carovillano. At Dallas 11 years ago there was considerable controversy
on whether the cusp extends over most of the dayside or is confined to
near local noon; now it is clear that there are in fact two distinct
regions, one broad and one narrow in local time, and to which one we
give the name "cusp" can be viewed as partly a matter of choice (albeit
one with significant consequences for how one visualizes the cusp's
relation to outer magnetosphere structures). What is not so clear is
the nature of the broad region - is it merely an extension and continuation of the auroral oval into the dayside, or is it really
distinct from the oval (and divided from it by some identifiable
boundary in particle precipitation)?

A feature of particle precipitation not necessarily directly
related to the cusp but nevertheless interesting is a persistent enhancement or "hot spot" at high latitudes near 14 hours local time,
studied in detail by Evans. The feature is apparently always there but
is most pronounced during magnetically quiet conditions, tending to be
masked by enhanced precipitation elsewhere when activity increases. It
is associated with intense upward Birkeland (field-aligned) currents,
part of the well-known Region 1 current system, and its precipitating
electrons appear to have been accelerated out of a population with
energies typical of the magnetosheath and a somewhat lower density;
hence the suggestion that this region may be mapped from the dayside
magnetopause boundary layer. Another result on particle precipitation,
not directly related to the cusp but potentially of great significance
for understanding the over-all magnetospheric structure, was reported
by Carovillano: the "polar rain" electron precipitation over the polar
cap occurs predominantly when the interplanetary magnetic field has a
southward component, and it furthermore exhibits a north-south asymmetry controlled by the x component of the interplanetary magnetic
field, in the sense that the polar rain is enhanced in that hemisphere
whose open field lines connect toward the sun (similar to the well-
known north-south asymmetry of solar particle access).

The second most widely used method, after charged particle observations, of determing the location of the polar cusp is by observing the aurora and other optical emissions. Compared to its status at earlier meetings, this topic has greatly expanded and matured. Here, too, large amounts of new data have become available: observations from very high latitude stations, especially South Pole (Eather) and Spitsbergen (Sandholt et al., Henriksen and Stamnes), observations by rocket-borne instruments of optical and UV emissions either alone or in correlation with precipitating particle measurements (Feldman et al., McEwen), auroral imaging from satellites (the well-known DE observations reported by Craven and more recent Hilat observations reported by Meng), observations by aircraft-borne instruments (Weber), and finally tests of some newly developed techniques of ground-based optical imaging (Shepherd).

The general pattern of auroral emissions agrees, at least in broad outline, with what would be expected from the local-time pattern of particle precipitation described already. There is a localized region near local noon where the 6300 Å emissions (produced by soft electrons) are enhanced and other auroral emissions (produced by precipitation of more energetic electrons) are reduced in comparison to their levels elsewhere in the auroral oval, consistent with the picture of a small region of low-energy precipitation near noon and a broad region of harder precipitation elsewhere. Some observers of auroral emissions in lines produced by more energetic electrons even speak of a "noon gap," although that name is not quite appropriate since the emissions only become weaker but do not disappear. Craven reported that a thin arc can occasionally be seen stretching across this so-called gap ("the gap arcs over!" - an irresistible pun when one thinks of the magnetospheric electric potential applied between the morning and evening sides of the oval). Within the region of soft precipitation near noon, small-scale auroral structures moving rapidly poleward have been observed (Eather, Sandholt).

Further support for the general pattern is provided by observations of electromagnetic waves (Holtet, Rosenberg, Maynard) which presumably are also produced ultimately by charged particle precipitation of one type or another. A special region of wave emission near local noon is clearly evident, but whether it is precisely coincident with the corresponding auroral region is less clear. Holtet, for example, finds a region of strong micropulsation activity over a few hours local time centered at noon (possibly coincident with the localized soft precipitation region) and in addition a region of ELF hiss in the same general area but distinctly shifted somewhat toward local morning. (There is also a region of VLF hiss which, however, lies well to the evening side of the localized noon cusp and is probably not directly connected to it; more likely, the VLF hiss may be related to the "hot spot" of precipitation at 14 hours local time.) Such examples illustrate the difficulty of finding a unique definition for the polar cusp, a structure that manifests itself in various physical effects not all of which need to be spatially coincident.

The position of the cusp, as determined from auroral observations, is highly variable in latitude. During extremely quiet geomagnetic con-

ditions it can be, as pointed out by Eather, at invariant latitudes as
high as 83-84°, farther poleward than ordinarily assumed (and nearly
coincident with the cusp location in the Mead-Beard models of a closed
magnetosphere with no magnetotail). At the other extreme, during large
magnetic storms the cusp can move equatorward to the 61° latitude of
the South Pole station (reported, on the basis of ionospheric signatures, by Broom). There is a major controversy on what determines the
latitudinal location of the cusp: does it depend primarily on the
southward component B_z of the interplanetary magnetic field, the cusp
moving equatorward when southward B_z increases (supported by Sandholt
et al. on the basis of individual event studies, with additional data
reported by Meng), or is it controlled largely by the substorm currents
measured by the AE index, the cusp moving equatorward when AE increases
(argued by Eather on the basis of statistical studies)? The controversy
involves some fine points of technique (possible inaccuracies in the
AE index for very quiet and very disturbed conditions when the auroral
oval moves away from the station network, possible uncertainties in the
time shift between interplanetary magnetic field changes at the earth
and at the ISEE-3 spacecraft 200 R_E upstream) and remains unresolved.

The implications of this controversy are not as obvious as one
might think. Meng formulated what he took to be its basic question as
"Is the change of magnetospheric topology an internal or an external
process?" but the obvious answer to that is "It's really both" - the
complex topology of the magnetosphere arises from the interaction of
the internal dipole and the external interplanetary magnetic field.
If one assumes that the cusp is located at the boundary of open and
closed field lines (and that is an assumption, presupposing a specific
answer to what was noted earlier to be a major unanswered question),
then a change of cusp location results from an imbalance between the
magnetic merging rate on the dayside (which is strongly influenced,
at the least, by B_z) and the magnetic merging rate on the nightside
(which may very well involve magnetotail currents and substorms). Thus
no simple predictions can be made; the resolution of the controversy is
important but cannot be expected to lead to any simple conclusions on
internal or external magnetospheric dynamics. A further complication
arises from the mutual interdependence of the various quantities involved (e.g. AE may depend on B_z, the various solar wind parameters are
to some extent correlated with each other, and so on); a multiple
correlation analysis is mandatory for any statistical studies, and one
should not forget that two quantities may have no direct relation and
yet may vary together just because they are both related to a third
quantity.

Meng reported that the motion of the polar cusp in latitude is
well correlated with the motion of the nightside auroral oval in
latitude, indicating that (whatever the responsible governing factor
may be) the entire polar cap expands and contracts in size while
keeping more or less the same shape. The implications of this result
for magnetospheric theory, again, are significant but not immediately
obvious.

In addition to the morphological and dynamical aspects of the
aurora discussed so far, auroral optical observations also have im-

portant spectroscopic aspects, reviewed by Sivjee who showed how they
can be used for the study of basic aeronomic processes (e.g. charge
exchange, atomic abundances, radiative processes, and others). Such
studies are much simpler in the dayside cusp region where the energies
of precipitating electrons are so low that the electrons are stopped at
high levels where the atmospheric consitutents are predominently
atomic, whereas the more energetic electrons in the nightside auroral
oval penetrate to deeper levels where the atmospheric constituents are
mainly molecular (everyone knows that atomic physics is much simpler
than molecular physics!).

The topic in which perhaps the most progress has been made is
magnetospheric convection. New data sets have become available: there
are extensive direct observations of plasma flow above the ionosphere
from satellites (Heelis), supported by observations from rockets
(Marklund and Heelis), and in addition the theory of magnetosphere-
ionosphere coupling has been used to infer the global distribution of
plasma flow on the basis of the Birkeland currents deduced from
satellite observations of magnetic fields (Potemra) as well as on the
basis of equivalent ionospheric currents deduced from ground-level
magnetic disturbances (Baumjohann and Friis-Christensen). All three
methods yield mutually consistent global patterns of plasma flow. On a
semi-global scale, plasma flow has been studied by both incoherent and
coherent scatter methods (Kelly, Greenwald, Foster), with results that
are consistent with the global patterns. Further support is provided
by the (somewhat uncertain) technique of ionospheric drift measurements
(Brekke) and by various local-scale observations from satellites
(Berthelier) and rockets (Primdahl et al.).

An additional and relatively new source of data bearing on mag-
netopsheric convection is provided by observations of neutral atmospheric winds at F-region altitudes (Smith, Killeen). The neutral wind
patterns here reflect ion-drag effects from magnetospheric convection
and yield results in general agreement with the over-all picture deduced from other data sets. The magnetosphere is clearly important for
the thermosphere (but at still lower heights, in the mesosphere, a
study by Myrabö suggests that effects from above, from the magnetosphere, are less important than effects from below).

The main features of the picture of magnetospheric convection
derived from all this multiplicity of data are, first, that that convection does flow across the polar cusp into the polar cap - this re-
solves a major controversy of the Dallas symposium, where antisunward
convection flow along the poleward boundary of the auroral oval,
avoiding the polar cap, was still considered a possible alternative.
Second, within the cusp region the flow is generally westward (i.e.
dusk to dawn) and poleward; but it is doubtful whether a distinct
feature in the flow pattern can be found that would correspond to the
localized noon-sector cusp of particle and auroral observations (the
so-called "throat" region of the flow does not seem to be a permanent
feature) - in other words, it may not be possible to determine the
location of the cusp on the basis of flow observations alone. Third,
the over-all pattern of plasma flow in the cusp, the auroral regions,
and the polar cap - and this is another case where the polar cusp is

closely connected to other aspects of magnetospheric physics – exhibits
a strong control, at least on the average, by the interplanetary
magnetic field. In particular, there is a marked effect of the B_y
component, primarily apparent as a dawn-dusk asymmetry which reverses
between the northern and the southern hemispheres (i.e. for a given
sign of B_y, an effect that appears on the dawn side in the north will
appear on the dusk side in the south, and vice versa – a necessary
consequence of the inherent symmetries of the configuration). There is
also a B_z effect: for a southward B_z, there is the well-known 2-cell
convection which intensifies as the southward B_z increases. More
striking is the effect of a prolonged northward B_z: there appears a
qualitatively different and more complex convection pattern, with 3 or
more cells, reversed (sunward) flow in parts of the polar cap, and
marked B_y effects on this pattern as well.

The detailed study of magnetospheric convection during periods of
prolonged northward B_z (and hence very quiet magnetic activity level)
is a relatively new topic, made possible by data sets (DE, Sondrestrom,
Spitsbergen, and others) that allow the coverage of very high latitudes. We have learned that what used to be called the "quiet-time
magnetosphere" is not just a simple and perhaps somewhat dull baseline
but a complex system with its own distinctive physics of solar wind-
magnetosphere interaction; the Θ aurora (discussed at the Workshop by
Craven) and sun-aligned arcs (Weber) are among the more spectacular
manifestations of this different state of the magnetosphere.

There are various direct ionospheric effects of the polar cusp.
These include enhanced electron density in the cusp region ionosphere,
already seen a long time ago by topside sounders (reviewed here by
Muldrew), and drifting patches of electron density enhancements, observed to drift through the cusp into the polar cap (Foster, Weber);
these are usually interpreted as the result of ionization by sunlight
in regions equatorward of the cusp and subsequent drift poleward –
one may say that the ionosphere is processed through the cusp to
provide an input to the polar cap (Carlson) – but there is some
unresolved controversy on whether the electron density enhancements
could be produced locally in a dark region by particle precipitation
(Stamnes) without the need for transport from a sunlit region. Another
ionospheric signature of the polar cusp is provided by irregularities
and turbulence in the F region (Greenwald, Muldrew) as well as
irregularities and instabilities in the E region discussed by Stauning
(observations) and Primdahl and Bahnsen (theory).

Concluding this entire Workshop, one need that is apparent is
to digest the large amounts of new and different data that are now
available, both on the cusp itself and on related regions of the
magnetosphere. An important question is: what are the interrelation-
ships of the various observed phenomena in which the cusp can be
recognized (particle precipitation, aurora, wave emissions, electric
fields, plasma flow, etc.)? One needs to examine the observations in
order to decide to what extent these various aspects of the cusp do or
do not coincide. There are basic questions on the relation of the cusp
to the topology and mapping of the magnetosphere that remain to be
answered. We do, however, now possess a greatly sharpened description

of at least the average configuration (particularly for the convection
pattern), which presents a much better defined set of problems for
the theorist. Understanding the polar cusp is essential for (and
indeed is very nearly equivalent to) understanding the basic physics
of the dynamical interaction between the solar wind and the magnetosphere.

REFERENCES

Vasyliunas, V.M. 1974, EOS, 55, 60.
Vasyliunas, V.M. 1979, p. 387 in Proceedings of Magnetospheric Boundary
Layers Conference (Ed. B. Battrick) ESA SP-148, Noordwijk,
The Netherlands.

SUBJECT INDEX

Acceleration mechanisms 63, 75-79, 99, 106
AC fields, electric 305-321, 325
        - magnetic 237
Aurora, cusp 60, 111-124, 127-134, 137-146, 149-161, 163-173, 193-202,
        282-287
        - theta 31
Auroral dynamics (cusp) 149-161, 163-173, 184-187, 251-256
        - emission rates 142
        - intensity ratios 119-123
        - spectrum 111-124, 127-130, 137. 166-169, 194, 201

Birkeland currents 16, 24, 99, 106, 197, 172, 184, 203-221, 235-241,
        413
        - classification of 203, 211
Boundary layers 1-5, 9-31, 33, 52-60

Charge exchange 111, 113-115
Circulation patterns/models 246, 261-276, 293-302, 337-340, 368, 402
Cleft: see Polar Cleft
Convection 208, 210, 217, 229-233, 252, 293-302, 310, 311, 318-321,
        337-340, 344, 349-353, 368, 391
Convection electric fields 15, 213, 293, 305
Coupling, magnetosphere-ionosphere 5, 23
        - solar wind - magnetosphere 9-31, 33-45, 47-64, 68, 79-81,
        99, 149-160, 163-172, 178-187, 220, 350
Current systems, ionospheric 165, 173, 204, 212, 216-217, 223-233
Currents, field aligned: see Birkeland currents
Cusp: see Polar cusp

Dayside aurora 60, 111-124, 127-134, 137-146, 149-161, 163-173, 193-
        202, 282-287
Dayside auroral gap 60-61, 182, 329, 350
Diamagnetic effects 43
Discontinuity, rotational 314
        - tangential 33, 37
Dispersion, energy 71-72
DPY currents 163, 173, 208, 212, 226
DP2 currents 226, 228
Drifts, ion 243, 252-255, 283, 285, 298, 338-340
Dynamics, polar cusp 149-161, 163-173, 177, 184-187, 398, 399

419

---

* Since this book is focusing on the Polar Cusp, also other key
  words will be connected to Polar Cusp properties